Modern Experimental Biochemistry

Modern Experimental Biochemistry

Rodney F. Boyer

Hope College

 ADDISON-WESLEY PUBLISHING COMPANY

Reading, Massachusetts • Menlo Park, California
Don Mills, Ontario • Wokingham, England • Amsterdam
Sydney • Singapore • Tokyo • Mexico City
Bogota • Santiago • San Juan

to H.B.

Library of Congress Cataloging in Publication Data

Boyer, Rodney F.
　　Modern experimental biochemistry.

　　Bibliography: p.
　　Includes index.
　　1. Biological chemistry—Methodology.　2. Biological
chemistry—Technique.　I. Title.
QP519.7.B68　1986　　　　574.19′2　　　　85–4042
ISBN　0-201-10131-9

ABCDEFGHIJ-DO-898765

Preface

The introduction of modern instrumental methods has revolutionized experimental biochemistry. The training of future life scientists must include both a theoretical and practical exposure to current biochemical techniques and concepts. The purpose of this textbook is to give students a thorough experience in modern experimental biochemistry. The book is designed for a teaching laboratory, accompanying a lecture course for junior-senior students. Its primary use will be in the undergraduate curriculum for students who have had one year of organic chemistry with laboratory. These students are principally in chemistry, biochemistry, biology and preprofessional (premedical, predental, preveterinarian) programs. This type of laboratory is probably offered in every four-year college and university, but the differences among these courses are vast. Biochemical equipment, staff expertise, scheduling, and student backgrounds vary considerably; hence, present laboratory courses are tailored to match local realities. This laboratory textbook was designed to respond to the needs, desires, and equipment limitations present in most biochemical teaching laboratories. Every effort has been made to produce a book that not only trains students in modern biochemical techniques, but one that does so in a flexible and generally usable fashion.

The book is divided into two parts. Part I, Theory and Experimental Techniques, introduces students to background material that is usually not provided in lecture courses nor readily available in existing laboratory textbooks. Since some laboratory procedures are used in several experiments, all of this information is consolidated in Part I for easy reference. Basic theory and practical information are offered for each experimental technique. This section will also become a useful reference for students during future research work.

Twenty-six experiments representing all areas of biochemistry are included in Part II. This broad coverage should lead to a well-rounded experience for the student and also allow the instructor to choose experiments that are appropriate or adaptable under a variety of local conditions.

The general outline for each experiment is as follows.

INTRODUCTION, THEORY, AND OBJECTIVES

In this section it is assumed that a student has studied the general subject in lecture. Only a summary or review of significant aspects of background is included. The general discussion includes practical information that is generally not available in biochemistry textbooks. The general thrust of the experiment is explained, and a flow chart of the experiment is presented, if appropriate. This outline of the experiment allows the student to recognize the importance of each part of the experiment to the achievement of the overall objective.

EXPERIMENTAL

Following a list of all materials and supplies is a carefully written procedure for the experiment. It is divided into logical parts for ease of completion and to facilitate the interruption of an experiment, if necessary.

ANALYSIS OF RESULTS

In this section the student is instructed in the proper collection and handling of data from the experiment. Each table or graph that is to be constructed is explained and sample calculations are outlined. Typical data for the experiment may be disclosed to the student but only to aid the student in interpretation of results.

QUESTIONS AND PROBLEMS

Several questions and problems are provided at the end of each experiment. Some questions will deal with various details of the experiment, and numerical problems are emphasized.

In the development of each experiment, primary consideration was given to the use of modern procedures and techniques. Students will learn procedures that are now used in actual laboratory settings, whether academic or industrial. Each experiment presents a challenging laboratory situation or problem to be solved by the student. In each experiment, a technique is introduced that allows students to obtain and evaluate some property of a biomolecule or biosystem. For each set of experiments dealing with a particular class of biomolecules, students proceed through three levels of investigation:

1. Isolation and purification of a macromolecule or cell fraction/organelle
2. Physical and chemical characterization of the macromolecule or cell fraction
3. Biological characterization or study of dynamic function

If students proceed through these three levels, they will gain an understanding and appreciation of the essence of biochemistry—the relationship between chemical structure and biological function.

Sequences of experiments that illustrate this process are:

I. Structure and Function of α-Lactalbumin

EXPERIMENT 4: Isolation and Purification of α-Lactalbumin, a Milk Protein

EXPERIMENT 5: Characterization of α-Lactalbumin

EXPERIMENT 8: Evaluation of a Regulatory Enzyme System: Lactose Synthetase

II. Lipids

EXPERIMENT 9: Isolation, Purification and Partial Characterization of a Lipid Extract from Nutmeg

EXPERIMENT 10: Characterization and Positional Distribution of Fatty Acids in Triacylglycerols

EXPERIMENT 12: Characterization of Erythrocyte Membrane Using Lipolytic Enzymes

III. DNA

EXPERIMENT 15: DNA Extraction from Bacterial Cells

EXPERIMENT 16: Characterization of DNA by Ultraviolet Absorption Spectrophotometry

EXPERIMENT 21: Ethidium Fluorescence Assay of Nucleic Acids: Binding of Polyamines to DNA

IV. DNA Recombination Techniques

EXPERIMENT 17: Growth of Bacteria and Amplification of ColE1 Plasmids

EXPERIMENT 18: Preparative-Scale Isolation and Purification of ColE1 Plasmids

EXPERIMENT 19: The Action of Restriction Enzymes on a Bacterial Plasmid or Viral DNA

EXPERIMENT 20: Rapid, Microscale Isolation and Electrophoretic Analysis of Plasmid DNA

Following Parts I and II is an appendix containing useful information including physical properties of biomolecules and solutions to selected questions. An instructor's manual describing the preparation of all reagents is available from Addison-Wesley.

Holland, Michigan   *R. F. B.*

Acknowledgments

This book would not have been possible without the superior assistance of Mrs. Norma Plasman who typed several revisions and the final manuscript. I am grateful for the many hours she spent on the project.

I would also like to thank students, too numerous to mention, who struggled through the first drafts of these experiments and provided excellent recommendations for changes. I am especially indebted to Anna Kalmbach and Dean Welsch, who each spent a summer testing and retesting the experiments. I also relied on the expert advice of many reviewers, including Hugh Akers (Lamar University), John R. Cronin (Arizona State University), Patricia Dwyer-Hallquist (Appleton Paper Inc.), Robert N. Lindquist (San Francisco State University), Kenneth A. Marx (Dartmouth College), Richard A. Paselk (Humboldt State University), William M. Scovell (Bowling Green State University), Ev Trip (University of British Columbia), Dennis Vance (University of British Columbia), and Ronald S. Watanabe (San Jose State University). Bob Rogers, Stuart Johnson, and Emily Silverman at Addison-Wesley constantly provided excellent supervision and assistance with the project.

Contents

I

Theory and Experimental Techniques 1

II

Experiments 235

Modern Experimental Biochemistry

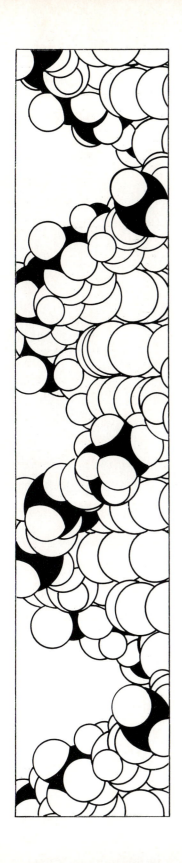

I

Theory and Experimental Techniques

Chapter **1**

Introduction to the Biochemistry Laboratory

Welcome to biochemistry laboratory! This is not the first chemistry laboratory course for most of you, but I believe you will find it to be among the most exciting and dynamic of those in which you have enrolled. Most of the experimental techniques and skills that you have acquired over the years will be of great value in this laboratory. However, you will be introduced to several new procedures. Your success in the biochemistry laboratory will depend, to a great extent, on your mastery of these specialized techniques and on your understanding of chemical-biochemical principles.

As you proceed through the schedule of experiments for this term, you will, no doubt, compare your work with previous laboratory experiences. In biochemistry laboratory you will seldom run reactions and isolate products on a gram-size scale as you did in the organic laboratory. Rather, you will work with milligram or even microgram quantities, and in most cases the biomolecules will be dissolved in solution so you never really "see" the materials under study. But, you will observe the dynamic chemical and biological changes brought about by biomolecules. The techniques and procedures introduced in the laboratory will be your "eyes" and will monitor the biochemical events occurring.

This chapter is an introduction to procedures that are of utmost importance for the safe and successful completion of a biochemical project. It is recommended that you become familiar with Sections A–F before you begin laboratory work.

A. SAFETY IN THE LABORATORY

The concern for laboratory safety can never be overemphasized. Most students have progressed through at least two years of laboratory work without even a minor accident. This record is, indeed, something to be proud of; however, it should not lead to overconfidence. You must always be aware that chemicals used in the laboratory are potentially toxic, irritating, or flammable. Such chemicals are a hazard, however, only when they are mishandled or improperly disposed. It is the author's experience that accidents happen least often to those students who come to each laboratory session mentally prepared and with a complete understanding of the experimental procedures to be followed. Since dangerous situations can develop unexpectedly, though, you must be familiar with general safety practices, facilities, and emergency action. Students must have a special concern for the safety of classmates. Carelessness on the part of one student can often cause injury to other students.

The experiments in this book are designed with an emphasis on safety. However, no amount of planning or pretesting of experiments substitutes for awareness and common sense on the part of the student. All chemicals used in the experiments outlined here must be handled with care and respect. Some chemicals and procedures will require special attention. When this is the case, a precautionary note will appear at the beginning of the experimental procedure. Here, the dangerous properties of the chemicals will be described along with recommendations for safe handling, first aid, and proper disposal.

So that all students are equally aware of potential laboratory hazards, it is imperative that a set of rules be developed for each laboratory situation. The following can serve as guidelines.[1]

1. Some form of eye protection is required at all times. Safety glasses with wide side shields are recommended, but normal eyeglasses with safety lenses may be permitted. *Contact lenses should never be worn in the laboratory.* Naturally, there is controversy surrounding the use of contacts (see Kingston, 1981). The major problem with contact lenses is that they reduce the rate of self-cleansing of the eye. If a chemical splatters into the eye of a contact lens wearer, he or she will possibly have greater injury because of less effective irrigation of the eye. Furthermore, the wearer may be reluctant to use an eye wash for fear of losing the contact lenses.

---

[1] Adapted from *Safety in Academic Chemistry Laboratories*, 3rd Edition, August, 1979. A publication of the American Chemical Society Committee on Chemical Safety.

2. Never work alone in the laboratory.

3. Be familiar with the physical properties of all chemicals used in the laboratory. This includes their flammability, reactivity, toxicity, and proper disposal.

4. Eating, drinking, and smoking in the laboratory are strictly prohibited.

5. Unauthorized experiments are not allowed.

6. Mouth suction should not be used to fill pipets or to start siphons.

7. Become familiar with the location and use of standard safety features in your laboratory. All chemistry laboratories should be equipped with fire extinguishers, eye washes, safety showers, fume hoods, chemical spill kits, first-aid supplies, and containers for chemical disposal. Any questions regarding the use of these features should be addressed to your instructor or teaching assistant.

Rules of laboratory safety are not designed to impede productivity nor should they instill a fear of chemicals or laboratory procedures. Rather, their purpose is to create a healthy awareness of potential laboratory hazards and to improve the efficiency of each student worker. The list of references at the end of this chapter includes books and manuals describing proper and detailed safety procedures.

B. THE LABORATORY NOTEBOOK

The biochemistry laboratory experience is not finished when you complete the actual experimental procedure and leave the laboratory. All scientists have the obligation to prepare written reports of the results of experimental work. Since this record may be studied by many individuals, it must be completed in a clear, concise, and accurate manner. This means that procedural details, observations, and results must be recorded in a laboratory notebook *while* the experiment is being performed. The notebook should be hardbound with quadrille ruled (gridded) pages, and used only for biochemistry laboratory. This provides a durable, permanent record and the potential for construction of graphs, charts, etc. It is recommended that the first one or two pages of the notebook be used for a constantly updated Table of Contents. Although your instructor may have his or her own rules for preparation of the notebook, the most readable notebooks are those in which only the right-hand pages are used for record-keeping. The left-hand pages may be used for your own notes, reminders and calculations. The following outline may be used for each experimental write-up. The basic outline follows that required by most biochemical research journals. Note that Parts I–IIb are labeled as Prelab and should be completed before you enter the laboratory.

Brief Outline of Experimental Write-up

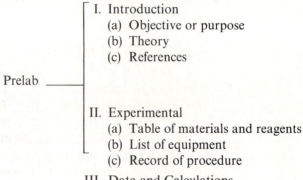

Prelab

I. Introduction
   (a) Objective or purpose
   (b) Theory
   (c) References

II. Experimental
   (a) Table of materials and reagents
   (b) List of equipment
   (c) Record of procedure

III. Data and Calculations
   (a) Record of all raw data
   (b) Method of calculation with statistical analysis
   (c) Present data in tables, graphs, or figures when appropriate

IV. Results and Discussion
   (a) Conclusions
   (b) Compare results with known values
   (c) Discuss the significance of the data
   (d) Was the original objective achieved?

Details of Experimental Write-up

I. Introduction

This section begins with a three- or four-sentence statement of the objective or purpose of the experiment. For preparing this statement, ask yourself, "What are the goals of this experiment?" This statement is followed by a brief discussion of the theory behind the experiment. If a new technique or instrumental method is introduced, give a brief description of the method. Include chemical or biochemical reactions when appropriate. Any references (textbooks, journals, etc.) used to prepare this section should be listed at the end.

II. Experimental

Begin this section with a list of all reagents and materials used in the experiment. The sources of all chemicals and the concentrations of solutions should be listed. Instrumentation is listed with reference to company name and model number.

The write-up to this point is to be completed as a Prelab. The experimental procedure followed is then recorded in your notebook as you proceed through the experiment. The detail should be sufficient so that a fellow student can use your notebook as a guide. You should include observations, such as color changes or gas evolution, made during the experiment.

III. Data and Calculations

All raw data from the experiment are to be recorded directly into your notebook, not on separate sheets of paper or paper towels. Calculations involving the data must be included for at least one series of measurements. Proper statistical analysis must be included in this section.

For many experiments, the clearest presentation of data is in a tabular or graphical form. The Analysis of Results section following each experimental procedure in this book describes the preparation of graphs and tables. These must all be included in your notebook.

IV. Results and Discussion

This may be the most important section of your write-up, because it answers the questions, "Did you achieve your proposed goals and objectives?" and, "What is the significance of the data?" Any conclusion that you make must be supported by experimental results. It is often possible to compare your data with known values and results from the literature. If this is feasible, calculate percentage error and explain any differences. If problems were encountered in the experiment, these should be outlined with possible remedies for future experimenters.

Everyone has his or her own writing style, some better than others. It is imperative that you continually try to improve your writing skills. When your instructor reviews your write-up, he or she should include helpful writing tips in the grading. For further instructions in scientific writing see Day (1983).

C. CLEANING LABORATORY GLASSWARE

The results of your experiments will depend, to a great extent, on the cleanliness of your equipment. There are at least two reasons for this. (1) Many of the chemicals and biochemicals will be used in milligram or microgram amounts. Any contamination, whether on the inner walls of a beaker, in a pipet, or in a glass cuvet, could possibly be a significant percentage of the total experimental sample. (2) Many biochemicals and biochemical processes are sensitive to one or more of the following common contaminants: metal ions, detergents, and organic residues. In fact, the objective of many experiments is to investigate the effect of a metal ion, organic molecule, or other chemical agent on a biochemical process. Contaminated glassware will virtually insure failure.

Cleaning Plasticware

Plastic containers made of polyethylene are increasingly being used in chemical procedures. Polyethylene containers, when used for the first time, should be cleaned in 8 M urea (adjusted to pH 1 with concentrated HCl). After rinsing it with distilled water, wash the plasticware with 1 N KOH, distilled water again, and then 10^{-3} M EDTA to remove metal-ion contamination. Finally, thoroughly rinse with glass-distilled water. After this initial treatment, a wash with 0.5% detergent solution followed by thorough glass-distilled water rinsing is recommended after each use. The base-urea-EDTA treatment may be repeated as desired.

Cleaning Glassware

Many contaminants, including organics and metal ions, will adhere to the inner walls of glass containers. Washing glassware with dilute detergent (0.5% in water) is probably sufficient for most purposes. However, if especially sensitive procedures are scheduled, special washing is essential. Soaking glassware in acid dichromate (see below) and rinsing with distilled water will free the glass of organic contaminants. A second wash with concentrated nitric acid followed by glass-distilled water will remove most metal ions. Dichromic acid solution is prepared as follows: Dissolve 5 g of sodium dichromate in 5 ml of water in a 250 ml beaker. Slowly add, with constant stirring, 100 ml of concentrated sulfuric acid. The temperature of the solution will rise to about 80°C. Allow the solution to cool to about 40° and transfer it to a dry bottle with glass stopper for storage. *Be cautious in the use of dichromate, as chromium has been determined to be carcinogenic.*

Cleaning Quartz and Glass Cuvets

Never clean cuvets or any optically polished glassware with ethanolic KOH or other strong base, as this will cause etching. All cuvets should be cleaned carefully with 0.5% detergent solution, using a cotton-tipped wood applicator. These may be purchased or made by twisting a small wad of cotton around a cylindrical wood applicator. Rinse with plenty of glass-distilled water after cleaning.

Cleaning Pipets

Pipets may require special cleaning treatment because reagents collect and dry in the tips. They should be soaked (tips up) in a 0.5% detergent solution and then rinsed with large amounts of tap water, followed by glass-distilled water.

Drying Plasticware and Glassware

Dry equipment is required for most processes carried out in biochemistry laboratory. When dry glassware was desired in organic laboratory, you probably rinsed the piece of equipment with acetone, which rapidly evaporates and leaves a dry surface. Unfortunately, that surface is coated with an organic residue consisting of nonvolatile contaminants in the acetone. Since this residue could interfere with your experiment, it is best to refrain from acetone washing. Glassware is best dried in an oven. Most plasticware may also be dried in an oven. *Never dry cellulose nitrate centrifuge tubes in an oven.* Cellulose nitrate is an explosive. If you are not sure of the type of plastic in a container, do not dry it in an oven.

D. QUANTITATIVE TRANSFER OF LIQUIDS

Practical biochemistry is highly reliant on analytical methods. Many analytical techniques must be mastered, but few are as important as the quantitative transfer of solutions. Some type of pipet will almost always be used in liquid transfer. Since students may not be familiar with the many types of pipets and the proper techniques in pipetting, this instruction is included here. Figure 1.1 illustrates the various types of pipets and fillers. Following is a discussion of the use of each.

Filling a Pipet

The use of any pipet requires some means of drawing reagent into the pipet. *Liquids should never be drawn into a pipet by mouth suction on the end of the pipet.* (One exception to this is the use of a micropipet, but here the mouth does not touch the glass pipet, but, rather a plastic mouthpiece attached to the pipet with latex rubber tubing.) Small latex bulbs are available for use with disposable pipets (Figure 1.1A). For volumetric and graduated pipets, two types of bulbs are available. One type (Figure 1.1B) features a special conical fitting that will accommodate common sizes of pipets. To use these, first place the pipet tip below the surface of the liquid. Squeeze the bulb with your left hand (if you are a right-handed pipettor) and then hold it tightly to the end of the pipet. Slowly release the pressure on the bulb to allow liquid to rise to 2 or 3 cm above the top graduated mark. Then, remove the bulb and quickly grasp the pipet with your index finger over the top end of the pipet, as shown in Figure 1.2. The level of solution in the pipet will fall slightly, but not below the top graduated mark. If it does fall too low, use the bulb to refill.

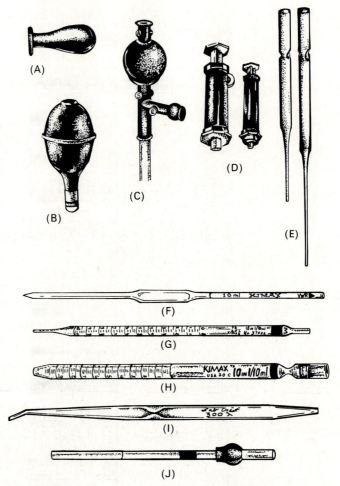

FIGURE 1.1
Examples of pipets and pipet-fillers. (A) Latex bulb courtesy of VWR Scientific, Division of Univar. (B) Pipet-filler courtesy of Curtin Matheson Scientific, Inc. (C) Mechanical pipet-filler courtesy of VWR Scientific, Division of Univar. (D) Pipetter pump courtesy of Cole-Parmer Instrument Co. (E) Pasteur pipet courtesy of VWR Scientific, Division of Univar. (F) Volumetric pipet courtesy of VWR Scientific, Division of Univar. (G) Mohr pipet courtesy of VWR Scientific, Division of Univar. (H) Serological pipet courtesy of VWR Scientific, Division of Univar. (I) Lang-Levy micropipet courtesy of VWR Scientific, Division of Univar. (J) Disposable micropipet courtesy of Curtin Matheson Scientific, Inc.

Mechanical pipet-fillers (sometimes called safety pipet-fillers, pro-pipets, or pi-fillers) are more convenient than latex bulbs (Figure 1.1C,D). Equipped with a system of hand-operated valves, these fillers can be used for the complete transfer of a liquid. The use of a safety pipet filler is outlined in Figure 1.3. *Never allow any solvent or solution to enter the pipet*

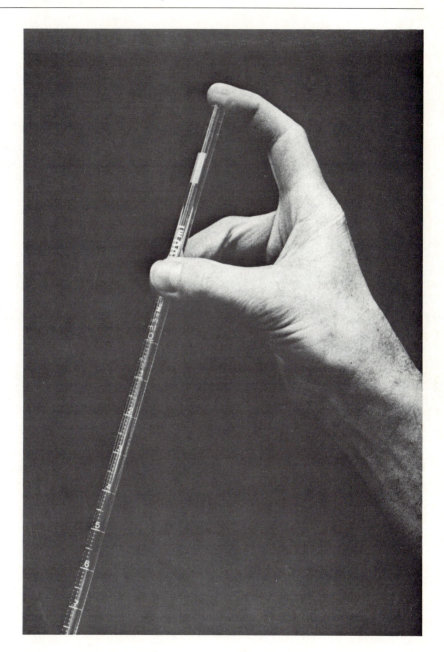

FIGURE 1.2
How to hold a pipet. Photo courtesy of Mr. Mark Billadeau.

bulb. To avoid this, two things must be kept in mind: (1) always maintain careful control while using valve (S) to fill the pipet, and (2) never use valve (S) unless the pipet tip is below the surface of the liquid. If the tip moves above the surface of the liquid, air will be sucked into the pipet and solution will be flushed into the bulb. Other pipet-fillers are used in a similar fashion.

1. Using thumb and forefinger,
   press on valve A and squeeze bulb with other fingers to
   produce a vacuum for aspiration.
   Release valve A leaving bulb compressed.

2. Insert pipette into liquid. Press on
   valve S. Suction draws liquid to desired
   level.

3. Press on valve E to expel liquid.

4. To deliver the last drop, maintain
   pressure on valve E, cover E inlet
   with middle finger, and squeeze the small bulb.

FIGURE 1.3
How to use a Spectroline[R] safety
pipette filler. Courtesy of Spec-
tronics Corporation, Westbury,
NY 11590.

Disposable Pasteur Pipets

Often it is necessary to perform a semiquantitative transfer of a small
volume (1 to 10 ml) of liquid from one vessel to another. Since pouring
is not efficient, a **Pasteur pipet** with a small latex bulb may be used

(Figure 1.1A,E). Pasteur pipets are available in two lengths (15 cm and 23 cm) and hold about 2 ml of solution. These are especially convenient for the transfer of nongraduated amounts to and from test tubes. Typical recovery while using a Pasteur pipet is 90 to 95%. If dilution is not a problem, rinsing the original vessel with a solvent will increase the transfer yield.

Calibrated Pipets

If a quantitative transfer of a specific and accurate volume of liquid is required, some form of calibrated pipet must be used. **Volumetric pipets** (Figure 1.1F) are used for the delivery of liquids required in whole milliliter amounts (1, 2, 3, 4, 5, 10, 15, 20, 25, 50, and 100 ml). To use these pipets, draw liquid with a latex bulb or mechanical pipet-filler to a level 2 to 3 cm above the "fill-line." Wipe the outside of the pipet with a lint-free tissue. Touch the tip of the pipet to the inside of the glass wall of the container from which it was filled. Release liquid from the pipet until the meniscus is directly on the "fill-line." Transfer the pipet to the inside of the second container and release the liquid. Hold the pipet vertically, allow the solution to drain until the flow stops, and then wait an additional 5 to 10 seconds. Touch the tip of the pipet to the inside of the container to release the last drop from the outside of the tip. Remove the pipet from the container. Some liquid may still remain in the tip. Most volumetric pipets are calibrated as "TD" (to deliver), which means the intended volume is transferred *without final blow-out*, i.e., the pipet delivers the correct volume.

Fractional volumes of liquid are transferred with **graduated pipets**, which are available in two types—**Mohr** and **serological**. Mohr pipets (Figure 1.1G) are available in long or short tip styles. Long-tip pipets are especially attractive for transfer to and from vessels with small openings. Virtually all Mohr pipets are "TD" and are available in many sizes (0.1 to 10.0 ml). The marked subdivisions are usually 0.01 or 0.1 ml, and the markings end a few centimeters from the tip. Selection of the proper size of pipet is especially important. For instance, do not try to transfer 0.2 ml with a 5 or 10 ml pipet. Use the smallest pipet that is practical. Class A pipets have tolerance limits ranging from ± 0.005 ml for a 0.1 ml pipet to ± 0.02 ml for a 10 ml pipet.

The use of a Mohr pipet is similar to that of a volumetric pipet. Draw the liquid into the pipet with a pipet-filler to a level about 2 cm above the "0" mark. Clean the tip with a tissue, and lower the liquid level to the "0" mark. Remove the last drop from the tip by touching the inside of the glass container. Transfer the pipet to the receiving container and release the desired amount of solution. The solution should not be allowed to move below the last graduated mark on the pipet. Touch off the last drop. If an additional transfer of the same solution is to be done, the same pipet may be used, but be sure to wipe the tip with a tissue before inserting it into the reagent.

Serological pipets (Figure 1.1H) are similar to Mohr pipets, except that they are graduated downward to the very tip and are designed for blow-out. Their use is identical to that of a Mohr pipet except that the last bit of solution remaining in the tip must be forced out into the receiving container with a rubber bulb. This final blow-out should be done after 15 to 20 seconds of draining. Serological pipets are used less often than Mohr pipets in biochemical procedures. Tolerance limits are in the same range as for Mohr pipets.

Micropipets

Long-tip Mohr pipets are useful for the transfer of volumes from 0.1 to 10 ml but of little value for volumes less than 0.1 ml. Small volume transfers (0.001 to 0.500 ml; 1 to 500 μl) are best achieved with volumetric micropipets (Lang-Levy or lambda (λ) pipets, Figure 1.1I). A lambda refers to 1 μl; hence, a 100 lambda pipet transfers 100 μl. Micropipets are available in several capacities from 1 to 500 μl. Liquid reagents are drawn into these pipets by one of two methods. Mechanical syringes that attach to the pipet are available for capacities up to 50 μl. A simpler method is to use a latex rubber tube with an adapter at one end for the pipet and a plastic mouthpiece at the opposite end. *Radioactive or other hazardous solutions must never be mouth pipetted, even with the latex tube.* The pipet is filled by gentle suction until the solution is slightly above the constriction. If air bubbles form in the pipet, drain and refill. (Bubbles sometimes indicate a dirty pipet.) Wipe the tip-end with a lint-free tissue. Drain excess liquid by touching the tip to the side of the container from which the pipet was filled. If the liquid does not drain, touch the tip to a piece of filter paper and allow the solution to be absorbed until the level reaches the constriction. Transfer the pipet to the receiving container. The receiving vessel should contain buffer or other solution required for the assay. The tip of the filled pipet is touched to the inner wall of the vessel just above the liquid level, and the reagent is dispensed with gentle air pressure. The pipet should be cleaned as soon as possible, as described later in this section.

Disposable micropipets (Figure 1.1J) are available in very small sizes (1 to 20 μl). These may be used to transfer small volumes and to apply solutions to thin-layer or paper chromatography sheets.

Automatic Pipetting Systems

If many identical small-volume transfers must be made, a mechanical microliter pipettor (Eppendorf type) is ideal. This allows for accurate, precise, and rapid dispensing of fixed volumes from 1 to 100 μl. The

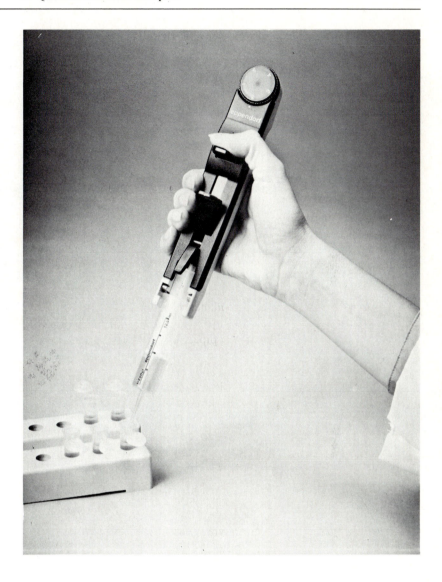

FIGURE 1.4
Eppendorf pipet with tip. Photo courtesy of Brinkmann Instruments Co., Westbury, NY.

pipet's push-button system can be operated with one hand, and it is fitted with detachable polypropylene tips (Figure 1.4). The advantage of polypropylene tips is that the reagent film remaining in the pipet after delivery is much less than for glass tips. Mechanical pipettors are available in up to 25 different sizes. Newer models offer continuous volume adjustment so a single model can be used for delivery of specific volumes within a certain range.

To use the pipettor, choose the proper size and place a polypropylene pipet tip firmly onto the cone as shown in Figure 1.5. Tips for pipets are

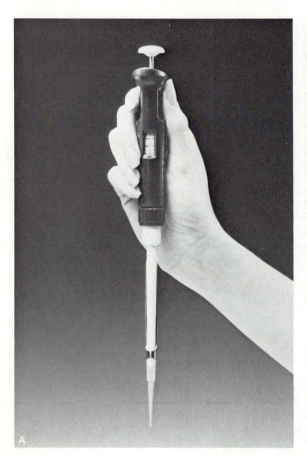
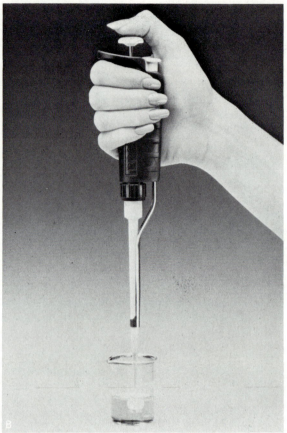

FIGURE 1.5
How to use an adjustable pipet device. (A) Set the digital micrometer to the desired volume. (B) Depress pushbutton to the first positive stop. Immerse pipet tip into sample fluid to several millimeters. (C) Allow pushbutton to return slowly to UP position. Wait 1 second, then withdraw pipet tip from fluid. (D) Transfer pipet to receiving container. Depress pushbutton to first positive stop to remove most of the liquid. (E) Depress pushbutton to second stop to achieve final blow out. Touch the pipet to the inside wall of the receiving container to remove last drop. Courtesy of Rainin Instrument Company, Inc.

available in two sizes, for 1 to 100 μl and for 200 to 1000 μl capacity. For adjustable pipets, set the correct volume to be delivered. The push button on the top has two depressed positions. The first setting (down) is to fill or empty the tip, and the fully depressed position (second position) is for final blow-out by compressed air. Depress the button to the first stop. Immerse the tip into the reagent. Release the push button and allow it to return to the up position. You will see liquid being drawn into

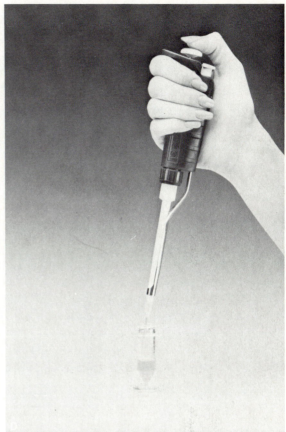

FIGURE 1.5 Continued

the tip. Transfer the pipet into the receiving vessel. Touch the tip to the inner wall just above any liquid already in the vessel, and push the button to the first stop. Wait 1 or 2 seconds and then depress the button to the second position (blow-out). Remove the pipet tip from the vessel with a slight twist while touching the side to remove the final drop from the outside of the tip. The same pipettor and tip may be used to transfer identical amounts of a single reagent to many containers. The polypropylene tips may be washed and autoclaved at 120°C for sterile conditions.

For rapid and accurate transfer of larger volumes, automatic repetitive dispensers are commercially available (Figure 1.6). These are particularly useful for the transfer of corrosive materials. The dispensers, which are available in several sizes, are simple to use. The volume of liquid to be dispensed is mechanically set; the syringe plunger is lifted for filling and pressed downward for dispensing. Hold the receiving container under

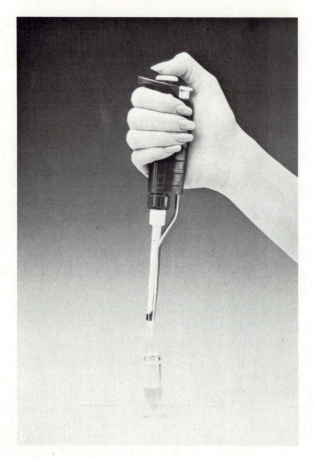

FIGURE 1.5   Continued

the spout while depressing the plunger. Touch-off the last drop on the inside wall of the receiving container. The set volumes are dispensed to an accuracy of $\pm 1\%$ and with a reproducibility of $\pm 0.1\%$.

Cleaning and Drying Pipets

Special procedures are required for cleaning pipets. Immediately after use, every pipet should be placed, tip up, in a vertical cylinder containing a dilute detergent solution (less than 0.5%). The pipet must be completely covered with solution. This insures that any reagent remaining in the pipet is forced out through the tip. If reagents are allowed to dry inside a pipet, the tips can easily become clogged and are very difficult to open. After several pipets have accumulated in the detergent solution, the pipets

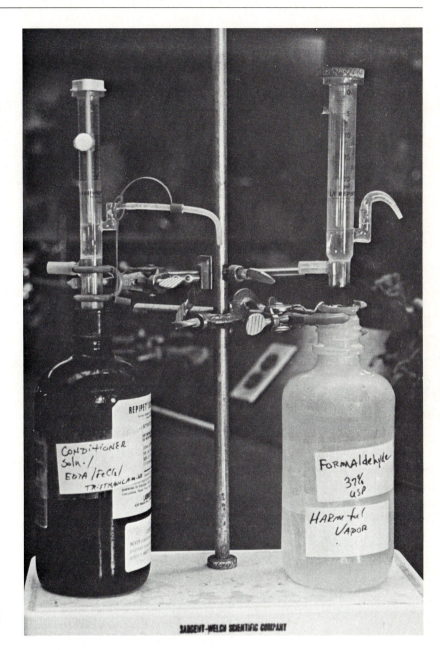

FIGURE 1.6
Automatic repetitive dispenser.
Photo courtesy of Mr. Mark
Billadeau.

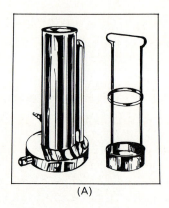

(A)

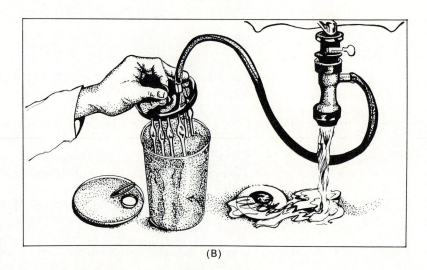

FIGURE 1.7
Pipet washers. (A) Pipet washer
with basket. (B) Micropipet
washer with aspirator.

(B)

should be transferred to a pipet rinser (Figure 1.7A). Pipet rinsers continually cycle fresh water through the pipets. Immediately after detergent
wash, tap water may be used to rinse the pipets, but distilled water should
be used for the final rinse. Pipets may then be dried in an oven.

Volumetric and graduated pipets are cleaned according to the above
procedure, but micropipets (lambda type) require special attention. Rinsing
water will not readily flow by gravity or capillary action through the
constrictions. Commercially available micropipet rinsers draw water or
other cleaning solutions through the pipets by suction (Figure 1.7B). The
pipets may be dried by ethanol rinsing or, preferably, in an oven at
100°C.

E. PREPARATION AND STORAGE OF SOLUTIONS

Water Quality

The most common solvent for solutions used in the biochemical laboratory is water. Tap water should never be used for the preparation of any reagent solutions. Some form of purified water (deionized or distilled) should be used for all procedures. The water quality necessary will depend on the solution to be prepared and on the biochemical procedure to be investigated. Most academic laboratories are equipped with "in house" water, which is either deionized or distilled. Water that is purified by passing through a mixed bed ion-exchange resin (deionized) will be low in the concentration of metal ions, but may contain organics that are washed from the polymeric resin. These contaminants will increase the ultraviolet absorbance properties of water. If sensitive ultraviolet absorbance measurements are to be made, distilled water is better than deionized.

Distilled water may be produced in a metal still or one in which it contacts only glass (glass-distilled water). Clearly, water in contact with metals will contain mineral ions that may intefere with reagents or biomaterials. This water will, however, be very low in organic contamination if it is distilled at a relatively slow rate.

For most procedures in the biochemistry laboratory, glass-distilled water is the choice. Several varieties of glass stills are commercially available. A glass still should operate only on deionized or distilled water. Use of tap water will cause a build-up of solid salts, which leads to "bumping" in the boiling flask. The still must be drained and cleaned with 10% hydrochloric acid almost daily if the tap water has a high mineral content. Organic contaminants can be removed during glass-distillation of water by maintaining a slow flow rate of cooling water through the condenser coils. The organics are "steam-distilled" and are carried out in the vapor escaping from the condenser. The instructions for glass stills give the most efficient flow rate for entry of water into the still and for condenser-cooling water.

All water will become contaminated upon storage. Ultraviolet-absorbing organics will be leached from some plastic containers, and metal ions may be present on glass surfaces. Bacterial contamination of glass-distilled water is not uncommon. A good practice is to tightly cap the purified water and store it in a refrigerator. If there is obvious bacterial growth (turbidity) in stored water, discard it.

Ultrapure water is often required for absorbance or fluorescence measurements. Distillation over acid permanganate (1 liter of water, 1 g of potassium permanganate, and 1 ml of 85% phosphoric acid in a glass still) will destroy organic contaminants by oxidation.

In summary, it is difficult, if not impossible, to remove all contaminants from large volumes of water. Compromises will have to be made in the choice of water. When deciding the method of water purification and storage, one must determine which type of contaminant has the least potential for interference. For all procedures described in this book, glass-distilled water is of sufficient purity.

Reagent and Solution Preparation

With only a few exceptions, reagents and chemicals used in this manual can be obtained from commercial sources and used without further purification. If a specific grade of chemical is required for an experiment, this will be stated in the instructions for solution preparation.

The concentrations for most biochemical solutions are in units of molarity, % (weight or volume), or weight per volume (mg/ml, g/liter, etc.). Solids should be weighed on weighing paper or plastic weighing boats, using an analytical or top-loading balance. Liquids are more conveniently dispensed by volumetric techniques; however, this assumes that the density is known. If a small amount of a liquid is to be weighed, it should be added to a tared flask by means of a disposable Pasteur pipet with a latex bulb. The hazardous properties of all materials should be known before use and the proper safety precautions obeyed.

The storage conditions of reagents and solutions are especially critical. Although some will remain stable indefinitely at room temperature, it is good practice to store all solutions in a closed container and in a refrigerator if they are prepared in advance. This inhibits bacterial growth and decomposition of the reagents. Some solutions may require storage below 0°C. If these are aqueous solutions or others that will freeze, be sure there is room for expansion inside the container.

All stored containers, whether at room temperature, 4°C, or below freezing, must be properly sealed. This reduces contamination by bacteria and vapors in the laboratory air (CO_2, NH_3, HCl, etc.). Volumetric flasks, of course, have glass stoppers, but test tubes, Erlenmeyers, bottles and other containers should be sealed with screw caps, corks, or hydrocarbon foil (Parafilm). Remember that hydrocarbon foil, a wax, is dissolved by solutions containing hydrocarbon and other organic solvents.

Bottles of pure chemicals and reagents should also be properly stored. Many manufacturers now include the best storage conditions for a reagent on the label. The common conditions are: store at room temperature, store at 0–4°C, store below 0°C, and store in a desiccator at 0–4°C or 0°C. Most biochemical reagents form hydrates by taking up moisture from the air. If the water content of a reagent increases, the molecular weight and purity of the reagent change. For example, when NAD^+ is purchased, the label usually reads: "Anhydrous molecular weight = 663.5;

when assayed, contained 3 H_2O per mole." The actual molecular weight that should be used for solution preparation is $663.5 + (18)(3) = 717.5$. However, if this reagent is stored in a moist refrigerator or freezer outside of a desiccator, the moisture content will increase to an unknown value.

If the proper method of storage for a specific reagent is in doubt, you are usually safe by refrigerating it at $4°C$ in a desiccator containing a drying agent.

F. THE BIOCHEMICAL LITERATURE

Experimental biochemists do not spend all their working time in the laboratory. An important component of a biochemistry research project is a search of the biochemical literature. The library should be considered as a tool for experimental biochemistry in the same way as any scientific instrument. The use of the biochemical literature by the student in biochemistry laboratory is not as extensive as that of a full-time researcher, but you must be aware of what is available in the library and how to use it.

The library is used in all stages of research. Before an investigator can begin experimentation, a research idea must be generated. This idea develops only after extensive reading and study of the literature. A research project usually begins in the form of a problem to be solved. For ease of solution, a major project is subdivided into questions that may be answered by experimentation. Before laboratory work can begin, the researcher must have a knowledge of the past and current literature dealing with the research area. In a sense, this can be reduced to two questions: What is the current state of knowledge in the area? What are the significant unknowns? These questions can be answered only by developing a familiarity with the biochemical literature. The researcher will find that this knowledge of the literature is also invaluable for the design of experiments. The development of experiments requires a knowledge of techniques and laboratory procedures. Excellent methods books and journals are available that provide experimental details. Finally, during the performance of experiments, there is often the need for physical and chemical constants and miscellaneous information. Various handbooks and encyclopedias are excellent for this purpose.

The beginning student in biochemistry laboratory will not be expected to proceed through all the above stages in the design of an experiment. However, a familiarity with the literature will increase your understanding of the experiment and may aid in the development of more effective methods. When you do begin a research program, you will be able to use the library to fullest advantage.

The Literature of Biochemistry

The biochemical literature is massive. It is almost a full-time job just to maintain a current awareness of a specialized research area. There are no discipline boundaries in the study of biochemistry. The biochemical literature overlaps into the biological sciences, the physical sciences, and the basic medical sciences. The intent of this section is to bring some order to the many textbooks, reference books, research journals, computer information retrieval services, and handbooks that are available. Each component of the biochemical literature will be dealt with individually.

Textbooks. The student's first exposure to biochemistry is probably a lecture course accompanied by the reading of a general textbook of biochemistry. By providing an in-depth survey of biochemistry, textbooks allow students to build a strong foundation of important principles and concepts. By the time most books are in print, the information is two to three years old, but textbooks still should be considered the starting point for mastery of biochemistry. The textbooks published since 1980 that have been most widely adopted for the 1-year undergraduate biochemistry course are:

R.C. Bohinski, *Modern Concepts in Biochemistry*, 4th Edition (1983), Allyn and Bacon (Boston).

A.L. Lehninger, *Biochemistry* (1982), Worth Publishers, Inc. (New York).

J.D. Rawn, *Biochemistry* (1983), Harper and Row (New York).

E.L. Smith, R.L. Hill, I.R. Lehman, R.J. Lefkowitz, P. Handler, and A. White, *Principles of Biochemistry*, 7th Edition (1983), McGraw-Hill (New York). This textbook is available in two volumes—*General Aspects* and *Mammalian Biochemistry*.

L. Stryer, *Biochemistry*, 2nd Edition (1981), W.H. Freeman (San Francisco).

G. Zubay, *Biochemistry* (1983), Addison-Wesley (Reading, MA).

Reference Books and Review Publications. For more specialized and detailed biochemical information that is not offered by textbooks, reference books must be used. Reference works range from general surveys to specialized series. The best works are those multivolume sets that continue publication of volumes on a periodical basis. Each volume usually covers a specialized area with articles written by recognized authorities in the field. Some of the more familiar reference and review publications are listed in Table 1.1. It should be noted that reference articles of interest to biochemists are often found in publications that are not strictly biochemical.

Research Journals. The main core of the biochemical literature consists of research journals. It is essential for a practicing biochemist to

TABLE 1.1
Reference Books in Biochemistry and Related Areas

General Reference/Survey Books

P. Boyer, *The Enzymes*, 3rd Edition, Academic Press, more than sixteen volumes.

C. Cantor and P. Schimmel, *Biophysical Chemistry*, Freeman, three volumes.

Volume I: The Conformation of Biological Macromolecules
Volume II: Techniques for the Study of Biological Structure and Function
Volume III: The Behavior of Biological Macromolecules

M. Florkin and E. Stolz, *Comprehensive Biochemistry*, Elsevier, more than thirty-five volumes.

Reviews

Advances in Cancer Research

Advances in Carbohydrate Chemistry

Advances in Cell Biology

Advances in Comparative Physiology and Biochemistry

Advances in Enzyme Regulation

Advances in Enzymology and Related Areas of Molecular Biology

Advances in Immunology

Advances in Lipid Research

Advances in Pharmacology and Chemotherapy

Advances in Protein Chemistry

Annals of the New York Academy of Science

Annual Reports in Medicinal Chemistry

Annual Reports of the Chemical Society

Annual Review of Biochemistry

Annual Review of Biophysics and Bioengineering

Annual Review of Genetics

Annual Review of Microbiology

Annual Review of Pharmacology

Annual Review of Physiology

Annual Review of Plant Physiology

Bacteriological Reviews

Biochemical Society Symposia

Biological Reviews

Chemical Reviews

Cold Spring Harbor Symposia in Quantitative Biology

Essays in Biochemistry

Harvey Lectures

International Review of Cytology

Physiological Reviews

Progress in Biophysics and Biophysical Chemistry

Progress in the Chemistry of Fats and Other Lipids

Progress in Nucleic Acid Research and Molecular Biology

Quarterly Review of Biology

Quarterly Reviews of Biophysics

Quarterly Review of the Chemical Society

Trends in Biochemical Sciences

Vitamins and Hormones

maintain a knowledge of biochemical advances in his or her field of research and related areas. Scores of research journals are published with the intent of keeping scientists "up-to-date." With the expansion of scientific information has come the need for efficient storage and use of research journals. Many publishers are now providing journals in forms such as microcards, microfilm, and microfiche. Some of the major research journals in biochemistry and related fields are listed in Table 1.2.

TABLE 1.2
Research Journals in Biochemistry and Related Areas

Acta Chemica Scandinavia	Inorganica Chemica Acta
Accounts of Chemical Research	International Journal of Biochemistry
American Journal of Physiology	Journal of the American Chemistry Society
Analytical Biochemistry	Journal of the American Oil Chemists' Society
Angewandte Chemie (English International Edition)	Journal of Applied Biochemistry
Archives of Biochemistry and Biophysics	Journal of Bacteriology
Biochemical and Biophysical Research Communications	Journal of Biochemistry (Tokyo)
Biochemical Education	Journal of Bioenergetics
Biochemical Journal	Journal of Biological Chemistry
Biochemistry	Journal of Cell Biology
Biochimica et Biophysica Acta	Journal of Cellular Physiology
Biokhimiya (Russian, translated into English)	Journal of Chemical Education
Bioelectrochemistry and Bioenergetics	Journal of Chromatography
Bioorganic Chemistry	Journal of Clinical Investigation
Biophysical Journal	Journal of Colloid and Interface Science
Biopolymers	Journal of Electron Microscopy
Biotechniques	Journal of Experimental Biology
Biotechnology and Bioengineering	Journal of General Microbiology
Biopolymers	Journal of General Virology
Brain Research	Journal of Histochemistry
Bulletin de la Societe Chimie Biologique	Journal of Immunology
Canadian Journal of Biochemistry	Journal of Inorganic Biochemistry
Canadian Journal of Microbiology	Journal of Lipid Research
Cell and Tissue Kinetics	Journal of Medicinal Chemistry
Chemistry and Physics of Lipids	Journal of Membrane Biology
Comparative Biochemistry and Physiology	Journal of Molecular Biology
Comptes Rendus de l'Academie des Sciences (Paris), Parts C and D	Journal of Neurochemistry
	Journal of Nutrition
Comptes Rendus des Seances de la Societe de Biologie	Journal of Organic Chemistry
European Journal of Biochemistry	Journal of Pharmacology and Experimental Therapeutics
Experimental Cell Research	
Experientia	Journal of Physical Chemistry
Federation Proceedings	Journal of Physiology
Federation of European Biochemical Societies (FEBS Letters)	Journal of Protein Chemistry
	Journal of the Chemical Society (London)
General Cytochemical Methods	Journal of Theoretical Biology
Hoppe-Seyler's Zeitschrift für Physiologische Chemie	Journal of Ultrastructure Research
Immunology	Journal of Virology
Indian Journal of Biochemistry	

TABLE 1.2 Continued
Research Journals in Biochemistry and Related Areas

Lipids	Plant Physiology
Macromolecules	Proceedings of the National Academy of Sciences (U.S.)
Metabolism—Clinical and Experimental	Proceedings of the Royal Society (London), Part B
Microchemical Journal	Proceedings of the Society for Experimental Biology and Medicine
Molecular and General Genetics	
Molecular Pharmacology	Prostaglandins
Mutation Research	Protoplasma
Nature	Science
Naturwissenschaften	Steroids
New England Journal of Medicine	Tetrahedron Letters
Physiological Chemistry and Physics	Toxicology Letters
Phytochemistry	Zeitschrift für Naturforschung

Methodology References. The active researcher continually has a need for new methods and techniques. Several publications specialize in providing details of research methods. Some of the major biochemical methodology publications are:

Analytical Biochemistry, a monthly journal.

Analytical Chemistry, a monthly journal.

Biochemical Preparations, an annual volume that contains laboratory methods for the preparation of biochemically significant materials.

Laboratory Techniques in Biochemistry and Molecular Biology, T.S. Work and R.G. Burdon, Eds. (formerly T.S. Work and E. Work). Each volume in the series is concentrated in an area of biochemistry and written by recognized authorities.

Methods of Biochemical Analysis, D. Glick, Ed. Contains techniques for the analysis and preparation of biochemicals.

Methods of Enzymatic Analysis (3rd Edition), H. Bergmeyer, Ed. Contains methods for enzyme purification and assay, in eleven volumes.

Methods in Enzymology, various editors. The most valuable methods series available. Each volume contains numerous articles describing biochemical techniques. The series is well indexed and easy to use. Over one hundred volumes.

Techniques of Organic Chemistry, A. Weissberger, Ed. Chemical techniques are covered, but of some value to biochemists.

Techniques in Protein Chemistry, J. Bailey. An excellent reference for the protein chemist, with coverage of all techniques for the study of amino acids and proteins.

Collections of Chemical and Biochemical Data. Students in introductory biochemistry laboratory may use books in this category more than any other type. While doing biochemical experiments, you will need physical, chemical, and biochemical information such as R_f values, K_M values, molecular weights, amino acid sequences of peptides and proteins, and pK values. This information is most easily found in the many handbooks and collections of biochemical data. You should be introduced to the use of these references early in your training. Some of the most useful handbooks are listed below with a brief description of contents.

CRC Handbook of Biochemistry and Molecular Biology, 3rd Edition, G.D. Fasman, Ed. (1977). This handbook consists of seven volumes covering physical and chemical data of proteins, nucleic acids, lipids, and carbohydrates.

Concise Encyclopedia of Biochemistry, T. Scott and M. Brewer, Eds. (1983), Walter De Gruyter, Inc. (Hawthorne, NY). Contains 4200 entries of biochemical terms.

Dictionary of Biochemistry, J. Stenesh (1975), Wiley Interscience. Contains definitions of 12,000 biochemical terms.

The Merck Index, 10th Edition, M. Windholz (1983), Merck and Co., Inc. (Rahway, NJ). An encyclopedia of chemicals, drugs and biological substances.

Worthington Enzyme Manual, L. Decker (1977), Worthington Biochemical Corp. (Freehold, NJ). Contains detailed assay information and references for over 100 enzymes.

Brochures of Commercial Suppliers. Some of the most up-to-date information on biochemical techniques, methods, and instrumentation is available from the many companies that manufacture and supply equipment and reagents. There is usually no charge for these brochures.

Aids for Literature Search. You will often need to complete a thorough literature search on some specialized area or topic. It is not practical to survey every reference book or journal for articles related to the topic. Two publications that provide brief summaries of each published article are *Chemical Abstracts* and *Biological Abstracts*. These periodical indexes provide rapid access to research articles that have been published by the many journals. These abstracts also survey patents and articles in new reference books. If you are not familiar with the use of these abstracts, ask your instructor or librarian for assistance.

Research articles of interest to biochemists may appear in many types of research journals. Research libraries do not have the funds necessary to subscribe to every journal, nor do scientists have the time to survey every current journal copy for articles of interest. Two publications that assist scientists in current awareness of published articles are *Chemical*

Titles (published every 2 weeks by the American Chemical Society) and the weekly *Current Contents* (published by the Institute of Science Information). *Chemical Titles* uses a keyword listing of article titles, which is indexed to a list of articles in a current journal. *Current Contents* lists the Table of Contents from most current journals. The Life Science edition of *Current Contents* is the most useful for biochemists.

The computer revolution has reached into the chemical and biochemical literature. Many college and university libraries now subscribe to computer bibliographic search services. By placing a telephone call to these computer banks, it is possible to obtain a complete listing of journal articles pertaining to a particular chemical structure, subject area, or topic. A proper choice of keywords and cross-references is essential in order to achieve a selective and useful search. Two information retrieval services available at the present time are CAS ONLINE by Chemical Abstracts (American Chemical Society) and Lockheed Dialog.

In conclusion, training in biochemical experimentation is not complete without some experience in the library. A familiarity with the biochemical literature will assist you in many ways. You will gain a perspective on the early development of biochemistry, an awareness of the current state of our knowledge, and an insight into the future of biochemistry.

REFERENCES

Safety

Safety in Academic Chemistry Laboratories, A Publication of the American Chemical Society Committee on Chemical Safety, 3rd Edition (1979).

L. Bretherick, *Handbook of Reactive Chemical Hazards*, 2nd Edition (1979), Butterworths (London).

Guide for Safety in the Chemical Laboratory, 2nd Edition (1972), Van Nostrand Reinhold Co., Litton Educational Publishing, Inc. (New York).

"Laboratory Health and Safety Seminar," *J. Chem. Ed.*, *55*, 140–150 (1972). A series of papers on academic laboratory safety.

D. Kingston, *J. Chem. Ed.*, *58*, A289 (1981). "Contact Lenses in the Laboratory."

Prudent Practices for Handling Hazardous Chemicals in Laboratories, (1981), National Research Council, National Academy Press (Washington, D.C.).

Writing Laboratory Reports

Robert A. Day, *How to Write and Publish a Scientific Paper*, 2nd Edition (1983), ISI Press (Philadelphia).

Chapter 2

General Laboratory Procedures

All biochemical laboratory activities, whether in teaching, research, or industry, are replete with techniques that must be carried out almost on a daily basis. This chapter outlines the theoretical and practical aspects of some of these general and routine procedures, including use of buffers, pH and other electrodes, dialysis, membrane filtration, lyophilization, and methods for protein measurement.

A. pH MEASUREMENTS, BUFFERS, AND ELECTRODES

Biochemical processes occurring in cells and tissues depend upon strict regulation of the hydrogen ion concentration. Biological pH is maintained at a constant value by natural buffers. When biological processes are studied *in vitro*, artificial media must be prepared that mimic the cell's natural environment. Because of the dependence of biochemical reactions on pH, the accurate determination of hydrogen ion concentration has always been of major interest. Today, we consider the measurement and control of pH to be a simple and rather mundane activity. However, an inaccurate pH measurement or a poor choice of buffer can lead to failure in the biochemistry laboratory. You should become familiar with several aspects of pH measurement, electrodes, and buffers.

Measurement of pH

A pH measurement is usually taken by immersing a glass combination electrode into a solution and reading the pH directly from a meter. At

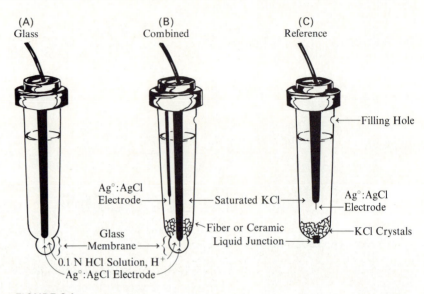

(A)
Glass

(B)
Combined

(C)
Reference

Filling Hole

Ag°:AgCl
Electrode

Saturated KCl

Ag°:AgCl
Electrode

Glass
Membrane

Fiber or Ceramic
Liquid Junction

KCl Crystals

0.1 N HCl Solution, H$^+$
Ag°:AgCl Electrode

FIGURE 2.1
pH electrodes. From "The Tools of Biochemistry," by T. G. Cooper. Copyright 1977, Wiley-Interscience. (A) pH-dependent glass electrode. (B) Combination electrode. (C) Calomel reference electrode.

one time, pH measurements required two electrodes, a pH-dependent glass electrode sensitive to H$^+$ ions and a pH-independent calomel reference electrode (Figure 2.1A,C). The potential difference that develops between the two electrodes is measured as a voltage defined by Equation 2.1.

$$V = E_{\text{constant}} + \frac{2.303RT}{F}\, \text{pH} \qquad \text{(Equation 2.1)}$$

where

$$V = \text{voltage of the completed circuit}$$
$$E_{\text{constant}} = \text{potential of reference electrode}$$
$$R = \text{the gas constant}$$
$$T = \text{the absolute temperature}$$
$$F = \text{the Faraday constant}$$

A pH meter is standardized with buffer solutions of known pH before a measurement of an unknown solution is taken. It should be noted from Equation 2.1 that the voltage depends on temperature. Hence, all pH meters have a knob labeled "temperature control," which is adjusted for the temperature of the measured solution.

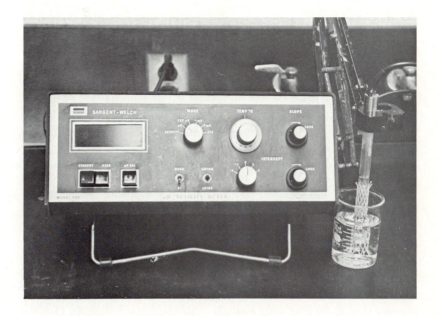

FIGURE 2.2
pH meter. Photo courtesy of Mr. Mark Billadeau.

Most pH measurements today are obtained using a single **combination electrode** (Figure 2.1B). Both the reference and the pH-dependent electrode are contained in a single glass tube. Although these are more expensive than dual electrodes, they are much more convenient to use, especially for smaller volumes of solution. The use of a pH meter with a combination electrode is relatively easy, but certain guidelines must be followed (see Figure 2.2). A pH meter not in use is left in a "standby" position. Before use, check the level of saturated KCl in the electrode (Figure 2.1B). If it is low, check with your instructor for the filling procedure. Turn the temperature control to the temperature of the standard calibration buffers and the test solutions. Be sure the function dial is set to pH. While not in use, the electrode will be stored in distilled water or buffer. Lift the electrode out of the storage solution and rinse it with distilled water from a wash bottle, and *gently* clean and dry the electrode with a tissue. Immerse the electrode into a standard buffer. Common standard buffers are pH 4, 7, and 10 with accuracy of ± 0.02 pH unit. The standard buffer should have a pH within two pH units of the expected pH of the test solution. The bulb of the electrode must be completely covered with solution. Turn the pH meter to "on" or "read" and adjust the meter with the "calibration dial" (sometimes called "intercept") until the proper pH of the standard buffer is indicated on the dial. Turn the pH meter to "standby" position. Remove the electrode and again rinse with distilled water and carefully blot dry with tissue. Immerse the electrode into a standard buffer of different pH and turn the pH meter to "read." The dial should read within ± 0.05 pH units of the known value. If not, adjust to the proper pH and again check the first standard pH buffer. Clean the electrode and immerse it into the test solution. Record the pH of the test solution.

As with all delicate equipment, the pH meter and electrode must receive proper care and maintenance. All glass electrodes should be stored in distilled water or buffer and kept clean. Normally, rinsing with distilled water and careful blotting with tissue paper is sufficient. Glass electrodes are fragile and expensive, so they must be handled with care. If pH measurements of protein solutions are often taken, a protein film may develop on the electrode; it can be removed by soaking in 0.1 N HCl, cleaning with dilute detergent, and rinsing well with water.

Measurements of pH are always susceptible to experimental errors. Some common problems are:

1. The Sodium Error. Most glass combination electrodes are sensitive to Na^+ as well as H^+. The sodium error can become quite significant at high pH values, where 0.1 N Na^+ may decrease the measured pH by 0.4 to 0.5 units. Several things may be done to reduce the sodium error. Some commercial suppliers of electrodes provide a standard curve for sodium error correction. Newer glass electrodes that are virtually Na^+-impermeable are now commercially available. If neither a standard curve nor a special glass electrode is available, potassium salts may be substituted for sodium salts.

2. Concentration Effects. The pH of a solution will vary with the concentration of buffer ions or other salts in the solution. This is because the pH of a solution depends on the **activity** of an ionic species, not on the concentration. **Activity**, you may recall, is a thermodynamic term used to define species in a nonideal solution. At infinite dilution, the activity of a species is equivalent to its concentration. At finite dilutions, however, the activity of a solute and its concentration are not equal.

It is common practice in biochemical laboratories to prepare concentrated "stock" solutions and buffers. These are then diluted to the proper concentration when needed. Because of the concentration effects described above, it is important to adjust the pH of these solutions *after* dilution.

3. Temperature Effects. The pH of a buffer solution is influenced by temperature. This effect is due to a temperature-dependent change of the dissociation constant (pK_a) of ions in solution. The pH of the commonly used buffer "Tris" is greatly affected by temperature changes, with a $\Delta pK_a/C°$ of -0.031. This means that a pH 7.0 Tris buffer made up at 4°C would have a pH of 5.95 at 37°C. The best way to avoid this problem is to prepare the buffer solution at the temperature at which it will be used, and to standardize the electrode with buffers at the same temperature as the solution you wish to measure.

Biochemical Buffers

Buffer ions are used to maintain solutions at constant pH values. The selection of a buffer for use in the investigation of a biochemical process is of critical importance. Before the characteristics of a buffer system are discussed, we will review some concepts in acid-base chemistry.

Weak acids and bases do not completely dissociate in solution but exist as equilibrium mixtures (Reaction 2.1).

$$HA \underset{k_2}{\overset{k_1}{\rightleftharpoons}} H^+ + A^- \qquad \text{(Reaction 2.1)}$$

HA represents a weak acid and A^- represents its conjugate base; k_1 represents the rate constant for dissociation of the acid and k_2 the rate constant for association of the conjugate base and hydrogen ion. The equilibrium constant, K_a, for the weak acid HA is defined by Equation 2.2,

$$K_a = \frac{k_1}{k_2} = \frac{[H^+][A^-]}{[HA]} \qquad \text{(Equation 2.2)}$$

which can be rearranged to define $[H^+]$ (Equation 2.3).

$$[H^+] = \frac{K_a[HA]}{[A^-]} \qquad \text{(Equation 2.3)}$$

The $[H^+]$ is often reported as pH, which is $-\log[H^+]$. In a similar fashion, $-\log K_a$ is represented by pK_a. Equation 2.3 can be converted to the $-\log$ form by substituting pH and pK_a:

$$pH = pK_a + \log \frac{[A^-]}{[HA]} \qquad \text{(Equation 2.4)}$$

Equation 2.4 is the familiar Henderson-Hasselbalch equation, which defines the relationship between pH and the ratio of acid and conjugate base concentrations. The Henderson-Hasselbalch equation is of great value in buffer chemistry because it can be used to calculate the pH of a solution if the molar ratio of buffer ions ($[A^-]/[HA]$) and the pK_a of HA are known. Also, the molar ratio of HA to A^- that is necessary to prepare a buffer solution at a specific pH can be calculated if the pK_a is known.

A solution containing both HA and A^- has the capacity to resist changes in pH, i.e., it acts as a **buffer**. If acid (H^+) were added to the buffer solution, it would be neutralized by A^- in solution:

$$H^+ + A^- \longrightarrow HA \qquad \text{(Reaction 2.2)}$$

Base (OH^-) added to the buffer solution would be neutralized by reaction with HA:

$$OH^- + HA \longrightarrow A^- + H_2O \qquad \text{(Reaction 2.3)}$$

The most effective buffering system contains equal concentrations of the acid, HA, and the conjugate base, A^-. According to the Henderson-Hasselbalch equation (2.4), when $[A^-]$ is equal to $[HA]$, pH equals pK_a. Therefore, the pK_a of a weak acid-base system represents the center of the buffering region. The effective range of a buffer system is generally two pH units, centered at the pK_a value (Equation 2.5).

$$\text{Effective pH range for a buffer} = pK_a \pm 1 \qquad \text{(Equation 2.5)}$$

Selection of a Biochemical Buffer

Virtually all biochemical investigations must be carried out in buffered aqueous solutions. The natural environment of biomolecules and cellular organelles is under strict pH control. When these components are extracted from cells, they are most stable if maintained in their normal pH range of 6 to 8. An artificial buffer system is found to be the best sub-

FIGURE 2.3
Effective buffering ranges of several common buffers.

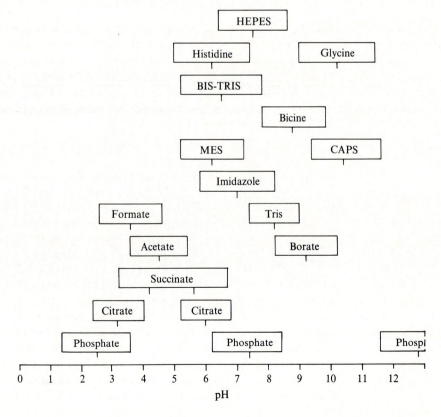

stitute for the natural cell milieu. It should also be recognized that many biochemical processes (especially some enzyme systems) produce or consume hydrogen ions. The buffer system neutralizes these solutions and maintains a constant chemical environment.

Although most biochemical solutions require buffer systems effective in the pH range of 6 to 8, there is occasionally the need for buffering over the pH range of 2 to 12. Obviously no single acid-conjugate base pair will be effective over this entire range, but several buffer systems are available that may be used in discrete pH ranges. Figure 2.3 compares the effective buffering ranges of common biological buffers. It should be noted that some buffers (phosphate, succinate, and citrate) have more than one pK_a value, so they may be used in different pH regions. Many buffer systems are effective in the usual biological pH range (6 to 8); however, there may be major problems in their use. Several characteristics of a buffer must be considered before a final selection is made. Following is a discussion of the advantages and disadvantages of the commonly used buffers.

Phosphate Buffers

The phosphates are among the most widely used buffers. These solutions have high buffering capacity and are very useful in the pH range of 6.5 to 7.5. Sodium or potassium phosphate solutions of all concentrations are easy to prepare. The major disadvantages of phosphate solutions are (1) precipitation or binding of common biological cations (Ca^{2+} and Mg^{2+}) and (2) inhibition of some biological processes, including some enzymes.

Tris Buffer

The use of the synthetic buffer Tris, [tris(hydroxymethyl)aminomethane], is now probably greater than that of phosphate. It is useful in the pH range of 7.5 to 8.5. Tris is available in a basic form as highly purified crystals, which makes buffer preparation especially convenient. To prepare solutions, the appropriate amount of Tris base is weighed and dissolved in water. For one liter of a 0.1 M solution, 12.11 g (0.1 mole) of Tris base is weighed and dissolved in 950 to 975 ml of distilled water. The pH is adjusted by addition of acid, with stirring (concentrated hydrochloric if Tris-HCl is desired), until the appropriate pH is attained. Water is added to a final volume of 1 liter and a final pH check is made. Although Tris is a primary amine, it causes minimal interference with biochemical processes and does not precipitate calcium ions. However, Tris has several disadvantages, including (1) pH dependence on concentration, since the pH decreases 0.1 pH unit for each 10-fold dilution,

(2) interference with some pH electrodes, and (3) a large $\Delta pK_a/C°$ compared to most other buffers. Most of these drawbacks can be minimized by (1) adjusting the pH *after* dilution to the appropriate concentration, (2) purchasing electrodes that are compatable with Tris, and (3) preparing the buffer at the same temperature at which it will be used.

Carboxylic Acid Buffers

The most widely used buffers in this category are acetate, formate, citrate, and succinate. This group is useful in the pH range of 3 to 6, a region that offers few buffer choices. All of these acids are natural metabolites, so they may interfere with the biological processes under investigation. Also, citrate and succinate may interfere by binding transition metal ions (Fe^{3+}, Zn^{2+}, Mg^{2+}, etc.). Formate buffers are especially useful because they are volatile and can be removed by evaporation under reduced pressure.

Borate Buffers

Buffers of boric acid are useful in the pH range of 8.5 to 10. Borate has the major disadvantage of complex formation with many metabolites, especially carbohydrates.

Amino Acid Buffers

The most commonly used amino acid buffers are glycine (pH 2 to 3 and 9.5 to 10.5) histidine (pH 5.5 to 6.5), glycine amide (pH 7.8 to 8.8), and glycylglycine (pH 8 to 9). These provide a more "natural" environment to cellular components and extracts; however, they may interfere with some biological processes as do the carboxylic acid and phosphate buffers.

Zwitterionic Buffers (Good's Buffers)

In the mid-1960s, N.E. Good and his colleagues recognized the need for a set of buffers specifically designed for biochemical studies (Good et al., 1966, 1972). He and others noted major disadvantages of the established buffer systems. Good outlined several characteristics essential in a biological buffer system. They are:

1. pK_a between 6 and 8.
2. Highly soluble in aqueous systems.
3. Exclusion or minimal transport by biological membranes.

TABLE 2.1
Structure and Properties of Several Synthetic Zwitterionic Buffers

Buffer Name	Abbreviation	Structure (All structures shown in salt form)	pK_a (20°C)	Useful pH Range	$\Delta pK_a/C°$	Concentration of a Saturated Solution, M, 0°C
N-2-acetamido-2-amino-ethanesulfonic acid	ACES	$H_2NCOCH_2\overset{+}{N}H_2CH_2CH_3SO_3^-$	6.9	6.4–7.4	−0.020	0.22
N-2-acetamidoiminodiacetic acid	ADA	$H_2NCOCH_2\overset{+}{N}\!\!\underset{H}{\overset{CH_2COO^-}{\diagdown}}{}_{CH_2COO^-}$	6.6	6.2–7.2	−0.011	—
N,N-bis(2-hydroxyethyl)-2-aminoethanesulfonic acid	BES	$(HOCH_2CH_2)_2\overset{+}{N}HCH_2CH_2SO_3^-$	7.15	6.6–7.6	−0.016	3.2
N,N-bis(2-hydroxyethyl)-glycine	Bicine	$(HOCH_2CH_2)_2\overset{+}{N}HCH_2COO^-$	8.35	7.5–9.0	−0.018	1.1
3-(cyclohexylamino)-propanesulfonic acid	CAPS	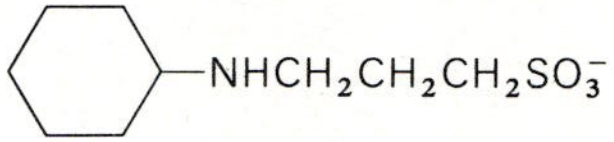 $-\overset{+}{N}HCH_2CH_2SO_3^-$ (cyclohexyl)	10.4	10.0–11.0	−0.009	0.85
cyclohexylaminoethane-sulfonic acid	CHES	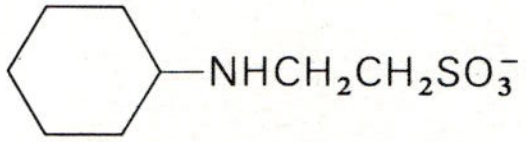 $-\overset{+}{N}HCH_2CH_2SO_3^-$ (cyclohexyl)	9.5	9.0–10.0	−0.009	0.85
glycylglycine	gly-gly	$H_3^+NCH_2CONHCH_2COO^-$	8.4	7.5–9.5	−0.028	1.1
N-2-hydroxyethyl-piperazine-N'-2-ethanesulfonic acid	HEPES	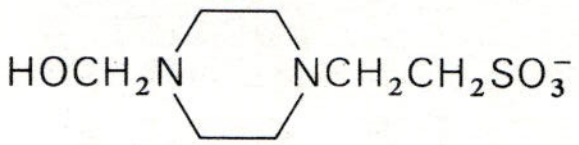 $HOCH_2N\!\!\diagup\!\!\diagdown\overset{+}{N}CH_2CH_2SO_3^-$	7.55	7.0–8.0	−0.014	2.25
2-(N-morpholino)ethane-sulfonic acid	MES	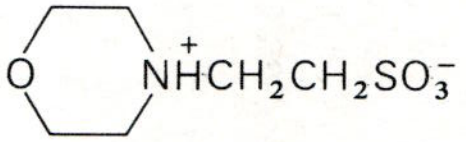 $O\!\!\diagup\!\!\diagdown\overset{+}{N}HCH_2CH_2SO_3^-$	6.15	5.8–6.5	−0.011	0.65
3-(N-morpholino)-propanesulfonic acid	MOPS	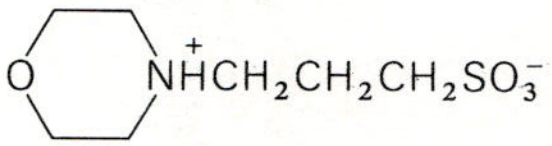 $O\!\!\diagup\!\!\diagdown\overset{+}{N}HCH_2CH_2CH_2SO_3^-$	7.20	6.5–7.9	—	—
piperazine-N,N'-bis-2-ethanesulfonic acid	PIPES	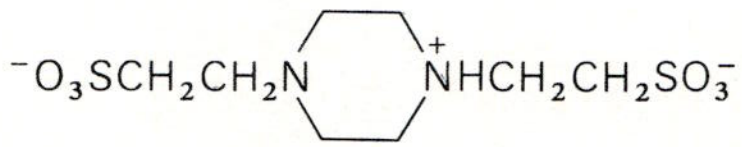 $^-O_3SCH_2CH_2\overset{+}{N}\!\!\diagup\!\!\diagdown\overset{+}{N}HCH_2CH_2SO_3^-$	6.8	6.4–7.2	−0.0085	—
N-tris(hydroxymethyl)methyl-2-aminoethane-sulfonic acid	TES	$(HOCH_2)_3\overset{+}{N}HCH_2CH_2SO_3^-$	7.5	7.0–8.0	−0.020	2.6
N-tris(hydroxymethyl)methylglycine	Tricine	$(HOCH_2)_3C\overset{+}{N}H_2CH_2COO^-$	8.15	7.5–8.5	−0.021	0.8
tris(hydroxymethyl)-aminomethane	Tris	$(HOCH_2)_3C\overset{+}{N}H_3$	8.3	7.5–9.0	−0.031	2.4

4. Minimal salt effects.

5. Minimal effects on dissociation due to ionic composition, concentration, and temperature.

6. Buffer–metal ion complexes nonexistent or soluble and well-defined.

7. Chemically stable.

8. Insignificant light absorption in the UV and VIS regions.

9. Readily available in purified form.

Good investigated a large number of synthetic zwitterionic buffers and found many of them to meet these criteria. Table 2.1 lists several of these buffers and their properties. Good's buffers are widely used, but their main disadvantage is high cost.

The Oxygen Electrode

Second in popularity only to the pH electrode is the oxygen electrode. This device is a polarographic electrode system that can be used to measure oxygen concentration in a liquid sample or to monitor oxygen uptake or evolution by a chemical or biological system. The only known interfering materials are H_2S and SO_2.

The basic electrode was developed by L.C. Clark, Jr. (Clark, 1953). The popular Clark-type oxygen electrode with biological oxygen monitor is shown in Figure 2.4. Two views of the oxygen electrode are shown in Figure 2.5. The electrode system, which consists of a platinum cathode and silver anodes, is molded into an epoxy block and surrounded by a

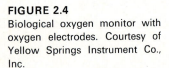

FIGURE 2.4
Biological oxygen monitor with oxygen electrodes. Courtesy of Yellow Springs Instrument Co., Inc.

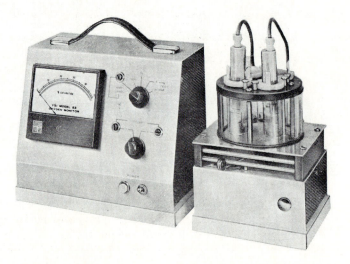

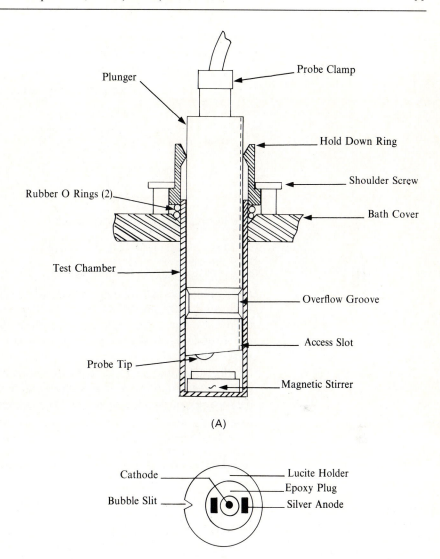

Plunger

Probe Clamp

Hold Down Ring

Shoulder Screw

Rubber O Rings (2)

Bath Cover

Test Chamber

Overflow Groove

Access Slot

Probe Tip

Magnetic Stirrer

(A)

Cathode

Lucite Holder

Epoxy Plug

Bubble Slit

Silver Anode

(B)

FIGURE 2.5
Two views of the Clark oxygen electrode. (A) Side view showing lucite plunger and probe tip courtesy of Yellow Springs Instrument Co., Inc. (B) End view showing anode and cathode from "The Tools of Biochemistry," by T. G. Cooper. Copyright 1977, Wiley-Interscience.

lucite holder. A thin Teflon membrane is stretched over the electrode end of the probe and held in place with an O-ring. The membrane separates the electrode elements, which are bathed in an electrolyte solution (saturated KCl), from the test solution in the sample chamber. The membrane is permeable to oxygen and other gases, which then come into contact with the surface of the platinum cathode. If a suitable polarizing voltage (about 0.8 volt) is applied across the cell, oxygen will be reduced at the cathode:

$$O_2 + 2e^- + 2H_2O \longrightarrow [H_2O_2] + 2OH^- \qquad \text{(Reaction 2.4)}$$

$$\underline{[H_2O_2] + 2e^- \longrightarrow 2OH^-} \qquad \text{(Reaction 2.5)}$$

$$\text{Total} \quad O_2 + 4e^- + 2H_2O \longrightarrow 4OH^- \qquad \text{(Reaction 2.6)}$$

The reaction occurring at the anode is:

$$4Ag^\circ + 4Cl^- \longrightarrow 4AgCl + 4e^- \qquad \text{(Reaction 2.7)}$$

The overall electrochemical process is:

$$4Ag^\circ + O_2 + 4Cl^- + 2H_2O \longrightarrow 4AgCl + 4OH^- \qquad \text{(Reaction 2.8)}$$

The current flowing through the electrode system is, therefore, directly proportional to the amount of oxygen passing through the membrane. According to the laws of diffusion, the rate of oxygen flowing through the membrane depends on the concentration of dissolved oxygen in the sample. Therefore, the magnitude of the current flow in the electrode is directly related to the concentration of dissolved oxygen. The function of the electrode can also be explained in terms of oxygen pressure. The oxygen concentration at the cathode (inside the membrane) is virtually zero because oxygen is rapidly reduced. Oxygen pressure in the sample chamber (outside the membrane) will force oxygen to diffuse through the membrane at a rate that is directly proportional to the pressure (or concentration) of oxygen in the test solution. The current generated by the electrode system is electronically amplified and transmitted to a meter, which usually reads in units of % oxygen saturation.

Many biochemical processes consume or evolve oxygen. The oxygen electrode is especially useful for monitoring such changes in the concentration of dissolved oxygen. If oxygen is consumed, for example, by a suspension of mitochondria or by an oxygenase enzyme system, the rate of oxygen diffusion through the probe membrane will decrease, leading to less current flow. A linear relationship exists between the electrode current and the amount of oxygen consumed by the biological system. Measurements of these changes in O_2 concentration are most conveniently made with a biological oxygen monitor equipped with a strip chart recorder.

The use of an oxygen electrode is not free of experimental and procedural problems. Probably the most significant source of trouble is the Teflon membrane. A torn, damaged, or dirty membrane is easily recognized by recorder noise or electronic "spiking." The membrane must not be touched with the fingers, contaminated with chemicals, or creased when it is stretched over the electrode. The membrane can also become coated with chemical reagents and biomolecules (especially proteins) present in the sample chamber. It is good practice to routinely inspect the membrane for damage and contamination with a low-power microscope. The silver anode should be cleaned frequently with dilute ammonium hydroxide in order to remove silver sulfide and maintain a bright surface.

Temperature control of the sample chamber and electrode is especially important since the rate of oxygen diffusion through the membrane depends on temperature. Commercial sample chambers are equipped with a bath assembly for temperature control. The bath assembly also contains a magnetic stirrer. It is essential that the solution in the sample chamber be constantly stirred in order to prevent oxygen depletion at the cathode.

Air bubbles in the sample chamber or in the KCl electrolyte solution inside the membrane lead to considerable error in measurements. When an air bubble (which may contain up to 20 times more oxygen than air-saturated water) is in the vicinity of the probe, erratic readings result from this rapid increase in oxygen concentration. If bubbles are present in the electrolyte solution inside the membrane, the membrane must be removed and carefully replaced to avoid bubbles. Air bubbles are easily removed from the sample chamber because the tightly fitting lucite plunger is slanted and contains a small groove or access slot. When the plunger is slowly pushed into the sample, air bubbles can be directed out this small groove. Reactants can also be added to the sample chamber through the access slot without removing the probe.

Successful results with an oxygen electrode will require some practice and an understanding of the limitations and characteristics of the instrument. Further reading and instructions are available from several sources (Fork, 1972, and Yellow Springs Instrument Co. Inc., 1976).

Ion Selective Electrodes

Although electrodes that respond to H^+ and O_2 are the most widely used, other ions and gases may be measured with specific electrodes and potentiometric methods. Table 2.2 lists some of the electrodes that are valuable in biochemical measurements. Each type of electrode is made from a specially prepared glass that has an increased permeability toward a single type of ion. Ion specific electrodes are sensitive (some are useful in the parts per billion range), require simple, inexpensive equipment (may be hooked to a pH meter set in the millivolt mode), and save time (after electrode calibration, an analysis takes about 1 minute). A major difficulty that has slowed the adoption of ion selective electrodes is interference by other ions. In most cases, the serious interfering ions are known (see Table 2.2), and procedures are available for masking or eliminating some competing ions (Orion Research, 1982).

TABLE 2.2
Ion Specific Electrodes

Ion	Concentration Range (M)	Interferences[1]
ammonium	10^0 to 10^{-6}	volatile amines
bromide	10^0 to 5×10^{-6}	S^{2-}, I^-
cadmium	10^0 to 10^{-7}	Ag^+, Hg^{2+}, Cu^{2+}, Pd^{2+}, Fe^{3+}
carbon dioxide	10^{-2} to 10^{-4}	Sr^{2+}, Mg^{2+}, Ba^{2+}, Ni^{2+}, volatile weak acids
chloride	10^0–8×10^{-6}	ClO_4^-, I^-, NO_3^-, SO_4^{2-}, Br^-, OH^-, OAc^-, HCO_3^-, F^-
cupric	saturated to 10^{-8}	S^{2-}, Hg^{2+}, Ag^+
cyanide	10^{-2} to 10^{-6}	S^{2-}, I^-, Br^-, Cl^-
divalent cations	10^0 to 6×10^{-6}	Na^+, Cu^{2+}, Zn^{2+}, Fe^{2+}, Ni^{2+}, Sr^{2+}, Bi^{2+}
fluoride	saturated to 10^{-6}	OH^-
iodide	10^0 to 2×10^{-5}	S^{2-}
lead	10^0 to 10^{-7}	Ag^+, Hg^{2+}, Cu^{2+}, Cd^{2+}, Fe^{3+}
nitrate	10^0 to 6×10^{-6}	I^-, Br^-, NO_2^-
nitrite	10^{-2} to 5×10^{-7}	CO_2
potassium	10^0 to 10^{-5}	Cs^+, NH_4^+, H^+
silver/sulfide	10^0 to 10^{-7}	Hg^{2+}
sodium	saturated to 10^{-6}	Cs^+, Li^+, K^+
thiocyanate	10^0 to 5×10^{-6}	OH^-, Br^-, Cl^-

[1] The presence of these ions does not rule out the use of a specific ion electrode, since many ions interfere only at relatively high concentrations.

B. DIALYSIS (McPHIE, 1971)

One of the oldest procedures applied to the purification and characterization of biomolecules is **dialysis**, an operation used to separate dissolved molecules on the basis of their molecular size. The technique involves sealing an aqueous solution containing both macromolecules and small molecules in a porous membrane. The sealed membrane is placed in a large container of low ionic strength buffer (Figure 2.6). The membrane pores are too small to allow diffusion of macromolecules greater than about 10,000 in molecular weight. Smaller molecules freely diffuse through the openings (Figure 2.7). The passage of small molecules continues until their concentrations inside the dialysis tubing and outside in the large volume of buffer are equal. Thus, the concentration of small molecules inside the membrane is reduced. Equilibrium is reached after 4 to 6 hours.

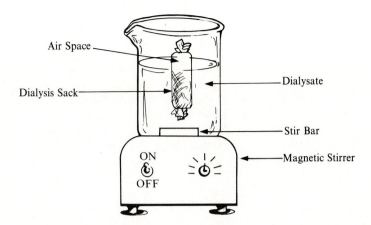

FIGURE 2.6
Typical set-up for dialysis.

Of course, if the outside solution (dialysate) is replaced with fresh buffer after equilibrium is reached, the concentration of small molecules inside the membrane will be further reduced by continued dialysis.

Dialysis membranes are available in a variety of materials and sizes. The most common materials are collodion, cellophane, and cellulose. Recent modifications in membrane construction make a range of pore sizes available. Spectrum Medical Industries sells Spectropor cellulose tubings with complete molecular weight cutoff ranging from 1000 to 50,000.

Commercial dialysis tubing is usually contaminated with glycerol, heavy metal ions, and traces of sulfur-containing compounds, and occasionally with hydrolytic enzymes (proteinases and nucleases), which must be removed before use. Typical preparation begins with cutting the tubing

FIGURE 2.7
Diffusion of smaller molecules through dialysis membrane pores.

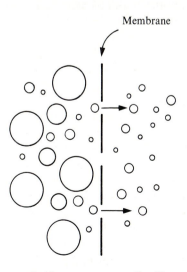

Inside   Outside

into convenient lengths (6″ to 8″), soaking it in 1% acetic acid for 1 hour, and then transferring it to glass-distilled water. To remove metal ions, the tubing is then boiled in basic EDTA solution (1% sodium carbonate, 10^{-3} M EDTA) for 1 hour. This step is repeated with a fresh portion of basic EDTA. The EDTA solution is poured off and heating is repeated about five times with distilled water. Recommended storage is in distilled water at 4°C with a small amount of sodium azide or a few drops of chloroform added as a bacterial-growth inhibitor. The treated tubing should not be handled with bare fingers, since it may become contaminated by hydrolytic enzymes.

Applications of Dialysis

Dialysis is most commonly used to remove salts and other small molecules from solutions of macromolecules. During the separation and purification of biomolecules, small molecules are added to selectively precipitate or dissolve the desired molecule. For example, proteins are often precipitated by addition of organic solvents or salts such as ammonium or sodium sulfate. Since the presence of organics or salts will usually interfere with further purification and characterization of the molecule, they must be removed. Dialysis is a simple, inexpensive, and effective method for removing all small molecules, ionic or nonionic.

Dialysis is also useful for the removal of small ions and molecules that are weakly bound to biomolecules. Protein cofactors such as NAD, FAD, and metal ions can often be dissociated by dialysis. The removal of metal ions is facilitated by the addition of a chelating agent (EDTA) to the dialysate.

Several precautions should be observed during dialysis set-up. Since biomolecules are usually less stable at elevated temperatures, dialysis is done in a refrigerated chamber at 3 to 4°C. The dialysate containing the sealed tube should be stirred very gently so that the small molecules are evenly dispersed throughout the solution and not concentrated in the vicinity of the dialysis bag. It is good practice to leave a small air bubble (about 10% of the total volume) in the dialysis sack when tying the final end. This insures that the sack will float and not be damaged by the rotating stir bar.

Although dialysis is still a popular method, it is being replaced by two recently developed techniques—gel filtration (see Chapter 3) and hollow-fiber filter dialysis (see below). The major disadvantage of dialysis that is overcome by the newer methods is that it may take several days of dialysis to attain a suitably low level of salts. The other methods require only 1 to 2 hours.

Hollow-fiber dialysis is more rapid because a bundle of very small-diameter porous tubing is used; this greatly increases the surface-to-volume ratio (see Figure 2.8A). The solution to be dialyzed is placed in the outside chamber, and water or low-ionic-strength buffer is slowly circulated through the tubes (see Figure 2.8B). Because of the concentration gradient, small molecules diffuse from the sample solution into the moving

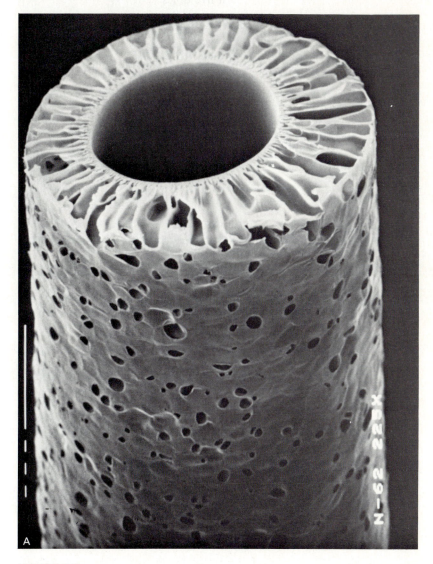

FIGURE 2.8
Hollow fiber-filter dialysis. (A) Electron micrograph of single hollow fiber courtesy of Amicon Corporation, Scientific Systems Division, Danvers, MA. (B) Apparatus for hollow fiber dialysis. Photo courtesy of Mr. Mark Billadeau.

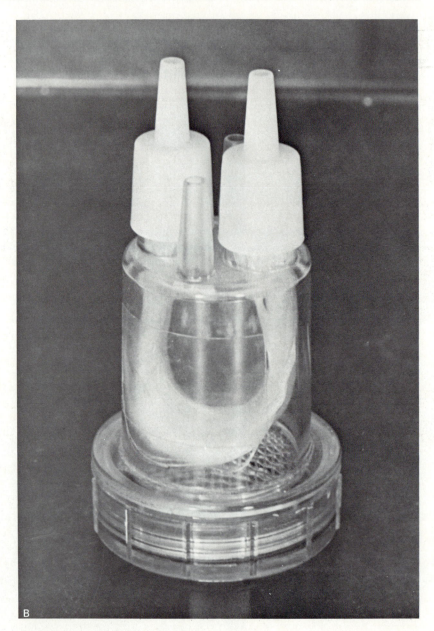

FIGURE 2.8   Continued

"dialysate" and are swept from the cell. The principles behind hollow-fiber dialysis and regular dialysis are identical except that the bundle of small tubes offers a much larger surface area of contact between the sample and the dialysate.

C. MICROFILTRATION

A modern variation of dialysis that is rapidly growing in popularity is microfiltration. This technique involves the removal of very small particles (micron or submicron size) from a solution by passage through a porous membrane. Two applications of membrane filtration are obvious: (1) clarification of turbid solutions by removal of small particles and (2) collection of fine precipitates for analysis. Further applications will be discussed below.

Filter Types

Membrane filters are divided into two major classes: **depth** and **screen**. **Depth filters**, which may be composed of paper, cotton, or fiberglass, function by trapping particles primarily within the "depths" of the filter matrix. The interior of these filters, as shown in Figure 2.9, is a random arrangement of fiber material forming tiny channels. Particles that are larger than the passages are retained in the filter by entrapment in the matrix. Since they are thick, depth filters have a high load capacity, retaining particles both on the surface and within the matrix. In addition, they have relatively high flow rates, are inert to most solvents, and are inexpensive. However, they have several disadvantages, including (1) ill-defined and variable pore sizes due to a random matrix, (2) extensive absorption and loss of liquid filtrate, and (3) loss of filter fragments that contaminate the filtrate.

Many of these disadvantages are overcome by **screen filters**, which have uniform pore size. The screen filters function by retaining particles on their surfaces rather than within the matrix (Figure 2.10). The most widely used screen-type filters are composed of cellulose esters (cellulose

FIGURE 2.9
Photomicrograph showing fiberglass depth filter material magnified 2600 × . Courtesy of Millipore Corporation.

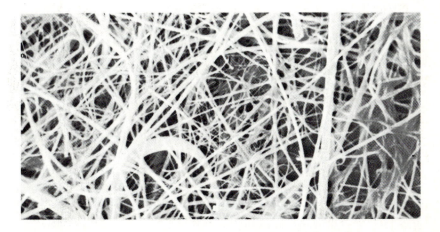

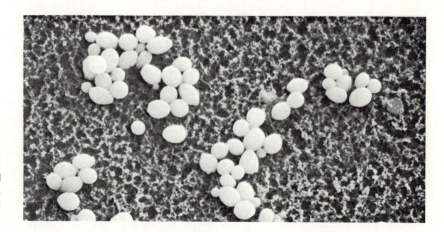

FIGURE 2.10
Photomicrograph of a typical screen filter magnified 100×. Courtesy of Millipore Corporation.

nitrate and acetate). Membrane filters of these materials can be manufactured with a predetermined and accurately controlled pore size. Filters are available with a mean pore size ranging from 0.025 to 8 μm. These filters clog more readily than do depth filters and require suction or pressure for liquid flow. A typical flow rate for the commonly used 0.45 μm membrane is 57 ml min^{-1} cm^{-2} at 10 psi. Clogging can be reduced by combining depth and screen filters. The depth filter serves as a "prefilter" to remove particles that would rapidly clog the screen filter.

Cellulose ester screen filters are hydrophobic and must be treated during manufacturing with a surfactant (Triton). This improves the wettability of the filter, but the wetting agent must be prewashed to avoid contamination of the filtrate.

A relatively new addition to the class of screen filters is the polycarbonate filter. These polymeric filters have pores of uniform size that are formed by bombardment with nuclear particles. They are naturally hydrophilic, so no wetting agent is required.

The principles behind microfiltration are sometimes misunderstood. The nomenclature implies that separations are the result of physical trapping of the particles by the filter. With polycarbonate and fiberglass filters, separations are made primarily on the basis of physical size. Other filters (cellulose nitrate, polyvinylidene fluoride, and to a lesser extent cellulose acetate) trap particles that cannot pass through the pores, but also retain macromolecules by adsorption. In particular, these materials have protein and nucleic acid binding properties. Each type of membrane displays a different affinity for various molecules. For protein, the relative binding affinity is polyvinylidene fluoride > cellulose nitrate > cellulose acetate. We can expect to see many applications of the "affinity membranes" in the future as the various membrane surface chemistries are altered and made more specific. Obvious applications are the purification

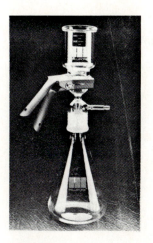

FIGURE 2.11
Assembly for membrane filtration. Courtesy of Millipore Corporation.

of macromolecules and quantitative binding assays (see application section below). More detailed information on the theory and application of membrane filtration is available in "A Guide to Membrane Separation Technology" (Millipore, 1982).

Applications of Microfiltration

1. Clarification of Solutions

Because of low solubility and denaturation, solutions of biomolecules or cellular extracts are often turbid. This is a particular disadvantage if spectrophotometric analysis is desired. The transmittance of turbid solutions can be greatly increased by passage through a membrane filter system. A typical filter set-up is shown in Figure 2.11. Since the pore sizes are so small, liquids will not enter the filter materials by gravity; suction (water aspirator) or pressure must be applied.

This simple technique may also be applied to the sterilization of non-autoclavable materials such as protein solutions or heat labile reagents. Bacterial contamination can be removed from these solutions by passing them through filter systems that have been sterilized by autoclaving. Most membrane filters can be autoclaved, but it is a good idea to read the accompanying literature before doing so.

2. Collection of Precipitates for Analysis

The collection of small amounts of very fine precipitates is the basis for many chemical and biochemical analytical procedures. Membrane filtration is an ideal method for sample collection. This is of great advantage in the collection of radioactive precipitates. Cellulose nitrate and fiberglass filters are often used to collect radioactive samples because they can be analyzed by direct suspension in an appropriate scintillation cocktail (see Chapter 6).

3. Harvesting of Bacterial Cells from Fermentation Broths

The collection of bacterial cells from nutrient broths is typically done by batch centrifugation. This time-consuming operation can be replaced by membrane filtration. Filtration is at least 10 times faster than centrifugation, and it allows for extensive cell washing.

4. Concentration of Protein Solutions

Protein solutions obtained by extraction or various purification steps are often too dilute for further investigation. Since they cannot be concentrated by high-temperature evaporation of solvent, more gentle methods have been developed. One of the most effective is the use of microfiltration pressure cells as shown in Figure 2.12. A membrane filter is placed in the

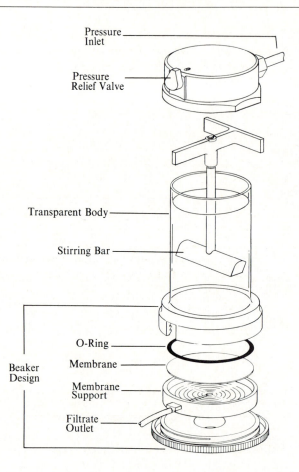

Pressure Inlet

Pressure Relief Valve

Transparent Body

Stirring Bar

O-Ring

Beaker Design

Membrane

Membrane Support

Filtrate Outlet

FIGURE 2.12
Schematic of an ultrafiltration cell. Courtesy of Amicon Corporation, Scientific Systems Division, Danvers, MA.

bottom and the protein solution is poured into the cell. High pressure, exerted by compressed nitrogen (air could cause oxidation and denaturation of protein), forces the flow of small molecules, including solvent, through the filter. Membranes are available in a number of sizes to allow for a large variety of molecular weight cutoffs. Larger molecules that cannot pass through the pores are, of course, concentrated in the sample chamber. This method of concentration is rapid and gentle, and can be performed at cold temperatures to insure minimal inactivation of the proteins. One major disadvantage is potential clogging of the pores, which reduces the flow rate through the filter; this can be lessened by constant but gentle stirring of the protein solution.

5. Binding Assays

The applications of membrane filters in binding assays are numerous. As was described earlier in this section, the surface chemistry of membranes can be altered to allow selectivity in the collection of macromolecules. Membrane filters have been applied to drug/hormone binding assays, ligand/protein binding assays, RNA/DNA hybrid formation, and antigen/antibody interactions (Millipore, 1982).

D. LYOPHILIZATION (FREEZE-DRYING)

Although microfiltration is being used more and more for the concentration of biological solutions, the older technique of lyophilization is still used. There are some situations (storing or transporting biological materials) where lyophilization is better. The technique of freeze-drying involves the removal of solvent from a frozen sample. This is one of the most effective methods for drying or concentrating heat-sensitive materials. In practice, a biological solution to be concentrated is "shell-frozen" on the walls of a round-bottom or freeze-drying flask. Freezing of the biological material is accomplished by placing the flask ($\frac{1}{2}$ full with sample) in a dry ice-acetone bath and slowly rotating it as it is held at a 45° angle. The biological material will freeze in layers on the wall of the flask. This provides a large surface area for evaporation of water. The flask is then connected to the lyophilizer, which consists of a refrigeration unit and a vacuum pump (see Figure 2.13). The combined unit maintains the sample at −40°C for stability of the biological materials, and applies a vacuum of approximately 5 to 25 microns on the sample. Ice formed from the aqueous solution sublimes and is pumped from the sample vial. In fact,

FIGURE 2.13
Schematic showing the components of a freeze dryer. Courtesy of FTS Systems, Inc., Stone Ridge, NY 12484.

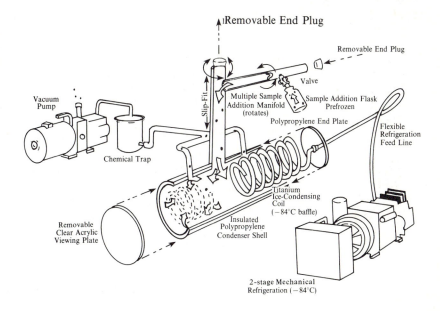

all materials that are volatile under these conditions ($-40°C$, 5 to 25 microns) will be removed, and nonvolatile materials (proteins, buffer salts, etc.) will be concentrated into a light, fluffy precipitate. Most freeze-dried biological materials are stable for long periods of time and some remain viable for many years. This stabilization is probably due to the fact that the various biological components in the mixture cannot interact because they are locked in an ice matrix.

As with any laboratory method, there are precautions and limitations of lyophilization that must be understood. Only aqueous solutions should be lyophilized. Organic solvents lower the melting point of aqueous solutions and increase the chances that the sample will melt and become denatured during freeze-drying. There is also the possibility that organic vapors will pass through the cold trap into the vacuum pump, where they may cause damage.

The freeze-drying of buffered solutions involves special problems. All nonvolatile materials in the aqueous solution, including inorganic or buffer salts, will crystallize during the freeze-drying. Biological materials generally can tolerate high salt concentration; however, freezing of a buffered solution may cause large changes in pH. When pH 7.0 phosphate buffer is frozen, disodium phosphate crystallizes before monosodium phosphate. The pH of the solution just before freezing is about 3.5. Problems associated with buffered solutions can be minimized by very rapid freezing of the solution before lyophilization.

FIGURE 2.14
Freeze dryer. Photo courtesy of Mr. Mark Billadeau.

Some biological samples, especially dilute protein solutions, form a very light and fluffy material upon lyophilization. Unless there is some barrier (a small piece of filter paper or glass wool) between the sample and the vacuum hook-up, sample may be lost by suction.

Freeze dryers are now available in several models and styles. The most versatile is a compact bench top unit that contains a refrigerator and has the capability to freeze-dry up to 12 samples (Figure 2.14). More details on the theory and techniques of lyophilization are available in Everse and Stolzenbach (1971).

E. MEASUREMENT OF PROTEIN SOLUTIONS

Practical biochemistry often requires the rapid measurement of microgram quantities of protein. The research literature contains many procedures for the determination of proteins in solution. Three older, classical methods and one recently developed method are in widespread use. The four methods will not work all the time with every type of protein. Furthermore, a single protein solution measured by the four procedures will probably give four different results. It is important to recognize that there is no "perfect" method for protein measurement. Each method has advantages and disadvantages that must be considered in selecting the best procedure. Important factors are the sensitivity and accuracy desired, the nature of the protein, the presence of interfering substances in the solution, and the time available for the assay. The four most widely used methods, compared in Table 2.3, will be described below. (For further discussion, see E. Layne, 1957; M. Bradford, 1976; and Bio-Rad, 1984.)

The Biuret Assay

When substances containing two or more peptide bonds react with alkaline copper sulfate, a purple complex is formed. The colored product is the result of coordination of peptide nitrogen atoms with Cu^{2+}. The amount of product formed depends on the concentration of protein.

In practice, a calibration curve must be prepared by using a standard protein solution. An aqueous solution of bovine serum albumin is an appropriate standard. Various known amounts of this solution are treated with the biuret reagent and the color is allowed to develop. Measurements of absorbance at 540 nm (A_{540}) are made against a blank containing biuret reagent and buffer or water. The A_{540} data are plotted versus protein concentration (mg/ml). Unknown protein samples are treated with biuret reagent and the A_{540} measured after color development. The protein concentration is determined from the standard curve.

TABLE 2.3
Methods for Protein Measurement

Method	Sensitivity	Time	Principle	Interferences	Comments
Biuret	Low sensitivity 1–20 mg	Moderate 20–30 min	Peptide bonds + alkaline $Cu^{2+} \rightarrow$ purple complex	Good's buffers, Tris, some amino acids	Useful for fast determination, but not especially sensitive. Similar color development with different proteins.
Lowry	High sensitivity ~5 μg	Slow ~40–60 min	(1) Biuret reaction (2) reduction of phosphomolybdate-phosphotungstate by tyr and trp	ammonium sulfate, glycine, Good's buffers, mercaptans	Time-consuming. Amount of color varies with different proteins. Critical timing of procedure. Useful for changes in protein conc.
Bradford	High sensitivity ~5 μg	Rapid ~15 min	λ_{max} of Coomassie dye shifts from 465 nm to 595 nm when bound to protein	strongly basic buffers, Triton X-100, SDS	Excellent method, only slight interferences, stable color. Amount of color varies with different proteins. Reagents are commercially available.
Warburg–Christian (spectrophotometric)	Moderate sensitivity 50–100 μg	Rapid 5–10 min	Absorption of 280 nm light by tyr and trp residues in protein	purines, pyrimidines nucleic acids	Useful for monitoring column eluents. Nucleic acid absorption can be corrected.

Few substances interfere with the biuret test. There is abnormal color development with amino acid buffers, Good's buffers, and Tris buffer.

The biuret assay has several advantages including speed, similar color development with different proteins, and few interfering substances. Its primary disadvantage is its lack of sensitivity. Its lower level of sensitivity is about 1 mg.

The Lowry Assay

This protein assay is one of the most sensitive and, hence, widely used. The Lowry procedure can detect protein levels as low as 5 μg. The principle behind color development is identical to that of the biuret assay except that a second reagent (Folin-Ciocalteu) is added to increase the amount of color development. Two reactions account for the intense blue color that develops, (1) the coordination of peptide bonds with alkaline copper (biuret reaction) and (2) the reduction of the Folin-Ciocalteu reagent (phosphomolybdate-phosphotungstate) by tyrosine and tryptophan residues in the protein.

The procedure is similar to that of the biuret assay. A standard curve is prepared with bovine serum albumin or other pure protein, and the concentration of unknown protein solutions determined from the graph.

The obvious advantage of this assay is its sensitivity, which is up to 100 times greater than the biuret assay; however, more time is required for the Lowry assay. The Lowry assay, unfortunately, is greatly affected by the presence of interfering substances, and critical timing is necessary for addition of reagents. Good's buffers result in a high blank absorbance and may not be present in the assay. Other interfering substances are ammonium sulfate (concentration greater than 0.15%), glycine (greater than 0.5%) and mercaptans. Since all proteins have varying contents of tyrosine and tryptophan, the amount of color development changes with different proteins, including the bovine serum albumin standard. Because of this, the Lowry protein assay should be used only for the measurement of changes in protein concentration, not absolute values of protein concentration. This does not decrease the value of the Lowry assay, because for many biochemical procedures (such as protein purification), relative measurements of protein concentration changes are suitable.

The Bradford Assay

The many limitations of the biuret and Lowry assays have encouraged researchers to seek better methods for quantitation of protein solutions. A recently developed procedure, based on protein binding of a dye, provides numerous advantages over other methods (Bradford, 1976; Bio-Rad, 1984).

The binding of Coomassie Brilliant Blue dye to protein in acidic solution causes a shift in λ_{max} of the dye from 465 nm to 595 nm. The absorption at 595 nm is directly related to the concentration of protein.

In practice, a calibration curve is prepared using bovine plasma gamma globulin or bovine serum albumin as a standard. The assay requires only a single reagent, an acidic solution of Coomassie Brilliant Blue G-250. After addition of dye solution to a protein sample, color development is complete in two minutes and remains stable for up to one hour. The sensitivity of the Bradford assay rivals and may surpass that of the Lowry assay. Using a microassay procedure, the Bradford assay can be used to determine proteins in the range of 1 to 20 μg. The Bradford assay shows significant variation with different proteins, but this also occurs with the Lowry assay. Not only is the Bradford method rapid, but it has very few interferences by nonprotein components. The only known interfering substances are detergents, Triton X-100 and sodium dodecyl sulfate. The many advantages of the Bradford assay have led to its wide adoption in biochemical research laboratories.

The Spectrophotometric Assay

Most proteins have relatively intense ultraviolet light absorption centered at 280 nm. This is due to the presence of tyrosine and tryptophan residues in the protein. However, the amount of these amino acid residues varies in different proteins, as was pointed out earlier. If certain precautions are taken, the A_{280} value of a protein solution is proportional to the protein concentration. The procedure is simple and rapid. A protein solution is transferred to a quartz cuvet and the A_{280} is read against a reference cuvet containing the protein solvent only (buffer, water, etc.)

Cellular extracts contain many other compounds that display absorbance in the vicinity of 280 nm. Nucleic acids, which would be common contaminants in a protein extract, absorb strongly at 280 nm ($\lambda_{max} = 260$). Warburg and Christian developed a method to correct for this interference by nucleic acids. Mixtures of pure protein and pure nucleic acid were prepared, and the ratio A_{280}/A_{260} was experimentally determined. They then calculated a factor (F) to correct for the presence of nucleic acid. Table 2.4 lists the correction factor for several A_{280}/A_{260} measurements. The following equation may be used for protein solutions containing up to 20% nucleic acids:

$$\text{protein concentration (mg/ml)} = A_{280} \cdot F$$

A modified equation from Layne (1957) that will yield similar results is:

$$\text{protein concentration (mg/ml)} = 1.55\, A_{280} - 0.76\, A_{260}$$

Table 2.4 can also be used to estimate the amount of nucleic acid in a protein solution.

TABLE 2.4

Protein Estimation by UV Absorption[1]

A_{280}/A_{260}	% Nucleic Acid	F
1.75	0.00	1.12
1.63	0.25	1.08
1.52	0.50	1.05
1.40	0.75	1.02
1.36	1.00	0.99
1.30	1.25	0.97
1.25	1.50	0.94
1.16	2.00	0.90
1.09	2.50	0.85
1.03	3.00	0.81
0.95	3.50	0.78
0.94	4.00	0.74
0.87	5.00	0.68
0.85	5.55	0.66
0.82	6.00	0.63
0.80	6.50	0.61
0.78	7.00	0.59
0.77	7.50	0.57
0.75	8.00	0.55
0.73	9.00	0.51
0.71	10.00	0.48
0.67	12.00	0.42
0.64	14.00	0.38
0.62	17.00	0.32
0.60	20.00	0.28

[1] Adapted from E. Layne in *Methods in Enzymology*, S. P. Colowick and N. O. Kaplan, Editors, Vol. III, p. 447 (1957). Copyright 1957 Academic Press.

Although the spectrophotometric assay of proteins is fast, relatively sensitive, and requires only a small sample size, it is nothing more than an estimate of protein concentration. It has certain advantages over the colorimetric assays in that most buffers and ammonium sulfate do not interfere.

The spectrophotometric assay is particularly suited to the rapid measurement of protein elution from a chromatography column, where only protein concentration changes are required.

REFERENCES

Bio-Rad, *Bio-Rad Protein Assay*, Bulletin 1069 (1984). 2200 Wright Avenue, Richmond, CA 94804.

M. Bradford, *Anal. Biochem.*, *72*, 248 (1976). "A Rapid and Sensitive

Method for the Quantitation of Microgram Quantities of Protein Utilizing the Principle of Protein-Dye Binding."

L.C. Clark, Jr., R. Wold, G. Granger, and Z. Taylor. *J. Appl. Physiol.*, 6, 189 (1953). "Continuous Recording of Blood Oxygen Tensions."

J. Everse and F. Stolzenbach in *Methods in Enzymology*, W. Jakoby, Editor, Vol. XXII, pp. 33–39 (1971), Academic Press (New York). "Lyophilization."

D.C. Fork in *Methods in Enzymology*, A. San Pietro, Editor, Vol. XXIV, pp. 113–122 (1972), Academic Press (New York). "Oxygen Electrode."

N.E. Good, G.D. Winget, W. Winter, T.N. Connolly, S. Izawa, and R.M.M. Singh, *Biochem.*, 5, 467 (1966). "Hydrogen Ion Buffers for Biological Research."

N.E. Good and S. Izawa in *Methods in Enzymology*, A. San Pietro, Editor, Vol. XXIV, pp. 53–68 (1972), Academic Press (New York). "Hydrogen Ion Buffers for Photosynthesis Research."

D.E. Gueffroy, Editor, *A Guide for the Preparation and Use of Buffers in Biological Systems* (1975). Calbiochem, 10933 N. Torrey Pines Road, La Jolla, CA 92037.

E. Layne in *Methods in Enzymology*, S.P. Colowick and N.O. Kaplan, Editors, Vol. III, pp. 447–454 (1957), Academic Press (New York). "Spectrophotometric and Turbidimetric Methods for Measuring Protein."

P. McPhie in *Methods in Enzymology*, W. Jakoby, Editor, Vol. XXII, pp. 23–32 (1971), Academic Press (New York). "Dialysis."

Millipore Corporation, *A Guide to Membrane Separation Technology* (1982). Ashby Road, Bedford, MA 01730.

Orion Research Incorporated, *Analytical Methods Guide* (1982). 380 Putnam Ave., Cambridge, MA 02139.

Yellow Springs Instrument Co., *Instructions for YSI Model 53 Biological Oxygen Monitor* (1976). Yellow Springs, Ohio 45387.

Chapter **3**

Separation and Identification of Biomolecules by Chromatography

The molecular details of a biochemical process cannot be fully elucidated until the reacting molecules have been isolated and characterized. Therefore, our understanding of biochemical principles has increased at about the same pace as the development of techniques for the separation and identification of biomolecules. Chromatography has been and will continue to be the most effective technique for isolating and purifying all types of biomolecules. In addition, it is widely used as an analytical tool.

A. INTRODUCTION TO CHROMATOGRAPHY

All types of chromatography are based on a very simple principle. The sample to be examined (called the **solute**) is allowed to interact with two physically distinct entities—**a mobile phase** and **a stationary phase** (see Figure 3.1). The mobile phase, which may be a gas or liquid, moves the sample through a region containing the solid or liquid stationary phase called the **sorbent**. The stationary phase will not be described in detail at this time, since it varies from one chromatographic method to another. However, it may be considered as having the ability to "bind" some types of solutes. The sample, which may contain one or many molecular components, comes into contact with the stationary phase. The components distribute themselves between the mobile and stationary phases. If some of the sample components are preferentially bound by the stationary

61

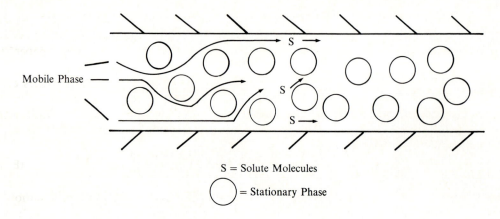

S = Solute Molecules

◯ = Stationary Phase

FIGURE 3.1
A representation of the principles of chromatography.

phase, they spend more time in the stationary phase and, hence, will be retarded in their movement through the chromatography system. Molecules that show weak affinity for the stationary phase will spend more time with the mobile phase and are more rapidly removed or **eluted** from the system. The many interactions that occur between solute molecules and the stationary phase bring about a separation of molecules because of different affinities for the stationary phase. The general process of moving a solute mixture through a chromatographic system is called **development**.

The mobile phase can be collected as a function of time at the end of the chromatographic system. The mobile phase, now called the **effluent**, contains the solute molecules. If the chromatographic process has been effective, fractions or "cuts" that are collected at different times will contain the different components of the original sample. In summary, molecules are separated because they differ in the extent to which they are distributed between the mobile phase and the stationary phase.

Throughout this chapter and others, biochemical techniques will be designated as **preparative** or **analytical**, or both. A preparative procedure is one that can be applied to the purification of a relatively large amount of a biological material. The purpose of such an experiment would be to obtain purified material for further characterization and study. Analytical procedures are used most often for the determination of purity of a biological sample; however, they may be used to evaluate any physical, chemical, or biological characteristic of a biomolecule or biological system.

Partition vs Adsorption Chromatography

Chromatographic methods are divided into two types according to how solute molecules bind to or interact with the stationary phase. **Parti-**

tion chromatography is the distribution of a solute between two liquid phases. This may involve direct extraction using two liquids, or it may use a liquid immobilized on a solid support as in the case of paper, thin-layer, and gas-liquid chromatography. For partition chromatography, the stationary phase in Figure 3.1 consists of inert solid particles coated with liquid adsorbent. The distribution of solutes between the two phases is based primarily on solubility differences. The distribution may be quantified by using the **partition coefficient**, K_D (Equation 3.1).

$$K_D = \frac{\text{concentration of solute in mobile phase}}{\text{concentration of solute in stationary phase}} \quad \text{(Equation 3.1)}$$

Adsorption chromatography refers to the use of a stationary phase or support, such as an ion exchange resin, that has a finite number of relatively specific binding sites for solute molecules. There is not a clear distinction between the processes of partition and adsorption. All chromatographic separations rely, to some extent, on adsorptive processes. However, in some methods (paper chromatography, TLC, GC) these specific adsorptive effects are minimal and the separation is based primarily on nonspecific solubility factors. Adsorption chromatography relies on relatively specific interactions between the solute molecules and binding sites on the surface of the stationary phase. The attractive forces between solute and support may be ionic, hydrogen bonding, or hydrophobic interactions. Binding of solute is, of course, reversible.

Because of the different interactions involved in partition and adsorption processes, they may be applied to different separation problems. Partition processes are most effective for the separation of small molecules, especially those in a homologous series. Partition chromatography has been widely used for the separation and identification of amino acids, carbohydrates, and fatty acids. Adsorption techniques, represented by ion-exchange chromatography, are most effective when applied to the separation of macromolecules including proteins and nucleic acids.

The rest of the chapter will be directed toward a discussion of the various chromatographic methods. You should recognize that no single chromatographic technique relies solely on adsorption or partition effects. Therefore, little emphasis will be placed on a classification of the techniques, but, rather, theoretical and practical aspects will be discussed.

B. PAPER AND THIN-LAYER CHROMATOGRAPHY

Because of the similarities in the theory and practice of these two procedures, they will be considered together. Both are examples of partition chromatography. In paper chromatography, the cellulose support is extensively hydrated, so distribution of the solutes occurs between the immobilized water (stationary phase) and the mobile developing solvent. The initial stationary liquid phase in thin layer chromatography (TLC) is the solvent used to prepare the thin layer of adsorbent. However, as developing solvent molecules move through the stationary phase, polar solvent molecules may bind to the immobilized support and become the stationary phase.

FIGURE 3.2
The procedure of paper and thin-layer chromatography. (A) Application of the sample. (B) Setting plate in solvent chamber. (C) Movement of solvent by capillary action. (D) Detection of separated components and calculation of R_f.

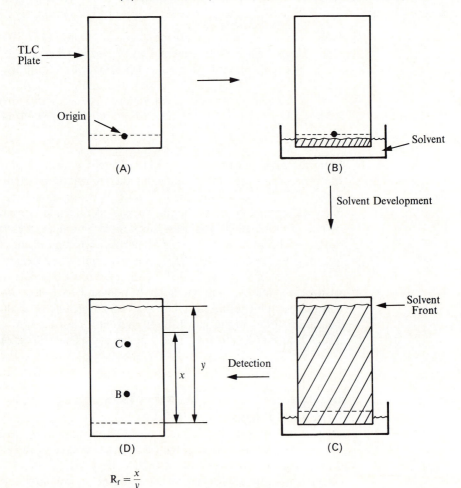

$$R_f = \frac{x}{y}$$

Preparation of the Stationary Support

The support medium may be a sheet of cellulose, or a glass or plastic plate covered with a thin coating of silica gel, alumina, or cellulose. Large sheets of cellulose chromatography paper are available in different porosities. These may be cut to the appropriate size and used without further treatment. The paper should never be handled with bare fingers, especially if amino acids are to be analyzed. Although thin layer plates can easily be prepared, it is much more convenient to purchase ready-made plates. These are available in a variety of sizes, materials, and thicknesses of stationary support. They are relatively inexpensive and have a more uniform support thickness than hand-made plates.

Figure 3.2 outlines the application procedure. A very small drop of sample is spotted onto the plate with a disposable microcapillary and allowed to dry; then the spotting process is repeated by superimposing more drops on the original spot. The exact amount of sample applied is critical. There must be enough sample so the developed spots can be detected, but overloading will lead to "tailing" and lack of resolution. Finding the proper sample size is a matter of trial and error. It is usually recommended that two or three spots of different concentrations be applied for each sample tested. Spots should be applied along a very faint line drawn with a **pencil** and ruler. TLC plates should not be heavily scratched or marked. Very faint marks may be made every 2 to 4 cm along the edge, but be sure the sample is not applied on or near these marks.

Development of the Plate

A wide selection of solvent systems is available in the biochemical literature. If a new solvent system must be developed, a preliminary analysis must be done on the sample with a series of solvents. Solvents can be rapidly screened by developing several small chromatograms (2×6 cm) in small sealed bottles containing the solvents. For the actual analysis, the sample should be run on a larger plate with appropriate standards in a development chamber (Figure 3.3). The chamber must be air-tight and saturated with solvent vapors. Filter paper on two sides of the chamber, as shown in Figure 3.3, enhances vaporization of the solvent.

Paper chromatograms may be developed in either of two types of arrangements—ascending or descending solvent flow. Descending solvent flow leads to faster development because of assistance by gravity, and it can offer better resolution for compounds with small R_f values because the solvent can be allowed to run off the paper. R_f values cannot be determined under these conditions, but it is useful for quantitative separations.

Two-dimensional chromatography is used for especially difficult separations. The chromatogram is developed in one direction by a solvent system, air dried, turned 90°, and developed in a second solvent system.

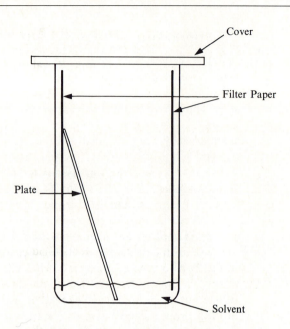

FIGURE 3.3
A typical chamber for paper and thin-layer chromatography.

Detection and Measurement of Components

Unless the components in the sample are colored, their location on a chromatogram will not be obvious after solvent development. Several methods can be used to locate the spots, including fluorescence, radioactivity, and treatment with chemicals that develop colors. Substances that are highly conjugated may be detected by fluorescence under a UV lamp. Chromatograms may be treated with different types of reagents to develop a color. **Universal reagents** produce a colored spot with any organic compound. When a solvent-developed plate is sprayed with concentrated H_2SO_4 and heated at 100°C for a few minutes, all organic substances will appear as black spots. A more convenient universal reagent is I_2. The solvent-developed chromatogram is placed in an enclosed chamber containing a few crystals of I_2. The I_2 vapor reacts with most organic substances on the plate to produce brown spots. The spots are more intense with unsaturated compounds.

Specific reagents react with a particular class of compound. For example, rhodamine B is often used to visualize lipids, ninhydrin for amino acids, and aniline phthalate for carbohydrates.

The position of each component of a mixture is quantified by calculating the distance traveled by a component relative to the distance traveled by the solvent. This is called **relative mobility** and symbolized by R_f. In Figure 3.2D, the R_f for component C is the ratio of distance X to distance Y. The R_f for a substance is a constant for a certain set of experimental conditions. However, the R_f value will vary with solvent, type of stationary support (paper, alumina, silica gel), temperature, humidity, and other environmental factors. R_f values are always reported along with solvent and temperature.

Applications of Paper and Thin-Layer Chromatography

Thin-layer chromatography is now more widely used than is paper chromatography. In addition to its greater resolving power, TLC is faster and plates are available with several sorbents (cellulose, alumina, silica gel).

Partition chromatography as described in this section may be applied to two major types of problems: (1) identification of unknown samples and (2) isolation of the components of a mixture. The first application is, by far, the more widely used. Paper and thin-layer chromatography require only a minute sample size, the analysis is fast and inexpensive, and detection is straightforward. Unknown samples are applied to a plate along with appropriate standards, and the chromatogram is developed as a single experiment. In this way any changes in experimental conditions (temperature, humidity, etc.) affect standards and unknowns to the same extent. It is then possible to compare the R_f values directly.

Purified substances can be isolated from developed chromatograms; however, only tiny amounts are present. In paper chromatography, the spot may be cut out with a scissors and the piece of paper extracted with an appropriate solvent. Isolation of a substance from a TLC plate is accomplished by scraping the solid support from the region of the spot with a knife edge or razor blade and extracting the sorbent with a solvent. "Preparative" thin-layer plates with a thick coating of sorbent (up to 2 mm) are especially useful because they have higher sample capacity.

C. GAS CHROMATOGRAPHY (GC)

When the mobile phase of a chromatographic system is gaseous, and the stationary phase is a liquid coated on inert solid particles, the technique is **gas-liquid chromatography**, or simply, gas chromatography. Separation by GC methods is based primarily on partitioning processes. The stationary phase, inert particles coated with a thin layer of liquid, is confined to a long stainless steel or glass tube, called the **column**, which is maintained at a suitable (usually elevated) temperature. A gaseous mobile phase under high pressure is continuously swept through the column. The sample to be analyzed is vaporized, introduced into the warm gaseous phase, and swept through the stationary phase. The vaporous chemical constituents in the sample then distribute themselves between the mobile phase and the stationary liquid film on the solid support. Components of the sample mixture that have affinity for the stationary phase are retarded in their movement through the column. Ideally, each component will have a different partition coefficient, K_D (see Equation 3.1), and each will pass through the column at a different rate.

Instrumentation

The essential components of a gas chromatography system are shown in Figure 3.4. The mobile phase (called the carrier gas) is inert, usually helium, nitrogen, or argon. The gas is directed past an injection port, the entry point of the sample. The sample, dissolved in a solvent, is injected with a syringe through a rubber septum into the injection port. The column, injection port, and detector are in individual ovens maintained at elevated temperatures so that the sample components remain vaporized throughout their residence time in the system.

Two types of columns are widely used. A **packed column** is one filled with inert, solid particles coated with a liquid stationary phase. Standard tubing is about 0.5 cm in diameter, with lengths ranging from 1 m to 20 m;

FIGURE 3.4
The essential components of a gas chromatography system.

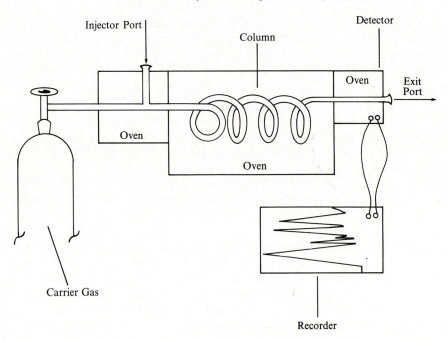

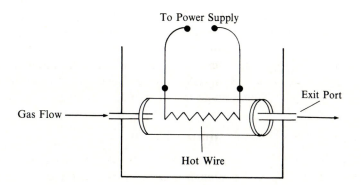

FIGURE 3.5
A thermal conductivity detector for gas chromatography.

however, columns for large-scale preparative work may be up to 5 cm in diameter and several meters long. Commonly used solid supports are diatomaceous earth, Teflon powder, and glass beads. The stationary liquid must be chosen on the basis of the compounds to be analyzed. A more recently developed type of column is the **open-tubular** or **capillary column**. This is prepared by coating the inner wall of the column with the stationary liquid phase. The inside diameter of a typical capillary tube is 0.25 mm, and the length ranges from 10 to 100 meters. The length of a column depends upon the degree of resolution required. Clearly, the longer the column, the better the separation; however, a long column increases the pressure differential from the beginning to the end of the column. A large pressure change will hamper analysis by increasing retention time and causing peak broadening. Stainless steel columns are still in widespread use, but they cannot be used with sensitive compounds. Many biochemical substances decompose when they contact warm metal surfaces. For example, derivatives of amino acids can be analyzed only in gas chromatographs with glass injection ports and columns.

After passage through the column, the carrier gas and separated components of the mixture are directed through a detector. Several types of detectors are available, but the two most commonly used for bioanalytical purposes are the **thermal conductivity cell** (TC cell) and the **flame ionization detector** (FID). The TC cell functions by measuring the temperature-dependent electrical resistance of a hot wire. The detector unit, as diagrammed in Figure 3.5, consists of a platinum or platinum-alloy wire through which an electric current is passed. The hot wire is cooled as the carrier gas passes over it. The extent of cooling depends on the gas flow rate and the thermal conductivity of the gas. Since the rate of carrier gas flow remains constant during a typical GC analysis, only changes in the thermal conductivity of the vapor will change the temperature and, therefore, the resistance of the wire. Organic vapors usually have lower thermal conductivities than that of the carrier gas. When carrier gas containing an organic vapor exits the column and passes over the wire, the electrical resistance of the wire decreases. The changing current in the wire is amplified and monitored by a recorder. The extent of resistance change depends upon the amount of organic vapor in the carrier gas. Therefore, the size of the recorder signal is a measure of the amount of that chemical constituent in the sample. The thermal conductivity cell has a poor level of sensitivity (about 5 μg), but it is widely used because it responds to all organic compounds. Since it does not destroy the sample, a TC cell can be used when samples are to be collected from the column.

In contrast to the TC detector, the flame ionization detector (FID) is sensitive to about 10^{-5} μg. Figure 3.6 shows the main components of the FID. A flame supported by hydrogen gas and air is used to burn organic vapors as they leave the column. Electrons and ionic fragments are produced upon combustion of the organics. A wire loop (called the ion collector) collects the charged particles and produces an electrical current, which is fed into a recorder. The FID is sensitive only to oxidizable compounds and does not respond to water vapor or CO_2. Since the detector response is proportional to the amount of organic material in the carrier

FIGURE 3.6
A flame ionization detector for gas chromatography.

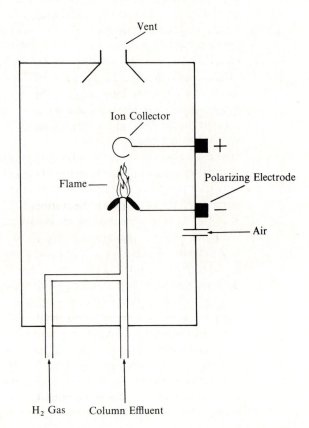

gas, quantitative analysis can be done. Samples cannot be collected after passage through the FID; however, it is possible to split the carrier gas flow before it enters the detector. Only a small fraction of the carrier gas with organic vapor is directed through the detector, and the rest is collected for further analysis.

The electrical signal from a detector is amplified and fed into a strip chart recorder. A typical recorder trace is shown in Figure 3.7. Each peak represents a component in the original mixture. A peak is identified by a *retention time*, the time lapse between injection of the sample and the maximum signal from the recorder. This number is a constant for a particular compound under specified conditions of carrier gas flow rate, temperature of the injector, column, and detector, and type of column. Retention time in GC analysis is analogous to the R_f value in thin-layer or paper chromatography.

Selection of Operating Conditions

Each type of gas chromatograph has its own set of operating instructions, but general experimental conditions are appropriate for all instruments. Three important factors must always be considered when a GC analysis is to be completed. These are (1) selection of the proper column, (2) choice of temperatures for injector, oven, and detector, and (3) adjustment of gas flow. Because hundreds of stationary phases are available, it is impossible to outline the characteristics of each. Selecting the stationary phase requires some knowledge of the nature of the sample to be analyzed. Books and journal articles on gas chromatography and commercial catalogs of GC supplies are the best sources of information to help select the right column.

The operating temperature of the column oven is critical, as it will greatly affect the resolving ability of the column. If the temperature is too high, low boiling components will be swept rapidly through the column without equilibration with the stationary phase. A temperature that is too low will cause condensation of some organic compounds in the liquid stationary phase, resulting in very extended retention times. A column temperature set near the boiling point of the primary component in the sample is often the best. If a sample containing a mixture of compounds with a wide range of boiling points is to be analyzed, a temperature gradient may be required. Research grade chromatographs are equipped with temperature programmers that gradually increase the column temperature at a preselected rate.

The detector and injector temperatures should be maintained about 10° above the column temperature. This insures against condensation of vapor and causes rapid vaporization of the sample upon injection.

The separation of a mixture by GC also depends upon the flow rate of the carrier gas. A slow rate allows for extensive equilibration of the

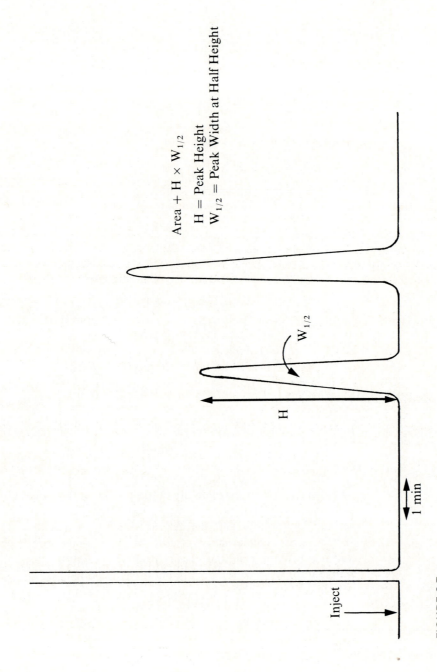

$\text{Area} + \text{H} \times \text{W}_{1/2}$

$\text{H} = \text{Peak Height}$
$\text{W}_{1/2} = \text{Peak Width at Half Height}$

$\text{W}_{1/2}$

H

1 min

Inject

FIGURE 3.7
A recorder trace from gas chromatography.

sample between the stationary phase and the gaseous phase, and leads to better separation; however, the recorder peaks become very broad. A fast flow rate greatly decreases column efficiency by decreasing equilibration time. A suitable flow rate for a 0.5 cm packed column is between 50 and 120 ml/min.

Analysis of GC Data

Gas chromatography can be applied to both qualitative and quantitative analyses. There are two general methods for qualitative identification of eluted compounds. A compound may be identified by **retention time** or by **peak enhancement**. If you have some idea of the identity of an unknown peak or if you know it is one of three or four compounds, then you can inject the known compounds into the GC under conditions identical for the unknown sample. Compounds with the same retention times (± 2 or 3%) can generally be considered to be identical. Alternatively, you can add pure sample of a suspected component to the unknown sample and inject the mixture into the GC. If a recorder peak is increased in size, this also provides evidence for the identity of an unknown. The latter technique is called **peak enhancement** or **spiking**.

Identification of unknown compounds by retention times or peak enhancement is not conclusive or absolute proof of identity. It is possible that two different substances will have identical retention times under the same experimental conditions. For positive identification, the sample must be collected at the exit port and characterized by mass spectrometry, infrared, nmr, or chemical analysis.

A graphical identification method can be used if compounds in a homologous series are to be analyzed. This is most often applied to the analysis of fatty acid methyl esters. A plot of log retention time vs. the number of carbon atoms is usually linear (Figure 3.8). Experiment 10 illustrates this technique.

The response from most recorders and detectors is proportional to the amount of compound in the original sample. In other words, the area of each recorder peak is a relative measure of the concentration of that substance in the sample. GC then becomes an important tool for quantitative analysis. Methods for the measurement of peak area range from sophisticated to primitive. The most ideal method is to use an electronic integrator, which automatically measures peak area and can be programmed to yield % composition of each component. Alternatively, one can use the method of triangulation. Here the peaks are assumed to be triangles with an area calculated according to Equation 3.2, where H is height and $W_{1/2}$ is the width at one-half height.

$$\text{Area} = HW_{1/2} \qquad \text{(Equation 3.2)}$$

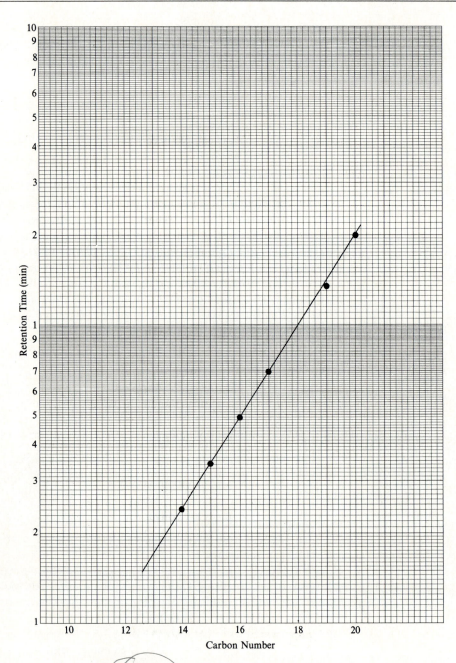

FIGURE 3.8
A plot of log retention time vs. carbon number for a series of saturated fatty acid methyl esters.

This method gives acceptable results if the peaks are symmetrical. If the area of each peak is measured, the % composition of the sample can be estimated using Equation 3.3.

$$\%x = \frac{\text{Area}_x}{\text{Area}_a + \text{Area}_b + \text{Area}_x} \times 100 \qquad \text{(Equation 3.3)}$$

The percent of x in the sample is equal to the area represented by component x divided by the sum of all peak areas. Area_a and Area_b refer to other components in the mixture.

These methods for quantitative analysis have a common limitation. Most detectors do not respond with equal sensitivity to all components in the sample. If samples that are isomeric or in a homologous series are analyzed, detector response is probably similar. However, if dissimilar substances are to be measured, response correction factors (RCF) must be determined. The procedure for measuring RCFs is outlined in Experiment 10.

Advantages and Limitations of GC

It is difficult to imagine another analytical technique that has all the advantages of gas chromatography. It provides for excellent separation of most organic molecules, it is simple, versatile, and rapid, it is highly sensitive, and finally, by most standards, it is relatively inexpensive.

The major limitation of GC is the requirement for heat stability and volatility of the sample. Obviously, compounds that decompose at elevated temperatures (below 250°C) cannot normally be subjected to GC analysis. Many compounds of biochemical interest are not volatile in the useful temperature range of GC (up to about 200–250°C). Such compounds can often be converted to volatile derivatives. Hydroxy groups in alcohols, carbohydrates, and sterols are converted to derivatives by trimethylsilylation or acetylation. Amino groups can also be converted to volatile derivatives by acetylation and silylation. Fatty acids are transformed to methyl esters for GC analysis.

D. ADSORPTION COLUMN CHROMATOGRAPHY

Adsorption chromatography in biochemical applications usually consists of a solid stationary phase and a liquid mobile phase. The most useful technique is column chromatography, in which the stationary phase is confined to a glass tube and the mobile phase (a solvent or buffer) is allowed to flow through the solid adsorbent. A small amount of the sample to be analyzed is layered on top of the column. The sample mixture enters the column of adsorbing material and is distributed between the mobile phase and the stationary phase. The various components in the sample will have different affinities for the two phases and will move through the column at different rates. Collection of the liquid phase emerging from the column yields separate fractions containing the individual components in the sample.

Specific terminology is used to describe various aspects of column chromatography. When the actual adsorbing material is made into a column, it is said to be **poured** or **packed**. Application of the sample to the top of the column is **loading** the column. Movement of solvent through the loaded column is called **developing** or **eluting** the column. The **bed volume** is the total volume of solvent and adsorbing material taken up by the column. The volume taken up by the liquid phase in the column is the **void volume**. The **elution volume** is the amount of solvent required to remove a particular solute from the column. This is analogous to R_f values in thin-layer or paper chromatography, or to retention time in GC.

In adsorption chromatography, solute molecules take part in specific interactions with the stationary phase. Herein lies the great versatility of adsorption chromatography. Many varieties of adsorbing materials are available, so a specific sorbent can be chosen that will effectively separate a mixture. There is still an element of trial and error in the selection of an effective stationary phase. However, experiences of many investigators are recorded in the literature and are of great help in choosing the proper system. Table 3.1 lists the most common stationary phases employed in adsorption column chromatography.

Adsorbing materials come in various forms and sizes. The most suitable forms are dry powders or a slurry form of the material in an aqueous buffer. Alumina, silica gel, and fluorisil do not normally need special pretreatment. The size of particles in an adsorbing material is defined by *mesh size*. This refers to a standard sieve through which the particles can pass. A 100 mesh sieve has 100 small openings per square inch. Adsorbing material with high mesh size (400 and greater) is extremely fine

TABLE 3.1

Adsorbents Useful in Biochemical Applications

Adsorbing Material	Uses
Alumina	Small organics, lipids
Silica gel	Amino acids, lipids, carbohydrates
Fluorisil (magnesium silicate)	Neutral lipids
Calcium phosphate (hydroxyapatite)	Proteins, polynucleotides, nucleic acids

TABLE 3.2
Mesh Sizes of Adsorbents and Typical Applications

Mesh Size	Applications
20–50	Crude preparative work, very high flow rate
50–100	Preparative applications, high flow rate
100–200	Analytical separations, medium flow rate
200–400	High resolution analytical separations, slow flow rate

and is most useful for very high resolution chromatography. Table 3.2 lists standard mesh sizes and the most appropriate application. For most biochemical applications, 100 to 200 mesh size is suitable.

Operation of a Chromatographic Column

A typical column set-up is shown in Figure 3.9. The heart of the system is, of course, the column of adsorbent. In general, the longer the column, the better the resolution of components. However, a compromise must be made because flow rate decreases with increasing column length. The actual size for a column depends upon the nature of the adsorbing material and the amount of chemical sample to be separated. For preparative purposes, column heights of 20 to 50 cm are usually sufficient to achieve acceptable resolution. Column inside diameters may vary from 2 to 5 cm. Increasing the inside diameter of a column does not improve resolution but it does increase the capacity.

Packing the Column

Once the adsorbing material and column size have been selected, the column is poured. If the glass tube does not have a fritted disc in the bottom, then a small piece of glass wool or cotton should be used to support the column. Most columns are packed by pouring a slurry of the sorbent into the glass tube and allowing it to settle by gravity into a tight bed. The slurry is prepared with the solvent or buffer that will be used as the initial developing solvent. Pouring of the slurry must be continuous to avoid formation of sorbent layers. Excess solvent is eluted from the bottom of the column while the sorbent is settling. The column must never run dry. Additional slurry is added until the column bed reaches the desired height. The top of the settled adsorbent is then covered with a small circle of filter paper or glass wool to protect the surface while the column is loaded with sample or the eluting solvent is changed.

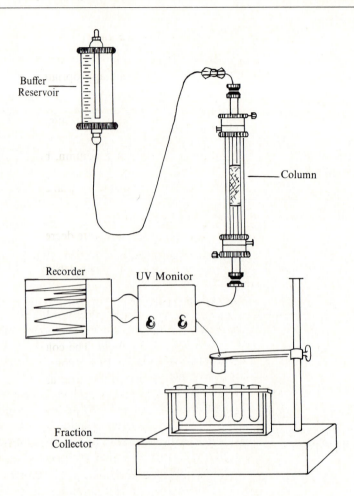

Buffer Reservoir

Column

Recorder

UV Monitor

Fraction Collector

FIGURE 3.9
Set-up for the operation of a chromatography column.

Sometimes it is necessary to pack a column under pressure (5 to 10 psi). This leads to a tightly packed bed that yields more reproducible results, especially with gradient elution (see above).

Loading the Column

The sample to be analyzed by chromatography should be applied to the top of the column in a concentrated form. If the sample is solid, it is dissolved in a minimum amount of solvent; if already in solution, it may be concentrated as described in Chapter 2. After the sample is loaded onto the column with a graduated or disposable pipet, it is allowed to percolate into the adsorbent. A few milliliters of solvent are then carefully added to wash the sample into the column material. The glass column is then filled with eluting solvent.

Eluting the Column

The chromatography column is developed by continuous flow of a solvent. Maintaining the appropriate flow rate is important for effective separation. If the flow rate is set too fast, there is not sufficient time for complete equilibration of the sample components with the two phases. Too slow a flow rate will allow diffusion of solutes, which leads to poor resolution and broad elution peaks. It is difficult to give guidelines for the proper flow rate of a column, but, in general, a column should be adjusted to a rate slightly less than "free flow." Sometimes it is necessary to find the proper flow rate by trial and error. One problem encountered during column development is a changing flow rate. As the solvent height above the column bed is reduced, there will be less of a "pressure head" on the column, so the flow rate decreases. This can be avoided by storing the developing solvent in a large reservoir and allowing it to enter the column at the same rate that it is emerging from the column (see Figure 3.9).

Adsorption columns are eluted in one of three ways. All components may be eluted by a single solvent or buffer. This is referred to as **continual elution**. In contrast, **stepwise elution** refers to an incremental change of solvent to aid development. The column is first eluted with a volume of one solvent, and then with a second solvent. This may continue with as many solvents or solvent mixtures as desired. In general, the first solvent should be the least polar of any used in the analysis, and each additional solvent should be of greater polarity or ionic strength. Finally, adsorption columns may be developed by **gradient elution** brought about by a gradual change in solvent composition. The composition of the eluting solvent can be changed by the continuous mixing of two different solvents to gradually change the ratio of the two solvents. Alternatively, the concentration of a component in the solvent can be gradually increased. This is most often done by addition of a salt (KCl, NaCl, etc.). Devices are commercially available to prepare predetermined, reproducible gradients.

Collecting the Eluent

The separated components emerging from the column in the eluent are usually collected as discrete fractions. This may be done manually by collecting specified volumes of eluent in Erlenmeyer flasks or test tubes. Alternatively, if many fractions are to be collected, a mechanical fraction collector is convenient and even essential. An automatic fraction collector (see Figure 3.9) directs the eluent into a single tube until a predetermined volume has been collected or until a preselected time period has elapsed; then the collector advances another tube for collection. Specified volumes are collected by a drop counter activated by a photocell, whereas a timer can be set to collect a fraction over a specific period. The fraction size collected is sometimes a critical factor in effective separations. If the solutes to be separated are very similar in structure and properties, it is best to collect relatively small fractions (1 to 3 ml).

Detection of Eluting Components

The completion of a chromatographic experiment calls for a means to detect the presence of solutes in the collected fractions. The detection method used will depend on the nature of the solutes. Smaller molecules such as lipids, amino acids, and carbohydrates can be detected by spotting fractions on a thin-layer plate or a piece of filter paper and treating them with a chemical reagent that produces a color. The same reagents that are used to visualize spots on a thin-layer or paper chromatogram are useful for this. Proteins and nucleic acids are conveniently detected by spectroscopic absorption measurements at 280 nm and 260 nm, respectively. Enzymes can be detected by measurements of catalytic activity associated with each fraction. Research-grade chromatographic systems are equipped with detectors that continuously monitor some physical property of the eluent (Figure 3.9). Most often the eluent is directed through a flow cell where absorbance or fluorescence characteristics can be measured. The detector is connected to a strip chart recorder for a permanent record of spectroscopic changes. When the location of the various solutes is determined, fractions containing identical components are pooled and stored for later use.

Advantages and Limitations of Adsorption Column Chromatography

Column procedures are relatively straightforward and trouble-free; however, several precautions should be taken. Most problems are the result of human error, and therefore can be avoided if the experimenter is careful, alert, and knowledgeable.

Proper column conditions in terms of adsorbent preparation, flow rate, and sample loading have already been discussed. A column must never be allowed to go dry during packing or development. This will lead to irregular flow or channeling and column cracking. The glass tube must be clamped to a ring stand in a level position before pouring. The top of the column bed must be level and undisturbed during the complete operation.

E. ION-EXCHANGE CHROMATOGRAPHY

Ion-exchange chromatography is a form of adsorption chromatography in which ionic solutes display reversible electrostatic interactions with a charged stationary phase. The chromatographic set-up is identical to that described in the last section and Figure 3.9. The column is packed with a stationary phase consisting of a synthetic resin that is tagged with ionic functional groups. The steps involved in ion-exchange chromatography are outlined in Figure 3.10. In Stage 1, the insoluble resin material (positively charged) in the column is surrounded by buffer counter-ions. Loading of the column in Stage 2 brings solute molecules of different charge into the ion-exchange medium. Solutes entering the column may be negatively charged, positively charged, or neutral under the experimental conditions. Solute molecules that have a charge opposite to that of the resin will bind tightly but reversibly to the stationary phase (Stage 3). The strength of binding depends on the size of the charge and the charge density (amount of charge per unit volume of molecule) of the solute. The greater the charge or the charge density, the stronger the interaction. Neutral solute molecules (Δ) or those with a charge identical to that of the resin will show little or no affinity for the stationary phase and will move with the eluting buffer. The bound solute molecules can be released by eluting the column with a buffer of increased ionic strength or pH (Stage 4). An increase in buffer ionic strength releases bound solute molecules by displacement. Increasing the buffer pH decreases the strength of the interaction by reducing the charge on the solute or on the resin (Stage 5).

FIGURE 3.10
Illustration of the principles of ion-exchange chromatography. See text for explanation.

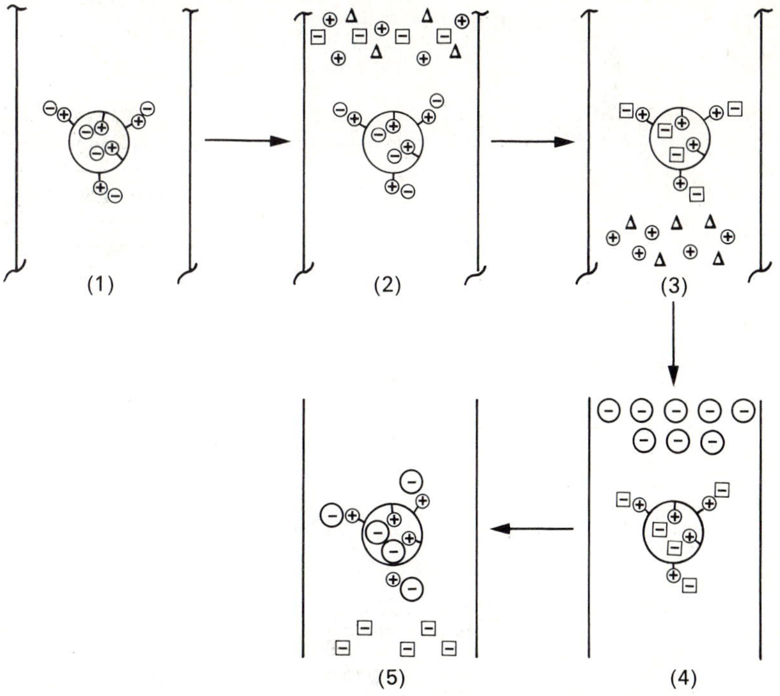

Summarizing the action of ion-exchange chromatography, solute molecules are separated on the basis of their affinity for a charged stationary phase. The following sections will focus on the properties of ion-exchange resins, selection of experimental conditions, and applications of ion-exchange chromatography.

Ion-Exchange Resins

Ion exchangers are made up of two parts—an insoluble, three-dimensional matrix and chemically bonded charged groups within and on the surface of the matrix. The resins are prepared from a variety of materials, including polystyrene, acrylic resins, polysaccharides (dextrans), and celluloses. An ion exchanger is classified as **cationic** or **anionic** depending on whether it will exchange cations or anions. A resin that has negatively charged functional groups will exchange positive ions and is a **cation exchanger**. Each type of exchanger is also classified as **strong** or **weak** according to the ionizing strength of the functional group. An exchanger with a quaternary amino group is, therefore, a **strongly basic anion exchanger**, whereas primary or secondary aromatic or aliphatic amino groups would lead to a **weakly basic anion exchanger**. A **strongly acidic**

TABLE 3.3

Ion Exchange Resins

Name			
Anion Exchangers	Functional Group	Matrix	Class
AG 1	tetramethylammonium	polystyrene	strong
AG 3	tertiary amine	polystyrene	weak
DEAE-cellulose	diethylaminoethyl	cellulose	weak
PEI-cellulose	polyethyleneimine	cellulose	weak
DEAE-Sephadex	diethylaminoethyl	dextran	weak
QAE-Sephadex	diethyl-(2-hydroxyl-propyl)-aminoethyl	dextran	strong
Cation Exchangers			
AG 50	sulfonic acid	polystyrene	strong
Bio-Rex 70	carboxylic acid	acrylic	weak
CM-Cellulose	carboxymethyl	cellulose	weak
P-Cellulose	phosphate	cellulose	intermediate
CM-Sephadex	carboxymethyl	dextran	weak
SP-Sephadex	sulfopropyl	dextran	strong

cation exchanger contains the sulfonic acid group. Table 3.3 lists the more common ion exchangers according to each of these classifications.

The ability of an ion exchanger to adsorb counter-ions is defined quantitatively by *capacity*. The *total capacity* of an ion exchanger is the quantity of charged and potentially charged groups per unit weight of dry exchanger. It is usually expressed as milliequivalents of ionizable groups per milligram of dry weight, and it can be experimentally determined by titration. The capacity of an ion exchanger is a function of the porosity of the resin. The resin matrix contains covalent crosslinking that creates a "molecular sieve." Ionized functional groups within the matrix are not readily accessible to large molecules that cannot fit into the pores. Only surface charges would be available to these molecules for exchange. The purely synthetic resins (polystyrene and acrylic) have crosslinking ranging from 2 to 16%, with 8% being the best for general purposes. The cellulosic ion exchangers have little or no crosslinking, while the dextran-based ion exchangers are available in two crosslink sizes.

With so many different experimental options and resin properties to consider, it is difficult to select the proper conditions for a particular separation. This section will outline the choices and offer guidelines for proper experimental design.

Selection of the Ion Exchanger

Before a proper choice of ion exchanger can be made, the nature of the solute molecules to be separated must be considered. For relatively small, stable molecules (amino acids, lipids, nucleotides, carbohydrates, pigments, etc.) the synthetic resins based on polystyrene are most effective. They have relatively high capacity for small molecules because the extensive crosslinking still allows access to the interior of the resin beads. For separations of peptides, proteins, nucleic acids, polysaccharides, and other large biomolecules, one must consider the use of fibrous cellulosic ion exchangers and low percent-crosslinked dextran or acrylic exchangers. The immobilized functional groups in these resins are readily available for exchange even to larger molecules. In addition, most macromolecules are found to be relatively stable on cellulose, dextran, and acrylic-based resins but are sometimes denatured on polystyrene-based exchangers.

The choice of ion exchanger has now been narrowed considerably. The next decision is whether to use a cationic or anionic exchanger. If the solute molecule has only one type of charged group, the choice is simple. A solute that has a positive charge will bind to a cationic exchanger and vice versa. However, many biomolecules have more than one type of ionizing group and may have both negatively and positively charged groups (they are amphoteric). The net charge on such molecules depends on pH. At the isoelectric point, the substance has no net charge and would not bind to any type of ion exchanger.

In principle, amphoteric molecules should bind to both anionic and cationic exchangers. However, when one is dealing with large biomolecules, the pH "range of stability" must also be evaluated. The "range of stability" refers to the pH range in which the biomolecule is not denatured. Figure 3.11 shows how the net charge of a hypothetical protein changes as a function of pH. Below the isoelectric point, the molecule has a net positive charge and would be bound to a cation exchanger. Above the isoelectric point, the net charge is negative, and the protein would bind to an anion exchanger. Superimposed on this graph is the pH range of stability for the hypothetical protein. Because it is stable in the range of pH 7.0–9.0, the ion exchanger of choice is an anionic exchanger. In most cases, the isoelectric point of the protein is not known. The type of ion exchanger must be chosen by trial and error as follows. Small samples of the solute mixture in buffer are equilibrated for 10 to 15 minutes in separate test tubes, one with each type of ion exchanger. The tubes are then

FIGURE 3.11
The effect of pH on the net charge of a protein.

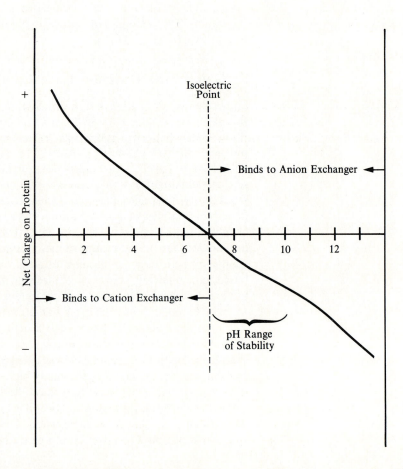

centrifuged or let stand to sediment the ion exchanger. Check each super-natant for the presence of solute (A_{260} for nucleic acids, A_{280} for proteins, catalytic activity for enzymes, etc.). If a supernatant has a relatively low level of added solute, that ion exchanger would be suitable for use. This simple test can also be extended to find conditions for elution of the desired macromolecule from the ion exchanger. The ion exchanger charged with the macromolecule is treated with buffers of increasing ionic strength or pH. The supernatant after each treatment is analyzed as before for release of the macromolecule.

The influence of porosity on ion exchange function has been mentioned. The general rule to follow is that ion exchangers with a high degree of crosslinking (small pore size) are best for the separation of small molecules. Since cellulose ion exchangers are fibrous and have little or no cross-linking, they are appropriate for large and small molecules. The dextran and acrylic-based ion exchangers are available in various pore sizes. These should be used according to the general guidelines furnished by the supplier.

Choice of Buffer

This decision includes not just the buffer substance, but also the pH and the ionic strength. Buffer ions will, of course, interact with ion exchange resins. Buffer ions with a charge opposite to that on the ion exchanger will compete with solute for binding sites and greatly reduce the capacity of the column. The following guidelines should be used: (1) cationic buffers should be used with anionic exchangers; (2) anionic buffers should be used with cationic exchangers.

The pH chosen for the buffer depends first of all on the range of stability of the macromolecule to be separated (see Figure 3.11). Secondly, the buffer pH should be chosen so that the desired macromolecule will bind to the ion exchanger. In addition, the ionic strength should be relatively low to avoid "damping" of the interaction between solute and ion exchanger. Buffer concentrations in the range of 0.05 to 0.1 M are recommended.

Preparation of the Ion Exchanger

The commercial suppliers of ion exchangers provide detailed instructions for the preparation of the adsorbents. Failure to pretreat ion exchangers will greatly reduce the capacity and resolution of a column. Because of the extensive use of the cellulose ion exchangers in biochemical research, a suitable method of preconditioning will be described here. This will also provide an example of the care and planning that must go into the design of a chromatography experiment. Most cellulosic ion exchangers are furnished in a dry powder form and must be swollen in aqueous solution to "uncover" ionic functional groups. The two most common cellulosic resins are diethylaminoethyl (DEAE) and carboxymethyl (CM) cellulose. To pretreat the anion exchanger, DEAE cellulose, 100 g is first suspended in 1 liter of 0.5 M HCl and stirred for 30 min. The cation exchanger, CM cellulose, is suspended in 1 liter of 0.5 M NaOH for the same period. This initial treatment puts a like charge on all the ionic functional groups, causing them to repel each other; thus, the exchanger swells. The cellulose is then allowed to settle (or filtered) and washed with water to a neutral pH. DEAE cellulose is then washed for 30 min. with one liter of 0.5 M NaOH (CM cellulose with 0.5 M HCl), followed by water until a neutral pH is attained. This treatment puts the ion exchanger in the free base or free acid form, and it is ready to be converted to the ionic form necessary for the separation. An additional step should be included at this point to remove trace metal ions from the ion exchanger. The presence of heavy metals in the column will interfere with the ion exchange process and also may denature sensitive macromolecules. Treatment of the swollen cellulose with 0.01 M EDTA will remove metal ions.

The above washing treatments do not remove small particles that are present in most ion exchange materials. If left in suspension, these particles, called fines, will result in decreased resolution and slow column flow rates. Removing the fines from an exchanger is done by suspending the swollen adsorbent in a large volume of water in a graduate cylinder and allowing at least 90% of the exchanger to settle. The cloudy supernatant containing the fines is decanted. This process is repeated until the supernatant is completely clear. The number of washings necessary to remove most fines is variable, but for a typical cellulose exchanger, 8 to 10 times is probably sufficient.

The washed, metal-free, defined cellulose is now ready for equilibration with the starting buffer. The equilibration is accomplished by washing the exchanger with a large volume of the desired buffer, either in a packed column or batchwise.

Using the Ion-Exchange Resin

Ion exchangers are most commonly used in a column form. The column method discussed earlier in this chapter can be directly applied to ion-exchange chromatography.

An alternate method of ion exchange is **batch separation**. This involves mixing and stirring equilibrated exchanger directly with the solute mixture to be separated. After an equilibration time of approximately one hour, the slurry is filtered and washed with buffer. The ion exchanger can be chosen so that the desired solute is adsorbed onto the exchanger or remains unbound in solution. If the latter is the case, then the desired material is in the filtrate. If the desired solute is bound to the exchanger, it can be removed by suspending the exchanger in a buffer of greater ionic strength or different pH. Batch processes have some advantages over column methods. They are rapid, and the problems of packing, channeling, and dry columns are avoided.

A new development in ion-exchange column chromatography allows for the separation of proteins according to their isoelectric points. This technique, **chromatofocusing**, involves the formation of a pH gradient on an ion exchange column. If a buffer of a specified pH is passed through an ion exchange column that was equilibrated at a second pH, a pH gradient is formed on the column. Proteins bound to the ion exchanger are eluted in the order of their isoelectric points. In addition, protein band concentration (focusing) takes place during elution. Chromatofocusing is similar to isoelectric focusing, introduced in Chapter 4, in which a column pH gradient is produced by an electric current. The requirements for chromatofocusing include specialized exchangers and buffers (available from Pharmacia under the tradename Polybuffer). Column equipment is appropriate for chromatofocusing experiments. Further information is available in a booklet, "Chromatofocusing," published by Pharmacia.

Storage of Resins

Most ion exchangers in the dry form are stable for many years. Aqueous slurried ion exchangers are still useful after several months. One major storage problem of a wet exchanger is microbial growth. This is especially true for the cellulose and dextran exchangers. If it is necessary to store pretreated exchangers, an antimicrobial agent must be added to the slurry. Sodium azide (0.02%) is suitable for cation exchangers and phenyl mercuric salts (0.001%) are effective for anion exchangers.

F. GEL EXCLUSION CHROMATOGRAPHY

The chromatographic methods discussed up to this point allow for the separation of molecules according to polarity, volatility, and charge. The method of **gel exclusion chromatography** (also called gel filtration, molecular sieve chromatography, or gel permeation chromatography) exploits the physical property of molecular size to achieve separation. The molecules of nature range in molecular weight from less than 100 to as large as several million. It should be obvious that a technique capable of separating molecules of 10,000 molecular weight from those of 100,000 would be very popular among research biochemists. Gel chromatography has been of major importance in the purification of thousands of proteins, nucleic acids, enzymes, polysaccharides, and other biomolecules. In addition, the technique may be applied to molecular weight determination and quantitative analysis of molecular interactions. In this section the theory and practice of gel filtration will be introduced and applied to several biochemical problems.

Theory of Gel Filtration

The operation of a gel filtration column is illustrated in Figure 3.12. The stationary phase consists of inert particles that contain small pores of a controlled size. Microscopic examination of a particle reveals an interior resembling a sponge. A solution containing solutes of various molecular sizes is allowed to pass through the column under the influence of continuous solvent flow. Solute molecules that are larger than the pores cannot enter the interior of the gel beads, so they are limited to the space between the beads. The volume of the column that is accessible to very large molecules is, therefore, greatly reduced. As a result, they are not slowed in their progress through the column and will rapidly elute in a single zone. Small molecules that are capable of diffusing in and out of the beads have a much larger volume available to them. Therefore, they will be delayed in their journey through the column bed. Molecules of intermediate size will migrate through the column at a rate somewhere between those for large and small molecules. Therefore, the order of elution of the various solute molecules is directly related to their molecular dimensions.

Physical Characterization of Gel Chromatography

Several physical properties must be introduced to define the performance of a gel and solute behavior. Some important properties are:

1. Exclusion Limit. This is defined as the molecular weight of the smallest molecule that cannot diffuse into the inner volume of the gel matrix. All molecules above this limit elute rapidly in a single zone. The

FIGURE 3.12
Separation of molecules by gel filtration. (A) Application of sample containing large and small molecules. (B) Large molecules cannot enter gel matrix, so they move more rapidly through the column. (C) Elution of the large molecules.

(A)   (B)   (C)

exclusion limit of a typical gel, Sephadex G-50, is 30,000. All solute molecules having a molecular weight greater than this value would pass directly through the column bed without entering the gel pores.

2. Fractionation Range. Sephadex G-50 has a fractionation range of 1500 to 30,000. Solute molecules within this molecular weight range would be separated in a somewhat linear fashion (see next page).

3. Water Regain and Bed Volume. Gel chromatography media are most often supplied in dehydrated form and are swollen in a solvent, usually water, before use. The weight of water taken up by 1 g of dry gel is known as the water regain. For G-50, this value is 5.0 ± 0.3 g. This value does not include the water surrounding the gel particles, so it cannot be used as an estimate of the final volume of a packed gel column. Most commercial suppliers of gel materials provide, in addition to water regain, a bed volume value. This is the final volume taken up by 1 g of dry gel when swollen in water. For G-50, bed volume is 9 to 11 ml/g dry gel.

4. Gel Particle Shape and Size. Ideally, gel particles should be spherical to provide for a uniform bed with a high density of pores. Particle size is defined either by mesh size or bead diameter (μm). The degree of resolution afforded by a column and the flow rate both depend on particle size. Larger particle sizes (50 to 100 mesh, 100 to 300 μm) offer high flow rates but poor chromatographic separation. The opposite is true for very small particle sizes ("superfine," 400 mesh, 10 to 40 μm). The most useful particle size, which represents a compromise between resolution and flowrate, is 100 to 200 mesh (50 to 150 μm).

5. Void Volume. This is the total space surrounding the gel particles in a packed column. This value is determined by measuring the volume of solvent required to elute a solute that is completely excluded from the gel matrix. Most columns can be calibrated for void volume with a dye, blue dextran, which has an average molecular weight of 2,000,000.

6. Elution Volume. This is the volume of eluting buffer necessary to remove a particular solute from a packed column.

Chemical Properties of Gels

Four basic types of gels are available: **dextran, polyacrylamide, agarose**, and **combined polyacrylamide-agarose**. The first gels to be developed were those based on a natural polysaccharide, dextran. These are supplied by Pharmacia Fine Chemicals, Inc. under the tradename of Sephadex. Table 3.4 gives the physical properties of the various sizes of Sephadex. The number given each gel refers to the water regain multiplied by 10. Sephadex is available in various particle sizes labeled coarse, medium, fine, and superfine. Dextran-based gels cannot be manufactured with

TABLE 3.4
Properties of Gel Filtration Media

Name	Fractionation Range for Proteins	Water Regain (ml/g dry gel)	Bed Volume (ml/g dry gel)
Dextran (Sephadex)[1]			
G-10	0–700	1.0 ± 0.1	2–3
G-15	0–1500	1.5 ± 0.2	2.5–3.5
G-25	1000–5000	2.5 ± 0.2	4–6
G-50	1500–30,000	5.0 ± 0.3	9–11
G-75	3000–80,000	7.5 ± 0.5	12–15
G-100	4000–150,000	10 ± 1.0	15–20
G-150	5000–300,000	15 ± 1.5	20–30
G-200	5000–600,000	20 ± 2.0	30–40
Polyacrylamide (Bio-Gels)[2]			
P-2	100–1800	1.5	3.0
P-4	800–4000	2.4	4.8
P-6	1000–6000	3.7	7.4
P-10	1500–20,000	4.5	9.0
P-30	2500–40,000	5.7	11.4
P-60	3000–60,000	7.2	14.4
P-100	5000–100,000	7.5	15.0
P-150	15,000–150,000	9.2	18.4
P-200	30,000–200,000	14.7	29.4
P-300	60,000–400,000	18.0	36.0
(Sephacryl)[1]			
S-200	5000–250,000	—	—
S-300	10,000–1,500,000	—	—
Agarose (Sepharose)[1]			
2B	70,000–40,000,000	—	—
4B	60,000–20,000,000	—	—
6B	10,000–4,000,000	—	—
Vinyl (Fractogel TSK)[3]			
HW-40	100–10,000	—	—
HW-55	1000–700,000	—	—
HW-65	50,000–5,000,000	—	—
HW-75	500,000–50,000,000	—	—
(Bio-Gel)[2]			
A-0.5 m	10,000–500,000	—	—
A-1.5 m	10,000–1,500,000	—	—
A-5 m	10,000–5,000,000	—	—
A-15 m	40,000–15,000,000	—	—
A-50 m	100,000–50,000,000	—	—
A-150 m	1,000,000–150,000,000	—	—

Combined Polyacrylamide-Agarose[4]

Gel Type	Acrylamide %(w/v)	Agarose %(w/v)	Fractionation Range
Ultragel AcA 202	20	2	1,000– 20,000
Ultragel AcA 54	5	4	5,000– 70,000
Ultragel AcA 44	4	4	10,000– 130,000
Ultragel AcA 34	3	4	20,000– 350,000
Ultragel AcA 22	2	2	100,000–1,200,000

[1] Pharmacia Fine Chemicals   [3] Pierce Chemical Co.

[2] Bio-Rad Laboratories   [4] LKB

an exclusion limit greater than 600,000 because the small extent of cross-linking is not sufficient to prevent collapse of the particles. If the dextran is crosslinked with N,N′-methylenebisacrylamide, gels for use in higher fractionation ranges are possible. Table 3.4 lists these gels, which are called Sephacryl and are also supplied by Pharmacia.

Polyacrylamide gels are produced by the copolymerization of acrylamide and the crosslinking agent, N,N′-methylenebisacrylamide. These are supplied by Bio-Rad Laboratories (Bio-Gel P). The Bio-Gel media are available in 10 sizes with exclusion limits ranging from 1800 to 400,000. Table 3.4 lists the acrylamide gels and their physical properties.

The agarose gels have the advantage of very high exclusion limits. Agarose, the neutral polysaccharide component of agar, is composed of alternating galactose and anhydrogalactose units. The gel structure is stabilized by hydrogen bonds rather than by covalent crosslinking. Agarose gels, supplied by Bio-Rad Laboratories (Bio-Gel A) and by Pharmacia (Sepharose), are listed in Table 3.4.

The combined polyacrylamide-agarose gels are commercially available under the tradename Ultragel (LKB). These consist of crosslinked polyacrylamide with agarose trapped within the gel network. The polyacrylamide gel allows for a high degree of separation and the agarose maintains gel rigidity, so high flow rates may be used. Typical Ultragels are listed in Table 3.4.

Selecting a Gel

The selection of the proper gel is a critical stage in successful gel chromatography. Most gel chromatographic experiments can be classified as either **group separations** or **fractionations**. Group separations involve dividing a solute sample into two groups, a fraction of relatively low molecular weight solutes and a fraction of relatively high molecular weight solutes. Specific examples of this are desalting a protein solution or removing small contaminating molecules from protein or nucleic acid extracts. For group separations, a gel should be chosen that allows for the complete exclusion of the high molecular weight molecules in the void volume. Sephadex G-25, Bio-Gel P-6, or Ultragel AcA 202 are recommended for most group separations. The particle size recommended is 100 to 200 mesh or 50 to 150 μm diameter.

Gel fractionation involves separation of groups of solutes of similar molecular weights in a multicomponent mixture. In this case, the gel should be chosen so that the fractionation range includes the molecular weights of the desired solutes. If the solute mixture contains macromolecules up to 120,000 in molecular weight, then Bio-Gel P-150, Ultragel AcA 44, or Sephadex G-150 would be most appropriate. If P-100, G-100 or AcA 54 were used, some of the higher molecular weight proteins in the sample would elute in the void volume. On the other hand, if P-200,

P-300, or G-200 were used, there would be a decrease in both resolution and flow rate. If the molecular weight range of the solute mixture is unknown, then empirical selection is necessary. The recommended gel grade for most fractionations is 100–200 or 200–400 mesh (20–80 μm or 10–40 μm). The finest grade that allows a suitable flow rate should be selected. For very critical separations, superfine grades offer the best resolution but with very slow flow rates.

Gel Preparation and Storage

The dextran and acrylamide gel products are supplied in a dehydrated form. They must be allowed to swell in water before use. The swelling time required differs for each gel, but the extremes are 3 to 4 hours at 20°C for highly crosslinked gels and up to 72 hours at 20°C for P-300 or G-200. The swelling time can be shortened if a boiling water bath is used. Agarose gels and combined polyacrylamide-agarose gels are supplied in a hydrated state, so there is no need for swelling.

Before a gel slurry is packed into the column, it should be defined and deaerated. Defining is necessary to remove very fine particles which would reduce flow rates. To define, pour the gel slurry into a graduated cylinder and add water equivalent to two times the gel volume. Invert the cylinder several times and allow the gel to settle. After 90 to 95% of the gel has settled, decant the supernatant, add water, and repeat the settling process. Two or three defining operations are usually sufficient to remove most small particles.

Deaerating (removal of dissolved gases) should be done on the gel slurry and all eluting buffers. Gel particles that have not been deaerated tend to float and form bubbles in the column bed. Dissolved gases are removed by placing the gel slurry in a side-arm vacuum flask and applying a vacuum from a water aspirator. The degassing process is complete when no more small air bubbles are released from the gel (usually 1 to 2 hours).

Antimicrobial agents must be added to stored, hydrated gels. One of the best agents is sodium azide (0.02%).

Operation of a Gel Column

The procedure for gel column chromatography is very similar to the general description given earlier. The same precautions must be considered in packing, loading, and eluting the column. A brief outline of important considerations follows.

1. Column Size. For fractionation purposes, it is usually not necessary to use columns greater than 100 cm in length. The ratio of bed length to width should be between 25 and 100. For group separations, columns less than 50 cm long are sufficient, and appropriate ratios of bed length to width are between 5 and 10.

2. Eluting Buffer. There are fewer restrictions for buffer choice in gel chromatography than in ion-exchange chromatography. Dextran and polyacrylamide gels are stable in the pH range of 1.0 to 10.0, whereas agarose gels are limited to pH 4 to 10. Since there is such a wide range of stability for the gels, the buffer pH should be chosen on the basis of the range of stability of the macromolecules to be separated.

3. Sample Volume. The sample volume is a critical factor in planning a gel chromatography experiment. If too much sample is applied to a column, resolution is decreased; if the sample size is too small, the solutes will be greatly diluted. For group separations, a sample volume of 10 to 25% of the column total volume is suitable. The sample volume for fractionation procedures should be between 1 and 5% of the total volume. Column total volume is determined by measuring the volume of water in the glass column that is equivalent to the height of the packed bed.

4. Column Flow Rate. The flow rate of a gel column depends on many factors, including length of column and the type and size of gel. It is generally safe to elute a gel column at a rate slightly less than free flow. A high flow rate will reduce sample diffusion or zone broadening, but may not allow complete equilibration of solute molecules with the gel matrix.

A specific flow rate cannot be recommended, since each type of gel requires a different range. The average flow rate given in literature references for small pore-size gels is 8 to 12 ml/cm² of cross-sectional bed area per hour (15 to 25 ml/hr). For large pore-size gels, a value of 2 to 5 ml/cm² of cross-sectional bed area per hour (5 to 10 ml/hr) is average.

Eluent can be made to flow through a column by either of two methods, gravity or pump elution. Gravity elution is most often used because no special equipment is required. It is quite acceptable for developing a column used for group separations and fractionations when small pore-sized gels are used. However, if the flow rate must be maintained at a constant value throughout an experiment, or if large pore gels are used, pump elution is recommended.

One variation of gel chromatography is ascending eluent flow. Some investigators report more reproducible results, better resolution, and a more constant flow rate if the eluting buffer is pumped backward through the gel. This type of experiment requires special equipment, including a specialized column and a peristaltic pump.

Applications of Gel Exclusion Chromatography

Several experimental applications of gel chromatography have already been mentioned, but more detail will be given here.

1. Desalting. Inorganic salts, organic solvents, and other small molecules are used extensively for the purification of macromolecules. Gel chromatography provides an inexpensive, simple, and rapid method for removal of these small molecules.

2. *Purification of Biomolecules.* This is probably the most popular use of gel chromatography. Because of the ability of a gel to fractionate molecules on the basis of size, gel filtration complements other purification techniques that separate molecules on the basis of polarity and charge.

3. *Estimation of Molecular Weight.* The elution volume for a particular solute is proportional to the molecular size of that solute. This indicates that it is possible to estimate the molecular weight of a solute on the basis of its elution characteristics on a gel column. An elution curve for several standard proteins is shown in Figure 3.13. This curve, a plot of protein concentration vs. volume collected, is representative of data obtained from a gel filtration experiment. The elution volume, V_e, for each protein can be estimated as shown in the figure. A plot of log molecular weight vs. elution volume for the proteins is shown in Figure 3.14. Note that the linear portion of the curve in Figure 3.14 covers the molecular weight range of 3000 to 100,000. Molecules below 3000 in molecular weight are not eluted in an elution volume proportional to size. Since they readily diffuse into the gel particles, they will be retarded. Molecules greater than 100,000 in molecular weight are all excluded from the gel in the void volume. A solution of the unknown protein is chromatographed through the calibrated column under conditions identical to the standards and the elution volume measured. The unknown molecular weight is then read directly from the graph. This method of molecular weight estimation is widely used because

FIGURE 3.13

Elution curve for several proteins. A = Hemoglobin, B = Egg Albumin, C = Chymotrypsinogen, D = Myoglobin, E = Cytochrome *c*.

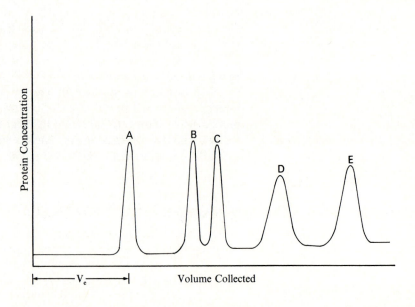

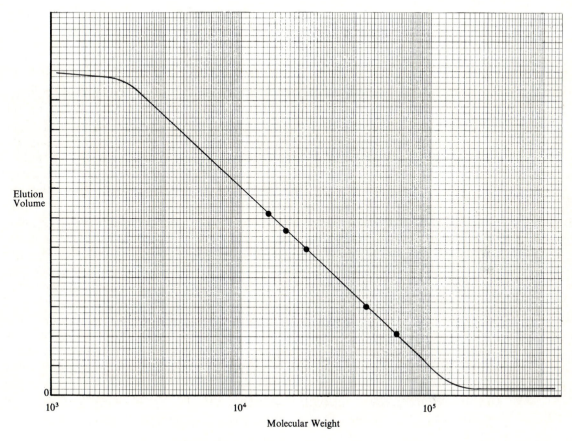

FIGURE 3.14
A plot of log molecular weight vs. elution volume for the proteins in Figure 3.13.

it is simple, accurate, inexpensive, and fast. It can be used with highly purified or impure samples. There are, of course, some limitations to consider. The gel must be chosen so that the molecular weight of the unknown is within the linear section of the curve. The method assumes that only steric and partition effects influence the elution of the standards and unknown. If a protein interacts with the gel by adsorptive or ionic processes, the estimate of molecular weight is lower than the true value. The assumption is also made that the unknown molecules have a general spherical shape, not an elongated or rod shape.

If only small samples of a particular protein are available, the molecular weight can be estimated by **thin-layer gel filtration**. A glass plate is coated with an appropriate hydrated gel (usually superfine grade) and the protein sample is applied along with standards just as in thin-layer chromatography. The loaded plate is set in a special chamber and developed with descending buffer at a rate of 1 to 2 cm/hour. This is particularly useful for molecular weight determination, but can also be used for analytical purposes.

4. Gel Chromatography in Organic Solvents. The gels discussed so far in this section are hydrophilic, and the inner matrix retains its integrity only in aqueous solvents. Because there is a need for gel chromatography of nonhydrophilic solutes, gels have been produced that can be used with organic solvents. The trade names for some of these products are Sepharose CL and Sephadex LH (Pharmacia), Bio Beads S (Bio-Rad) and Styragel (Dow Chemical). Organic solvents that are of value in gel chromatography are ethanol, acetone, dimethylsulfoxide, dimethylformamide, tetrahydrofuran, chlorinated hydrocarbons, and acetonitrile. For specific instructions in the use of these gels, brochures from the individual suppliers should be studied.

G. HIGH PERFORMANCE LIQUID CHROMATOGRAPHY (HPLC)

The previous discussions on the theory and practice of the various chromatographic methods should convince you of the tremendous influence chromatography has had on our biochemical understanding. It is tempting, but unfair, to make comparisons of the methods; however, each serves a specific purpose. There will, for example, always be a need for fast, inexpensive, and qualitative analyses as afforded by TLC. Traditional column chromatography will probably always be preferred in large-scale protein purification.

However, during the past 15 years, an analytical method has been developed that currently rivals and may soon surpass the traditional liquid chromatographic techniques in importance for analytical separations. This new technique, high performance liquid chromatography (HPLC), is ideally suited for the separation and identification of amino acids, carbohydrates, lipids, nucleic acids, proteins, pigments, steroids, pharmaceuticals, and many other biologically active molecules.

The future promise of HPLC is indicated by its classification as "modern liquid chromatography" when compared to other forms of column-liquid chromatography, now referred to as "classical" or "traditional." Compared to the classical forms of liquid chromatography (paper, TLC, column), HPLC has several advantages:

1. Resolution and speed of analysis far exceed the classical methods.
2. HPLC columns can be reused without repacking or regeneration.

3. Reproducibility is greatly improved because the parameters affecting the efficiency of the separation can be closely controlled.

4. Instrument operation and data analysis are easily automated.

5. HPLC is adaptable to large-scale, preparative procedures.

Gas chromatography has had and will continue to have a major impact in biochemical research. However, the major disadvantage of GC is that many biochemical samples either are not volatile or are thermally unstable. HPLC, which can be operated at ambient temperatures, does not have this limitation.

The advantages of HPLC are the result of two major advancements: (1) the development of very small particle-size stationary supports with large surface areas, and (2) the improvement of elution rates by applying high pressure to the solvent flow.

The great versatility of HPLC is evidenced by the fact that all chromatographic modes, including partition, adsorption, ion-exchange, chromatofocusing, and gel exclusion, are possible. In a sense, HPLC can be considered as automated liquid chromatography. The theory of each of these chromatographic modes has been discussed and needs no modification for application to HPLC. However, there are unique theoretical and practical characteristics of HPLC that should be introduced.

The **retention time** of a solute in HPLC (t_R) is defined as the time necessary for maximum elution of the particular solute. This is analogous to retention time measurements in GC. **Retention volume** (V_R) of a solute is the solvent volume required to elute the solute and is defined by Equation 3.5, where F is the flow rate of the solvent.

$$V_R = Ft_R$$
(Equation 3.5)

In all forms of chromatography, a measure of column efficiency is **resolution**, R. Resolution indicates how well solutes are separated; it is defined by Equation 3.6, where t_R and t'_R are the retention times of two solutes and w and w' are the base peak widths of the same two solutes.

$$R = 2 \frac{t_R - t'_R}{w + w'}$$
(Equation 3.6)

Instrumentation

The increased resolution achieved in HPLC compared to classical column chromatography is primarily the result of adsorbents of very small particle size (less than 20 μm) and large surface areas. The smallest gel beads used in gel exclusion chromatography are "superfine" grade with diameters of 20 to 50 μm. Recall that the smaller the particle size, the lower the flow rate; therefore, it is not feasible to use very small gel beads

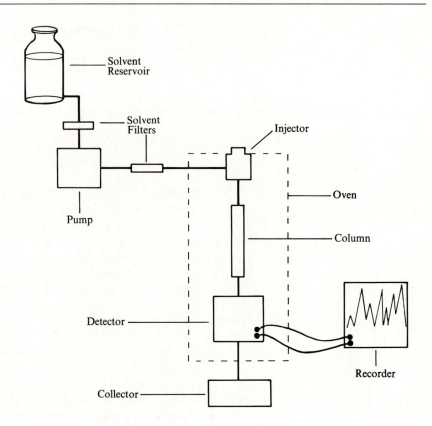

FIGURE 3.15
A schematic diagram of a high
pressure liquid chromatograph.

in liquid column chromatography because slow flow rates lead to solute diffusion and because the time necessary for completion of an analysis would be impractical. In HPLC, increased flow rates are obtained by applying a pressure differential across the column. A combination of high pressure and adsorbents of small particle size leads to the high resolving power and short analysis times characteristic of HPLC.

A schematic diagram of a typical high pressure liquid chromatograph is shown in Figure 3.15. The basic components are a solvent reservoir, high pressure pump, packed column, detector, and recorder. The similarities between a gas chromatograph and an HPLC are obvious. The tank of carrier gas in GC is replaced by the solvent reservoir and high pressure pump in HPLC.

Solvent Reservoir

The solvent chamber should have a capacity of at least 500 ml for analytical applications, but larger reservoirs are required for preparative work. In order to avoid bubbles in the column and detector, the solvent must be degassed. Several methods may be used to remove unwanted gases, including refluxing, filtration through a vacuum filter, ultrasonic vibration, and purging with an inert gas. The solvent should also be filtered to remove particulate matter that would be drawn into the pump and column.

Pumping Systems

The purpose of the pump is to provide a constant, reproducible flow of solvent through the column. Two types of pumps are available—constant pressure and constant volume. Typical requirements for a pump are:

1. It must be capable of pressure outputs of at least 500 psi and preferably up to 5000 psi.
2. It should have a controlled, reproducible flow delivery of at least 3 ml/min for analytical applications and up to 100 ml/min for preparative applications.
3. It should yield pulse-free solvent flow.
4. It should have a small hold-up volume.

Although neither type of pump meets all these criteria, constant volume pumps maintain a more accurate flow rate and a more precise analysis is obtained.

Injection Port

A sample must be introduced onto the column in an efficient and reproducible manner. One of the most popular injectors is the syringe injector shown in Figure 3.16. The sample, in a microliter syringe, is

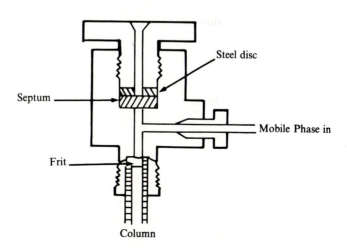

FIGURE 3.16
A syringe injector for HPLC. From *Introduction to High Performance Chromatography* by R. Hamilton and P. Sewell, Chapman and Hall. Copyright 1977.

Steel disc

Septum

Mobile Phase in

Frit

Column

injected through a neoprene/Teflon septum. This type of injector can be used at pressures up to 3000 psi.

Columns

HPLC columns are prepared from stainless steel or glass-Teflon tubing. Typical column inside diameters are 2.1 mm, 3.2 mm, or 4.5 mm for analytical separations and up to 30 mm for preparative applications. The length of the column can range from 5 to 100 cm, but 10 to 20 cm columns are common.

Detector

Liquid chromatographs are equipped with a means to continuously monitor the column effluent and recognize the presence of solute. Only small sample sizes are used with most HPLC columns, so a detector must have high sensitivity. The type of detector that has the most universal application is the **differential refractometer**. This device continuously monitors the refractive index difference between pure solvent (mobile phase) and the mobile phase containing sample (column effluent). The sensitivity of this detector is on the order of 0.1 μg, which, compared to other detectors, is only moderately sensitive. The major advantage of the refractometer detector is its versatility; its main limitation is that there must be at least 10^{-7} refractive index units between the mobile phase and sample. However, the refractometer cannot be used with gradient elution (see below); this presents a major limitation to its use.

The most widely used HPLC detectors are the **photometric detectors**. These detectors measure the extent of absorption of ultraviolet or visible radiation by a sample. Since few compounds are colored, visible detectors are only of limited value. Ultraviolet detectors, however, are the most widely used in HPLC. The typical UV detector functions by focusing radiation from a low pressure mercury lamp on a flow cell that contains column effluent. The mercury lamp provides a primary radiation at 254 nm. The use of filters or other lamps provides radiation at 220, 280, 313, 334, and 365 nm. Many compounds absorb strongly in this wavelength range, and sensitivities on the order of 1 ng are possible. The extent of absorption is measured with a photocell. The ultraviolet detector is suitable for compounds containing π-bonding and nonbonded electrons. Most biochemicals are detected, including proteins, nucleic acids, pigments, vitamins, some steroids, and aromatic amino acids. Aliphatic amino acids, carbohydrates, lipids, and other biochemicals that do not absorb UV can be detected by chemical derivatization with UV-absorbing functional groups. UV detectors have many positive characteristics, including high sensitivity, small sample volumes, linearity over wide concentration ranges, nondestructiveness to sample, and suitability for gradient elution.

A third type of detector that has only limited use is the **fluorescence detector**. This type of detector is extremely sensitive; its use is limited to samples containing trace quantities of biological materials. Its response is not linear over a wide range of concentrations, but it may be up to 100 times more sensitive than the UV detector.

Collection of Eluent

All of the detectors described here are nondestructive to the samples, so column effluent can be collected for further chemical and physical analysis.

Analysis of HPLC Data

The response from the detector is directed to a strip chart recorder for a trace of solute elution vs. time. These data are useful for qualitative or quantitative analysis. As in GC, sample components may be identified by **retention time**, **peak enhancement**, or **physical/chemical characterization** of collected samples.

Most HPLC instruments are on-line with an integrator and a microprocessor for data handling. For quantitative analysis of HPLC data, operating parameters such as rate of solvent flow must be controlled. In modern instruments, the whole system (including the pump, injector, detector, and data system) is under the control of a central microprocessor.

Figure 3.17 illustrates the separation by HPLC of several phenylhydantoin derivatives of amino acids.

Stationary Phases in HPLC

The adsorbents in HPLC are typically small-diameter, porous materials. Two types of stationary phases are illustrated in Figure 3.18. **Porous layer beads** (Figure 3.18A) have an inert solid core with a thin porous outer shell of silica, alumina, or ion-exchange resin. The average diameter of the beads ranges from 20 to 45 μm. They are especially useful for analytical applications, but, because of their short pores, their capacities are too low for preparative applications.

Microporous particles are available in two sizes: 20 to 40 μm diameter with longer pores and 5 to 10 μm with short pores (see Figure 3.18B). These are now more widely used than the porous layer beads because they offer greater resolution and faster separations with lower pressures. The microporous beads are prepared from alumina, silica, ion-exchanger resins, and chemically bonded phases (see next page).

The HPLC can function in several chromatographic modes. Each type of chromatography will be discussed below, with information about stationary phases.

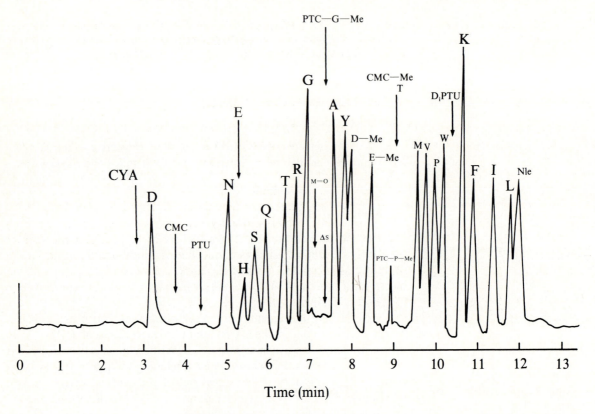

FIGURE 3.17
The separation of several amino acid phenylhydantoins. Abbreviations: D = aspartic acid, N = asparagine, H = histidine, S = serine, Q = glutamine, T = threonine, R = arginine, G = glycine, A = alanine, Y = tyrosine, D-Me = aspartic acid, methylester, E-Me = glutamic acid, methyl ester, M = methionine, V = valine, P = proline, W = tryptophan, K = lysine, F = phenylalanine, I = isoleucine, L = leucine, Nle = norleucine. From S. Black and M. Coon, *Anal, Biochem., 121*, 281 (1982).

FIGURE 3.18
Adsorbents used in HPLC. (A) Porous layer with short pores. (B) Microporous particle with longer pores.

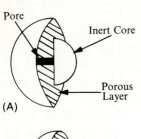

(A)

(B)

Liquid-Solid (Adsorption) Chromatography

HPLC in the adsorption mode can be carried out with silica or alumina porous-layer-bead columns. Small glass beads are often used for the inert core. Some of the more widely used packings are μPorasil (Waters Associates), BioSil A (Bio-Rad Laboratories), LiChrosorb Si-100 Partisil, Vydac, ALOX 60D (several suppliers), and Supelcosil (Supelco).

In high pressure adsorption chromatography, solutes adsorb, with different affinities, to binding sites in the solid stationary phase. Separation of solutes in a sample mixture occurs because polar solutes will adsorb more strongly than nonpolar solutes. Therefore, the various components in a sample will be eluted with different retention times from the column. This form of HPLC is usually called **normal phase** (polar stationary phase and a nonpolar mobile phase).

Liquid-Liquid (Partition) Chromatography

In the early days of HPLC (1970–78), solid supports were coated with a liquid stationary phase as in gas chromatography. Columns with these packings were found to have short lifetimes and a gradual decrease in resolution because there was continuous loss of the liquid stationary phase with use of the column.

This problem was remedied by the discovery of methods for chemically bonding the active stationary phase to the inert support. Most chemically bonded stationary phases are produced by covalent modification of the surface silica. Three modification processes are shown in Reactions 3.1 to 3.3.

Silicate Esters

$$-Si-OH + ROH \longrightarrow -Si-O-R \qquad \text{(Reaction 3.1)}$$

Silica-Carbon

$$-Si-Cl + \begin{matrix} RMgBr \\ \text{or} \\ RLi \end{matrix} \longrightarrow -Si-R \qquad \text{(Reaction 3.2)}$$

Siloxanes

$$-Si-OH + \begin{matrix} ClSiR_3 \\ \text{or} \\ ROSiR_3 \end{matrix} \longrightarrow -Si-O-SiR_3 \qquad \text{(Reaction 3.3)}$$

The major advantage of the bonded stationary phases is stability. Since it is chemically bonded, there is very little loss of stationary phase with column use. The siloxanes are the most widely used silica supports. Functional groups that can be attached as siloxanes are alkyl nitriles ($-Si-CH_2CH_2-CN$), phenyl ($-Si-C_6H_5$), alkylamines ($-Si(CH_2)_n-NH_2$), and alkyl side chains ($-Si-C_8H_{17}$; $-Si-C_{18}H_{37}$).

The use of nonpolar chemically bonded stationary phases with a polar mobile phase is referred to as **reverse phase HPLC**. This technique separates sample components according to hydrophobicity. It is widely used for the separation of smaller biomolecules, including peptides, nucleotides, carbohydrates, and derivatives of amino acids. Typical solvent systems are water-methanol, water-acetonitrile, and water-tetrahydrofuran mixtures.

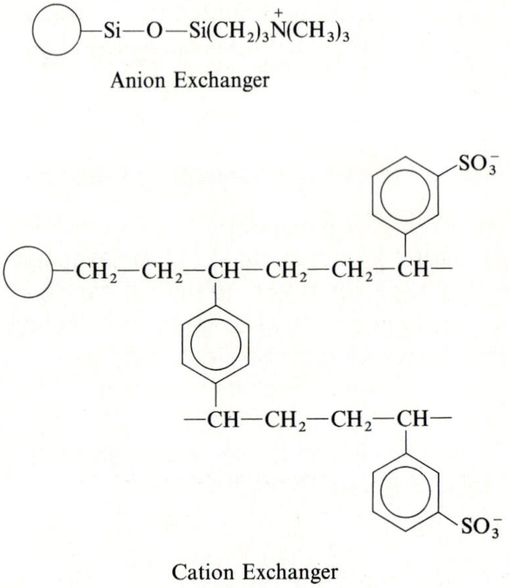

Anion Exchanger

Cation Exchanger

FIGURE 3.19
Ion exchangers used in HPLC.

Ion-Exchange Chromatography

Ion-exchange HPLC uses column packings with charged functional groups. Structures of typical ion exchangers are shown in Figure 3.19. They are prepared by chemically bonding the ionic groups to the support via silicon atoms or by using polystyrene-divinylbenzene resins. These stationary phases may be used for the separation of proteins, peptides, and other charged biomolecules. Two widely used phases are DEAE-silica and CM-silica. Both of these columns can be used with aqueous buffers in the pH range of 2.5 to 7.5. Chromatofocusing columns are available from commercial suppliers.

Gel Exclusion Chromatography

The combination of HPLC and gel exclusion chromatography is used extensively for the separation of large biomolecules, especially proteins and nucleic acids.

The exclusion gels discussed in Section F are not appropriate for HPLC use because they are soft and, except for small-pore beads (G-25 and less), will collapse under high pressure conditions. Semirigid gels based on crosslinked styrene-divinylbenzene, polyacrylamide, and vinyl-acetate copolymer are available with various fractionation ranges useful for the separation of molecules up to a molecular weight of 10,000,000. The semirigid gels are useful only with aqueous-based solvents and a few organic solvents.

Rigid packings for HPLC gel exclusion are prepared from porous glass or silica. They have several advantages over the semirigid gels, including several fractionation ranges, ease of packing and compatibility with water and organic solvents.

The Mobile Phase

The selection of a column packing that is appropriate for a given analysis does not ensure a successful HPLC separation. A suitable solvent system must also be chosen. Several critical solvent properties will be considered here.

Purity. Very high purity solvents with no particulate matter are required. Many laboratory workers do not purchase expensive prepurified solvents, but rather they purify lesser grade solvents by microfiltration through a millipore system or distillation in glass.

Reactivity. The mobile phase must not react with the analytical sample or column packing. This does not present a major limitation since many relatively unreactive hydrocarbons, alkyl halides, and alcohols are suitable.

Detector Compatibility. A solvent must be carefully chosen to avoid interference with the detector. Most UV detectors monitor the column effluent at 254 nm. Any UV-absorbing solvent, such as benzene or olefins, would be unacceptable because of high background. Since refractometer detectors monitor the difference in refractive index between solvent and column effluent, a greater difference leads to greater ability to detect the solute.

Solvents for HPLC Operation

It has long been recognized that the eluting power of a solvent is related to its polarity. Chromatographic solvents have been organized into a list according to their ability to displace adsorbed solutes (eluting power, ε°). This list of solvents, called an **eluotropic series**, is shown in Table 3.5. It should be noted that ε° increases with an increase in polarity. Using the eluotropic series makes solvent choice less a matter of trial and error. Quite often, trial experiments can be performed on a silica gel thin-layer plate spotted with the analytical sample. A solvent, ethyl acetate, for example, is used to develop the plate. If ethyl acetate is too nonpolar to cause suitable movement and separation of the sample components, a solvent of higher eluent strength (ethanol, for example) is tried. If the solutes move too far, a solvent of lower ε° is used (toluene, for example).

Occasionally, a single solvent does not provide suitable resolution of solutes. Solvent **binary mixtures** can be prepared with eluent strengths intermediate between the ε° for the individual solvents.

TABLE 3.5
Eluotropic Series of HPLC Solvents

Solvent	ε° [1]	n_0 [2]	UV cutoff (nm)
n-Pentane	0.00	1.358	210
n-Hexane	0.01	1.375	210
Cyclohexane	0.04	1.427	210
Carbon tetrachloride	0.18	1.466	265
2-Chloropropane	0.29	1.378	225
Toluene	0.29	1.496	285
Ethyl ether	0.38	1.353	220
Chloroform	0.40	1.443	245
Tetrahydrofuran	0.45	1.408	230
Acetone	0.56	1.359	330
Ethyl acetate	0.58	1.370	260
Dimethylsulfoxide	0.62	1.478	270
Acetonitrile	0.65	1.344	210
2-Propanol	0.82	1.380	210
Ethanol	0.88	1.361	210
Methanol	0.95	1.329	210
Ethylene glycol	1.11	1.427	210

[1] eluent strength

[2] refractive index

Gradient Elution in HPLC

Figure 3.20A illustrates the separation of a multicomponent sample using a single solvent and a silica column. Note that some earlier components (1 to 5) are poorly resolved and later peaks (9 to 11) display extensive broadening. This common difficulty encountered in the analysis of multicomponent samples is referred to as the **general elution problem**. It is due to the fact that the components have a wide range of K_D values and no single solvent system is equally effective in efficiently displacing all components from the column.

The general elution problem is solved by the use of **gradient elution**. This is achieved by varying the composition of the mobile phase during elution. The similarity between gradient elution and temperature programming in gas chromatography should be readily apparent. In practice, gradient elution is performed by beginning with a weakly eluting solvent (low ε°) and gradually increasing the concentration of a more strongly eluting solvent (higher ε°). The weaker solvent is able to improve resolution of those components having low K_D values (peaks 1 to 5, Figure 3.20B). In addition, the gradual increase in ε° of the solvent mixture decreases line broadening of components 9 to 11 and provides a more effective separation.

The choice of solvents to use for gradient elution is still somewhat empirical; however, using the data from Table 3.5 narrows the choices. Modern HPLC instruments are equipped with **solvent programming units** which control gradient elution in a stepwise or continuous manner.

Gradient elution cannot be used with the differential refractometer detector because it is difficult to match the refractive indices of the column

FIGURE 3.20
The general elution problem. (A) Normal elution. (B) Gradient elution. From L. R. Snyder, *J. Chromatogr. Sci., 8,* 692 (1970).

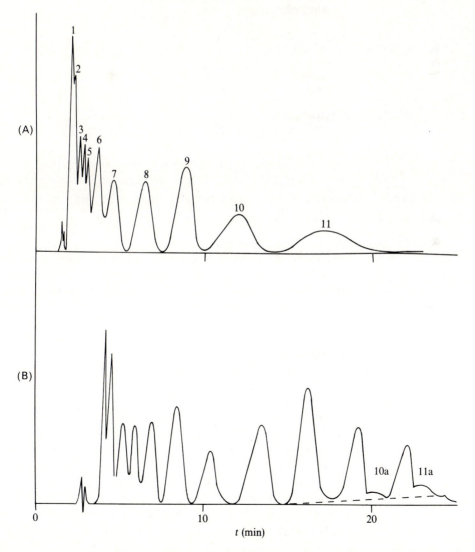

effluent with the reference (a steady base-line is difficult to achieve). Despite this limitation, gradient elution provides the conditions for maximum resolution for most multicomponent samples.

Selection of HPLC Operating Conditions

Each type of HPLC instrument has its own characteristics and operating directions. It is not feasible to describe those here. However, it is appropriate to outline the general approach taken when an HPLC analysis is desired. The following items must be considered:

1. Chemical nature of the sample.
2. Selection of type of chromatography (partition, adsorption, ion exchange, gel exclusion).
3. Choice of solvent system and mode of elution.
4. Selection of column packing.
5. Choice of equipment (type of detector).

H. AFFINITY CHROMATOGRAPHY

The more conventional chromatographic procedures that we have studied up to this point rely on rather nonspecific physico-chemical interactions between a stationary support and solute. The molecular characteristics of net charge, size, and polarity do not provide a basis for high selectivity in the separation and isolation of biomolecules. The desire for more specificity in chromatographic separations led to the discovery of **affinity chromatography**. This technique offers the ultimate in specificity—separation on the basis of biological interactions. The biological function displayed by most macromolecules (antibodies, transport proteins, enzymes, nucleic acids, polysaccharides, receptor proteins, etc.) is a result of a recognition and interaction with specific molecules called **ligands**. This can be illustrated by Reaction 3.4, where A represents a macromolecule and B is a smaller molecule or ligand. The two molecules interact in a specific manner to form a complex, A:B.

$$A + B \longrightarrow A:B \longrightarrow \text{Biological Response} \qquad \text{(Reaction 3.4)}$$

In a biological system, the formation of the complex often triggers some response such as immunological action, control of a metabolic process, hormone action, catalytic breakdown of a substrate, or membrane transport. The biological response depends upon proper molecular recognition and binding as shown in the reaction. The most common example of Reaction 3.4 is the interaction that occurs between an enzyme molecule, E, and a substrate, S, with reversible formation of an ES complex. The biological event resulting from this interaction is the transformation of S

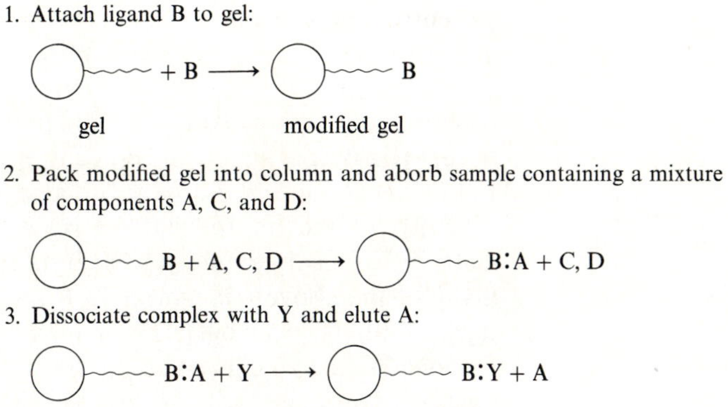

1. Attach ligand B to gel:

gel   modified gel

2. Pack modified gel into column and aborb sample containing a mixture of components A, C, and D:

3. Dissociate complex with Y and elute A:

FIGURE 3.21
The steps of affinity chromatography.

to a metabolic product, P. Only the first step in Reaction 3.4, formation of the complex, is of concern in affinity chromatography.

In practice, affinity chromatography requires the preparation of an insoluble stationary phase, to which appropriate ligand molecules (B) are covalently affixed. Thus, ligand molecules are immobilized on the stationary support. The affinity support is packed into a column through which a mixture containing the desired macromolecule, A, is allowed to percolate. There are many types of molecules in the mixture, especially if it is a crude cell extract, but only macromolecules that recognize and bind to immobilized B will be retarded in their movement through the column. After the nonbinding molecules have washed through the column, the desired macromolecules, B, are eluted by gentle disruption of the A:B complex. Study Figure 3.21 for an illustration of affinity chromatography.

Affinity chromatography can be applied to the isolation and purification of virtually all biological macromolecules. It has been used to purify enzymes, transport proteins, antibodies, hormone receptor proteins, drug binding proteins, neurotransmitter proteins, and many others.

Successful application of affinity chromatography requires careful design of experimental conditions. The essential components, which will be outlined below, are (1) creation and preparation of a stationary matrix with immobilized ligand, and (2) design of column development and eluting conditions.

Chromatographic Media

Selection of the matrix used to immobilize a ligand requires consideration of several properties. The stationary supports used in gel exclusion chromatography are found to be quite suitable for affinity chromatography because: (1) they are physically and chemically stable under most experimental conditions, (2) they are relatively free of nonspecific adsorptive effects, (3) they display satisfactory flow characteristics, (4) they are available with very large pore sizes, and (5) they have reactive functional groups to which an appropriate ligand may be attached.

Four types of media possess most of these desirable characteristics: agarose, polyvinyl, polyacrylamide, and controlled porosity glass (CPG) beads. Highly porous agarose beads such as Sepharose 4B (Pharmacia Fine Chemicals) and Bio-Gel A-150 m (Bio-Rad Laboratories) possess virtually all the above characteristics and are the most widely used matrices. Polyacrylamide gels such as Bio-Gel P-300 (Bio-Rad) display many of the above recommended features; however, the porosity is not especially high. Fractogel HW-65F, a vinyl polymer supplied by Pierce Chemical Company, has many of the above characteristics and a protein fractionation range of 50,000 to 5,000,000. CPG beads have not yet found wide acceptance, primarily because of extensive nonspecific protein adsorption and small numbers of reactive functional groups for ligand attachment. Future developments in the manufacture of these beads should minimize these difficulties. The advantages of the CPG beads are mechanical strength and chemical stability.

Selection of the Immobilized Ligand

The ligand (B in Reaction 3.4 and Figure 3.21) can be selected only after the nature of the macromolecule to be isolated is known. When a hormone receptor protein is to be purified by affinity chromatography, the hormone itself is an ideal candidate for the ligand. For antibody isolation, an antigen or hapten may be used as ligand. If an enzyme is to be purified, then a substrate analog, inhibitor, cofactor, or effector may be used as the immobilized ligand. The actual substrate molecule may be used as a ligand, but only if column conditions can be modified to avoid catalytic transformation of the bound substrate.

In addition to the above requirements, the ligand must display a strong, specific, but reversible interaction with the desired macromolecule and it must have a reactive functional group for attachment to the matrix. It should be recognized that several types of ligand may be used for affinity purification of a particular macromolecule. Of course, some ligands will work better than others, and empirical binding studies can be performed to select an effective ligand.

Attachment of Ligand to Matrix

Several procedures have been developed for the covalent attachment of the ligand to the stationary support. All procedures for gel modification proceed in two separate chemical steps: (1) activation of the functional groups on the matrix and (2) joining the ligand to the functional group on the matrix.

A wide variety of activated gels is now commercially available. The most widely used are:

1. Cyanogen Bromide-Activated Agarose. This gel is especially versatile since all ligands containing primary amino groups are easily attached to the agarose. It is available under the tradename CNBr-activated Sepharose 4B (Pharmacia Fine Chemicals). Since the gel is extremely reactive, very gentle conditions may be used to couple the ligand. One disadvantage of CNBr activation is that small ligands are coupled very closely to the matrix surface; macromolecules, because of steric repulsion, may not be able to interact fully with the ligand. The procedure for CNBr activation and ligand coupling is outlined in Figure 3.22A.

2. 6-Aminohexanoic Acid (CH)-Agarose and 1,6-Diaminohexane (AH)-Agarose. These activated gels overcome the steric interference problems stated above by positioning a six-carbon spacer arm between the ligand and the matrix. Ligands with free primary amino groups can be covalently attached to CH-agarose, whereas ligands with free carboxyl groups can be coupled to AH-agarose. CH- and AH-agarose are marketed by several companies including Pharmacia Fine Chemicals, Pierce Chemical Co.,

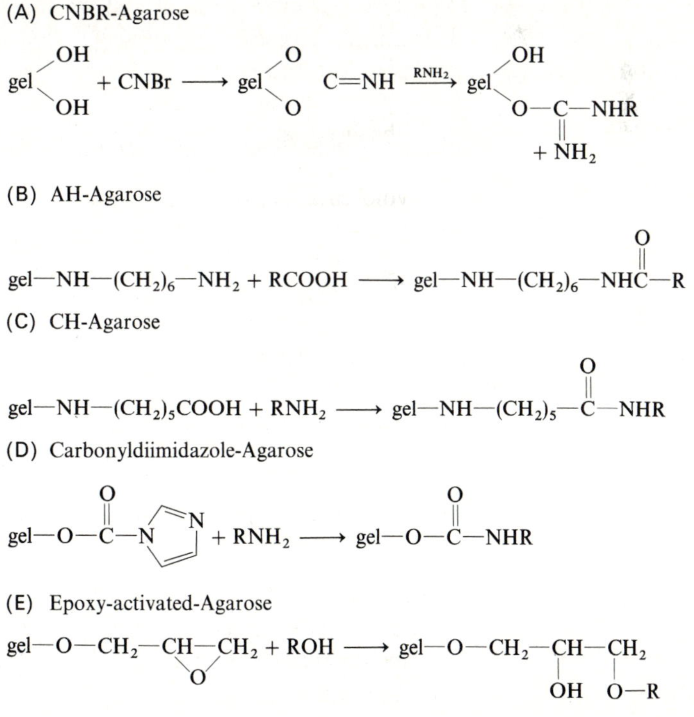

FIGURE 3.22
Attachment of specific ligands to activated gels; R = Ligand.

and Bio-Rad Laboratories. The attachment of ligands to AH and CH gels is outlined in Figure 3.22B,C.

3. Carbonyldiimidazole (CDI)-Activated Supports. Reaction with CDI produces gels that contain uncharged N-alkylcarbamate groups (see Figure 3.22D). CDI-activated agarose, dextran, and polyvinyl acetate are sold by Pierce Chemical Co. under the tradename Reacti-Gel.

4. Epoxy-Activated Agarose. The structure of this gel is shown in Figure 3.22E. It provides for the attachment of ligands containing hydroxyl, thiol, or amino groups. The hydroxyl groups of mono-, oligo-, and polysaccharides can readily be attached to the gel. Epoxy-activated Sepharose 6B is available from Pharmacia Fine Chemicals.

5. Group-Specific Adsorbents. The affinity materials described up to this point are modified with a ligand that displays specificity for a particular macromolecule. Therefore, each time a biomolecule is to be isolated by affinity chromatography, a new adsorbent must be designed and prepared. Ligands of this type are called substance-specific. In contrast, group-specific adsorbents contain ligands that have affinity for a class of biochemically related substances. Table 3.6 provides a list of several commercially available group-specific adsorbents and the specificity of each.

One interesting and versatile affinity material from this list is thiopropyl agarose. This gel contains 2-thiopyridyl functional groups on a short hydrophilic spacer arm (Figure 3.23). It is especially attractive for

TABLE 3.6

Group Specific Adsorbents Useful in Biochemical Applications

Group Specific Adsorbent	Group Specificity
Concanavalin A-agarose	macromolecules with glucopyranosyl and other carbohydrates (glycoproteins and glycolipids)
Protein A-agarose	IgG-type antibodies
Poly(U)-agarose	nucleic acids containing poly(A) sequences, poly(U)-binding proteins
Poly(A)-agarose	nucleic acids containing poly(U) sequences, *m*-RNA binding proteins
Cibracron blue-agarose	enzymes with nucleotide cofactors (dehydrogenases, kinases, DNA polymerase); serum albumin
5′-AMP-agarose	enzymes that have NAD^+ as cofactor; ATP-dependent kinases
Boronic acid-agarose	compounds with cis-diol groups; sugars, catecholamines, ribonucleotides
Iminodiacetate-agarose	proteins with affinity for heavy metals

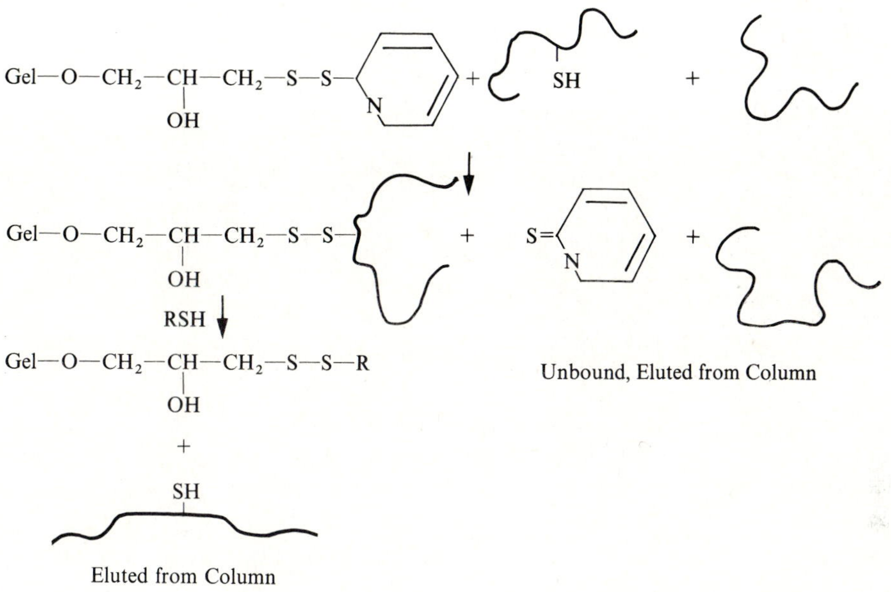

FIGURE 3.23
Affinity chromatography of a thioprotein on thiopropylagarose.

reversible coupling of small thiol compounds and thiol-proteins. Substances with the —SH group can be coupled to the gel by thiol-disulfide exchange and then removed by disulfide reduction. The use of a group-specific support, boronate-agarose, is illustrated in Experiment 13.

Experimental Procedure for Affinity Chromatography

Although the procedure will be different for each type of substance isolated, a general experimental plan is outlined here. Figure 3.24 provides a step-by-step plan in a flow-chart form. Many types of matrix-ligand systems are commercially available and the costs are reasonable, so it is not always necessary to spend valuable laboratory time for affinity gel preparation. Even if a specific gel is not available, time can be saved by purchasing preactivated gels for direct attachment of the desired ligand. Once the gel is prepared, the procedure is similar to that described earlier. The major difference, however, is the use of shorter columns. Most affinity gels have high capacities, and column beds less than 10 cm in length are commonly used. A second difference is the mode of elution. Ligand-macromolecule complexes immobilized on the column are held together by hydrogen bonding, ionic interactions, and hydrophobic effects. Any agent

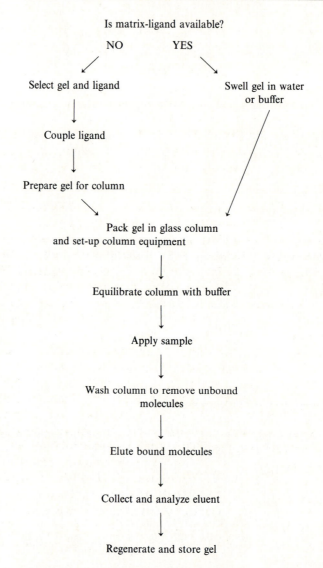

FIGURE 3.24
Experimental procedure for affinity chromatography.

that diminishes these forces causes the release and elution of the macromolecule from the column. The common methods of elution are change of buffer pH, increase of buffer ionic strength, affinity elution, and chaotropic agents. The choice of elution method depends on many factors, including the types of forces responsible for complex formation and the stability of the ligand matrix and isolated macromolecule.

Buffer pH or Ionic Strength. If ionic interactions are important for complex formation, a change in pH or ionic strength weakens the interaction by altering the extent of ionization of ligand and macromolecule.

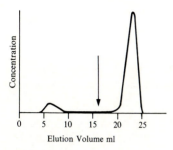

FIGURE 3.25
Purification of α-chymotrypsin by affinity chromatography on immobilized D-tryptophan methyl ester. From *Affinity Chromatography: Principles and Methods,* Pharmacia Fine Chemicals AB, Uppsala, Sweden.

In practice, either a decrease in pH or a gradual increase in ionic strength (continual or stepwise gradient) is used.

Affinity Elution. In this method of elution, a selective substance added to the eluting buffer competes for binding to the ligand or for binding to the adsorbed macromolecule. An example of the first case is the elution of glycoproteins from a boronic acid-agarose gel by addition of a free sugar, sorbitol, to the eluting buffer (Experiment 13). The latter type of affinity elution is illustrated by the elution of lactate dehydrogenase isoenzymes from a $2',5'$-ADP-agarose gel with a gradient of NADH (Brodelius and Mosbach, 1973).

Chaotropic Agents. If gentle and selective elution methods do not release the bound macromolecule, then mild denaturing agents can be added to the buffer. These substances deform protein and nucleic acid structure and decrease the stability of the complex formed on the affinity gel. The most useful agents are urea, guanidine·HCl, CNS^-, ClO_4^-, and CCl_3COO^-. These substances should be used with care, because they may cause irreversible structural changes in the isolated macromolecule.

The application of affinity chromatography is limited only by the imagination of the investigator. Every year literally hundreds of research papers appear with new and creative applications of affinity chromatography. Figure 3.25 illustrates the purification of α-chymotrypsin by affinity chromatography on immobilized D-tryptophan methyl ester. α-Chymotrypsin can recognize and bind, but not chemically transform, D-tryptophan methyl ester. The enzyme catalyzes the hydrolysis of L-tryptophan methyl ester. The impure α-chymotrypsin mixture was applied to the gel, D-tryptophan methyl ester coupled to CH-Sepharose 4B, and the column washed with Tris buffer. At the point shown by the arrow, the eluent was changed to 0.1 M acetic acid. The decrease in pH caused release of α-chymotrypsin from the column.

The best source of further information and more applications is the large number of excellent brochures available from commercial suppliers of affinity gels. See the following references for a list.

REFERENCES

General

R.C. Bohinski, *Modern Concepts in Biochemistry*, 4th Edition (1983), Allyn and Bacon (Boston), pp. 119–122. Brief introduction to chromatography.

J. Brewer, A. Pesce, and R. Ashworth, *Experimental Techniques in Biochemistry* (1974), Prentice-Hall (Englewood Cliffs, NJ), pp. 32–96. Excellent theoretical background and applications of chromatography.

T. Cooper, *The Tools of Biochemistry* (1977), John Wiley (New York). Chapters on ion-exchange, gel permeation, and affinity chromatography.

D. Freifelder, *Physical Biochemistry*, 2nd Edition (1982). W.H. Freeman (San Francisco), pp. 216–275. Theory and applications of chromatography.

R.L. Grob, Editor, *Modern Practice of Gas Chromatography* (1977), Wiley-Interscience (New York). Excellent coverage of theory, instrumentation, and application of GC.

R.J. Hamilton and P.A. Sewell, *Introduction to High Performance Liquid Chromatography* (1977), Chapman and Hall (London). Good coverage of theory and practice of HPLC.

W.B. Jakoby, Editor, *Methods in Enzymology*, Vol. XXII (1971), Academic Press (New York), pp. 273–378. Chapters on ion exchange, gel filtration, and hydroxyapatite chromatography.

W.B. Jakoby, Editor, *Methods in Enzymology*, Vol. 34B (1974), Academic Press (New York). Entire volume is devoted to affinity chromatography.

W. Jennings, *Gas Chromatography with Glass Capillary Columns*, 2nd Edition (1980), Academic Press (New York). Detailed coverage of GC.

A.M. Krstulovic' and P.R. Brown, *Reversed Phase High Performance Liquid Chromatography* (1982), Wiley-Interscience (New York). Theory, practice, and biomedical applications.

C.R. Lowe and P.D.G. Dean, *Affinity Chromatography* (1974), Wiley-Interscience (New York). Theory, practice, and applications.

A. Niederwieser in *Methods in Enzymology*, C.H.W. Hirs and S.N. Timasheff, Editors, Vol. XXV (1972), Academic Press (New York), pp. 60–99. Thin-layer chromatography of amino acids.

J. Pisano, *ibid.* pp. 27–44. GC of amino acid derivatives.

R.D. Plattner in *Methods in Enzymology*, J.M. Lowenstein, Editor, Vol. 72 (1981), Academic Press (New York), p. 21–33. HPLC of triglycerides.

N.A. Porter, *ibid.* pp. 34–40. HPLC of complex lipids.

J.D. Rawn, *Biochemistry* (1983), Harper and Row (New York), pp. 163–192. Chapter on isolation and characterization of biological molecules.

W.H. Scouten, *Affinity Chromatography* (1981), Wiley-Interscience (New York). Theory, practice, and application of affinity chromatography.

L.R. Snyder and J.J. Kirkland, *Introduction to Modern Liquid Chromatography*, 2nd Edition (1979), Wiley-Interscience (New York). Extensive coverage of HPLC.

J.C. Touchstone and M.F. Dobbins, *Practice of Thin Layer Chromatography* (1978), Wiley-Interscience (New York). Practical coverage of TLC.

Specific

P. Brodelius and K. Mosbach, *FEBS Lett.*, *35*, 223–225 (1973). "Affinity Chromatography of Enzymes on an AMP-Analogue."

Technical Bulletins from Suppliers of Chromatography Materials

Bio-Rad Laboratories
    A Laboratory Manual on Gel Chromatography
    Affi-Gel Anthology
    HPLC Application Sheets

Pierce Chemical Company
    Immobilized Ligands for Affinity Chromatography
    Reacti-Gel
    Fractogel TSK

Pharmacia Fine Chemicals
    Affinity Chromatography, Principles and Methods
    Chromatofocusing
    Gel Filtration, Theory and Practice
    Sephadex Ion Exchangers, A Guide to Ion Exchange Chromatography

Chapter *4*

Characterization of Biomolecules by Electrophoretic Methods

Electrophoresis is the study of the movement of charged molecules in an electric field. The technique, which has been used by biochemists for at least 50 years, is especially applicable to characterization and analysis of biological polymers. Molecular migration in an electric field is influenced by the size, shape, charge, and chemical nature of the molecule.

Electrophoretic techniques can be classified as either **moving boundary electrophoresis** (solution electrophoresis) or **zone electrophoresis**. In moving boundary electrophoresis, an aqueous buffered solution of macromolecules, in an enclosed cell, is subjected to an electric field. The rate of migration of the macromolecules is measured by direct optical monitoring at the boundary separating the zone of macromolecules from the free solvent. Because of severe experimental difficulties, including sample mixing due to convection currents and vibration, sample diffusion, and tedious sample application and removal, moving boundary electrophoresis does not lend itself well to routine analysis or preparative applications, and it is rarely used today.

Zonal electrophoretic techniques overcome most of these problems by employing a buffer-saturated solid matrix as the electrophoretic support medium. The sample to be analyzed is applied to the medium as a spot or thin band. The migration of molecules is influenced by the applied electric field *and* the rigid, mazelike matrix of the support. Although it is difficult to quantify the electrophoretic movement of molecules in a solid support, zonal electrophoresis has gained widespread use in analysis of the purity of biomolecules and determination of molecular size.

A. INTRODUCTION TO ELECTROPHORESIS

The movement of a charged molecule subjected to an electric field is represented by Equation 4.1.

$$Eq = fv \qquad \text{(Equation 4.1)}$$

where

E = the electric field in volts/cm

q = the net charge on the molecule

f = the frictional coefficient, which depends upon the mass and shape of the molecule

v = the velocity of the molecule

The charged particle moves at a constant velocity that depends directly on the electrical force (Eq) but inversely on a counteracting force, **viscous drag** (f). The velocity of the charged molecule is defined by Equation 4.2.

$$v = \frac{Eq}{f} \qquad \text{(Equation 4.2)}$$

The applied field (voltage or current) represented by E in Equation 4.2 is usually held constant during electrophoresis. Under these conditions, the movement of the charged molecule actually depends on the charge-to-mass ratio (q/f). Recall that the frictional coefficient of a molecule is a measure of its mass and shape.

The movement of a charged particle in an electric field is often defined in terms of **mobility**, μ, the velocity per unit of electric field (Equation 4.3).

$$\mu = \frac{q}{f} \qquad \text{(Equation 4.3)}$$

In theory, if the net charge, q, on a molecule is known, it should be possible to measure f and, therefore, obtain information about the hydrodynamic size and shape of that molecule by investigating its mobility in an electric field. Attempts to define f by electrophoresis have not been successful, primarily because Equation 4.3 does not adequately describe the electrophoretic process. Important factors that are not accounted for in the equation are interaction of migrating molecules with the support medium and shielding of the molecules by buffer ions. This means that electrophoresis is not useful for describing specific details of the size and shape of a molecule. Instead, it has been applied to the analysis of purity and identification of macromolecules. Each molecule in a mixture is expected to have a unique charge and size, and its mobility in an electric field will therefore be unique. This expectation forms the basis for analysis and separation by all electrophoretic methods. The technique is especially useful for the analysis of amino acids, carbohydrates, peptides, proteins, nucleotides, and nucleic acids.

B. METHODS OF ZONE ELECTROPHORESIS

All types of zone electrophoresis are based on the principles stated above. The variable factor in each method is the type of support medium. Electrophoresis support media are either papers or gels. Following is a description of the most popular and effective methods.

Paper and Cellulose Acetate Electrophoresis

A typical experimental arrangement for paper or cellulose acetate electrophoresis is shown in Figure 4.1. In a manner similar to paper chromatography, the sample to be analyzed is applied as a spot or a band near the middle of the sheet. Two techniques may be used for sample application to paper, **dry application**, where an aqueous buffered sample is applied to dry paper, or **wet application**, consisting of application of the sample to buffer-soaked paper. The dry method results in smaller spots and better resolution than the wet method; however, the latter method is more easily performed. Wet application is used on cellulose acetate strips because the support must be presoaked to induce an inner matrix structure.

After sample application, the paper or cellulose acetate strip is placed in the electrophoresis chamber such that opposite ends are in separate buffer reservoirs in order to complete electrical contact (see Figure 4.1). A power supply is connected to the chamber and voltage is applied to the strip. Voltages between 6 and 8 volts/cm are appropriate for most separations of biological samples.

The length of time required for electrophoresis depends on the support medium, the nature of the sample, the voltage, the dimensions of the medium, and many other factors. Trial and error is sometimes necessary to find the correct time period. For most applications, 3/4 to 1 hour is sufficient.

When electrophoresis is complete, visualization of the sample components is most readily accomplished by spraying or dipping the solid support into a dye. The reagent used will, of course, depend on the chemical nature of the sample. Specific dyeing procedures are discussed later in this chapter.

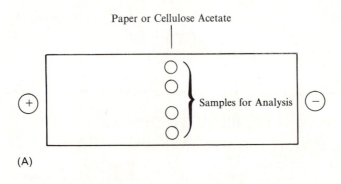

(A)

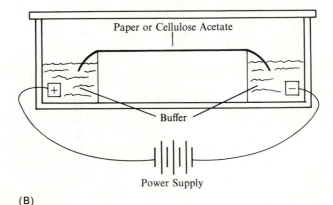

(B)

FIGURE 4.1

Experimental arrangement for paper or cellulose acetate electrophoresis. (A) Top view of the support with samples. (B) Side view of the electrophoresis chamber with support.

Although paper and cellulose acetate supports are both discussed in this section, each type of medium has its own characteristics and limitations, and different results are obtained for each method. Paper electrophoresis is seldom used today except for analysis of small, charged molecules such as amino acids, peptides, carbohydrates, and nucleotides. Because these molecules have small charges, electrophoretic mobility is low unless high voltages (about 200 volts/cm) are used. High-voltage electrophoresis produces considerable heat, which causes evaporation of buffer and drying of the paper strip. This is remedied by sandwiching the strip between two glass plates, immersing the paper in a nonconducting liquid, or performing the experiment in a refrigerated chamber.

The many hydroxy groups on cellulose provide for extensive interaction between polar macromolecules and the paper electrophoresis strip. Therefore, hydrophilic proteins and nucleic acids tend to have low electrophoretic mobility. Acetylation of the cellulose hydroxy groups produces a medium (cellulose acetate) that greatly improves electrophoretic separations of polar macromolecules. Cellulose acetate has several other advantages over cellulose, including more rapid analysis with better resolution, detection of smaller samples because of low background staining, and the use of lower voltages. The apparatus necessary for a paper or cellulose acetate electrophoresis is relatively simple and may be constructed for student use. However, inexpensive and versatile models are commercially available (Gelman, Bio-Rad).

Gel Electrophoresis

Cellulose acetate electrophoretic separations are characterized by ease of operation and high resolution. If even greater resolution is required for biochemical analysis, some type of gel support should be chosen.

Starch Gels. The first gel medium to receive attention was starch. A slab gel is prepared from potato starch paste layered on a horizontal glass plate. The sample is placed in a well cut in the slab, and the plate is subjected to a voltage. After electrophoresis, the starch slab is stained for visualization of the sample components. The enhanced resolution achieved by starch gels compared to cellulose-based media is probably due to a molecular sieving effect and to reduced diffusion because of a more rigid gel network. Starch gels are still used for some protein and isoenzyme analyses.

Polyacrylamide Gels. Starch gels have rather limited use because pore sizes in the matrix cannot be controlled. Recall from the earlier discussion of gel filtration (Chapter 3) that gels can be prepared with different pore sizes by changing the concentration of crosslinking agents. Electrophoresis through polyacrylamide gels leads to enhanced resolution of sample components because the separation is based on both molecular sieving and electrophoretic mobility. Gels have other advantages over starch, including minimal protein adsorption, rapid analysis, ease of preparation and staining, and reproducible, preselected pore sizes. A major difference between gel filtration and polyacrylamide gel electrophoresis must be emphasized at this time. In gel filtration, the rate of movement is large molecules > small molecules. The opposite is the case for gel electrophoresis. In gel electrophoresis, there is no void volume in the system, only a continuous matrix network throughout the gel. The electrophoresis gel is comparable to a single bead in gel filtration. Therefore, large molecules do not move easily through the medium.

Polyacrylamide gels are prepared by the free radical polymerization of acrylamide and the crosslinking agent, N,N'-methylene-bis-acrylamide (Figure 4.2). Chemical polymerization is controlled by an initiator/catalyst system, ammonium persulfate/N,N,N',N'-tetramethylenediamine

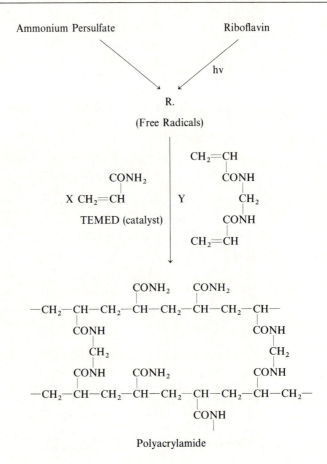

FIGURE 4.2
Chemical reactions to illustrate the copolymerization of acrylamide and N,N'-methylene-bis-acrylamide.

(TEMED). Photochemical polymerization may be brought about by riboflavin in the presence of UV radiation. The standard gel is 7.5% polyacrylamide. It can be used over the molecular weight range of 10,000 to 1,000,000; however, the best resolution is obtained in the range of 30,000 to 300,000. Gel that is 3.5% polyacrylamide can be used for macromolecules in the range of 1,000,000 to 5,000,000. Above 5,000,000, agarose/polyacrylamide mixtures or pure agarose gels are used.

Polyacrylamide electrophoresis can be set up in one of two experimental arrangements—**column gel** or **slab gel**. Figure 4.3 shows the typical arrangement for column gel. Glass tubes (10 cm × 6 mm ID) are filled with a mixture of acrylamide, N,N'-methylene-bis-acrylamide, buffer, and free radical initiator/catalyst. Polymerization occurs in 30 to 40 minutes. The gel column is inserted between two separate buffer reservoirs; the upper reservoir contains the cathode, and the lower one contains the

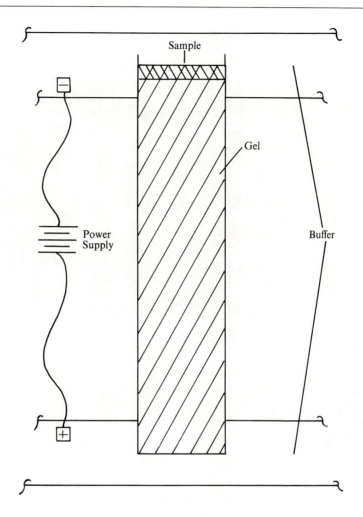

FIGURE 4.3

A column gel for polyacrylamide electrophoresis.

anode. Gel electrophoresis is usually carried out at basic pH, where most biological polymers are anionic; hence they move down toward the anode. The sample to be analyzed is layered on top of the gel and voltage is applied to the system. A "tracking dye" is also applied, which moves more rapidly through the gel than the sample components. When the dye band has moved to the opposite end of the column, the voltage is turned off, and the gel is removed from the column and stained with a dye. Chambers for gel electrophoresis are commercially available (Figure 4.4) or can be constructed from inexpensive materials. Up to 12 samples can be run in the illustrated apparatus, one in each glass tube.

The slab gel arrangement is shown in Figure 4.5. The polyacrylamide slab is prepared between glass plates that are separated by spacers. The

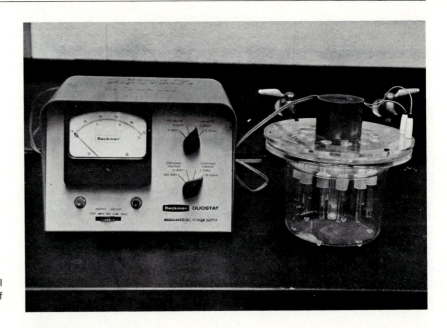

FIGURE 4.4
A typical chamber for column gel electrophoresis. Photo courtesy of Mr. Mark Billadeau.

FIGURE 4.5
Experimental arrangement for horizontal slab get electrophoresis. Courtesy of Bio-Rad Laboratories, Richmond, CA.

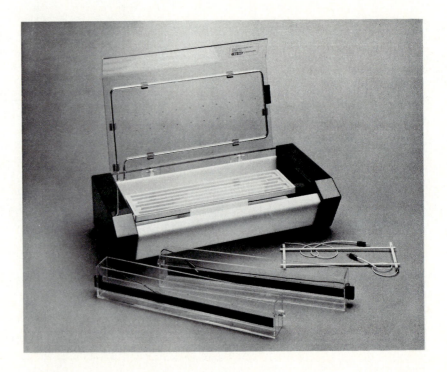

spacers typically allow a uniform slab thickness of 1.75 mm or 1.5 mm, which is appropriate for analytical procedures. Slab gels are usually prepared in sizes of 12×12 cm or 14×16 cm. The slab top is equipped with a plastic mold accessory called a "comb." When in place during polymerization of the acrylamide, the comb forms indentations in the gel that serve as sample wells. Up to 20 wells can be etched in the gel so that many samples can be run simultaneously. The samples are loaded on the gel through a thin opening at the top, and the voltage is applied. For visualization, the slab is removed and stained with an appropriate dye. ·

Two modifications of polyacrylamide gel electrophoresis have greatly increased its versatility.

Discontinuous Gel Electrophoresis (Shuster, 1971). The experimental arrangement for "disc" gel electrophoresis is shown in Figure 4.6. The three significant differences in this method are (1) there are now two gel layers, a lower or **resolving gel** and an upper or **stacking gel**; (2) the buffers used to prepare the two gel layers are of different ionic strengths and pH; and (3) the stacking gel has a lower acrylamide concentration, so that its pore sizes are larger. These three changes in the experimental conditions cause the formation of highly concentrated bands of sample in the stacking gel and greater resolution of the sample components in the lower gel. Sample concentration in the upper gel occurs in the following manner. The sample is usually dissolved in glycine-chloride buffer, pH 8 to 9, before loading on the gel. Glycine exists primarily in two forms at this pH, a zwitterion and an anion (Reaction 4.1).

$$\overset{+}{H_3}NCH_2COO^- \rightleftharpoons H_2NCH_2COO^- + H^+ \qquad \text{(Reaction 4.1)}$$

The average charge on glycine anions at pH 8.5 is about -0.2. When the voltage is turned on, buffer ions (glycinate and chloride) and protein or nucleic acid sample move into the stacking gel, which has a pH of 6.9. Upon entry into the upper gel, the equilibrium of Reaction 4.1 shifts toward the left, increasing the concentration of glycine zwitterion, which has no net charge and hence no electrophoretic mobility. In order to maintain a constant current in the electrophoresis system, a flow of anions must be maintained. Since most proteins and nucleic acid samples are still anionic at pH 6.9, they replace glycinate as mobile ions. Therefore, the relative ion mobilities in the stacking gel are chloride > protein or nucleic acid sample > glycinate. The sample will tend to accumulate and form a thin, concentrated band sandwiched between the chloride and glycinate as they move through the upper gel. Since the acrylamide concentration in the stacking gel is low (2 to 3%), there is little impediment to the mobility of the large sample molecules.

Now, when the ionic front reaches the lower gel with pH 8 to 9 buffer, the glycinate concentration increases and anionic glycine and chloride

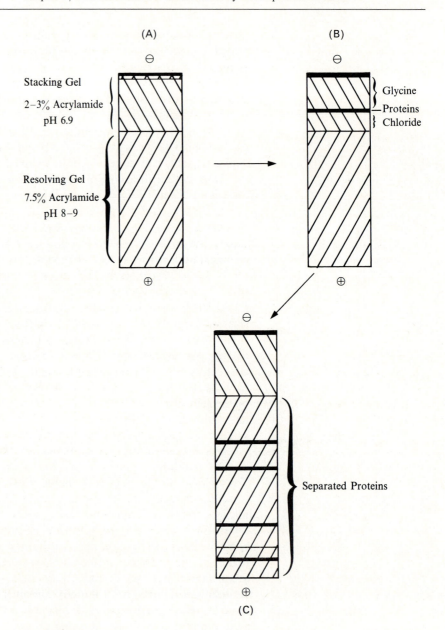

FIGURE 4.6

The process of disc gel electrophoresis. (A) Before electrophoresis. (B) Movement of chloride, glycinate, and protein through the stacking gel. (C) Separation of protein samples by the resolving gel.

carry the majority of the current. The protein or nucleic acid sample molecules, now in a narrow band, encounter both an increase in pH and a decrease in pore size. The increase in pH would, of course, tend to increase electrophoretic mobility, but the smaller pores will decrease mobility. The relative rate of movement of anions in the lower gel is chloride > glycinate > protein or nucleic acid sample. The separation of sample components in the resolving gel occurs as described in an earlier section on gel electrophoresis. Each component has a unique charge/mass ratio and a discrete size and shape, which will directly influence its mobility.

Disc gel electrophoresis yields excellent resolution and is the method of choice for analysis of proteins and nucleic acid fragments. Protein or nucleic acid bands containing as little as 1 or 2 μg can be detected by staining the gels after electrophoresis. Disc gel is not limited to the column arrangement, but is also adaptable to slab gel electrophoresis.

Sodium Dodecyl Sulfate–Polyacrylamide Electrophoresis. The electrophoretic techniques previously discussed are not applicable to the measurement of the molecular weights of biological molecules because mobility is influenced by both charge and size. If protein samples are treated so that they have a uniform charge, electrophoretic mobility then depends primarily on size (see Equation 4.3). The molecular weights of proteins may be estimated if they are subjected to electrophoresis in the presence of a detergent, sodium dodecyl sulfate (SDS), and a disulfide reducing agent, mercaptoethanol (Shapiro, *et al.*, 1967; Weber and Osborn, 1969).

When protein molecules are treated with SDS, the detergent disrupts the secondary, tertiary, and quarternary structure to produce polypeptide chains in a random coil. The presence of mercaptoethanol assists in protein denaturation by reducing all disulfide bonds. The detergent binds to hydrophobic regions of the denatured protein chain in a constant ratio of about 1.4 gram of SDS per gram of protein. The bound detergent molecules carrying negative charges mask the native charge of the protein. In essence, polypeptide chains of a constant charge/mass ratio and uniform shape are produced. The electrophoretic mobility of the SDS-protein complexes will be influenced primarily by molecule size; the larger molecules will be retarded by the molecular sieving effect of the gel, while the smaller molecules will have greater mobility. Empirical measurements have shown a linear relationship between the log molecular weight and the electrophoretic mobility (Figure 4.7).

In practice, a protein of unknown molecular weight and subunit structure is treated with 1% SDS and 0.1 M mercaptoethanol in electrophoresis buffer. A standard mixture of proteins with known molecular weights must also be subjected to electrophoresis under the same conditions. Two sets of standards are available, one for low molecular weight proteins (MW range 14,000 to 100,000) and one for high molecular weight proteins

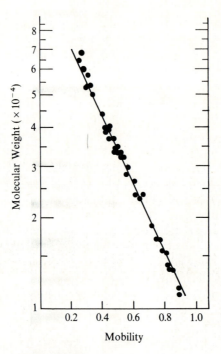

FIGURE 4.7

A graph illustrating the linear relationship between electrophoretic mobility of a protein and its molecular weight. Thirty-seven different polypeptide chains with a molecular weight range of 11,000 to 70,000 are shown. From K. Weber and M. Osborn, *J. Biol. Chem.*, *244*, 4406 (1969). By permission of the copyright owner, The American Society of Biological Chemists, Inc.

(45,000 to 200,000). Figure 4.8 shows a stained gel after electrophoresis of a protein standard mixture. After electrophoresis and dye staining, the molecular weight is determined graphically.

SDS-gel electrophoresis is extremely valuable for estimating the molecular weight of protein subunits. This modification of gel electrophoresis finds its greatest use in characterizing the sizes and different types of subunits in oligomeric proteins. SDS polyacrylamide gel electrophoresis is limited to a molecular weight range of 10,000 to 200,000. Gels of less than 2.5% acrylamide must be used for molecular weight determinations above 200,000, but these gels do not set well because of minimal crosslinking. A recent modification using gels of agarose/acrylamide mixtures allows for the measurement of molecular weights above 200,000. It is not necessary to treat nucleic acids with SDS, since they already have a uniform shape and charge.

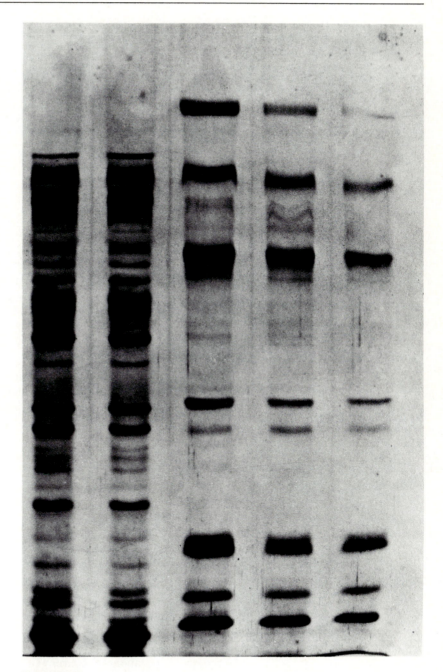

FIGURE 4.8
A silver-stained gel obtained by electrophoresis of a standard protein mixture. Courtesy
of Bio-Rad Laboratories, Richmond, CA.

Agarose Gel Electrophoresis

The electrophoretic techniques discussed up to this point are readily adapted to the analysis of proteins and small fragments of nucleic acids up to 3,500,000 in molecular weight; however, the small pore sizes in the gel are not appropriate for analysis of larger nucleic acid fragments or intact DNA. The standard method used to characterize RNA and DNA is electrophoresis through agarose gels.

The mobility of nucleic acids in agarose gels is influenced by the *agarose concentration*, the *molecular size of the nucleic acid*, and the *molecular shape of the nucleic acid*. Agarose concentrations of 0.5 to 3% are most effective for nucleic acid separation. Gels with agarose concentrations less than 0.5% are rather fragile and must be used in a horizontal slab arrangement or in a refrigerated chamber. As do proteins, nucleic acids migrate at a rate that is inversely proportional to the logarithm of their molecular weights; hence, molecular weights can be estimated from electrophoresis results using standard nucleic acids of known molecular weight. A standard plot of mobility vs. log MW of standard DNA fragments is shown in Experiment 19.

Several different conformations of double-stranded DNA have been identified. The forms of DNA most frequently encountered are supercoiled closed circular (Form I), nicked circular (Form II), and linear (Form III). Passage of a molecule through a gel is influenced by the shape of conformation of that molecule. A small, compact molecule would be expected to have a greater mobility than rodlike, linear molecules. The electrophoretic mobility of the three forms of DNA depends on the agarose concentration in the gel, but the general order of mobility is not exactly as predicted. The relative order for most gels is supercoiled $\geq$ linear $>$ nicked circular. The reasons for these observations are not entirely clear.

Most agarose gel electrophoresis experiments are carried out with horizontal slab gels as shown in Figure 4.5. This method is chosen over the vertical mode because low agarose concentrations can be used. The sample to be separated is placed in a sample well made with the comb, and voltage is applied until the separation is complete. For maximum resolution, the applied voltage should be no more than 5 volts/cm.

Nucleic acids can be visualized on the slab after separation by using ethidium bromide, a dye that displays enhanced fluorescence when intercalated between stacked nucleic acid bases.

Specific Applications of Agarose Gel Electrophoresis

The versatility of agarose gels is obvious when one reviews their many applications in nucleic acid analysis. In the past 15 years, the rapid advancement in our understanding of DNA structure and function is primarily due to the development of agarose gel electrophoresis. Two of the many applications of agarose gel will be described here. For further examples, see Freifelder (1982).

Analysis of DNA Fragments after Restriction Endonuclease Action (Sharp, et al., 1973; Southern, 1979). Many prokaryotes produce enzymes that catalyze the hydrolysis of phosphodiester bonds in the DNA backbone. Each enzyme, called a **restriction endonuclease**, recognizes a specific base sequence in double-stranded DNA and causes cleavage in or near that specific region. (See Experiment 19.) Most viral, bacterial, or animal DNA molecules are substrates for the enzymes. When each type of DNA is treated with a restriction endonuclease, a specific number of DNA fragments is produced. The base sequence recognized by the enzyme occurs only a few times in any particular DNA molecule; therefore, the smaller the DNA molecule, the fewer the number of specific sites. Viral or phage DNA, for example, is cleaved into up to 50 fragments depending on the enzyme used, whereas larger bacterial or animal DNA may be cleaved into hundreds of fragments. Smaller DNA molecules, upon cleavage with a particular enzyme, will produce a unique set of fragments. It is unlikely that this set of fragments will be the same for any two different DNA molecules, so the fragmentation pattern can be considered a "fingerprint" of the DNA substrate. The **restriction pattern** is produced by electrophoresis of the cleavage reaction mixture through agarose gels, followed by staining with ethidium bromide (Figure 4.9). The separation of the fragments is based on molecular size. In addition to the characterization of

FIGURE 4.9
Restriction patterns produced by agarose electrophoresis of DNA fragments after restriction endonuclease action. Courtesy of Bio-Rad Laboratories, Richmond, CA.

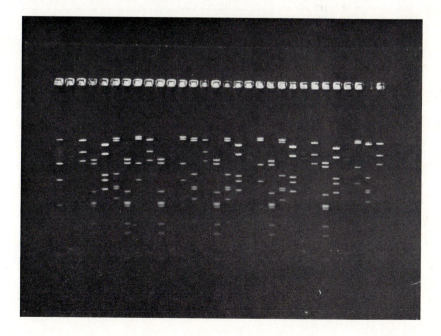

DNA structure, endonuclease action coupled with agarose gel electrophoresis is a valuable tool for sequencing large DNA molecules, for DNA recombination experiments, and for the isolation of genes (see Chapter 9).

Characterization of Superhelical Structure of DNA by Agarose Gel Electrophoresis. The structure of native bacterial or viral DNA is often closed circular with negative superhelical turns (see Experiment 21). It is possible under various experimental conditions to induce reversible changes in the conformation of DNA (Bauer and Vinograd, 1968). The intercalating dye, ethidium bromide, causes an unwinding of supercoiled DNA that affects its centrifugal sedimentation rate and electrophoretic mobility. Electrophoresis of DNA on agarose in the presence of increasing concentrations of ethidium bromide provides an unambiguous method for identifying closed circular and other DNA conformations.

Closed circular, negative supercoiled DNA (Form I) usually has the greatest electrophoretic mobility of all DNA forms. This would be predicted because the supercoiled DNA molecule is extremely compact. If ethidium bromide is added to Form I DNA, the dye intercalates between the stacked DNA bases, causing unwinding of some of the negative supercoils. As the concentration of ethidium bromide is increased, more and more of the negative supercoils are removed until no more are present in the DNA. The conformational change of the DNA supercoil can be monitored by electrophoresis because the mobility decreases with each unwinding step. With increasing concentration of ethidium bromide, the negative supercoils are progressively unwound and the electrophoretic mobility decreases to a minimum (Johnson and Grossman, 1977; Espejo and Lebowitz, 1976). This minimum represents the free dye concentration necessary to remove all negative supercoils. (The free dye concentration at this minimum has been shown to be related to the **superhelix density**, which is a measure of the extent of supercoiling in a DNA molecule). The circular DNA at this point is said to be in the "relaxed form." If more ethidium bromide is added to the relaxed DNA, positive superhelical turns are induced in the structure and the electrophoretic mobility increases. Forms II and III of DNA, under the same conditions of increasing ethidum bromide concentration, show a gradual decrease in the electrophoretic mobility throughout the entire concentration range.

Agarose gel electrophoresis is able to resolve native DNA that differs only in the degree of supercoiling. This technique has proved useful in the analysis and characterization of a new class of enzymes that catalyzes changes in the conformation or topology of native DNA. These enzymes, called **topoisomerases**, have been isolated from bacterial and mammalian cells. They change DNA conformations by catalyzing nicking and closing of phosphodiester bonds in circular duplex DNA. Agarose gel electrophoresis is an ideal method for identifying and assaying topoisomerases because the intermediate DNA molecules can be resolved on the basis of the extent of supercoiling. Topoisomerases may be assayed by incubating native DNA with an enzyme preparation. Aliquots are removed after various periods of time and subjected to electrophoresis on agarose gel with standard supercoiled and relaxed DNA.

C. ISOELECTRIC FOCUSING OF PROTEINS (VESTERBERG, 1971; WRIGLEY, 1971)

Another effective method of electrophoresis is **isoelectric focusing (IEF)**, which is the study of electrophoretic mobility as a function of pH. The net charge on a protein molecule depends on pH. Proteins below their isoelectric pH (pH_I, pH of zero net charge) are positively charged and will migrate, in a medium of fixed pH, toward the cathode. At a pH above its isoelectric point, a protein will migrate toward the anode. If the pH of the electrophoretic medium is identical to the pH_I of a protein, that protein will have a net charge of zero and will not migrate toward either electrode. Theoretically, it should be possible to separate protein molecules and to estimate the pH_I of a protein by investigating the electrophoretic mobility in a series of separate experiments in which the pH of the medium is changed. The pH at which there is no protein migration should coincide with the pH_I of the protein. Since it is rather tedious and time consuming to determine the pH_I by trial and error, an alternative is to perform a single electrophoresis in a medium of gradually changing pH, i.e., a pH gradient. This is essentially what is done in isoelectric focusing. Before the practical aspects of IEF are discussed, the principles behind the techniques will be outlined.

Imagine an electrophoresis system as shown in Figure 4.10. Acid, usually phosphoric, is placed at the cathode; base, triethanolamine, at the anode. Between the electrodes is a medium in which the pH gradually

FIGURE 4.10
Illustration of isoelectric focusing. See text for details.

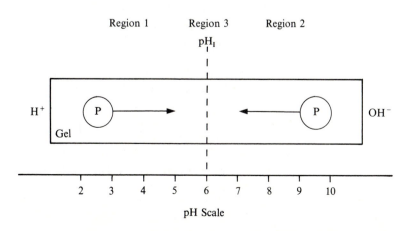

increases from 2 to 10. The pH values are designated for each region. Now, envision protein molecules (P) in two different regions of the pH gradient. Assume that the pH at region 1 is less than the pH_I of the protein, and the pH of region 2 is greater than the pH_I of the protein. Molecules of P near region 1 will be positively charged and will migrate in the applied electric field toward the cathode. As P migrates, it will encounter an increasing pH, which will influence its net charge. (The net charge will decrease toward zero.) When P reaches a region where its net charge is zero (region 3), it will stop. This region of the electrophoretic medium will have a pH coinciding with the pH_I of the protein. P molecules in region 2 will be negatively charged and will migrate toward the anode. The net charge on P will gradually approach zero as P moves through the gradient, and P molecules originally in region 2 will approach region 3 and come to rest. The P molecules move in opposite directions, but the final outcome of IEF is that P molecules located anywhere in the gradient will migrate toward region 3 and will eventually come to rest in a sharp band; that is, they will "focus" at a point corresponding to the pH_I.

It should be clear from this explanation that, since different protein molecules in a mixture have unique pH_I values, it is possible to apply IEF to protein separation problems. In addition, the pH_I of each protein in the mixture can be determined by measuring the pH of the region where the protein is focused.

The experimental set-up for IEF is shown in Figure 4.11. The pH gradient is prepared in a horizontal glass tube or slab. Special precautions must be taken so the pH gradient remains stable and is not disrupted by diffusion or convective mixing during the electrophoresis experiment. The most common stabilizing technique is to form the gradient in a polyacrylamide, agarose, or dextran gel. The pH gradient is formed in the gel by electrophoresis of synthetic polyelectrolytes, called **ampholytes**. These materials are low molecular weight polymers that have a wide range of isoelectric points as the result of numerous amino and carboxyl or sulfonic acid groups. The polymer mixtures are available in specific pH ranges (pH 5–7, 6–8, 3.5–10, etc.) under the tradenames LKB-Ampholine, Bio-Rad-Lyte, and Pharmacia-Pharmalyte. It is critical to select the appropriate pH range for the ampholyte so that the proteins to be studied have pH_I values in that range. The best resolution is, of course, achieved with an ampholyte mixture over a small pH range (about two units) encompassing the pH_I of the sample proteins. If the pH_I values for the proteins under study are unknown, an ampholyte of wide pH range (pH 3–10) should be used first and then a narrower pH range selected for use.

The gel medium is prepared as previously described except that the appropriate ampholyte is mixed prior to polymerization. The gel mixture is poured into the desired form (column tubes, horizontal slabs, etc.) and allowed to set. Immediately after casting of the gel, the pH is constant

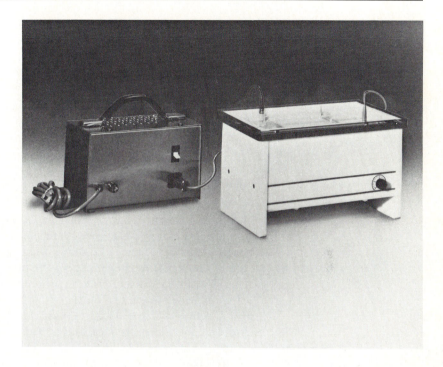

FIGURE 4.11
Typical apparatus for isoelectric focusing. Courtesy of Bio-Rad Laboratories, Richmond, CA.

throughout the medium, but application of voltage will induce migration of ampholyte molecules to form the pH gradient. The standard gel for proteins with a molecular weight up to 100,000 is 7.5% polyacrylamide; however, if larger proteins are of interest, gels with larger pore size must be prepared. Such gels can be prepared with a lower concentration of acrylamide (about 2%) and 0.5 to 1% agarose to add strength.

Loading the protein sample on the gel can be done in one of two ways. A concentrated, salt-free sample can be layered on top of the gel as previously described for traditional gel electrophoresis. The second method is simply to add the protein to the gel preparation, resulting in an even distribution of protein throughout the medium. The protein molecules move more slowly than the ampholyte molecules, so the pH gradient is set up before significant migration of the proteins occurs.

Very small protein samples can be evaluated by IEF. For analytical purposes, 10 to 50 μg is a typical sample size. Larger sample sizes (up to 20 mg) may be used for preparative purposes.

Electrofocusing is carried out under an applied current of 2 mA per tube and usually takes from 30 to 120 minutes. The time period for electrofocusing is not as critical as for traditional electrophoresis. Trial and error can be used to find a period that results in an effective separation; however, longer periods do not move the protein samples out of the medium as in gel electrophoresis.

After IEF electrophoresis, gels cannot be stained directly for protein detection because the commonly used stains are bound by the ampholytes. Carrier ampholytes are removed from gels by soaking in 5% trichloroacetic acid, and then the protein zones are stained with Coomassie blue dye.

SDS-Isoelectric Focusing Gel Electrophoresis

The separation of proteins by IEF is based on pH_I value and charge. SDS-gel electrophoresis separates molecules on the basis of molecular size. A combination of the two methods should lead to enhanced resolution of protein mixtures. Such an experiment was first reported by O'Farrell (1975) and has since become a routine and powerful technique (Garrels, 1983). Figure 4.12 shows the results of O'Farrell's analysis of total *E. coli* protein. The sample was first separated by isoelectric focusing. The protein zones were then transferred to an SDS gel. At least 1000 discrete protein spots are visible.

Electrofocusing is now established as the best method for analysis of proteins, including enzymes, and is considered the standard criterion of purity. Calibration kits are available for the determination of pH_I values of proteins.

FIGURE 4.12
SDS-isoelectric focusing gel electrophoresis of total *E. coli* protein. Photo courtesy of Dr. P. O'Farrell.

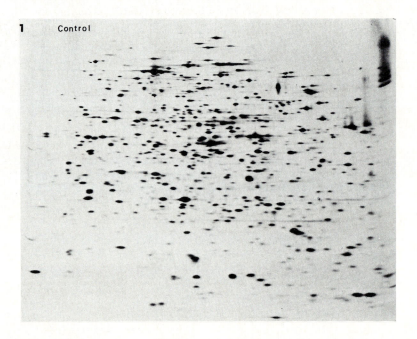

D. PRACTICAL ASPECTS OF ELECTROPHORESIS

Instrumentation

Only two pieces of equipment are required for electrophoresis—a power supply and an electrophoresis chamber. A power supply that provides a constant current is suitable for most electrophoresis experiments. If isoelectric focusing experiments are planned, a power supply that furnishes a constant voltage is necessary. Power supplies that provide both constant voltage (up to 500 volts) and constant current (up to 20 mA) are available commercially or may be constructed. Inexpensive power supply kits are available from Heath Co., Benton Harbor, MI.

Electrophoresis chambers in numerous designs are available from manufacturers, or they may be constructed from inexpensive and readily available materials. Table 4.1 lists literature references for the construction of electrophoresis chambers. Commercial suppliers include Bio-Rad Laboratories, Pharmacia, LKB, Gelman, Hoefer, Sargent-Welch, VWR Scientific, CM Scientific, and others.

Reagents

High quality chemicals must be used, since impurities may influence both the gel polymerization process and electrophoretic mobility. The reagents used for gel formation should be stored in a refrigerator and recrystallized before use. Acrylamide is recrystallized from chloroform, whereas N,N'-methylene-bis-acrylamide may be recrystallized from acetone. **CAUTION:** *Acrylamide, N,N'-methylene-bis-acrylamide, N,N,N',N'-tetramethylenediamine, and ammonium persulfate are toxic and must be used with care. Acrylamide is a neurotoxin and a potent skin irritant, so gloves and a mask must be worn while handling it in the unpolymerized form.*

TABLE 4.1
Construction of Electrophoresis Chambers

Starch Gel
> H. Tsuyuki, *Anal. Biochem.*, 6, 205 (1963).

Acrylamide Gel
> B.J. Davis, *Ann. N.Y. Acad. Sci.*, *121*, 404 (1964).
> S. Raymond, *Clin. Chem.*, *8*, 455 (1962).

Slab Gel—Horizontal
> T. Maniatis, E.F. Fritsch, and J. Sambrook, *Molecular Cloning—A Laboratory Manual* (1982). Cold Spring Harbor Press, p. 153.

Slab Gel—Vertical
> E. Southern, *Methods in Enzymology 68*, 152 (1979).

Several buffer systems appropriate for gels are Tris-glycine, Tris-phosphate, and Tris-borate at concentrations of about 0.05 M. Phosphate buffers are used for paper electrophoresis, while 0.1 M barbital buffer, pH 8.6, is standard for cellulose acetate electrophoresis.

Stains and Staining Techniques

During the electrophoretic process, it is important to know when to stop applying the voltage or current. If the process is run for too long, the desired components may pass entirely through the medium and into the buffer; if too short a period is used, the components may not be completely resolved. It is common practice to add a "tracking dye," usually bromphenol blue, to the sample mixture. The dye, which is small and anionic, moves rapidly through the gel ahead of most proteins or nucleic acids. After electrophoresis, the protein bands have the tendency to widen by diffusion processes. Because this may decrease resolution, the gels must be treated immediately with an agent that "fixes" the proteins in their final positions. The fixing process is often combined with staining.

Reagent dyes that are suitable for visualization of biomolecules after electrophoresis have already been briefly mentioned. The most commonly used stain for proteins is *Coomassie brilliant blue*. This dye can be used as a 0.25% aqueous solution. It is followed by destaining (removing excess background dye) by repeated washing of the paper or gel with 7% acetic acid. Alternatively, gels may be stained by soaking in 0.25% dye in H_2O: methanol:acetic acid (5:5:1) followed by repeated washings with the same solvent. Lipoproteins may be stained with Oil Red O (saturated solution in methanol) or Sudan Black B prepared by dissolving, with heating and stirring, 0.5 ml of the dye in 50 ml of ethylene glycol.

The search for more rapid and sensitive methods of protein detection after electrophoresis led to the development of fluorescent staining techniques. Two commonly used fluorescent reagents are fluorescamine and anilinonaphthalene sulfonate. New dyes based on silver salts (silver diamine or silver-tungstosilicic acid complex) have recently been developed for protein staining (see Experiment 5). They are 10 to 100 times more sensitive than Coomassie blue. Very recently Eastman Kodak has developed a new rapid method for fixing and staining.

The most time-consuming procedure in the visualization process is destaining, which often requires days of washing. A rapid destaining of gels may be brought about electrophoretically. The gel, after soaking in the stain, is replaced in the glass tube and subjected to electrophoresis again, using a buffer of higher concentration.

There is often the need for a visualization procedure that is specific for a certain biomolecule, for example, an enzyme. If the enzyme remains in an active form while in the gel, any substrate that produces a colored product could be used to locate the enzyme on the gel. Several reactions have been developed that may be used as specific assays for enzymes (Gabriel, 1971).

Although it is less desirable for detection, the electrophoresis support medium may be cut into small segments, and each part extracted with buffer and analyzed for the presence of the desired component.

Nucleic acids are visualized in agarose and polyacrylamide gels with the fluorescent dye ethidium bromide. The gel is soaked in a solution of the dye and washed to remove excess dye. Illumination of the washed slab with UV light reveals red-orange stains where nucleic acids are located. Although ethidium bromide stains both single- and double-stranded nucleic acids, the fluorescence is much greater for double-stranded molecules. The detection limit is about 10 ng of DNA. **CAUTION:** *Ethidium bromide must be used with great care as it is a potent mutagen. Gloves should be worn at all times while using dye solutions or handling the gels.* The electrophoresis may be performed with the dye incorporated in the gel and buffer. This has the advantage that the gel can be illuminated with UV light during electrophoresis to view the extent of separation. However, the mobility of double-stranded DNA is reduced by 10 to 15% in the presence of ethidium bromide.

Analysis of Electrophoresis Results

Each experimenter has specific reasons for performing electrophoresis and will view the results accordingly. The information sought often includes the following.

(1) What is the purity of the sample? This is indicated by the number of stained bands in the electrophoresis medium. One band usually means that only one detectable component is present, that is, the sample is homogeneous. The presence of two or more bands usually indicates that the sample contains two or more components, or is heterogeneous. There are, of course, exceptions to this description. Other proteins may be present in what appears to be a homogeneous sample, but they may be below the limit of detection of the dye. Occasionally a homogeneous sample may result in two or more bands because of degradation during the electrophoresis process. Information on purity may be obtained by all of the methods discussed. The specific electrophoretic method used will depend on the type of biomolecules to be analyzed. For small molecules (amino acids, peptides, carbohydrates, nucleotides, etc.), paper or cellulose acetate electrophoresis is best. For proteins, polyacrylamide disc gel or isoelectric focusing is the method of choice. For nucleic acid analysis, agarose gel electrophoresis is most often used.

(2) What is the molecular size of the protein or nucleic acid sample? SDS-gel electrophoresis may be used for protein measurement, but nucleic acid determination requires the larger matrix size provided by agarose.

REFERENCES

General

R.C. Bohinski, *Modern Concepts in Biochemistry*, 4th Edition (1983), Allyn and Bacon (Boston), pp. 69–71 and 122–123. Brief introduction to electrophoresis.

J. Brewer, A. Pesce, and R. Ashworth, *Experimental Techniques in Biochemistry* (1974), Prentice-Hall (Englewood Cliffs, NJ), pp. 128–160. Theoretical background and applications of electrophoresis.

T. Cooper, *The Tools of Biochemistry* (1977), John Wiley (New York), pp. 194–233. Theory and experiments in electrophoresis.

C.R. Cantor and P.R. Schimmel, *Biophysical Chemistry*, Part II (1980). W.H. Freeman (San Francisco), pp. 676–682. An excellent theoretical discussion of electrophoresis.

D. Freifelder, *Physical Biochemistry*, 2nd Edition (1982), W.H. Freeman (San Francisco), pp. 276–322. Theory and applications of electrophoresis especially with nucleic acids.

T. Maniatis, E.F. Fritsch, and J. Sambrook, *Molecular Cloning—A Laboratory Manual* (1982). Cold Spring Harbor Press. An excellent manual for techniques and experiments in molecular biology.

J.D. Rawn, *Biochemistry* (1983), Harper and Row (New York), pp. 184–188. Brief introduction to electrophoresis.

L. Stryer, *Biochemistry*, Second Edition (1981), W.H. Freeman (San Francisco), pp. 89–92. Electrophoretic analysis of hemoglobin.

G. Zubay, *Biochemistry* (1983), Addison-Wesley (Reading, MA), pp. 54–55. Brief introduction to electrophoretic techniques.

Specific

W. Bauer and J. Vinograd, *J. Mol. Biol.*, *33*, 141–171 (1968). "The Interaction of Closed Circular DNA with Intercalative Dyes."

R. Espejo and J. Lebowitz, *Anal. Biochem.*, *72*, 95–103 (1976). Determination of superhelix density using agarose gel electrophoresis and ethidium bromide.

O. Gabriel in *Methods in Enzymology*, W.B. Jakoby, Editor, Vol. XXII (1971), Academic Press (New York), pp. 578–604. "Locating Enzymes on Gels."

J. Garrels in *Methods in Enzymology*, R. Wu, L. Grossman, and K. Moldave, Editors, Vol. 100B (1983), Academic Press (New York), pp. 411–423. "Quantitative Two-Dimensional Gel Electrophoresis of Proteins."

P.H. Johnson and L.I. Grossman, *Biochemistry, 16,* 4217–4225 (1977). "Electrophoresis of DNA in Agarose Gels."

P. O'Farrell, *J. Biol. Chem., 250,* 4007–4021 (1975). "High Resolution Two-Dimensional Electrophoresis of Proteins."

A.L. Shapiro, E. Vinuela, and J. Maizel, *Biochem. Biophys. Res. Commun., 28,* 815–820 (1967). "Molecular Weight Estimation of Polypeptide Chains by Electrophoresis in SDS-Polyacrylamide Gels."

P.A. Sharp. B. Sugden, and J. Sambrook, *Biochemistry, 12,* 3055–3063 (1973). Detection of endonuclease activity using agarose gels.

L. Shuster in *Methods in Enzymology,* W.B. Jakoby, Editor, Vol. XXII (1971), Academic Press (New York), pp. 412–437. "Polyacrylamide Gel Electrophoresis."

E. Southern in *Methods in Enzymology,* R. Wu, Editor, Vol. 68 (1979), Academic Press (New York), pp. 152–176. "Gel Electrophoresis of Restriction Fragments."

O. Vesterberg in *Methods in Enzymology,* W.B. Jakoby, Editor, Vol. XXII (1971), Academic Press (New York), pp. 389–412. "Isoelectric Focusing of Proteins."

K. Weber and M. Osborn, *J. Biol. Chem., 244.* 4406–4412 (1969). "The Reliability of Molecular Weight Determination by Dodecyl Sulfate-Polyacrylamide Gel Electrophoresis."

C. Wrigley in *Methods in Enzymology,* W.B. Jakoby, Editor, Vol. XXII (1971), Academic Press (New York), pp. 559–564. "Gel Electrofocusing."

Chapter **5**

Spectrophotometry of Biomolecules

Some of the earliest experimental measurements made on biomolecules involved a study of their interactions with light. Early observations found that when light impinges on solutions of molecules, at least two distinct processes occur, light scattering and light absorption. Both processes have become the basis of useful techniques for characterizing and analyzing biomolecules. Some of these techniques will be described in this chapter.

With some molecules, the process of absorption is followed by the emission of light of a different wavelength. This process, called **fluorescence**, depends on molecular structure and environmental factors and serves as a valuable tool for the characterization and analysis of biologically significant molecules and dynamic processes.

Another technique involving molecules and light is **polarimetry**. Here, the interaction of plane-polarized light with optically active molecules is studied.

A. BASIC PRINCIPLES OF ABSORPTION SPECTROPHOTOMETRY

The electromagnetic spectrum, as shown in Figure 5.1, is composed of a continuum of waves with different properties. The regions of primary importance in biochemistry include ultraviolet (UV) and visible (VIS) light (180 to 800 nm). Light in these regions has sufficient energy to excite the valence electrons of molecules. Figure 5.2 shows that the propagation of light is due to an electrical field component, E, and a magnetic field component, H, which are perpendicular to each other. The **wavelength** of light, defined by Equation 5.1, is the distance between adjacent wave peaks as shown in Figure 5.2.

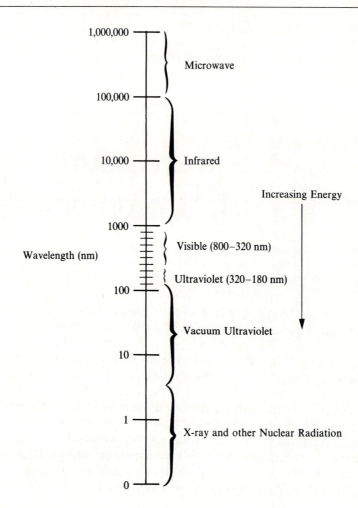

FIGURE 5.1
The electromagnetic spectrum.

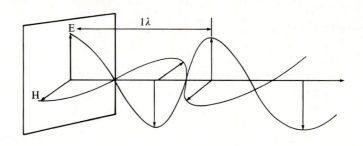

FIGURE 5.2
An electromagnetic wave, showing the *E* and *H* components.

$$\lambda = \frac{c}{v}$$ (Equation 5.1)

where

λ = wavelength

c = speed of light

v = frequency, the number of waves passing a certain point per unit time

Light also behaves as though it were composed of energetic particles. The amount of energy, E, associated with these particles (or photons) is given by Equation 5.2.

$$E = hv$$ (Equation 5.2)

where

h = Planck's constant

When a photon of specified energy interacts with a molecule, one of two processes may occur. The photon may be **scattered**, or it may transfer its energy to the molecule, producing an **excited state** of the molecule. The former process, called **Rayleigh scattering**, occurs when a photon collides with a molecule and is diffracted or scattered with unchanged frequency. This process results from the induction of an oscillating dipole in the molecule when it interacts with the electrical component of the photon. The oscillating dipole then acts as a radiation source, emitting light of the same frequency as the incident light. Rayleigh derived an equation describing the angular dependence of light scattering (Equation 5.3 and Figure 5.3). Note that the predominant characteristic of Rayleigh scattering is the large inverse dependence on wavelength.

$$\frac{I}{I_0} = \frac{8\pi^2 a\alpha^2}{\lambda^4 r^2} (1 + \cos^2 \theta)$$ (Equation 5.3)

FIGURE 5.3

The experimental set-up for measuring light scattering.

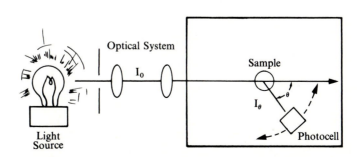

where

I = the intensity of the scattered light

I_0 = the intensity of the incident light

a = the number of scattering centers per cubic centimeter

α = the polarizability (induced dipole per unit field strength)

r = the distance between the molecules and detector

θ = the angle between the incident beam and the scattered photon

Light scattering is the physical basis of several experimental methods that are used to characterize macromolecules. Before the development of electrophoresis, light scattering techniques were used to measure the molecular weights of macromolecules. The widely used techniques of x-ray diffraction (crystal and solution), electron microscopy, laser-light scattering, and neutron scattering all rely in some way on the light scattering process. For more detailed information on scattering phenomena, see the references at the end of the chapter.

The other process mentioned above, the transfer of energy from a photon to a molecule, is **absorption**. In order for a photon to be absorbed, its energy must match the energy difference between two energy levels of the molecule.

Molecules possess a set of quantized energy levels, as shown in Figure 5.4. Although several states are possible, only two electronic states are shown, a **ground state**, G, and the **first excited state**, S_1. These two states differ in the distribution of valence electrons. When electrons are promoted from a ground state orbital in G to an orbital of higher energy in S_1, an **electronic transition** is said to occur. The energy associated with ultraviolet and visible light is sufficient to promote molecules from one

FIGURE 5.4
Energy-level diagram showing the ground state, G, and the first excited state, S_1.

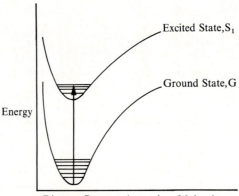

Distance Between Atoms in a Molecule

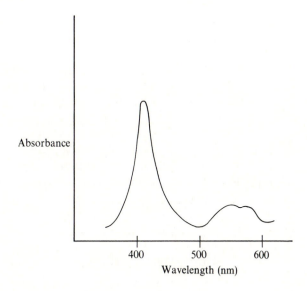

FIGURE 5.5
The visible absorbance spectrum
of hemoglobin.

electronic state to another, that is, to move electrons from one orbital
to another.

Within each electronic energy level is a set of **vibrational levels**. These
represent changes in the stretching and bending of covalent bonds. The
importance of these energy levels will not be discussed here, but transitions between these levels are the basis of infrared spectroscopy.

The electronic transition for a molecule from G to S_1, represented by
the vertical arrow in Figure 5.4, has a high probability of occurring if the
energy of the photon corresponds to the energy necessary to promote an
electron from energy level E_1 to energy level E_2:

$$E_2 - E_1 = \Delta E = \frac{hc}{\lambda}$$

(Equation 5.4)

A transition may occur from any vibrational level in G to some other
vibrational level in S_1, for example $v = 3$; however, all transitions do not
have equal probability. The probability of absorption is described by
quantum mechanics and will not be discussed here.

A UV-VIS spectrum is obtained by measuring the light absorbed by
a sample as a function of wavelength. Since only discrete packets of energy
(specific wavelengths) are absorbed by molecules in the sample, the spectrum theoretically should consist of sharp discrete lines. However, the
many vibrational levels of each electronic energy level increase the number
of possible transitions. This results in several spectral lines, which together
make up the familiar spectrum of broad peaks as shown in Figure 5.5.

An absorption spectrum can aid in the identification of a molecule
because the wavelength of absorption depends on the functional groups or

arrangement of atoms in the sample. The spectrum of oxyhemoglobin in Figure 5.5 is due to the presence of the iron porphyrin moiety and is useful for the characterization of heme derivatives or hemoproteins. Note that the spectrum in Figure 5.5 consists of several peaks at wavelengths where absorption reaches a maximum (415, 542, and 577 nm). These points, called λ_{max}, are of great significance in the identification of unknown molecules and will be discussed later in the chapter.

A second parameter that is evaluated in absorption spectroscopy is the efficiency or extent of absorption at selected wavelengths. This parameter is traditionally called the **extinction coefficient**. It is defined by the Beer-Lambert law:

$$A = Ebc \qquad \text{(Equation 5.5)}$$

where

$$A = \text{absorbance or } -\log \frac{I}{I_0}$$

$I_0 = $ the intensity of light irradiating the sample

$I = $ the intensity of light transmitted through the sample

$E = $ extinction coefficient of the absorbing material

$b = $ pathlength of light through the sample, or thickness of the cell

$c = $ concentration of the absorbing material in the sample

Since the absorbance, A, is derived from a ratio ($-\log I/I_0$), it is unitless; therefore, the units of E depend on the units of b and c in Equation 5.5. It is most conveniently used in the form of the **molar extinction coefficient**, ε, which is defined as the absorbance of a 1 M solution of pure absorbing material in a 1 cm cell under specified conditions of wavelength and solvent. To illustrate the use of Equation 5.5, consider the following calculation of ε from experimental data.

EXAMPLE 1 ■ The absorbance, A, of a 5×10^{-4} M solution of the amino acid tyrosine, at a wavelength of 280 nm, is 0.75. The pathlength of the cuvet is 1 cm. What is the molar extinction coefficient, ε?

$$A = \varepsilon bc$$

$$A = 0.75$$

$$b = 1 \text{ cm}$$

$$c = 5 \times 10^{-4} \text{ M}$$

$$\varepsilon = \frac{0.75}{(1 \text{ cm})(5 \times 10^{-4} \text{ mole/liter})}$$

$$\varepsilon = 1500 \frac{\text{liter}}{\text{mole cm}} = 1500 \text{ M}^{-1} \text{ cm}^{-1}$$

Notice that the units of ε are defined by the concentration units of the tyrosine solution (M) and the dimension units of the cuvet (cm). Although E is most often expressed as a molar extinction coefficient, you may encounter other units such as $E_\lambda^\%$, which is the absorbance of a 1% (w/v) solution of pure absorbing material in a 1 cm cuvet at a specified wavelength, λ.

B. INSTRUMENTATION FOR MEASURING THE ABSORPTION OF VISIBLE-ULTRAVIOLET LIGHT

The **spectrophotometer** is used to measure absorbance experimentally. This instrument produces light of a preselected wavelength, directs it through the sample (usually dissolved in a solvent and placed in a cuvet), and measures the intensity of light transmitted by the sample. The major components are shown in Figure 5.6. These consist of a light source, a monochromator (including various filters, slits, and mirrors), a sample chamber, a detector, and a meter or recorder.

Light Source. For absorption measurements in the ultraviolet region, a high-pressure hydrogen or deuterium lamp is used. These lamps produce radiation in the 200 to 320 nm range. The light source for the visible region is the tungsten lamp, with a wavelength range of 320 to 800 nm. Instruments with both lamps have greater flexibility and can be used for the study of most biologically significant molecules.

Monochromator. Both lamps discussed above produce continuous emissions of all wavelengths within their range. Therefore, a spectrophotometer must have an optical system to select monochromatic light (light of a specific wavelength). Modern instruments use a prism or, more often, a diffraction grating to produce the desired wavelengths. It should be

FIGURE 5.6
A diagram of a typical spectrophotometer.

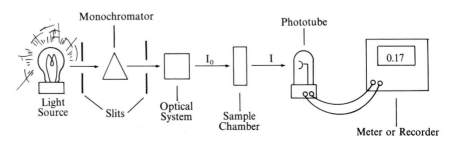

noted that light emitted from the monochromator is not entirely of a single wavelength, but is enhanced in that wavelength. That is, the majority of the light is of a single wavelength, but shorter and longer wavelengths are present.

Before the monochromatic light impinges on the sample, it passes through a series of slits, lenses, filters, and mirrors. This optical system concentrates the light, increases the spectral purity and focuses it toward the sample. The operator of a spectrophotometer has little control over the optical manipulation of the light beam, except for adjustment of slit width. Light passing from the monochromator to the sample encounters a "gate" or slit. On some instruments this is fixed to allow a light beam of a certain width to pass. More expensive instruments have variable slits, allowing the operator to adjust the slit width. The slit width determines both the intensity of light impinging on the sample and the spectral purity of that light. Decreasing the slit width increases the spectral purity of the light, but the amount of light directed toward the sample decreases. The efficiency or sensitivity of the detector then becomes a limiting factor. Setting the proper slit width requires finding a balance between spectral purity and detector sensitivity.

Sample Chamber. The processed monochromatic light is then directed into a sample chamber, which can accommodate a wide variety of sample holders. Most UV-VIS measurements on biomolecules are taken on solutions of the molecules. The sample is placed in a tube or cuvet made of glass, quartz, or other transparent material. Figure 5.7 shows the design of the most common sample holders and the transmission properties of several transparent materials used in cuvet construction. Glass cuvets are the least expensive, but, because they absorb UV light, they can be used only above 320 nm. Quartz or fused silica must be used in the UV range (200 to 320 nm). The quality and condition of cuvets are critical factors in spectrophotometric use. High quality cuvets are expensive and must be carefully maintained. The care and use of cuvets was discussed in Chapter 1.

Less expensive spectrophotometers are "single beam" instruments that allow insertion of one cuvet at a time in the sample holder. This is the case with the Coleman Model 6/35, the Bausch and Lomb Model 20, or the Sequoia-Turner Model 340. More sophisticated instruments are "double beam" and accept two cuvets, one containing sample dissolved in a solvent (sample position in Figure 5.6) and the other containing pure solvent. The use of both types of instruments will be outlined in the application section.

Spectrophotometers are often used to monitor the rates of chemical reactions. For these measurements, the sample cuvet must be held at a constant temperature. Many sample chambers allow for circulating a constant-temperature liquid around the sample holder.

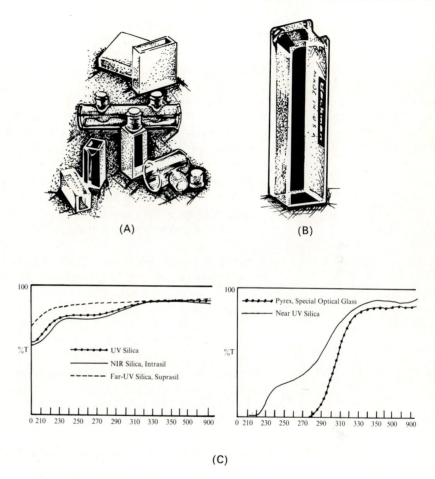

(A)

(B)

(C)

FIGURE 5.7

(A) An assortment of cuvets. Courtesy of VWR Scientific, Division of Univar. (B) A standard 3 ml cuvet. Courtesy of Beckman Instruments, Inc. (C) The transmission properties of several materials used in cuvets. Courtesy of Beckman Instruments, Inc.

Detector. The intensity of the light that passes through the sample under study depends upon the amount of light absorbed by the sample. Intensity is measured by a light-sensitive detector, usually a photomultiplier tube (PMT). The PMT detects a small amount of light energy, amplifies this by a cascade of electrons accelerated by dynodes, and converts it into an electrical signal that can be fed into a meter or recorder.

Meters or Recorders. Less expensive instruments give a direct readout of absorbance and/or transmittance in analog or digital form. These instruments are suitable for single wavelength measurements; however, if a scan of wavelength vs. absorbance (Figure 5.5) is desired, then a recorder must be used. Some spectrophotometers are equipped with pen recorders, but ordinary strip chart recorders can be connected to most spectrophotometers. The absorbance range for most recorders is 0 to 1, but more expensive instruments can be used at higher absorbance levels (1–2) or lower, more sensitive ranges (0–0.1, 0–0.01, 0.1–0.2, etc.).

C. APPLICATIONS OF ABSORPTION SPECTROPHOTOMETRY

Now that you are familiar with the theory and instrumentation of absorption spectrophotometry, you will more easily understand the actual operation and typical applications of a spectrophotometer. Since virtually all UV-VIS measurements are made on samples dissolved in solvents, only those applications will be described here. Although many different types of operations can be carried out on a spectrophotometer, all applications can be placed in one of two categories: (1) measurement of absorbance at a fixed wavelength, and (2) measurement of absorbance as a function of wavelength.

For fixed wavelength measurements with a single beam instrument, a cuvet containing solvent only is placed in the sample beam and the instrument is adjusted to read "zero" absorbance. A matched cuvet containing sample plus solvent is then placed in the sample chamber and the absorbance is read directly from the meter or recorder. The adjustment to zero absorbance with only solvent in the sample chamber allows the operator to obtain a direct reading of absorbance for the sample.

Fixed wavelength measurements using a double beam spectrophotometer are made by first zeroing the instrument with no cuvet in either the sample or reference holder. Alternatively, the spectrophotometer can be balanced by placing matched cuvets containing water or solvent in both sample chambers. Then, a cuvet containing pure solvent is placed in the reference position and a matched cuvet containing solvent plus sample is set in the sample position. The absorbance reading given by the instrument is that of the sample, that is, the absorbance due to solvent is subtracted by the instrument.

An **absorbance spectrum** of a compound is obtained by scanning a range of wavelengths and plotting the absorbance at each wavelength. Most double-beam spectrophotometers automatically scan the desired wavelength range and record the absorbance as a function of wavelength. If solvent is placed in the reference chamber and solvent plus sample in the sample position, the instrument will continuously and automatically subtract the solvent absorbance from the total absorbance (solvent plus sample) at each wavelength; hence, the recorder output is really a "difference spectrum" (absorbance of sample plus solvent, minus absorbance of solvent).

Both types of measurements (fixed wavelength and absorbance spectrum) are common in biochemistry, and you should be able to interpret results from each. The following examples are typical of the kinds of problems readily solved by spectrophotometry.

Measurement of the Concentration of a Solution

According to the Beer-Lambert law, the absorbance of a material in solution is directly dependent upon the concentration of that material. Two methods are commonly used for the measurement of concentration. If the extinction coefficient is known for the absorbing species, then the concentration can be calculated after experimental measurement of the absorbance of the solution.

EXAMPLE 2 ■ A solution of the nucleotide base uracil, in a 1 cm cuvet, has an absorbance at λ_{max} (260 nm) of 0.65. Pure solvent in a matched quartz cuvet has an absorbance of 0.07. What is the molar concentration of the uracil solution? Assume the molar extinction coefficient, ε, is 8.2×10^3 M^{-1} cm^{-1}.

$A = \varepsilon b c$

$A = (\text{absorbance of solvent} + \text{sample}) - (\text{absorbance of solvent})$

$A = 0.65 - 0.07 = 0.58$

$\varepsilon = 8.2 \times 10^3$ M^{-1} cm^{-1}

$b = 1$ cm

$c = \dfrac{A}{\varepsilon b} = \dfrac{0.58}{(8.2 \times 10^3 \ M^{-1} \ cm^{-1})(1 \ cm)}$

$c = 7.1 \times 10^{-5}$ M

If the extinction coefficient for an absorbing species is known, then the concentration of that species in solution can be calculated as outlined. However, there are limitations to this application. Most spectrophotometers are useful for measuring absorbances up to 1, although more sophisticated instruments can measure absorbances as high as 2. Also, some substances do not obey the Beer-Lambert law; that is, absorbance may not increase in a linear fashion with concentration. This may occur with substances that dissociate or associate in solution. Linearity is readily tested by preparing a series of concentrations of the absorbing species and measuring the absorbance of each. A plot of A vs. concentration should be linear if the Beer-Lambert law is valid. If the extinction coefficient for a species is unknown, its concentration in solution can be measured if the absorbance of a standard solution of the compound is known.

EXAMPLE 3 ■ The absorbance of a 1% (w/v) solution of the enzyme tyrosinase, in a 1 cm cell at 280 nm, is 24.9. What is the concentration of a tyrosinase solution that has an A_{280} of 0.25?

Since the extinction coefficient, $E^\%$, is the same for both solutions, the concentration can be calculated by a direct ratio:

$$\frac{A_{\text{std}}}{C_{\text{std}}} = \frac{A_x}{C_x}$$

A_{std} = absorbance of the 1% standard solution = 24.9

C_{std} = concentration of the standard solution = 1% (1 g/dl)

A_x = absorbance of the unknown solution = 0.25

C_x = concentration of the unknown solution in $\%$

$$\frac{24.9}{1\%} = \frac{0.25}{C_x}$$

$$C_x = 0.01\% = 0.01 \text{ g/dl} = 0.1 \text{ mg/ml}$$

Alternatively, the concentration of a species in solution can be determined by the preparation of a standard curve of absorbance vs. concentration.

EXAMPLE 4 ■ The Lowry protein assay is one of the most used spectrophotometric assays in biochemistry. (For a discussion of the Lowry assay, see Chapter 2.) Solutions of varying amounts of a standard protein are mixed with reagents that cause the development of a color. The amount of color produced depends upon the amount of protein present. The absorbance at 540 nm of each reaction mixture is plotted against the known protein concentration. A protein sample of unknown concentration is treated with the Lowry reagents and the color is allowed to develop.

The following absorbance measurements are typical for the standard curve of a protein:

Protein (μg per assay)	A_{540}
15	0.05
25	0.07
50	0.14
100	0.27
150	0.39
200	0.52
0.1 ml unknown protein solution	0.10
0.2 ml unknown protein solution	0.22

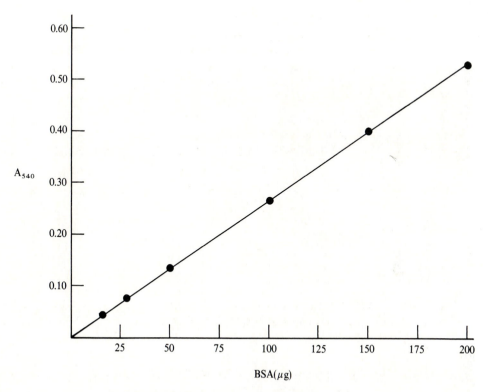

FIGURE 5.8
A standard curve obtained from the Lowry protein method.

A standard curve for the data is plotted in Figure 5.8. Note the linearity, indicating that the Beer-Lambert law is obeyed over this concentration range of standard protein. Two different volumes of unknown protein were tested. This is to ensure that one volume will be in the concentration range of the standard curve. Spectrophotometric readings of absorbance are most accurate in the range of 0.10 to 0.75, so the A_{540} for 0.2 ml of unknown protein will be used. From graphical analysis, an A_{540} of 0.22 corresponding to 85 μg of protein per 0.20 ml. This indicates that the original protein solution concentration was approximately 430 μg/ml or 0.43 mg/ml.

Identification of Unknown Biomolecules by Spectrophotometry

The UV-VIS spectrum of a biomolecule reveals much about its molecular structure. Therefore, a spectral analysis is one of the first experimental measurements made on an unknown biomolecule. Natural molecules often contain chromophoric (color-producing) functional groups that have characteristic spectra. Figure 5.9 displays spectra of well-characterized biomolecules including DNA, FMN, $FMNH_2$, NAD, NADH, and nucleotides.

FIGURE 5.9
Absorbance spectra of significant biomolecules. (A) DNA, (B) FMN, (C) NAD^+ and NADH, (D) d-GMP, (E) thymine.

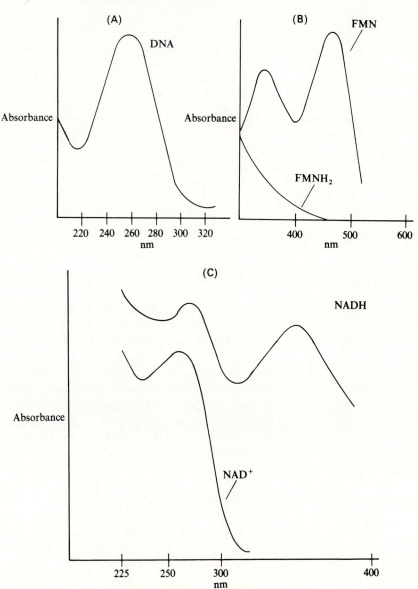

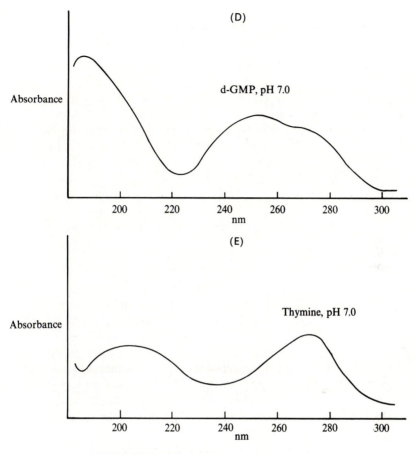

FIGURE 5.9 (*Continued*)

The procedure for obtaining a UV-VIS spectrum begins with the preparation of a solution of the species under study. A standard solution should be prepared in an appropriate solvent. An aliquot of the solution is transferred to a cuvet and placed in the sample chamber of a spectrophotometer. A cuvet containing solvent is placed in the reference holder. The spectrum is scanned over the desired wavelength range and an extinction coefficient is calculated for each major λ_{max}.

Kinetics of Chemical Reactions

Spectrophotometry is one of the best methods available to measure the rates of chemical reactions. Consider a general chemical reaction as shown in Reaction 5.1.

$$A + B \longrightarrow C + D$$

(Reaction 5.1)

If reactants A or B absorb in the UV-VIS region of the spectrum at some wavelength, λ_1, then the rate of the reaction can be measured by monitoring the decrease of absorbance at λ_1 due to loss of A or B. Alternatively, if products C or D absorb at a specific wavelength, λ_2, the kinetics of the reaction can be evaluated by monitoring the absorbance increase at λ_2. According to the Beer-Lambert law, the absorbance change of a reactant or product is proportional to the concentration change of that species occurring during the reaction. This method is widely used to assay enzyme-catalyzed processes. Experiment 7 utilizes spectrophotometry to characterize the kinetics of the tyrosinase-catalyzed oxidation of 3,4-dihydroxyphenylalanine.

Characterization of Macromolecule: Ligand Interactions by Difference Spectroscopy

Many biological processes depend upon a specific interaction between molecules. The interaction often involves a macromolecule (protein or nucleic acid) and a smaller molecule, a **ligand**. Specific examples include enzyme: substrate interactions and receptor protein: hormone interactions. One of the most effective and convenient methods for detecting and characterizing such interactions is **difference spectroscopy**. The interaction of small molecules with the transport protein, hemoglobin, is a classic example of the utility of difference spectroscopy. If a small molecule or ligand, such as inositol hexaphosphate, binds to hemoglobin, there is a change in the heme spectral properties (Giardina *et al.*, 1975). The spectral change is small and would be difficult to detect if the experimenter records the spectrum in the usual fashion. Normally, one would obtain a spectrum of a hemeprotein by placing a solution of the protein in a cuvet in the sample compartment of a spectrophotometer and the neat solvent in the reference compartment. Any absorption due to solvent is subtracted, because the solvent is present in both light beams, so the spectrum is that due to the hemeprotein. Then, the ligand to be tested would be added to the hemeprotein, and the spectrum would be obtained for this mixture vs. solvent. If the free or bound ligand molecule does not absorb light in the wavelength range studied, there would be no need to have ligand in the reference cell. The two spectra (hemeprotein in solvent vs. solvent, and hemeprotein and ligand in solvent vs. solvent) can then be compared and differences noted.

A difference spectrum is faster than the preceding method because only one spectral recording is necessary. Two cuvets are prepared in the following manner. The reference cuvet contains hemeprotein and solvent, whereas the sample cuvet contains hemeprotein, solvent, and ligand. There must be equal concentrations of the hemeprotein in the two cuvets.

(Why?) The two cuvets are placed into a double-beam spectrophotometer and the spectrum is recorded. If the spectrum of the hemeprotein is not influenced by the ligand, the result would be a zero difference spectrum, that is, a straight line (Figure 5.10A). Both samples have identical spectral properties, indicating that there is probably little or no interaction between hemeprotein and ligand. However, such data should be treated with caution because it is possible that the heme group is not affected by ligand binding. A nonzero difference spectrum indicates that the ligand interacts with the heme protein and induces a change in the environment of the heme group (Figure 5.10B).

A difference spectrum can be analyzed and used in several ways. This is a useful technique to qualitatively demonstrate whether an interaction occurs between a macromolecule and a ligand. Quantitative analysis of difference spectra requires measurement of λ_{max} and ΔA at λ_{max}. This method can be used to quantify the strength of ligand-protein interaction. You should note that every time you record a spectrum in a double-beam spectrometer, you are obtaining a difference spectrum between sample and reference. Although a hemeprotein was used in this example, this does not imply that only heme interactions can be characterized by difference

FIGURE 5.10
Difference spectroscopy. (A) Hemoglobin vs. hemoglobin. (B) Hemoglobin vs. hemoglobin + inositol hexaphosphate.

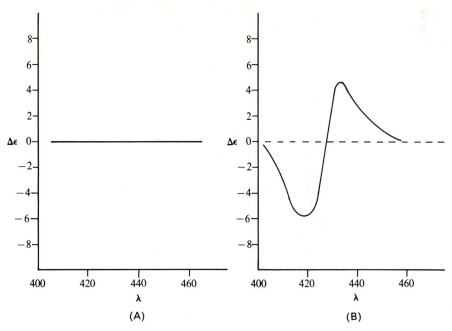

spectroscopy. Any protein that contains a chromophoric group, whether it be an aromatic amino acid, cofactor, prosthetic group, or metal ion, can be studied by difference spectroscopy (Herskovits, 1967).

Limitations and Precautions in Spectrophotometry

The use of a spectrophotometer is relatively straightforward and can be mastered in a short period of time. There are, however, difficulties that must be considered. A common problem encountered with biochemical measurements is turbidity or cloudiness of biological samples. This can lead to great error in absorbance measurements because much of the light entering the cuvet is not absorbed, but is scattered. This causes artificially high absorbance readings. Occasionally, absorbance readings on turbid solutions are desirable (as in measuring the rate of bacterial growth in a culture), but in most cases turbid solutions must be avoided or clarified by filtration or centrifugation.

A difficulty encountered in measuring the concentration of unknown absorbing species in solution is deviation from the Beer-Lambert law. For reasons stated earlier in this chapter, some absorbing species do not demonstrate an increase in absorbance that is proportional to an increase in concentration. (In reality, most compounds display the Beer-Lambert relationship over a relatively small concentration range.) When measuring solution concentration, adherence to the Beer-Lambert law must always be tested in the concentration range under study.

Finally, care in sample preparation and cleanliness are especially important in performing spectrophotometric measurements. The cause of many errors in absorbance measurements is improper preparation and dilution of solutions used to test the Beer-Lambert law. Cuvets used for UV-VIS measurements must be clean and lint-free. Always handle the cuvets on the sides, not in the light beam; better yet, use tissue paper to handle them.

D. BASIC PRINCIPLES OF FLUORESCENCE SPECTROSCOPY

In our discussion of absorption spectroscopy, we noted that the interaction of photons with molecules resulted in the promotion of valence electrons from ground state orbitals to higher energy level orbitals. The molecules were said to be in an excited state.

Molecules in the excited state do not remain there long, but spontaneously relax to the more stable ground state. With most molecules, the relaxation process is brought about by collisional energy transfer to sol-

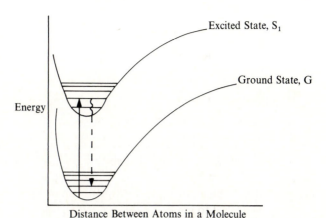

FIGURE 5.11
Energy level diagram describing the fluorescence process.

vent or other molecules in the solution. Some excited molecules, however, return to the ground state by emitting the excess energy as light. This process, called **fluorescence**, is illustrated in Figure 5.11. The solid vertical arrow in the figure indicates the photon absorption process in which the molecule is excited from G to some vibrational level in S. The excited molecule loses vibrational energy by collision with solvent and ground state molecules. This relaxation process, which is very rapid, leaves the molecule in the lowest vibrational level of S, as indicated by the wavy arrow. The molecule may release its energy in the form of light (fluorescence, dashed arrow) to return to some vibrational level of G.

Two important characteristics of the emitted light should be noted: (1) It is of longer wavelength (lower energy) than the excitation light. This is because part of the energy initially associated with the S state is lost as heat energy, and because the energy lost by emission may be sufficient only to return the excited molecule to a higher vibrational level in G. (2) The emitted light is composed of many wavelengths, which results in a fluorescence spectrum as shown in Figure 5.12. This is due to the fact that fluorescence from any particular excited molecule may return the molecule to one of many vibrational levels in the ground state. Just as in the case of an absorption spectrum, a wavelength of maximum fluorescence is observed, and the spectrum is composed of a wavelength distribution centered at this emission maximum. Note that all emitted wavelengths are longer and of lower energy than the initial excitation wavelength.

In our discussion above, it was pointed out that a molecule in the excited state can return to lower energy levels by collisional transfer or by light emission. Since these two processes are competitive, the **fluorescence intensity** of a fluorescing system depends upon the relative importance of each process. The fluorescence intensity is often defined in terms of **quantum yield**, represented by Q. This describes the efficiency or probability of the fluorescence process. By definition, Q is the ratio of the

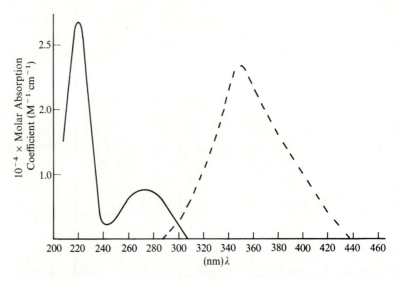

FIGURE 5.12

Absorption (————) and fluorescence (——————) spectra of tryptophan. From *Physical Biochemistry* by D. M. Freifelder, W. H. Freeman and Company. Copyright © 1982. All rights reserved.

number of photons emitted to the number of photons absorbed (Equation 5.6).

$$Q = \frac{\text{number of photons emitted}}{\text{number of photons absorbed}} \qquad \text{(Equation 5.6)}$$

The measurement of quantum yield is often the goal in fluorescence spectroscopy experiments. Q is of interest because it may reveal important characteristics of the fluorescing system. Two types of factors affect the intensity of fluorescence, internal and external (environmental) influences. Internal factors, such as the number of vibrational levels available for transition and the rigidity of the molecules, are associated with properties of the fluorescent molecules themselves. Internal factors will not be discussed in detail here because they are of more interest in theoretical studies. The external factors that affect Q are of great interest to biochemists because information can be obtained about macromolecule conformation and molecular interactions between small molecules (ligands) and larger biomolecules (proteins, nucleic acids). Of special value is the study of experimental conditions that result in **quenching** or **enhancement** of the quantum yield. Quenching in biochemical systems can be caused by chemical reactions of the fluorescent species with added molecules, transfer of energy to other molecules by collision (actual contact between molecules), and transfer of energy over a distance (no contact, resonance energy transfer). The reverse of quenching, enhancement of fluorescent intensity, is also observed in some situations. Several fluorescent dye molecules are quenched in aqueous solution, but their fluorescence is greatly enhanced in a nonpolar or rigidly bound environment (the interior of a protein, for example). This is a convenient method for the characterization of ligand binding.

Both fluorescence quenching and fluorescence enhancement studies can yield important information about biomolecular structure and function. Several applications will be described in a later section of this chapter and in Experiments 6 and 21.

E. INSTRUMENTATION FOR FLUORESCENCE MEASUREMENTS

The basic instrument for measuring fluorescence is the **spectrofluorometer**. It contains a light source, two monochromators, a sample holder, and a detector. A typical experimental arrangement for fluorescence measurement is shown in Figure 5.13. The set-up is similar to that for absorption measurements, with two significant exceptions. First, there are two monochromators, one for selection of the excitation wavelength, another for wavelength analysis of the emitted light. Second, the detector is at an angle (usually 90°) to the excitation beam. This is to eliminate interference by the light that is transmitted through the sample. Upon excitation of

FIGURE 5.13
A diagram of a typical fluorometer.

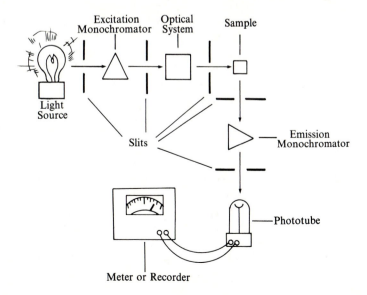

the sample molecules, the fluorescence will be emitted in all directions and is detected by a photocell at right angles to the excitation light beam.

The lamp source used in most instruments is a xenon arc lamp that emits radiation in the ultraviolet, visible, and near infrared regions (200 to 1400 nm). The light is directed by an optical system to the excitation monochromator, which allows either the preselection of a wavelength or scanning of a certain wavelength range. The exciting light then passes into the sample chamber, which contains a fluorescence cuvet with dissolved sample. Because of the geometry of the optical system, a typical fused absorption cuvet with two opaque sides cannot be used; instead, special fluorescence cuvets with four translucent quartz or glass sides must be used. When the excitation light beam impinges on the sample cell, molecules in the solution are excited, and some will emit light.

Light emitted at right angles to the incoming beam is analyzed by the emission monochromator. In most cases, the wavelength analysis of emitted light is carried out by measuring the intensity of fluorescence at a preselected wavelength (usually the wavelength of emission maximum). The analyzer monochromator directs emitted light of only the preselected wavelength toward the detector. A photomultiplier tube serves as a detector to measure the intensity of the light. The output current from the photomultiplier is fed to some measuring device that indicates the extent of fluorescence. The final read-out is not in terms of Q, but in units of the photomultiplier tube current (microamperes) or in relative units of percent of full scale. Therefore, the scale must be standardized with a known. This is described in Experiments 6 and 21. Some newer instruments provide, as output, the ratio of emitted light to incident light intensity. This type of information is advantageous because the xenon arc lamp is not a particularly stable light source, and its output may vary with time.

F. APPLICATION OF
FLUORESCENCE MEASUREMENTS

Two types of measurements are most common in fluorescence experiments, **measurements of relative fluorescence intensities** and **measurements of the quantum yield**. Experiments introduced in this book will require only relative fluorescence intensity measurements, and they proceed as follows. The fluorometer is set to "zero" or "full scale" fluorescence intensity (microamps or %) with the desired biochemical system under standard conditions. Some perturbation is then made in the system (pH change, addition of a chemical agent in varying concentrations, change of ionic strength, etc.) and the fluorescence intensity is determined relative to the standard conditions. This is a straightforward type of experiment because it consists of replacing one solution with another in the fluorometer and reading the detector output for each. For these experiments, the excitation wavelength and the emission wavelength are preselected and set for each monochromator.

The measurement of quantum yield is a more complicated process. Before these measurements can be made, the instrument must be calibrated. A thermopile or chemical actinometer may be used to measure the absolute intensity of incident light on the sample. Alternatively, quantum yields may be measured relative to some accepted standard. Two commonly used fluorescence standards are quinine sulfate in 1 N H_2SO_4 ($Q = 0.70$) and fluorescein in 0.1 N NaOH ($Q = 0.93$). The quantum yield of the unknown, Q_x, is then calculated by Equation 5.7.

$$\frac{Q_x}{Q_{std}} = \frac{F_x}{F_{std}}$$

(Equation 5.7)

where

Q_x = quantum yield of unknown

Q_{std} = quantum yield of standard

F_x = experimental fluorescence intensity of unknown

F_{std} = experimental fluorescence intensity of standard

Some biomolecules are **intrinsic fluors**, that is, they are fluorescent themselves. The amino acids with phenyl rings (phenylalanine, tyrosine, and tryptophan) are fluorescent; hence, proteins containing these amino acids have intrinsic fluorescence. The purine and pyrimidine bases in nucleic acids (adenine, guanine, cytosine, uracil) and some coenzymes (NAD, FAD) are also intrinsic fluors. Intrinsic fluorescence is most often used to study protein conformational changes and to probe the location of active sites and coenzymes in enzymes.

Valuable information can also be obtained by the use of **extrinsic fluors**. These are fluorescent molecules that are added to the biochemical system under study. Many fluorescent dyes have enhanced fluorescence when they are in a nonpolar solution or bound in a rigid hydrophobic environment. Some of these dyes bind to specific sites on proteins or nucleic acid molecules, and the resulting fluorescence intensity depends on the environmental conditions at the binding site. Extrinsic fluorescence is of value in characterizing the binding of natural ligands to biochemically significant macromolecules. This is because many of the extrinsic fluors bind in the same sites as do natural ligands. Extrinsic fluorescence has been used to study the binding of fatty acids to serum albumin (see Experiment 6), to characterize the binding sites for cofactors and substrates in enzyme molecules, to characterize the heme-binding site in various hemoproteins, and to study the intercalation of small molecules into the DNA double helix (Experiment 21).

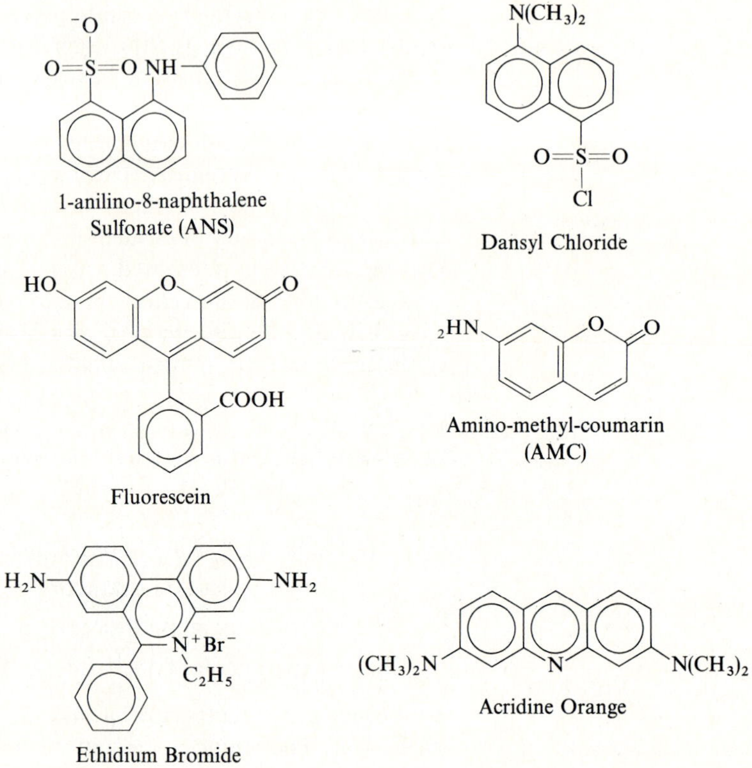

FIGURE 5.14
Fluorescent molecules useful in biochemical studies.

1-anilino-8-naphthalene Sulfonate (ANS)

Dansyl Chloride

Fluorescein

Amino-methyl-coumarin (AMC)

Ethidium Bromide

Acridine Orange

Figure 5.14 shows the structures of extrinsic fluors that have been of value in studying biochemical systems. ANS, dansyl chloride, and fluorescein are used for protein studies, whereas ethidium, proflavine, and various acridines are useful for nucleic acid characterization. Ethidium bromide has the unique characteristic of enhanced fluorescence when bound to double-stranded DNA, but not to single-stranded DNA. Aminomethyl coumarin (AMC) is of value as a fluorogenic leaving group in measuring peptidase activity (Castillo *et al.*, 1979).

The potential applications of fluorescence spectroscopy are too numerous to mention here. The use of a fluorometer is becoming routine in biochemical research and can be applied, as illustrated here, to many aspects of biochemistry.

Difficulties in Fluorescence Measurements

Fluorescence measurements have much greater sensitivity than absorption measurements. Therefore, the experimenter must take special precautions in making fluorescence measurements because any contaminant or impurity in the system can lead to poor results. The following factors must be considered when preparing for a fluorescence experiment.

Xenon Arc Lamp. As was noted earlier, this lamp is not particularly stable. It is best to make several measurements on a sample and average the readings. Instruments that provide data in terms of the ratio of light emitted to incident light intensity have a special advantage.

Preparation of Reagents and Solutions. Since fluorescence measurements are very sensitive, dilute solutions of biomolecules and other reagents are appropriate. Special precautions must be taken to maintain the integrity of these solutions. All solvents and reagents must be checked for the presence of fluorescent impurities, which can lead to large errors in measurement. "Blank" readings should be taken on all solvents and solutions, and any background fluorescence must be subtracted from the fluorescence of the complete system under study. Solutions should be stored in the dark, in clean glass-stoppered containers, in order to avoid photochemical breakdown of the reagents and contamination by corks and rubber stoppers. Some biomolecules, especially proteins, tend to adsorb to glass surfaces, which can lead to loss of fluorescent material or to contamination of fluorescence cuvets. All glassware must be scrupulously cleaned with a mild detergent. Solutions for use in fluorometry should not be stored for extended periods of time; ideally, they should be prepared just prior to use. All turbid solutions must be clarified by centrifugation or filtration.

Control of Temperature. Fluorescence measurements, unlike absorption, are temperature-dependent. All solutions, especially if relative fluorescent measurements are taken, must be thermostated at the same temperature.

G. BASIC PRINCIPLES OF POLARIMETRY

Both absorption and fluorescence spectroscopy provide information about the molecular structure and conformation of biomolecules. A spectroscopic technique that reveals even more detailed information about molecular structure involves **polarimetry**, the interaction of **plane-polarized light** with biomolecules. Polarimetry is used to measure optical activity or chirality in molecules. A high percentage of biochemically significant molecules display chirality. This last section of the chapter will introduce the principles of polarimetry and discuss its application to the characterization of biomolecules.

Recall, from an earlier discussion, that light is composed of electrical and magnetic vectors (Figure 5.2). If light is passed through a Polaroid

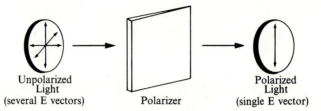

FIGURE 5.15
Production of polarized light.

Unpolarized Light (several E vectors)   Polarizer   Polarized Light (single E vector)

lens or a Nicol prism, only one plane of the electrical vector is transmitted (Figure 5.15). The transmitted light is said to be *plane-polarized*, since it has only a single plane of polarization in the *E* vector.

Plane-polarized light interacts in a unique and reproducible manner with molecules that are asymmetric. (Asymmetric molecules are those that cannot be superimposed on their mirror images). Such molecules are said to be optically active because their solutions change the orientation of plane-polarized light. The most often encountered optically active molecules are those with chiral centers, carbon atoms bonded to four different groups. In Figure 5.16 are shown two chiral molecules, the amino acids D- and L-phenylalanine and the carbohydrates D-glucose and L-glucose. Only one form of each of these occurs in nature (L-phenylalanine, D-glucose); therefore, when they are extracted from biological material they should show optical activity.

When plane-polarized light is transmitted through a solution of a chiral substance, the plane of light exiting the sample is rotated or changed. In polarimetry, the direction and extent of that rotation are measured (see

FIGURE 5.16
Optical isomers of phenylalanine and glucose.

$$\begin{array}{cc}
\text{COOH} & \text{COOH} \\
| & | \\
\text{H}_2\text{N}{-}\text{C}{-}\text{H} & \text{H}{-}\text{C}{-}\text{NH}_2 \\
| & | \\
\text{CH}_2 & \text{CH}_2
\end{array}$$

L-phenylalanine   D-phenylalanine

$$\begin{array}{cc}
\text{CHO} & \text{CHO} \\
| & | \\
\text{HCOH} & \text{HOCH} \\
| & | \\
\text{HOCH} & \text{HCOH} \\
| & | \\
\text{HCOH} & \text{HOCH} \\
| & | \\
\text{HCOH} & \text{HOCH} \\
| & | \\
\text{CH}_2\text{OH} & \text{CH}_2\text{OH}
\end{array}$$

D-glucose   L-glucose

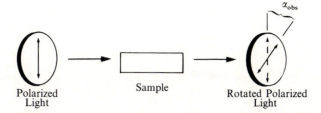

FIGURE 5.17
Diagram showing the basic principle of polarimetry.

Figure 5.17). The direction (+ or −) of the rotation depends upon the form of the chiral molecule under study. The extent of rotation is expressed in terms of the angle (α) and identified by $[\alpha]_\lambda$ defined by Equation 5.8:

$$[\alpha]_\lambda^T = \frac{\alpha_{obs}}{\ell c} \qquad \text{(Equation 5.8)}$$

where

$[\alpha]_\lambda^T$ = specific rotation of an asymmetric substance at temperature T, with light of wavelength λ

α_{obs} = experimental measurement of rotation in degrees

ℓ = pathlength of the cell holding the sample, in dm

c = concentration of the optically active sample in g/ml

The magnitude and sign of α_{obs} depend on the nature and concentration of the asymmetric substance, the temperature of the solution, the pathlength of the plane-polarized light through the sample, and the wavelength of the plane-polarized light. Most polarimetric measurements use the D line from a sodium lamp; hence, the **specific rotation** becomes $[\alpha]_D^T$. Some instruments allow for α_{obs} measurements at several different wavelengths. More expensive instruments plot optical rotation as a function of wavelength, which provides extensive information about molecular structure. (This technique, called optical rotatory dispersion, ORD, will not be discussed here.)

The size and sign of the specific rotation can be used to identify unknown asymmetric molecules, because each compound has a characteristic $[\alpha]_D^T$. This is a physical property of that substance, similar to boiling point, refractive index, and density.

Instrumentation for Measurement of $[\alpha]_D$

The essential components of a polarimeter are shown in Figure 5.18. Light from a sodium lamp is transmitted through a Polaroid lens or a Nicol prism. The plane-polarized light is directed through a glass cell containing a solution of the material under study. A second polarizer (called the analyzer) serves to measure the angle of rotation.

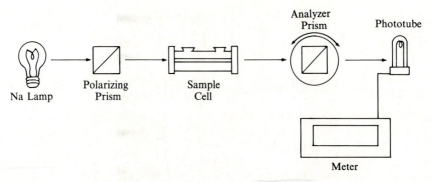

FIGURE 5.18
A diagram of a typical polarimeter.

In older instruments, the experimenter must adjust the analyzer manually and use the naked eye to detect the angle of rotation. More recent instruments provide for automatic measurement of the angle of rotation with digital readout of the magnitude and sign of α_{obs}.

Applications of Polarimetry

Polarimetry is a valuable tool for identifying and characterizing optically active biomolecules such as amino acids, carbohydrates, nucleic acids, and lipids. Once a molecule has been classified by other techniques (chromatography, titration, chemical reactivity, etc.) into one of these groups, its identity can be confirmed by polarimetric measurements.

The technique is simple, fast, and accurate and requires relatively inexpensive equipment. A standard solution of known concentration of the compound is prepared, and the optical rotation is measured relative to the optical rotation of pure solvent. The observed α is then used to calculate specific rotation, $[\alpha]_D^T$. Standard $[\alpha]_D^T$ values for biomolecules are readily available in common reference books. Since α_{obs} depends on temperature and wavelength, these variables must be known. Most modern polarimeters provide thermostated cells to maintain constant temperature (see Figure 5.19). When specific rotation measurements are reported, they must include the temperature of the solution, the λ of the plane-polarized light, the concentration of the optically active material, and the identity of the solvent.

More detailed information can be obtained from polarimetry experiments that are conducted under conditions that cause changes in optical rotation measurements (mutarotation). This technique is of particular value in identifying and characterizing carbohydrate structure (see Experiment 11).

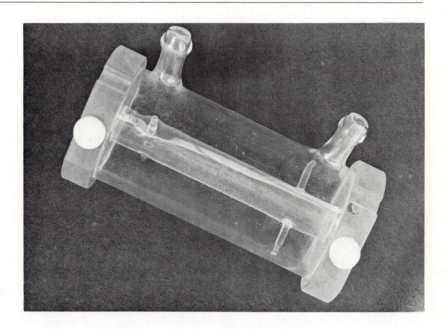

FIGURE 5.19
A polarimeter cell. Photo courtesy of Mr. Mark Billadeau.

REFERENCES

General

R.C. Bohinski, *Modern Concepts in Biochemistry*, 4th Edition (1983), Allyn and Bacon (Boston), pp. 495–497. Brief introduction to spectroscopy.

J. Brewer, A. Pesce, and R. Ashworth, *Experimental Techniques in Biochemistry* (1974), Prentice-Hall (Englewood Cliffs, NJ), pp. 216–262. Detailed theory and practice of absorption and fluorescence spectroscopy.

T. Copper, *The Tools of Biochemistry* (1977), John Wiley (New York), pp. 36–64. Theory and experiments in the use of the spectrophotometer.

C. Cantor and P. Schimmel, *Biophysical Chemistry*, Part II (1980), W.H. Freeman (San Francisco), pp. 349–480; 687–846. An excellent theoretical discussion of spectroscopy and scattering phenomena.

D. Freifelder, *Physical Biochemistry*, 2nd Edition (1982), W.H. Freeman (San Francisco), pp. 494–572; 573–579; 691–698. Theory and applications of absorption, fluorescence spectroscopy, and light scattering. Brief introduction to optical rotation.

J.D. Rawn, *Biochemistry* (1983), Harper and Row (New York), p. 274. A discussion of optical rotation of carbohydrates.

G. Zubay, *Biochemistry* (1983), Addison-Wesley (Reading, MA), pp. 61–63. A brief introduction to optical methods.

Specific

M. Castillo, K. Nakajima, M. Zimmerman, and J. Powers, *Anal. Biochem.*, *99*, 53 (1979). "Sensitive Substrates for Human Leukocyte and Porcine Pancreatic Elastase."

B. Giardina, F. Ascoli, and M. Brunori, *Nature*, *256*, 761 (1975). Spectroscopic study of the binding of inositol hexaphosphate to hemoglobin.

T. Herskovits in *Methods in Enzymology*, C.H.W. Hirs, Editor, Vol. XI (1967), Academic Press (New York), pp. 748–775. "Difference Spectroscopy."

Chapter **6**

Radioisotopes in Biochemistry

The use of radioactive isotopes in experimental biochemistry has provided us with a wealth of information about biological processes. In the earlier days of biochemistry, radioisotopes were primarily used to elucidate metabolic pathways. In modern biochemistry, there is probably no experimental technique that offers such a diverse range of applications as that provided by radioactive isotopes. The measurement of radioactivity is now a tool in all areas of experimental biochemistry, including enzyme assays, biochemical pathways of synthesis and degradation, analysis of biomolecules, measurements of antibodies, binding and transport studies, and many others. Radioisotope use has two advantages over other analytical techniques: (1) sensitive instrumentation allows the detection of minute quantities of radioactive material (some radioactive substances can be measured in the picomole or 10^{-12} M range), and (2) radioisotope techniques offer the ability to differentiate physically between substances that are chemically indistinguishable.

This chapter will explore the origin, properties, detection, and biochemical applications of radioisotopes.

A. ORIGIN AND PROPERTIES OF RADIOACTIVITY

Radioactivity results from the spontaneous nuclear disintegration of unstable isotopes. The hydrogen nucleus, consisting of a proton, is represented as $^{1}_{1}$H. Two additional forms of the hydrogen nucleus contain one and two neutrons; they are represented by $^{2}_{1}$H and $^{3}_{1}$H. These forms of hydrogen, which are commonly called deuterium and tritium, respectively, are **isotopes** of hydrogen. They have an identical number of protons (constant charge, $+1$) but they differ in the number of neutrons. Only tritium is radioactive.

The stability of a nucleus depends upon the ratio of neutrons to protons. Some nuclei are unstable, and undergo spontaneous nuclear disintegration accompanied by emission of particles. Unstable isotopes of this type are called **radioisotopes**. Three main types of radiation are emitted during nuclear decay, α particles, β particles and γ rays. The α particle, a helium nucleus, is emitted only by elements of mass number greater than 140. These elements are seldom used in biochemical research.

Most of the radioisotopes commonly used in biochemistry are β emitters. The β particles, which exist in two forms (β^+, positrons, and β^-, electrons), are emitted from a given radioisotope with a continuous range of energies. However, an average energy (called the *mean* energy) of β particles from an element can be determined. The mean energy is a characteristic of that isotope and can be used to identify one β emitter in the presence of a second (see Figure 6.1).

A few radioisotopes of biochemical significance are γ emitters. Emission of a γ ray (a photon of electromagnetic radiation) is often a secondary process occurring after initial decay by β emission. The disintegration of the isotope ^{131}I is an example of this multistep process. Each radioisotope emits γ rays of a distinct energy, which can be measured for identification of the isotope.

The spontaneous disintegration of a nucleus is, kinetically, a first-order process. That is, the rate of radioactive decay of N atoms ($-dN/dt$,

FIGURE 6.1

The energy spectra for the β emitters ^3H and ^{14}C. The dotted lines indicate the upper and lower limits for discrimination counting.

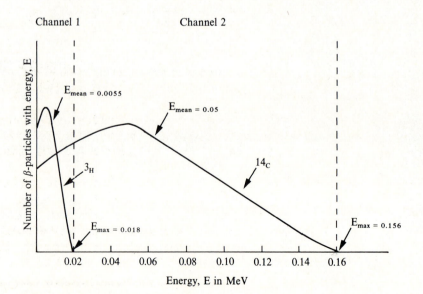

the change of N with time, t) is proportional to the number of radioactive atoms present (Equation 6.1).

$$-\frac{dN}{dt} = \lambda N \qquad \text{(Equation 6.1)}$$

The proportionality constant, λ, is called the **disintegration** or **decay constant**. Equation 6.1 can be transformed to a more useful equation by integration within the limits of $t = 0$ and t. The result is shown in Equation 6.2.

$$N = N_0 e^{-\lambda t} \qquad \text{(Equation 6.2)}$$

where

$N_0 =$ the number of radioactive atoms present at $t = 0$

$N =$ the number of radioactive atoms present at time $= t$

Equation 6.2 can be expressed in the natural logarithmic form as in Equation 6.3.

$$\ln \frac{N}{N_0} = -\lambda t \qquad \text{(Equation 6.3)}$$

Combination of Equations 6.1 and 6.2 leads to a relationship that defines the **half-life**. Half-life, $t_{1/2}$, is a term used to describe the time necessary for one-half of the radioactive atoms initially present in a sample to decay. At the end of the first half-life, N in Equation 6.3 becomes $\frac{1}{2}N_0$, and the result is shown in Equations 6.4 and 6.5.

$$\ln \frac{\frac{1}{2}N_0}{N_0} = -\lambda t_{1/2} \qquad \text{(Equation 6.4)}$$

$$\ln \tfrac{1}{2} = -\lambda t_{1/2}$$

$$t_{1/2} = \frac{0.693}{\lambda} \qquad \text{(Equation 6.5)}$$

Equations 6.3 and 6.5 allow the calculation of the ratio of N/N_0 at any time during an experiment. This calculation is especially critical when radioisotopes with short half-lives are used.

Units of Radiation Measurement

The basic unit of radioactivity is the **curie**, Ci. One curie is the amount of radioactive material that emits particles at a rate of 3.7×10^{10} disintegrations per second (dps), or 2.2×10^{12} min^{-1} (dpm). Amounts that large are seldom used in experimentation, so subdivisions are convenient. The **millicurie** (mCi, 2.2×10^9 min^{-1}) and **microcurie** (μCi, 2.2×10^6 min^{-1}) are standard units for radioactive measurement.

The number of disintegrations emitted by a radioactive sample depends upon the purity of the sample (number of radioactive atoms present) and the decay constant, λ. Therefore, radioactive decay is also expressed in terms of **specific activity**, the disintegration rate per unit mass of radioactive atoms. Typical units for specific activity are mCi/mmole and μCi/μmole.

Although radioactivity is defined in terms of nuclear disintegrations per unit of time, rarely does one measure this absolute number in the laboratory. Instruments that detect and count emitted particles respond to only a small fraction of the particles. Data from a radiation counter are in counts per minute (cpm), which can be converted to actual disintegrations per minute if the counting efficiency of the instrument is known. The percent efficiency of an instrument is determined by counting a standard compound of known radioactivity and using the ratio of detected activity (observed cpm) to actual activity (disintegrations per minute). Equation 6.6 illustrates the calculation using a standard of 1 μCi.

$$\% \text{ efficiency} = \frac{\text{observed cpm of standard}}{\text{dpm of 1 } \mu\text{Ci of standard}} \times 100$$

$$= \frac{\text{observed cpm}}{2.2 \times 10^6 \text{ dpm}} \times 100 \qquad \text{(Equation 6.6)}$$

Isotopes in Biochemistry

Table 6.1 lists the properties of several radioisotopes that are of value in biochemical research. Most of the biochemically significant isotopes

TABLE 6.1

Characteristics of Biochemically Significant Isotopes

Isotope	Particle Emitted	Energy of Particle (MeV)	Half-life, $t_{1/2}$
^{3}H	β^-	0.018	12.3 years
^{14}C	β^-	0.155	5,570 years
^{24}Na	β^-	1.39	14.97 hours
	γ	1.7, 2.75	
^{32}P	β^-	1.71	14.2 days
^{35}S	β^-	0.167	87.1 days
^{36}Cl	β^-	0.714	3×10^5 days
^{40}K	β^-	1.33	1.3×10^9 years
^{45}Ca	β^-	0.25	165 days
^{125}I	γ	0.035	60 days
	γ	0.027	
^{131}I	β^-	0.61, 0.33	8.1 days
	γ	0.637, 0.364, 0.284	

are β emitters, but a couple, ^{24}Na and ^{131}I, emit γ rays in addition to β particles. ^{125}I emits γ rays. Also note from Table 6.1 that there are no radioisotopes of oxygen and nitrogen, two elements common in biomolecules. Stable isotopes of these elements (^{18}O and ^{15}N) are available for labeling studies, but techniques such as mass spectrometry must be used for detection.

B. DETECTION AND MEASUREMENT OF RADIOACTIVITY

Since most of the radioisotopes used in biochemical research are β emitters, only those methods that detect and measure β particles will be emphasized. Two counting techniques are in current use, scintillation counting and Geiger-Müller counting.

Liquid Scintillation Counting

Samples for liquid scintillation counting consist of three components, (1) the radioactive material, (2) a solvent, usually aromatic, in which the radioactive substance is dissolved or suspended, and (3) one or more organic fluorescent substances. Components 2 and 3 make up the **liquid scintillation system**. The β particles emitted from the radioactive sample interact with the scintillation system, producing small flashes of light or scintillations. The light flashes are detected by a photomultiplier tube (PMT). Electronic pulses from the PMT are amplified and registered by a counting device called a scaler.

The scintillation process, in detail, begins with the collision of emitted β particles with solvent molecules, S (Reaction 6.1).

$$\beta + S \longrightarrow S^* + \beta \text{ (less energetic)} \qquad \text{(Reaction 6.1)}$$

Contact between the energetic β particles and S in the ground state results in transfer of energy and conversion of an S molecule into an excited state, S*. Aromatic solvents such as toluene or dioxane are most often used because their electrons are easily promoted to an excited state orbital (see discussion on fluorescence, Chapter 5). The β particle after one collision still has sufficient energy to excite several more solvent molecules. The excited solvent molecules normally return to the ground state by emission of a photon, $S^* \longrightarrow S + h\nu$. Photons from the typical aromatic solvent are of short wavelength and are not efficiently detected by photocells. A convenient way to resolve this technical problem is to add one or more fluorescent substances (fluors) to the scintillation mixture. Excited solvent molecules interact with a **primary fluor**, F_1, as shown in Reaction 6.2.

$$S^* + F_1 \longrightarrow S + F_1^* \qquad \text{(Reaction 6.2)}$$

Energy is transferred from S* to F_1, resulting in ground state S molecules and excited F_1 molecules, F_1*. F_1* molecules are fluorescent and emit light of a longer wavelength than S* (Reaction 6.3).

$$F_1^* \longrightarrow F_1 + hv_1 \qquad \text{(Reaction 6.3)}$$

If the light emitted during the decay of F_1* is still of a wavelength too short for efficient measurement by a PMT, a **secondary fluor**, F_2, that accepts energy from F_1* may be added to the scintillation system. Reactions 6.4 and 6.5 outline the continued energy transfer process and fluorescence of F_2*.

$$F_1^* + F_2 \longrightarrow F_1 + F_2^* \qquad \text{(Reaction 6.4)}$$

$$F_2^* \longrightarrow F_2 + hv_2 \qquad \text{(Reaction 6.5)}$$

The light, hv_2, from F_2* is of longer wavelength than hv_1 from F_1* and is more efficiently detected by a PMT. Two widely used primary and secondary fluors are 2,5-diphenyloxazole (PPO) with an emission maximum of 380 nm and 1,4-bis-2-(5-phenyloxazolyl)-benzene (POPOP) with an emission maximum of 420 nm.

The most basic elements in a liquid scintillation counter are the PMT, a pulse amplifier, and a counter, called a **scaler**. This simple assembly may be used for counting; however, there are many problems and disadvantages with this set-up. Many of these difficulties can be alleviated by more sophisticated instrumental features. Some of the problems and practical solutions are outlined below.

Thermal Noise in Photomultiplier Tubes. The energies of the β particles from most β emitters are very low. This, of course, leads to low energy photons emitted from the fluors and relatively low energy electrical pulses in the PMT. In addition, photomultiplier tubes produce thermal background noise with 25 to 30% of the energy associated with the fluorescence-emitted photons. This difficulty cannot be completely eliminated, but its effects can be lessened by placing the samples and the PMT in a freezer at -5 to $-8°C$ in order to decrease thermal noise.

A second method to help resolve the thermal noise problem is the use of two photomultiplier tubes for detection of scintillations. Each flash of light that is detected by the photomultiplier tubes is fed into a **coincidence circuit**. A coincidence circuit counts only those flashes that arrive simultaneously at the two photodetectors. Electrical pulses that are the result of simultaneous random emission (thermal noise) in the individual tubes are very unlikely. A schematic diagram of a typical scintillation counter with coincidence circuitry is shown in Figure 6.2. Coincidence circuitry has the disadvantage of inefficient counting of very low energy β particles emitted from ^{3}H and ^{14}C.

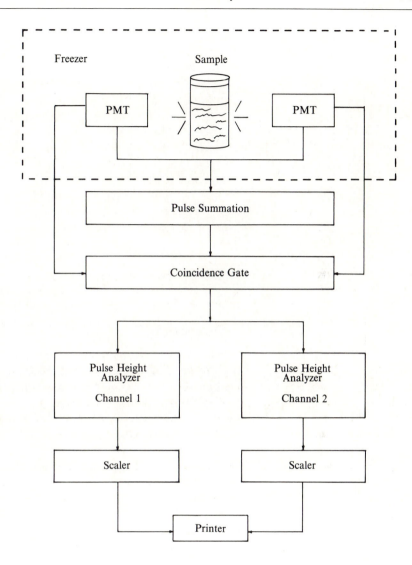

FIGURE 6.2
Diagram of a typical scintillation counter, showing coincidence circuitry.

Counting More Than One Isotope in a Sample. The basic liquid scintillation counter with coincidence circuitry can only be used to count samples containing one type of isotope. Many experiments in biochemistry require the counting of just one isotope; however, more valuable experiments can be performed if two radioisotopes can be simultaneously counted in a single sample (double-labeling experiments). The basic scintillation counter previously described has no means of discriminating between electrical pulses of different energies.

The size of the current generated in a photocell is nearly proportional to the energy of the β particle initiating the pulse. Recall that β particles from different isotopes have characteristic energy spectra with an average energy (Figure 6.1). Modern scintillation counters are equipped with **pulse height analyzers** that measure the size of the electrical pulse and count only those pulses within preselected energy limits set by **discriminators**. The circuitry required for pulse height analysis and energy discrimination of β particles is shown in Figure 6.2. Discriminators are electronic "windows" that can be adjusted to count β particles within certain energy or voltage ranges called **channels**. The channels are set to a **lower limit** and an **upper limit**, and all voltages within those limits are counted. Figure 6.1 illustrates the function of discriminators for the counting of ^{3}H and ^{14}C in a single sample. Discriminator channel number 1 is adjusted to accept typical β particles emitted from ^{3}H and channel 2 is adjusted to receive β particles of the energy characteristic of ^{14}C.

C. PRACTICAL ASPECTS OF SCINTILLATION COUNTING

Quenching

Any chemical agent or experimental condition that reduces the efficiency of the scintillation and detection process leads to a reduced level of counting, or **quenching**. There are four common origins of quenching.

Color quenching is a problem if chemical substances that absorb photons from the secondary fluors are present in the scintillation mixture. Since the secondary fluors emit light in the visible region between 410 and 420 nm, colored substances may absorb the emitted light before it is detected by the photocells. Radioactive samples must be treated to remove colored impurities before mixing with the scintillation solvent.

Chemical quenching occurs when chemical substances in the scintillation solution interact with excited solvent and fluor molecules and, hence, decrease the efficiency of the scintillation process. To avoid this type of quenching the sample can be purified or the fluors can be increased in concentration.

Point quenching occurs if the radioactive sample is not completely dissolved in the solvent. The emitted β particles may be absorbed before they have a chance to interact with solvent molecules. The addition of solubilizing agents such as Cab-O-Sil or Thixin decreases point quenching by converting the liquid scintillator to a gel.

Dilution quenching results when a large volume of liquid radioactive sample is added to the scintillation solution. In most cases, this type of quenching cannot be eliminated, but it can be corrected by one of the techniques discussed below.

Since quenching can occur during all experimental counting of radioisotopes, it is important to be able to determine the extent of the reduced counting efficiency. Two methods for quench correction are in common use.

In the **internal standard ratio** method, the sample, X, under study is first counted; then a known amount of a standard radioactive solution is added to the sample, and it is counted again. The absolute activity of the original sample, A_x, is represented by Equation 6.7.

$$A_x = A_s \frac{C_x}{C_s}$$ (Equation 6.7)

where

A_x = activity of the sample, X
C_x = radioactive counts from sample
A_s = activity of the standard
C_s = radioactive counts from standard

The absolute value for C_s is determined from Equation 6.8.

$$C_T = C_x + C_s$$ (Equation 6.8)

C_T = total radioactive counts from sample plus standard

Equation 6.7 can then be modified to Equation 6.9.

$$A_x = \frac{A_s C_x}{C_T - C_x}$$ (Equation 6.9)

The internal standard ratio method for quench correction is tedious and time-consuming, and it destroys the sample, so it is not an ideal method. Modern scintillation counters are equipped with a standard radiation source inside the instrument but outside the scintillation solution. The radiation source, usually a gamma emitter, is mechanically moved into a position next to the vial containing the sample, and the combined system of standard and sample is counted. Gamma rays from the standard will excite solvent molecules in the sample, and the scintillation process occurs as previously described. However, the instrument is adjusted to register only scintillations due to γ particle collisions with solvent molecules. This method for quench correction, called the **external standard method**, is fast and precise.

Scintillation Cocktails and Sample Preparation

Many of the quenching problems discussed above can be lessened if an experiment is carefully planned. A wide variety of sample preparation

TABLE 6.2
Useful Scintillation Systems

For Nonpolar Organic Compounds
1. Toluene-based: PPO, POPOP (or dimethyl-POPOP) in toluene.

For Aqueous Solutions
1. Toluene-based: PPO, POPOP, emulsifying agent (Aquasol, Biosolv, Cab-O-Sil or Triton-X100) in toluene.
2. Bray's Solution: PPO, POPOP, naphthalene, methanol, ethylene glycol in dioxane.
3. Dioxane-based: PPO, p-bis(O-methylstyryl)benzene, naphthalene in dioxane.

For CO_2 and Acidic Substances
1. Toluene-dioxane-based: PPO, POPOP, naphthalene in dioxane-toluene. Sample is prepared in a base such as hyamine hydroxide.

methods and scintillation liquids are available for use. Only a few of the more common techniques will be mentioned here.

A **scintillation cocktail** is a mixture of solvent and fluor(s) used for the scintillation solution. Table 6.2 lists some of the more popular scintillation systems. Two major types of solvent are noted in the table, those that are immiscible with water and those that dissolve in water. The toluene-based solution is most extensively used for nonaqueous radioactive samples. Many biological samples, however, contain water or are not soluble in the toluene solution. Dioxane-based cocktails have been developed that dissolve most hydrophilic samples; however, dioxane causes some quenching itself. Bray's solution, containing the polar solvents ethylene glycol, methanol, and dioxane, is especially suitable for aqueous samples. Toluene-based cocktails can be used with small amounts of aqueous samples if an emulsifying agent such as Aquasol, Biosolv, Cab-O-Sil, or Triton is added.

Radioactive samples that are insoluble in both toluene and water must be pretreated before analysis. Some insoluble samples can be chemically or physically transformed into a soluble form. A common method is to burn the sample under controlled conditions and collect the ^3H_2O and/or $^{14}CO_2$ in the scintillation solution. Alternatively, the radioactive substance can be collected on cellulose or fiberglass membrane filters (see Chapter 2). The filter with the imbedded radioactive sample is placed directly in a scintillation vial containing the proper cocktail. Acidic substances and CO_2 may be measured after treatment with a base such as hyamine.

The radioactive sample and scintillation solution are placed in a glass vial for counting. Special glass with a low potassium content must be used because natural potassium contains the radioisotope ^{40}K, a β emitter. Polyethylene vials are also in common use today.

Background Radiation

All radiation counting devices register counts even if there is no specific or direct source of radiation near the detectors. This is due to **background radiation**. It has many sources, including natural radioactivity, cosmic rays, radioisotopes in the construction materials of the counter, radioactive chemicals stored near the counter, and contamination of sample vials or counting equipment. Background count in a scintillation counter will depend on the scintillation solution used, but is usually 30 to 50 cpm. Background counts that are the result of natural sources cannot be eliminated; however, those due to contamination can be greatly reduced if the laboratory and counting instrument are kept clean and free of radioactive contamination. Since it is difficult to regulate background radiation, it is usually necessary to monitor its level. The typical procedure is to count a radioactive sample and then count a background sample containing the same scintillation system but no radioisotope. The actual or corrected activity is obtained by subtracting the background counts from the results of the radioactive sample.

Scintillation Counting of γ Rays

^{24}Na and ^{131}I are both β and γ emitters, and are often used in biochemical research. Gamma rays are more energetic than β particles, and denser materials are necessary for absorption. A gamma counter consists of a sample well, a sodium iodide crystal as fluor, and a photomultiplier tube (see Figure 6.3). The high energy γ rays are not absorbed by the scintillation solution or vial, but they interact with a crystal fluor, producing scintillations. The scintillations are detected by photomultiplier tubes and electronically counted.

FIGURE 6.3
Diagram of a γ counter.

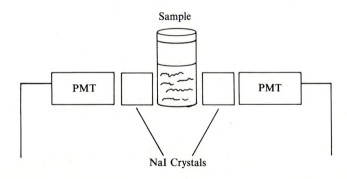

Sample

PMT

PMT

NaI Crystals

Geiger-Müller Counting of Radioactivity

Another method of radiation detection that has some applications in biochemistry is the use of gas ionization chambers. The most common device that uses this technique is the **Geiger-Müller tube** (G-M tube). When β particles pass through a gas, they collide with atoms and may cause ejection of an electron from a gas atom. This results in the formation of an ion pair made up of the negatively charged electron and the positively charged atom. If this ionization occurs between two charged electrodes (an anode and a cathode), the electron will be attracted to the anode and the positive ion toward the cathode. This results in a small current in the electrode system. If only a low voltage difference exists between the anode and cathode, the ion pairs will move slowly and will, most likely, recombine to form neutral atoms. Clearly, this would result in no pulse in the electric circuit because the individual ions did not reach the respective electrodes. At higher voltages, the charged particles are greatly accelerated toward the electrodes and collide many times with unionized gas atoms. This leads to extensive ionization and a cascade or avalanche of ions. If the voltage is high enough (1000 volts for most G-M tubes), all ions are collected at the electrodes. A Geiger-Müller counting system uses this voltage region for ion acceleration and detection. A typical G-M tube is diagramed in Figure 6.4. It consists of a mica window for entry of β particles from the radiation source, an anode down the center of the tube, and a cathode surface inside the walls. A high voltage is applied between the electrodes. Current generated from electron movement toward the anode is amplified, measured, and converted to counts per minute. The cylinder contains an inert gas that readily ionizes (argon, helium, or neon) plus a quenching gas (Q gas, usually butane) to reduce continuous ionization of the inert gas. Beta particles of high energy emitted from atoms such as ^{24}Na, ^{32}P, and ^{40}K have little difficulty entering the cylinder by penetrating the mica window. Particles from weak β emitters (especially ^{14}C and ^{3}H) are unable to efficiently pass through the window to induce ionization inside the chamber. Modified G-M tubes with thin mylar windows, called **flow window tubes**, may be used to count weak β emitters.

G-M counting has several disadvantages compared to liquid scintillation counting. The counting efficiency of a G-M system is not as high. The response time for G-M tubes is longer than for photomultiplier tubes; therefore, samples of high radioactivity are not efficiently counted by the G-M tube. In addition, sample preparation is more tedious and time consuming for G-M counting. The sample is prepared in a metal disc called a **planchet**. Liquid samples are allowed to dry, leaving a radioactive residue for counting on the metal disc. Alternatively, a sample may be collected on a membrane filter and inserted into the planchet. The thickness of the sample on the planchet is a significant factor to consider. The process of self-absorption of β particles increasingly becomes a prob-

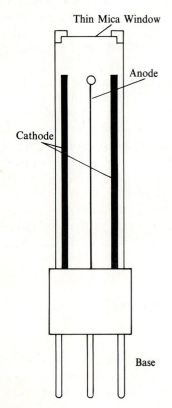

Thin Mica Window

Anode

Cathode

Base

lem as the thickness of the sample becomes greater. Self-absorption may be corrected by counting samples of different thicknesses and preparing a standard self-absorption curve of cpm_{obs}/cpm_{actual} vs. sample thickness.

If a choice of counting methods is available, a researcher will usually decide on liquid scintillation unless the sample is not amenable to such measurement. Insoluble materials or the presence of strong quenching agents would probably indicate the need for G-M counting.

Applications of Radioisotopes

The applications of radioisotopes in biochemical measurements are too numerous to outline here. Excellent reference books, some of which are listed at the end of this chapter, provide experimental procedures for radioisotope utilization. The technique of **autoradiography**, however, should be mentioned. This procedure allows the detection and localization of a radioactive substance (molecule or atom) in a tissue, cell, or cell organelle. Briefly, the technique involves placing a radioactive material close to a photographic emulsion. Radioisotopes in the material emit radiation that impinges on the photographic plate, activating silver halide crystals in the emulsion. Upon development of the plate, a pattern or image is displayed that yields information about the location and amount of the radioactive material in the sample. Autoradiography has been especially useful in concentration and localization studies of biomolecules in cells and cell organelles. For more detailed information, see the list of references.

Statistical Analysis of Radiation Measurements

Radioactive decay is a random process. It is impossible to predict when a radioactive event will occur. Therefore, counting measurements will only be average rates of decay; however, they may be treated by Gaussian distribution analysis. This allows the determination of an average counting rate, standard deviation, percent confidence level, and other statistical parameters. An introduction to statistical analysis of radioactivity counting data and other experimental measurements is given in Chapter 8.

D. SAFETY RULES FOR HANDLING RADIOACTIVE MATERIALS

Safety should be of major concern during the performance of any chemistry experiment, but when radioisotopes are involved, special precautions should be taken. Many chemistry and biochemistry departments have specially equipped laboratories for radioisotope work. Alternatively, your instructor may set aside a specific area of your laboratory for using radioactive materials. In either situation, specific guidelines should be followed.

Preparation for the Experiment

1. The first responsibility of the student is to become familiar with the properties and hazards of the radioactive substances to be used. You must know which radioisotopes are to be handled and the form, liquid or solid, of the material. It is also important to know whether the isotope is a β and/or γ emitter and whether it is "weak" or "strong." This information may be obtained from the instructor or a radiochemistry textbook before you begin the experiment.

2. Confine your work with radioisotopes to a small area in the laboratory. A convenient plan is to use a stainless steel tray lined with absorbent blotter paper coated on the bottom side with polyethylene. The paper must be replaced every day. If the radioactive materials are volatile, the work should be done in a fume hood. If spills occur, a small work area such as a tray is much easier to clean than a large lab bench.

3. Label all glassware and equipment that will be used with special adhesive tape labeled "Radioactive." Plan the experiment so that a minimal number of transfers of radioactive materials is required. This will reduce the amount of contaminated glassware.

4. Two types of containers should be available for disposal purposes. One should be labeled "Liquid Radioactive Waste" and used for all waste solutions; the other, "Solid Radioactive Waste," for blotter paper, broken glassware, etc. **Liquid wastes must not be poured down any drain, nor solid wastes in normal trash cans.**

5. Take all precautions against contamination of yourself or fellow workers. Always wear gloves and a lab coat. No smoking, eating, or drinking is allowed in the laboratory. It is especially hazardous to ingest radioactive materials.

6. The laboratory should be equipped with a portable G-M survey meter in order to check spills or possible self-contamination.

7. A specific area with a sink should be set aside for washing weakly contaminated glassware and equipment.

8. All radioactive materials should be stored in well-labeled, glass containers. The label must include your name, the type of isotope, the radioactive compound, the total amount of radioactivity, the specific activity, and the date of measurement.

Performing the Experiment

1. **Do not pipet any solutions by mouth; always use a mechanical pipet filler.**

2. Contaminated glassware should be kept separated from uncontaminated. Contaminated beakers and flasks are placed in the special sink or other container for washing. Clean and wash all equipment with soap and water immediately after the experiment has been completed. If water-insoluble materials are being used, the first washing should be done with an organic solvent such as acetone. Soak contaminated pipets in a container filled with water. All broken glassware is disposed in the "Solid Radioactive Waste" container.

3. Minor spills must be cleaned first with absorbent blotter paper and then *thoroughly* rinsed with water. Always wear gloves during clean-up. Dispose of the blotter paper in the "Solid Radioactive Waste" container. The spill area should then be checked with the portable G-M counter.

4. All skin cuts, accidental ingestion of radioactive materials, and major spills must be reported immediately to the instructor.

5. After final clean-up of all contaminated glassware and the work area, remove your gloves and thoroughly wash your hands with soap and warm water. Use the portable G-M counter to check for contamination on your hands, clothes, and work area in the laboratory. Report any areas of unusually high count (above 200 cpm) to your instructor.

REFERENCES

General

H.L. Andrews, *Radiation Biophysics*, 2nd Edition (1974), Prentice-Hall (Englewood Cliffs, NJ). The entire book is excellent for an introduction to radiochemistry theory and techniques.

J. Brewer, A. Pesce, and R. Ashworth, *Experimental Techniques in Biochemistry* (1974), Prentice-Hall (Englewood Cliffs, NJ), pp. 263–311. A useful discussion on theory and applications of radioisotopes.

G. Chase and J. Rabinowitz, *Principles of Radioisotope Methodology*, 3rd Edition (1967), Burgess Publishing Co. (Minneapolis). An excellent source for methodology, procedure, and other practical aspects of radiochemistry.

T. Copper, *The Tools of Biochemistry* (1977), John Wiley (New York), pp. 65–135. One chapter on the theory and application of radioisotopes.

D. Freifelder, *Physical Biochemistry*, 2nd Edition (1982), W.H. Freeman (San Francisco), pp. 129–192. Theory and applications of radioisotopes.

R. Lapp and H.L. Andrews, *Nuclear Radiation Physics*, 3rd Edition (1963), Prentice-Hall (Englewood Cliffs, NJ). Excellent source for discussion of theory and measurement of radiation.

K.Z. Morgan and J.E. Turner (Editors), *Principles of Radiation Protection* (1973), R.E. Krieger Publishers (Huntington, NY). Advanced textbook on safety procedures.

J. Shapiro, *Radiation Protection—A Guide for Scientists and Physicians* (1972), Harvard University Press (Cambridge MA). Safeguards and practical aspects of the use of radioisotopes.

Y. Wang (Editor), *CRC Handbook of Radioactive Nuclides* (1969), Chemical Rubber Co. (Cleveland). An excellent reference for properties and use of radioisotopes.

Specific Techniques and Experiments using Radioisotopes

D. Geller, in *Methods in Enzymology*, A. San Pietro, Editor, Vol. XXIVB (1972), Academic Press (New York), pp. 88–92. "ATP Formation."

N. Shavit, *ibid.* pp. 318–321. "Exchange Reactions (Plant)."

D. Canvin and H. Fock, *ibid.*, pp. 246–260. "Measurement of Photorespiration."

E.D. Bransome, Jr. in *Methods in Enzymology*, B.W. O'Malley and J.G. Hardman, Editors, Vol. XL (1975), Academic Press (New York), pp. 293–302. "The Design of Double Label Radioisotope Experiments."

H.J. Rajaniemi and A.R. Midgley, Jr., *ibid.*, pp. 145–167. "Autoradiographic Techniques for Localizing Protein Hormones in Target Tissue."

"Techniques for Isotope Studies" in *Methods in Enzymology*, S. Colowick and N.O. Kaplan, Editors, Vol. IV (1957), Academic Press (New York), pp. 425–905. Excellent introduction with many applications to the study of enzymes.

Chapter 7

Centrifugation

A centrifuge of some kind is found in nearly every biochemistry laboratory. Centrifuges have many uses, but they are used primarily for the preparation of biological samples and the analytical measurement of the hydrodynamic properties (shape, size, density, etc.) of purified macromolecules or cellular organelles. This is done by subjecting a biological sample to an intense force by spinning the sample at high speed. These experimental conditions cause sedimentation of particles, cell organelles, or macromolecules at a rate that depends upon their masses, sizes, and densities.

In this chapter we will explore the underlying principles of centrifugation and discuss the application of this technique to the isolation and characterization of biological molecules and cellular components.

A. BASIC PRINCIPLES OF CENTRIFUGATION

A particle, whether it is a precipitate, a macromolecule, or a cell organelle, is subjected to a centrifugal force when it is rotated at a high rate of speed. The **centrifugal force**, F, is defined by Equation 7.1.

$$F = m\omega^2 r \qquad \text{(Equation 7.1)}$$

where

F = intensity of the centrifugal force

m = effective mass of the sedimenting particle

ω = angular velocity of rotation in rad/sec

r = distance of the migrating particles from the central axis of rotation

The force on a sedimenting particle increases with the velocity of the rotation and the distance of the particle from the axis of rotation. A more common measurement of F, in terms of the earth's gravitation force, g, is **relative centrifugal force**, RCF, defined by Equation 7.2.

$$\text{RCF} = (1.119 \times 10^{-5})(\text{rpm})^2(r) \qquad \text{(Equation 7.2)}$$

This equation relates RCF to revolutions per minute of the sample. Equation 7.2 dictates that the RCF on a sample will vary with r, the distance of the sedimenting particles from the axis of rotation (see Figure 7.1). It is convenient to determine RCF by use of the nomogram in Figure 7.2. It should be clear from Figures 7.1 and 7.2 that since RCF varies with r, it is important to define r for an experimental run. Often an average RCF is determined using a value for r midway between the top and bottom of the sample container. The RCF value is reported as "a number times gravity, g." To illustrate this, consider a sample with an average r of 7 cm being rotated at 20,000 rpm. From Figure 7.2, RCF is 32,000 × g.

Although this simple introduction outlines the basic principles of centrifugation, it does not take into account other factors that influence the rate of particle sedimentation. Centrifuged particles migrate at a rate that depends upon the mass, shape, and density of the particle and the density of the medium. The centrifugal force felt by the particle is defined by Equation 7.1. The term m is the **effective mass** of the particle, that is, the actual mass, m_0, minus a correction factor for the weight of water displaced (buoyancy factor) (Equation 7.3).

FIGURE 7.1
A diagram illustrating the variation of RCF with r, the distance of the sedimenting particles from the axis of rotation. Courtesy of Beckman Instruments, Inc.

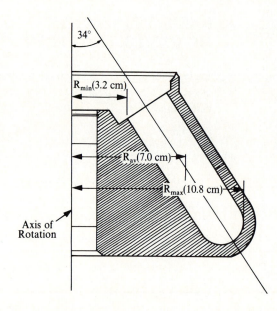

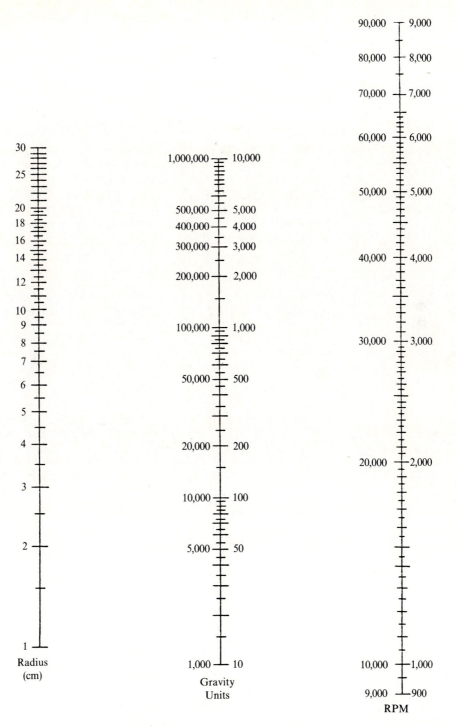

FIGURE 7.2
A nomogram for estimating RCF. To use, place a straight edge connecting the instrumental revolutions (right column) with the distance from the center of rotation (left column). The RCF is read where the straight edge crosses the middle column. Courtesy of Beckman Instruments, Inc.

193

$$m = m_0 - m_0 \bar{v} \rho \qquad \text{(Equation 7.3)}$$

where

m = effective mass of the sedimenting particle

m_0 = actual mass of the particle

$\bar{v}$ = partial specific volume, the volume change occurring when a particle is placed in a large excess of solvent

ρ = density of the solvent or medium

Equations 7.1 and 7.3 may be combined to describe the centrifugal force on a particle (Equation 7.4).

$$F = m_0(1 - \bar{v}\rho)\omega^2 r \qquad \text{(Equatior 7.4)}$$

As particles sediment under the influence of the centrifugal field, their movement is countered by a resistance force, the **frictional force**. The frictional force is defined by Equation 7.5.

$$\text{frictional force} = fv \qquad \text{(Equation 7.5)}$$

where

f = frictional coefficient

v = velocity of the sedimenting particle (sedimentation velocity)

The **frictional coefficient**, f, depends on the size and shape of the particle, as well as the viscosity of the solvent. The frictional force increases with the velocity of the particle until a constant velocity is reached. At this point, the two forces are balanced (Equation 7.6).

$$m_0(1 - \bar{v}\rho)\omega^2 r = fv = f\left(\frac{dr}{dt}\right) \qquad \text{(Equation 7.6)}$$

The rate of sedimentation, sometimes called **sedimentation velocity**, v, is defined by Equation 7.7.

$$v = \frac{dr}{dt} = \frac{m_0(1 - \bar{v}\rho)\omega^2 r}{f} \qquad \text{(Equation 7.7)}$$

It is cumbersome and sometimes impractical to express sedimentation velocity in terms of ρ, $\bar{v}$, and f, since these factors are difficult to measure. A new term, **sedimentation coefficient**, s (the ratio of sedimentation velocity to centrifugal force) is introduced by rearranging Equation 7.7 to Equation 7.8.

$$s = \frac{v}{\omega^2 r} = \frac{m_0(1 - \bar{v}\rho)}{f} \qquad \text{(Equation 7.8)}$$

The term s is usually defined under standard conditions, 20°C and water as the medium, denoted by s_{20w}. The s value is a physical characteristic commonly used to classify biological macromolecules, especially nucleic acids. Sedimentation coefficients for most macromolecules are in the range from 10^{-13} to 10^{-11} seconds. For convenience, these are expressed in Svedberg units, S, where $1\,S = 1 \times 10^{-13}$ seconds. The enzyme ribonuclease A has an s of 1.85×10^{-13} seconds, which would be reported as 1.85 S. The sedimentation coefficients for most proteins range from 1 S to 10 S; those for most nucleic acids are 30 S to 50 S, and those for subcellular organelles are 50 S to 100 S. The goal of many experiments using the centrifuge is the measurement of s. This value is significant because it can be related to the molecular weight of a macromolecule, as will be discussed later in this chapter. The units of the sedimentation coefficient (seconds) are not obvious. Dimensional analysis of Equation 7.8 provides the following units: v in cm/sec, ω in radians/sec, r in cm, m_0 in grams, $\bar{v}$ in cm^3/g, ρ in g/cm^3, and f in g/sec. Therefore, the unit of s is seconds.

B. INSTRUMENTATION FOR CENTRIFUGATION

The basic centrifuge consists of two components, an electric motor with drive shaft to spin the sample and a **rotor** to hold tubes or other containers of the sample. A wide variety of centrifuges is available, ranging from the simple design outlined above to sophisticated instruments that include accessories for making analytical measurements during centrifugation. Here we will describe three types, the benchtop or clinical centrifuge, the high speed centrifuge, and the ultracentrifuge.

Benchtop Centrifuges

Most laboratories have a standard low speed centrifuge used for routine sedimentation of relatively heavy particles. The common centrifuge has a maximum speed in the range of 4000 to 5000 rpm, with RCF values up to $3000 \times g$. These instruments usually operate at room temperature with no means for temperature control of the samples. A typical low speed centrifuge with rotor is pictured in Figure 7.3. Two types of rotors, **fixed-angle** and **swinging bucket**, may be used in the instrument. Centrifuge tubes or bottles that contain 12 or 50 ml of sample are commonly used. Low speed centrifuges are especially useful for the rapid sedimentation of coarse precipitates or red blood cells. The sample is centrifuged until the particles are tightly packed into a **pellet** at the bottom of the tube. The liquid portion, the **supernatant**, is then separated by decantation.

FIGURE 7.3
A typical low speed, benchtop centrifuge. Courtesy of Beckman Instruments, Inc.

High Speed Centrifuges

For more sensitive biochemical applications, higher speeds and temperature control of the rotor chamber are essential. Refrigerated, high speed centrifuges are capable of speeds up to 25,000 rpm. A typical high speed centrifuge is shown in Figure 7.4. The operator of this instrument can carefully control speed and temperature, which is especially important for carrying out reproducible centrifugations of temperature-sensitive biological samples. Rotor chambers in most instruments are maintained at or near 4°C. Both fixed-angle and swinging bucket rotors may be used; however, the former are more common. The two most useful rotors, shown in Figure 7.5, are (A) a rotor holding eight 50 ml tubes and (B) a rotor accommodating six 250 ml bottles. Rotor (A) is capable of generating an

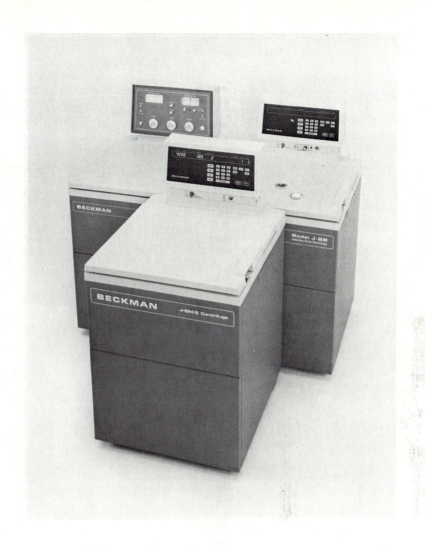

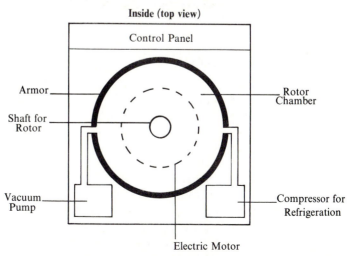

Inside (top view)

Control Panel

Armor

Rotor
Chamber

Shaft for
Rotor

Vacuum
Pump

Compressor for
Refrigeration

Electric Motor

FIGURE 7.4
Outside and inside views of a
typical high speed, refrigerated
centrifuge. Courtesy of Beckman
Instruments, Inc.

FIGURE 7.5
Rotors for a high speed centrifuge. (A) A fixed-angle rotor for holding eight 50 ml tubes. (B) A swinging bucket rotor. Courtesy of DuPont/Sorvall.

average RCF of approximately 30,000 × g, while rotor (B) exerts an RCF of approximately 20,000 × g. Centrifuge and rotor manufacturers usually supply a table of RCF values for various rotor speeds. Modern instruments are equipped with a brake to slow the rotor rapidly after centrifugation.

The preparation of biological samples will almost always require the use of a high speed centrifuge. Specific examples will be described later, but high speed centrifuges may be used to sediment (1) cell debris after cell homogenization, (2) ammonium sulfate precipitates of proteins, (3) microorganisms, and (4) cellular organelles such as chloroplasts, mitochondria, and nuclei.

Ultracentrifuges

The most sophisticated of the centrifuges are the **ultracentrifuges**. These instruments are capable of speeds up to 75,000 rpm with RCF values up to 500,000 × g. Because these high speeds will generate intense heat in the rotor, the chamber is refrigerated and placed under a high vacuum to reduce friction.

The sample in a cell or tube is placed in an aluminum or titanium rotor, which is then driven by an electric motor. Although it is relatively uncommon, these rotors, when placed under high stress, sometimes break into fragments. The rotor chamber on all ultracentrifuges is covered with protective steel armor plate. The drive shaft of the ultracentrifuge is constructed of a flexible material to accommodate any "wobble" of the rotor due to imbalance of the samples. It is still important to counterbalance samples as carefully as possible.

The two previously discussed centrifuges—the benchtop and the high speed—are of value only for preparative work, that is, for the isolation and separation of precipitates and biological samples. Ultracentrifuges can be used both for preparative work and for analytical measurements. Thus, two types of ultracentrifuges are available, **preparative models**, primarily used for separation and purification of samples for further analysis, and **analytical models**, which are designed for performing physical measurements on the sample during sedimentation. A typical preparative model is shown in Figure 7.6.

Analytical ultracentrifuges have the same basic design as preparative models except that they are equipped with optical systems to monitor directly the sedimentation of the sample during centrifugation (Figure 7.7). For analysis, the sample of 0.1 to 1.0 ml is placed in a special analytical rotor and rotated. Light is directed through the sample, parallel to the axis of rotation, and measurements of absorbance or transmittance of the sample are made. The optical system is capable of relating light measurements to the rate of movement of particles in the sample. The optical system actually detects and measures the **moving boundary** of the sedimenting particles. These measurements can lead to an analysis of concentration distributions within the centrifuge cell. Applications of these measurements will be discussed in the next section.

FIGURE 7.6
A preparative ultracentrifuge. Photo courtesy of Mr. Mark Billadeau.

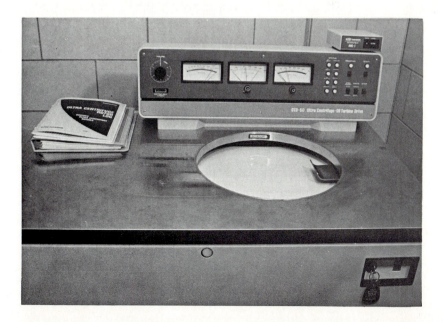

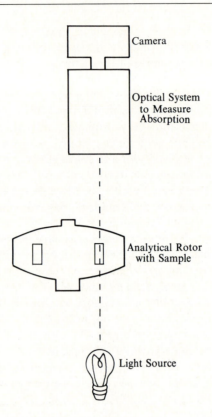

Camera

Optical System
to Measure
Absorption

Analytical Rotor
with Sample

Light Source

FIGURE 7.7
Optical system for monitoring sedimentation in an ultracentrifuge.

C. APPLICATIONS OF CENTRIFUGATION

Preparative Techniques

Centrifuges in undergraduate biochemistry laboratories are used most often for preparative scale separation procedures. This technique is quite straightforward, consisting of placing the sample in a tube or similar container, inserting the tube in the rotor, and spinning the sample for a fixed period. The sample is removed and the two phases, pellet and supernatant (which should be readily apparent in the tube), may be separated by careful decantation. Further characterization or analysis is usually carried out on the individual phases. This technique, called **velocity sedimentation centrifugation**, separates particles ranging in size from coarse precipitates to cellular ribosomes. Relatively heavy precipitates are sedimented in low speed, benchtop centrifuges, whereas lighter ribosomes require the high centrifugal forces of an ultracentrifuge.

Much of our current understanding of cell structure and function depends on separation of subcellular components by centrifugation. The specific method of separation, called **differential centrifugation**, consists of successive centrifugations at increasing rotor speeds. Figure 7.8 illustrates the differential centrifugation of a cell homogenate, leading to the separation and isolation of the common cell organelles. For most biochemical applications, the rotor chamber must be kept at low temperatures to maintain the native structure and function of each cellular organelle and its component biomolecules. A high speed centrifuge equipped with a fixed-angle rotor is most appropriate for the first two centrifugations at $600 \times g$ and $20,000 \times g$. After each centrifuge run, the supernatant is poured into another centrifuge tube, which is then rotated at the next higher speed. The final centrifugation at $100,000 \times g$ to sediment microsomes and ribosomes must be done in an ultracentrifuge. The $100,000 \times g$ supernatant, the **cytosol**, is the soluble portion of the cell and consists of soluble proteins and smaller molecules.

Analytical Measurements

A variety of analytical measurements can be made on biological samples during and after a centrifuge run. Most often, the measurements are taken to determine molecular weight, density, and purity of biological samples. All analytical techniques require the use of an ultracentrifuge and can be classified as **differential** or **density gradient**.

Differential Centrifugation

Differential methods involve sedimentation of particles in a medium of homogeneous density. Although the technique is similar to preparative differential centrifugation as previously discussed, the goal of an experiment is to measure the sedimentation coefficient of a particle. The underlying principles of this techniques are illustrated in Figure 7.9A. During centrifugation, a moving boundary is generated between pure solvent and sedimenting particles. An analytical ultracentrifuge is capable of detecting and measuring the rate of movement of the trailing boundary. Hence, the sedimentation velocity, v, can be experimentally determined. By using

FIGURE 7.8
Differential centrifugation of a cell homogenate. See text for description.

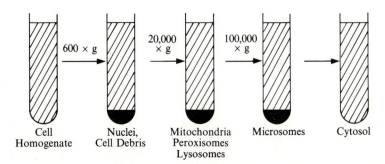

| Cell Homogenate | Nuclei, Cell Debris | Mitochondria Peroxisomes Lysosomes | Microsomes | Cytosol |

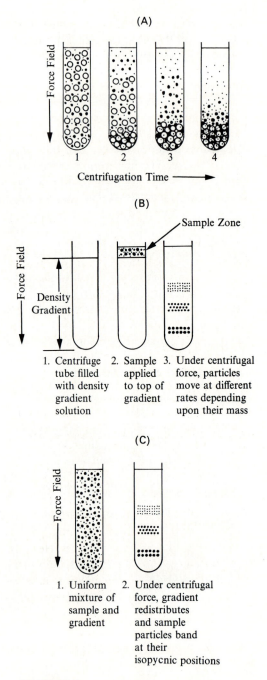

FIGURE 7.9
A comparison of differential and density gradient measurements. (A) Differential centri-fugation in a medium of unchanging density. (B) Zonal centrifugation in a prepared density gradient. (C) Isopycnic centrifugation; the density gradient forms during centri-fugation. Illustration courtesy of Beckman Instruments, Inc.

Equation 7.8 the sedimentation coefficient, s, can be calculated. The value of s for a sedimenting particle is related to the molecular weight of that particle by the Svedberg equation. The Svedberg equation is derived from Equation 7.8 by recognizing that the frictional force, f, may be defined by Equation 7.9.

$$f = \frac{RT}{ND}$$

(Equation 7.9)

where

R = the gas constant, 8.3×10^7 g cm²/sec/deg/mole

T = the absolute temperature

D = the diffusion coefficient for the solute in units of cm²/sec

N = Avogadro's number

Thus, Equation 7.8 may be transformed into Equation 7.10.

$$s = \frac{m_0(1 - \bar{v}\rho)}{RT/ND}$$

(Equation 7.10)

$$RTs = m_0ND(1 - \bar{v}\rho)$$

Since molecular weight, MW, is equal to m_0N, Equation 7.10 is converted to Equation 7.11, the Svedberg equation.

$$MW = \frac{RTs}{D(1 - \bar{v}\rho)}$$

(Equation 7.11)

This equation provides an accurate calculation of molecular weight and is applicable to macromolecules such as proteins and nucleic acids. However, its usefulness is limited because diffusion coefficients are difficult to measure and are not readily available in the literature.

An alternative method sometimes used to determine molecular weights of macromolecules is **sedimentation equilibrium**. In the previous example, using the Svedberg equation, the sample is rotated at a rate sufficient to sediment the particles. Here, the sample is rotated at a slower rate, and the particles sediment until they reach an equilibrium position at the point where the centrifugal force is equal to the frictional component opposing their movement (see Equation 7.6). The molecular weight is then calculated using Equation 7.12.

$$MW = \frac{RT}{(1 - \bar{v}\rho)\omega^2}\left(\frac{1}{rc}\right)\left(\frac{dc}{dr}\right)$$

(Equation 7.12)

where

MW, R, T, $\bar{v}$, ρ, ω and r have the same definitions previously given

$\dfrac{dc}{dr} =$ the change in particle concentration as a function of distance from the rotation center, as measured in the ultracentrifuge

This technique is time consuming, as it may require 1 to 2 days to reach equilibrium. Also, the use of Equation 7.12 is complicated by the difficulty of making concentration measurements in the ultracentrifuge.

Differential ultracentrifugation methods may also be applied to analysis of the purity of macromolecular samples. If one sharp moving boundary is observed in a rotating centrifuge cell, this indicates that the sample has one component, and therefore is pure. In an impure sample, each component would be expected to form a separate moving boundary upon sedimentation.

Density Gradient Centrifugation

In differential procedures, the sample is uniformly distributed in a cell before centrifugation, and the initial concentration of the sample is the same throughout the length of the centrifuge cell. Although useful analytical measurements can be made with this technique, it has disadvantages when applied to impure samples or samples with more than one component. Large particles that sediment faster pass through a medium consisting of solvent and particles of smaller size. Therefore, clearcut separations of macromolecules are seldom obtained. This can be avoided if the sample is centrifuged in a fluid medium that gradually increases in density from top to bottom. This technique, called **density gradient centrifugation**, permits the separation of multicomponent mixtures of macromolecules and the measurement of sedimentation coefficients. Two methods are used, **zonal centrifugation**, in which the sample is centrifuged in a preformed gradient, and **isopycnic centrifugation**, in which a self-generating gradient forms during centrifugation.

Zonal Centrifugation. Figure 7.9B outlines the procedure for zonal centrifugation of a mixture of macromolecules. A density gradient is prepared in a tube prior to centrifugation. This is accomplished with the use of an automatic gradient mixer. Solutions of low molecular weight solutes such as sucrose or glycerol are allowed to flow into the centrifuge cell. The sample under study is layered on top of the gradient and placed in a swinging bucket rotor. Sedimentation in an ultracentrifuge results in movement of the sample particles at a rate dependent upon their individual s values. As shown in Figure 7.9B, the various types of particles sediment as *zones* and remain separated from the other components. The centrifuge run is terminated before any particles reach the bottom of the gradient. The various zones in the centrifuged tubes are then isolated by collecting fractions from the bottom of the tube and analyzing them for the presence of macromolecules. The zones of separated macromolecules are relatively stable in the gradient because it slows diffusion and convection. The gradient conditions can be varied by using different ranges of sucrose concentration. Sucrose concentrations up to 60% can be used, with a density limit of 1.28 g/cm^3. However, solutions above 20% become sticky and viscous.

The zonal method can be applied to the separation and isolation of macromolecules (preparative ultracentrifuge) and to the determination of *s* (analytical ultracentrifuge).

Isopycnic Centrifugation. In the isopycnic technique, the density gradient is formed during the centrifugation. Figure 7.9C outlines the operation of isopycnic centrifugation. The sample under study is dissolved in a solution of a dense salt such as cesium chloride or cesium sulfate. The cesium salts may be used to establish gradients to an upper density limit of 1.8 g/cm^3. The solution of biological sample and cesium salt is uniformly distributed in a centrifuge tube and rotated in an ultracentrifuge. Under the influence of the centrifugal force, the cesium salt redistributes to form a continuously increasing density gradient from the top to the bottom. The macromolecules of the biological sample seek an area in the tube where the density is equal to their respective densities. That is, the macromolecules move to a region where the sum of the forces (centrifugal and frictional) is zero (Equation 7.6). The macromolecules will either sediment or float to this region of equal density. Stable zones or bands of the individual components will be formed in the gradient (see Figure 7.9C). These bands can be isolated as previously described. Up to two days of centrifugation may be required to completely establish a gradient of cesium salt.

Density gradients are widely used in the separation and purification of biological samples. In addition to this preparative application, measurements of *s* can also be made. Gradient techniques have been used to isolate and purify the subcellular components, microsomes, ribosomes, lysosomes, mitochondria, peroxisomes, chloroplasts, and others. After isolation, they have been biochemically characterized as to their protein, lipid, and enzyme contents.

Nucleic acids, in particular, have been extensively studied by density gradient techniques. Both RNA and DNA are routinely classified according to their *s* values. The different structural forms of DNA discussed in Chapter 4 can be determined by density gradient centrifugation.

Space does not allow an exhaustive review of centrifuge applications. Interested students should consult the references listed at the end of the chapter for recent developments.

Care of Centrifuges and Rotors

Centrifuge equipment can represent a sizable investment for a laboratory, so proper maintenance is essential. In addition, poorly maintained equipment is unsafe. Since many instruments are now available, specific instructions will not be given here, but general guidelines are outlined.

1. Carefully read the operating manual or receive proper instructions before you use any centrifuge.

2. Select the proper operating conditions on the instrument. If refrigeration is necessary, set the temperature to the appropriate level and allow 1 to 2 hours for temperature equilibration.

3. Check the rotor chamber for cleanliness and for damage. Clean with soap and warm water.

4. Select the proper rotor. Many sizes and types are available. Follow guidelines already stated in this chapter or consult your instructor.

5. Be sure the rotor is clean and undamaged. Observe any nicks, scratches, or other damage that may cause imbalance. If dirty, the rotor should be cleaned with warm water and a mild, nonbasic detergent. A soft brush can be used inside the cavities. Rinse well with distilled water and dry. Scratches should not be made on the surface coating, as corrosion may result.

6. Filled centrifuge tubes or bottles should be weighed carefully and balanced before centrifugation.

7. Rotor manufacturers provide a maximum allowable speed limit for each rotor. Do not exceed that limit.

8. Keep an accurate record of centrifuge and rotor use. Just as your automobile needs service after a certain number of miles, the centrifuge should be serviced after certain intervals of use. Centrifuge maintenance is usually determined by hours of use and total revolutions of the rotor. It is also essential to maintain a record of the use for each rotor. Rotors weaken with use, and the maximum allowable speed limit decreases. Rotor manufacturers usually provide guidelines for decreasing the allowable speed for a rotor.

9. If an unusual noise or vibration develops during centrifugation, immediately turn the centrifuge off.

10. Carefully clean the rotor chamber and rotor after centrifugation.

REFERENCES

General

K.C. Aune in *Methods in Enzymology*, C.H.W. Hirs and S.N. Timasheff, Editors, Vol. 48 (1978), Academic Press (New York), pp. 163–185. "Molecular Weight Measurements by Sedimentation Equilibrium: Some Common Pitfalls and How to Avoid Them."

R.C. Bohinski, *Modern Concepts in Biochemistry*, 4th Edition (1983), Allyn and Bacon (Boston), pp. 25–27; 172–173. Brief introduction to centrifugation.

J. Brewer, A. Pesce, and R. Ashworth, *Experimental Technique in Biochemistry* (1974), Prentice-Hall (Englewood Cliffs, NJ), pp. 161–215. Excellent theoretical and practical background to centrifugation.

C. Cantor and P. Schimmel, *Biophysical Chemistry*, Part II (1980), W.H. Freeman (San Francisco), pp. 591–641. Excellent but advanced treatment of theory and applications of ultracentrifugation.

T. Copper, *The Tools of Biochemistry* (1977), John Wiley (New York), pp. 309–354. Theory and applications of centrifugation.

D. Freifelder in *Methods in Enzymology*, C.H.W. Hirs and S.N. Timasheff, Editors, Vol. 27 (1973), Academic Press (New York), pp. 140–150. "Zonal Centrifugation."

D. Freifelder, *Physical Biochemistry*, 2nd Edition (1982), W.H. Freeman (San Francisco), pp. 362–454. Theoretical background with many applications, especially to the study of nucleic acids.

O.M. Griffith, *Techniques of Preparative, Zonal, and Continuous Flow Ultracentrifugation*, 2nd Edition (1976), Beckman Instruments, Inc. (Palo Alto, CA). Theory and practical aspects of ultracentrifugation.

J.D. Rawn, *Biochemistry* (1983), Harper and Row (New York), pp. 164–171. A brief introduction to theory and applications.

G. Zubay, *Biochemistry* (1983), Addison-Wesley (Reading, MA), pp. 57–59. A brief discussion of centrifuge applications.

Chapter 8

Statistical Analysis of Experimental Measurements

The purpose of each laboratory exercise in this book is to observe and measure characteristics of a biomolecule or a biological system. These measured characteristics may be the molecular weight of a protein, the pH of a buffer solution, the absorbance of a colored solution, the rate of an enzyme catalyzed reaction, or the radioactivity associated with a molecule. If you measure a single characteristic many times under identical conditions, a slightly different result will most likely be obtained each time. For example, if a radioactive sample is counted twice under identical experimental and instrumental conditions, the second measurement immediately following the first, the probability is very low that the numbers of counts will be identical. If the absorbance of a solution is determined several times at a specific wavelength, the value of each measurement will surely vary from the others. Which measurements, if any, are correct? Before this question can be answered, you must understand the source and treatment of variations in experimental measurements.

A. ANALYSIS OF EXPERIMENTAL DATA

An **error** in an experimental measurement is defined as a deviation of an observed value from the true value. There are two types of errors, **determinate** and **indeterminate**. Determinate errors are those that can be controlled by the experimenter and are associated with malfunctioning equipment, improperly designed experiments, and variations in experimental conditions. These are sometimes called human errors because they can be corrected or at least partially alleviated by careful design and performance of the experiment. Indeterminate errors are random or accidental and cannot be controlled by the experimenter. Specific examples of indeterminate errors are variations in radioactive counting and small differences in the successive measurements of glucose in a serum sample.

Two statistical terms involving error analysis that are often used and misused are **accuracy** and **precision**. Precision refers to the extent of agreement among repeated measurements of an experimental value. Accuracy is defined as the difference between the experimental value and the true value for the quantity. Since the true value is seldom known, accuracy is better defined as the difference between the experimental value and the accepted true value. Several experimental measurements may be precise (that is, in close agreement with each other) without being accurate.

If an infinite number of identical measurements could be made on a biosystem, this series of numerical values would constitute a **statistical population**. The average of all of these numbers would be the **true value** of the measurement. It is obviously not possible to achieve this in practice. The alternative is to obtain a relatively small **sample of data**, which is a subset of the infinite population data. The significance and precision of these data are then determined by statistical analysis.

This chapter will explore the mathematical basis for the statistical treatment of experimental data. Most measurements required for the completion of the experiments can be made in duplicate, triplicate, or even quadruplicate, but it would be impractical and probably a waste of time and materials to make numerous determinations of the same measurement. Rather, when you perform an experimental measurement in the laboratory, you will collect a small sample of data from the population of infinite values for that measurement. To illustrate, imagine that an infinite number of experimental measurements of the pH of a buffer solution are determined, written on slips of paper, and placed in a container. It is not feasible to calculate an average value of the pH from all of these numbers, but it is possible to draw five slips of paper, record these numbers, and calculate an average pH. By doing this, you have collected a sample of data. By proper statistical manipulation of this small sample, it is possible to determine whether it is representative of the total population and the amount of confidence you should have in these numbers.

The data analysis will be illustrated here primarily with the counting of radioactive materials, although it is not limited to such applications. Any replicate measurements made in the biochemistry laboratory can be analyzed by these methods. The objective of this chapter is to provide a practical knowledge of statistics; therefore, less emphasis will be placed on the derivation of statistical equations.

B. DETERMINATION OF THE MEAN, SAMPLE DEVIATION, AND STANDARD DEVIATION

Radioactive decay with emission of particles is a random process. It is impossible to predict with complete certainty when a radioactive event will occur. Therefore, a series of measurements made on a radioactive sample will result in a series of different count rates, but they will be centered around an **average** or **mean** value of counts per minute. Table 8.1 contains such a series of count rates obtained with a scintillation counter on a single radioactive sample. A similar table could be prepared for other biochemical measurements, including the rate of an enzyme catalyzed reaction or the protein concentration of a solution as determined by the Lowry or Bradford method. The arithmetic average or mean of the numbers is calculated by totaling all the experimental values observed for a sample (the counting rates, the velocity of the reaction, or protein concentration) and dividing the total by the number of times the measurement was made. The mean is defined by Equation 8.1.

$$\bar{x} = \frac{\sum\limits_{}^{n} x_i}{n}$$

(Equation 8.1)

$\bar{x}$ = arithmetic average or mean

x_i = the value for an individual measurement

n = the total number of experimental determinations

TABLE 8.1

The Observed Counts and Sample Deviation
from a Typical Radioactive Sample

Counts per Minute	Sample Deviation $x_i - \bar{x}$
1243	+21
1250	+28
1201	−21
1226	+4
1220	−2
1195	−27
1206	−16
1239	+17
1220	−2
1219	−3
Mean = 1222	

The mean counting rate for the data in Table 8.1 is 1222. If the same radioactive sample were again counted for a series of ten observations, that series of counts would most likely be different from those listed in the table, and a different mean would be obtained. If we were able to make an infinite number of counts on the radioactive sample, then a **true mean** could be calculated. The true mean would be the actual amount of radioactivity in the sample. Although it would be desirable, it is not possible to experimentally measure the true mean. Therefore, it is necessary to use the average of the counts as an approximation of the true mean, and to use statistical analysis to evaluate the precision of the measurements (that is, to assess the agreement among the repeated measurements).

Since it is not usually practical to observe and record a measurement many times as in Table 8.1, what is needed is a means to determine the reliability of an observed measurement. This may be stated in the form of a question. How close is the result to the true value? One approach to this analysis is to calculate the **sample deviation**, which is defined as the difference between the value for an observation and the mean value, $\bar{x}$ (Equation 8.2). The sample deviations are also listed for each count in Table 8.1.

$$\text{Sample deviation} = x_i - \bar{x} \qquad \text{(Equation 8.2)}$$

A more useful statistical term for error analysis is **standard deviation**, a measure of the spread of the observed values. Standard deviation, s, for a sample of data consisting of n observations may be estimated by Equation 8.3.

$$s = \sqrt{\frac{\sum(x_i - \bar{x})^2}{n - 1}} \qquad \text{(Equation 8.3)}$$

It is a useful indicator of the probable error of a measurement. Standard deviation is often transformed to **standard deviation of the mean** or **standard error**. This is defined by Equation 8.4, where n is the number of measurements.

$$s_m = \frac{s}{\sqrt{n}} \qquad \text{(Equation 8.4)}$$

It should be clear from this equation that as the number of experimental observations becomes larger, s_m becomes smaller, or the precision of a measurement is improved.

C. THE NORMAL DISTRIBUTION FUNCTION

Figure 8.1 shows the results from counting a radioactive sample several hundred times. The figure is obtained by plotting the frequency of occurrence of a given count against that count. Similar results could be obtained

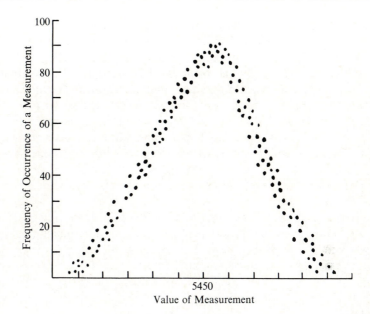

FIGURE 8.1

The distribution of observed counts from a radioactive sample.

by making replicate measurements of any quantitative characteristic of a material. To further illustrate the graphing procedure, the peak of the curve corresponds to the mean, 5450 cpm, and this number was observed 92 times.

The shape of the curve in Figure 8.1 is closely approximated by the **Gaussian distribution** or **normal distribution curve**. This mathematical treatment is based on the fact that a plot of relative frequency of a given event yields a dispersion of values centered about the mean, $\bar{x}$. The value of $\bar{x}$ is measured at the maximum height of the curve. The normal distribution curve shown in Figure 8.2 defines the spread or dispersion of the data.

FIGURE 8.2

The normal distribution curve.

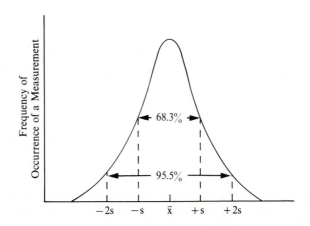

The probability that an observation will fall under the curve is unity or 100%. By using an equation derived by Gauss, it can be calculated that for a single set of sample data, 68.3% of the observed values will occur within the interval $\bar{x} \pm s$, 95.5% of the observed values within $\bar{x} \pm 2s$, and 99.7% of the observed values within $\bar{x} \pm 3s$. Stated in other terms, there is a 68.3% chance that a single observation will be in the interval $\bar{x} \pm s$.

For many experiments, a single measurement is made so a mean value, $\bar{x}$ is not known. In these cases, error is expressed in terms of s, but is defined as the **percentage proportional error**, %E_x in Equation 8.5.

$$\%E_x = \frac{100k}{\sqrt{n}} \qquad \text{(Equation 8.5)}$$

The parameter, k, is a proportionality constant between E_x and the standard deviation. The percent proportional error may be defined within several probability ranges. **Standard error** refers to a confidence level of 68.3%, that is, there is a 68.3% chance that a single measurement will not exceed the %E_x. For standard error, $k = 0.6745$. **Ninety-five hundredths error** means there is a 95% chance that a single measurement will not exceed the %E_x. The constant, k then becomes 1.45. As an example of the use of %E_x, consider an observation of 7640 counts per minute for a radioactive sample. Calculate the proportional error of this measurement at a percentage probability level of 95%.

$$E_x = \frac{100\ k}{\sqrt{n}} = \frac{(100)(1.45)}{\sqrt{7640}}$$
$$\%E_x = 1.65\%$$

At a confidence level of 95%, the percent proportional error is 1.65%. In words, there is a 1.65% chance that the observed count is outside the 95% confidence level or a 98.35% chance that the observed count lies inside the 95% confidence level.

D. CONFIDENCE LIMITS

The previous discussion of standard deviation and related statistical analysis placed emphasis on estimating the reliability or precision of experimentally observed values. However, standard deviation does not give specific information about how close an experimental mean is to the true mean. Statistical analysis may be used to estimate, within a given probability, a range within which the true value might fall. The range or confidence interval is defined by the experimental mean and the standard deviation. This simple statistical operation provides the means to determine quantitatively how close the experimentally determined mean is to the true mean.

Confidence limits (L_1 and L_2) are created for the sample mean as shown in Equation 8.6 and 8.7. The parameter t is calculated by

$$L_1 = \bar{x} + (t)(s_m) \qquad \text{(Equation 8.6)}$$

$$L_2 = \bar{x} - (t)(s_m) \qquad \text{(Equation 8.7)}$$

where

t = a statistical parameter that defines a distribution between a sample mean and a true mean

integrating the distribution between percent confidence limits. Values of t are tabulated for various confidence limits. Such a table is in Appendix VIII. Each column in the table refers to a desired confidence level (0.05 for 95%; 0.02 for 98% and 0.01 for 99% confidence). The table also includes the term **degrees of freedom** which is represented by $n - 1$, the number of experimental observations minus 1. The values of $\bar{x}$ and s_m are calculated as previously described in Equations 8.1 and 8.4.

E. STATISTICAL ANALYSIS IN PRACTICE

The equations for statistical analysis that have been introduced in this chapter are of little value if you have no understanding of their practical use, meaning, and limitations. A set of experimental data will first be presented, and then several statistical parameters will be calculated using the equations. This example will serve as a summary of the statistical formulas and will also illustrate their application.

Ten identical protein samples were analyzed by the Bradford method. The following values for protein concentration were obtained.

Observation Number	Protein Concentration (mg/ml), x
1	1.02
2	0.98
3	0.99
4	1.01
5	1.03
6	0.97
7	1.00
8	0.98
9	1.03
10	1.01

Sample Mean

$$\bar{x} = \frac{\sum x}{n} = \frac{10.02}{10} = 1.00 \text{ mg/ml}$$

Sample Deviation

Sample deviation $= x_i - \bar{x}$

Observation	$x_i - \bar{x}$
1	+0.02
2	−0.02
3	−0.01
4	+0.01
5	+0.03
6	−0.03
7	0.00
8	−0.02
9	+0.03
10	+0.01

Calculation of the sample deviation for each measurement gives an indication of the precision of the determinations.

Standard Deviation

$$s = \sqrt{\frac{\sum(x_i - \bar{x})^2}{n - 1}}$$

$$s = 0.02$$

The mean can now be expressed as $\bar{x} \pm s$ (for this specific example, 1.00 ± 0.02 mg/ml). The probability of a single measurement falling within these limits is 68.3%. For 95.5% confidence ($2s$), the limits would be 1.00 ± 0.04 mg/ml.

Standard Error of the Mean

$$s_m = \frac{s}{\sqrt{n}}$$

$$s_m = \frac{0.02}{\sqrt{10}}$$

$$s_m = 0.006$$

This statistical parameter can be used to gauge the precision of the experimental data.

Confidence Limits. The desired confidence limits will be set at the 95% confidence level. Therefore we will choose a value for t from Appendix VIII in the column labeled $t_{0.05}$ and $n - 1 = 9$.

$$L_1 = \bar{x} + (t_{0.05})(s_m)$$
$$L_2 = \bar{x} - (t_{0.05})(s_m)$$
$$s_m = 0.006$$

$$t_{0.05} = 2.262$$
$$\bar{x} = 1.00$$
$$L_1 = 1.00 + (2.262)(0.006)$$
$$L_1 = 1.01$$
$$L_2 = 0.99$$

We can be 95% confident that the true mean falls between 0.99 and 1.01 mg/ml.

REFERENCES

General

J. Brewer, A. Pesce, and R. Ashworth, *Experimental Techniques in Biochemistry* (1974), Prentice-Hall (Englewood Cliffs, NJ), pp. 10–31; 305–310. An excellent treatment of error analysis by Gaussian distribution. Also, an application of statistics to radioactive counting.

G. Chase and J. Rabinowitz, *Principles of Radioisotope Methodology*, 3rd Edition (1967), Burgess (Minneapolis), pp. 75–108. Application of statistics to radioactive counting analysis.

T. Copper, *The Tools of Biochemistry* (1977), John Wiley (New York), pp. 108–113. A brief discussion of the statistics of radioactive counting.

Chapter 9

Introduction to Molecular Cloning of DNA

Our knowledge of DNA structure and function has increased at a gradual pace over the past 40 years, but some discoveries have had a special impact on the progress and direction of DNA research. Although DNA was first discovered in cell nuclei in 1869, it was not confirmed as the carrier of genetic information until 1944. This major discovery was closely followed by the announcement of the double helix model of DNA in 1953. The more recent development of technology that allows manipulation by insertion of "foreign" DNA fragments into the natural, replicating DNA of an organism may well have a greater impact on the direction of DNA research than the previous two discoveries. In the short time since the first construction and replication of plasmid "recombinant DNA" (Cohen *et al.*, 1973; Morrow *et al.*, 1974; Hershfield *et al.*, 1974), several scientific and medical applications of the new technology have been announced. This new era of "genetic engineering" has captivated the general public and students alike. Some of the predicted achievements in recombinant DNA research have been slow in coming; however, future workers in biochemistry and related fields must be familiar with the principles and techniques of DNA recombination.

A. RECOMBINANT DNA

DNA recombination or molecular cloning consists of the covalent insertion of DNA fragments from one type of cell or organism into the replicating DNA of another type of cell. Many copies of the hybrid DNA

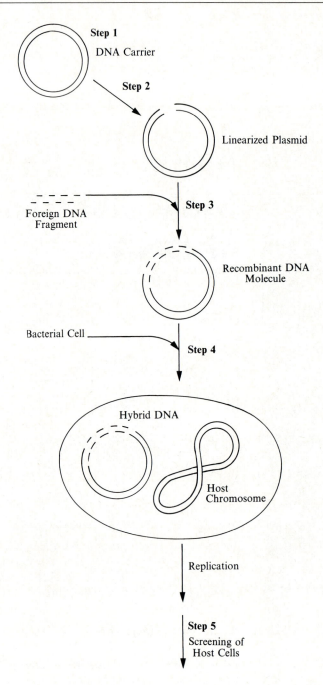

FIGURE 9.1
Outline of a typical procedure for
a DNA recombination experiment.

may be produced by the progeny of the recipient cells; hence, the DNA molecule is **cloned**. If the inserted fragment is a functional gene carrying the code for a specific protein, many copies of that gene and translated protein will be produced in the host cell. This process has become important for the large-scale production of proteins (insulin, somatostatin, and other hormones) that are of value in medicine and basic science but are difficult and expensive to obtain by other methods.

The basic steps involved in performing a recombinant DNA experiment are listed below and are outlined in Figure 9.1.

1. Selection and isolation of a natural DNA to serve as the carrier (sometimes called vehicle or vector) for the foreign DNA.
2. Cleavage of both strands of the DNA carrier.
3. Preparation and insertion of a foreign DNA fragment into the vehicle. This produces a **recombinant** or **hybrid** DNA molecule.
4. Introduction of the hybrid DNA into a host organism (usually a bacterial cell), where it can be replicated. This process is called **transformation**.
5. Development of a method for screening those host cells that have accepted and are replicating the hybrid DNA.

Our current state of knowledge regarding recombinant DNA is the result of several recent technological advances. The first major breakthrough was the isolation of mutant strains of *E. coli* that are not able to degrade or restrict foreign DNA. These strains are now used as host organisms for the replication of recombinant DNA. The second advance was the development of bacterial extrachromosomal DNA (plasmids) and bacteriophage DNA as cloning vehicles. The final, necessary advance was the development of methods for the insertion of the foreign DNA into the natural vehicle. The discovery of **restriction endonucleases**, enzymes that catalyze the hydrolytic cleavage of DNA at selective sites, provided a gentle and specific method for opening (linearizing) circular vehicles. Methods for covalent joining of the foreign DNA to the vehicle ends and final closure of the circular hybrid plasmid were then developed.

The following discussion reviews the current methods for making, analyzing, and using recombinant DNA.

B. CLONING ORGANISMS

The most important organisms for cloning hybrid DNA are various strains of *E. coli* K 12. The mutants of *E. coli* that are commonly used as host strains are HB101, RR1, GM48, SK1592, and χ^{1776}. Ideally, the host organism should not have restriction endonucleases capable of degrading foreign DNA. The use of yeast cells, higher plant cells, and mammalian cell cultures as host cells is still in its infancy, but advances in this area will be rapid in the future. Since *E. coli* cells are presently the most versatile host organisms, techniques for their use will be emphasized here.

C. CLONING VEHICLES

Three types of molecular vehicles are in common use for the preparation and replication of hybrid DNA in *E. coli* cells.

Plasmids. Many bacterial cells contain self-replicating, extrachromosomal DNA molecules called plasmids. This form of DNA is closed circular and double-stranded; its molecular weight ranges from 2×10^6 to 20×10^6, which corresponds to between 3000 and 30,000 base pairs. Bacterial plasmids normally contain genetic information for the translation of proteins that confer a specialized and sometimes protective characteristic (phenotype) to the organism. Examples of these characteristics are enzyme systems necessary for the production of antibiotics, enzymes that degrade antibiotics, and enzymes for the production of toxins.

Plasmids are replicated in the cell by one of two possible modes. **Stringent replicated** plasmids are present in only a few copies and **relaxed replicated** plasmids are present in many copies, sometimes up to 200. In addition, some relaxed plasmids continue to be produced even after the antibiotic chloramphenicol is used to inhibit chromosomal DNA synthesis in the host cell. Under these conditions, many copies of the plasmid DNA may be produced (up to 2000 or 3000), and may accumulate to 30 to 40% of the total cellular DNA.

The ideal plasmid cloning vehicle has the following properties.

1. The plasmid should replicate in a relaxed fashion so that many copies are produced.

2. The plasmid should be small; then it is easier to separate from the larger chromosomal DNA and easier to handle without physical damage, and it will probably contain very few sites for attack by restriction nucleases.

3. The plasmid should contain identifiable markers so that it is possible to screen progeny for the presence of the plasmid. At least two selective markers are desirable, a primary one to confirm the presence of the plasmid and a secondary marker to confirm the insertion of foreign DNA. Resistance to antibiotics is a convenient type of marker.

4. The plasmid should have only one cleavage site for a restriction endonuclease. This provides for only two "ends" to which the foreign DNA can be attached. Ideally, the single restriction site should be within a gene, so that the insertion of the foreign DNA will inactivate the gene (called insertional marker inactivation). (See below.)

Among the most widely used *E. coli* plasmids are derivatives of the replicon plasmid ColE1. This plasmid carries a resistance gene against the antibiotic colicin E. The plasmid is under relaxed control, and up to 3000 copies may be produced when the proper *E. coli* strain is grown in the presence of chloramphenicol (Clewell, 1972). One especially useful derivative plasmid of ColE1 is pBR322. It has all the properties previously outlined; in addition, its nucleotide sequence of about 4200 bases is known, and it contains six different restriction endonuclease cleavage sites where foreign DNA can be inserted.

Bacteriophage DNA. Two types of phage DNA have been widely used as cloning vehicles, λ phage and M13 phage DNA. λ DNA is a double-stranded molecule with approximately 50,000 base pairs. It has several advantages as a cloning vehicle:

1. Millions of copies of recombinant phage DNA can be replicated in the host cell.

2. The recombinant phage DNA may be efficiently packaged in the phage particle, which can be used to infect the host bacteria.

3. Recombinant phage DNA is easily screened for identification purposes.

4. λ phage DNA is larger than plasmid DNA; therefore, it is especially useful for insertion of larger fragments of eukaryotic DNA.

M13 phage DNA is single-stranded and contains approximately 6500 nucleotide bases. It is especially useful when DNA sequencing by the Sanger method is to be carried out on the recombinant DNA.

SV40 Virus. The cloning vehicle most often used to introduce recombinant DNA in eukaryotic cells is SV40 virus. Its capacity is rather small; it can accept foreign DNA fragments of only about 4300 base pairs. In contrast, plasmids can effectively be used to carry foreign DNA of up to 10,000 base pairs.

New cloning vehicles are continually being developed and studied. However, at the present time, plasmids are the most versatile. Methods for the production, isolation, purification, and analysis of plasmid DNA will be described in Experiments 17 through 20.

D. PREPARATION OF DNA FRAGMENTS FOR INSERTION INTO THE VEHICLE

The foreign, double-stranded DNA that is to be replicated may be prepared by one of several methods, including chemical synthesis, restriction endonuclease action on a larger fragment, reverse transcription of mRNA, and controlled mechanical shearing. For preparation of prokaryotic DNA, the most widely used method is cleavage of DNA into several fragments by restriction enzymes. The entire mixture of fragments may be inserted into the plasmid (shotgun method), or the fragments may be separated and purified by chromatography or electrophoretic techniques before insertion into the vehicle. If only a certain gene is to be inserted, it is essential to carefully prepare and purify the DNA fragment before joining it to the vehicle.

The preparation of eukaryotic, double-stranded DNA fragments for insertion is done using copies of mRNA. The process involves two steps:

1. Synthesis of one strand of the DNA by reverse transcriptase (RNA-dependent DNA polymerase) action on mRNA.

2. Synthesis of the complementary strand using DNA polymerase I.

E. JOINING THE DNA FRAGMENT TO THE VEHICLE

DNA fragments prepared for plasmid insertion by restriction endonuclease action will have either **cohesive ends** or **blunt ends**, as shown in Figure 9.2. The unpaired regions of cohesive ends, which may be up to five nucleotide bases in length, can be joined to a plasmid that also has been cleaved by the same restriction enzyme. This process is shown in Figure 9.3. However, this leads to an overlap region of only a few bases and may not yield a stable linkage. This can be remedied by using DNA ligase to catalyze the formation of phosphodiester linkages at the insertion sites. The reaction catalyzed by DNA ligase is shown in Figure 9.3.

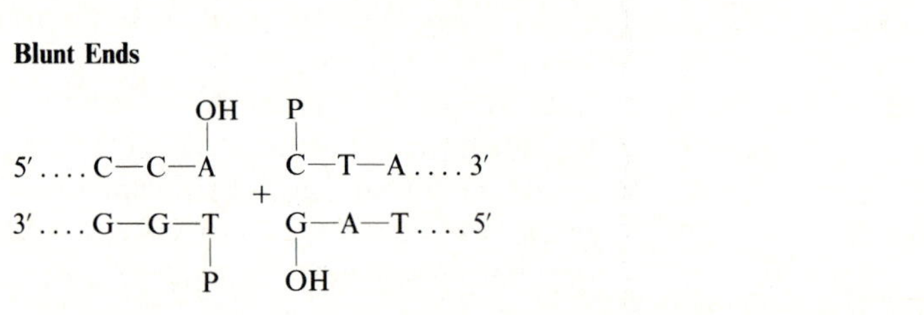

FIGURE 9.2
Result of the cleavage of DNA by a restriction endonuclease.

Cohesive Ends

$$
\begin{array}{l}
\quad\quad\quad\quad\quad \overset{\text{OH}}{\underset{|}{}} \quad\quad\quad\quad\quad\quad \overset{\text{P}}{\underset{|}{}} \\
5'\ldots C{-}G{-}T{-}C \quad\quad\quad\quad T{-}T{-}A{-}C{-}A{-}T{-}G\ldots 3' \\
\quad\quad\quad\quad\quad\quad\quad\quad\quad\quad\quad + \\
3'\ldots G{-}C{-}A{-}G{-}A{-}A{-}T{-}G \quad\quad\quad T{-}A{-}C\ldots 5' \\
\quad\quad\quad\quad\quad\quad\quad\quad \underset{\text{P}}{\underset{|}{}} \quad\quad\quad\quad\quad \underset{\text{OH}}{\underset{|}{}}
\end{array}
$$

Blunt Ends

$$
\begin{array}{l}
\quad\quad\quad\quad\quad \overset{\text{OH}}{\underset{|}{}}\quad \overset{\text{P}}{\underset{|}{}} \\
5'\ldots C{-}C{-}A \quad\quad C{-}T{-}A\ldots 3' \\
\quad\quad\quad\quad\quad\quad + \\
3'\ldots G{-}G{-}T \quad\quad G{-}A{-}T\ldots 5' \\
\quad\quad\quad\quad\quad \underset{\text{P}}{\underset{|}{}}\quad \underset{\text{OH}}{\underset{|}{}}
\end{array}
$$

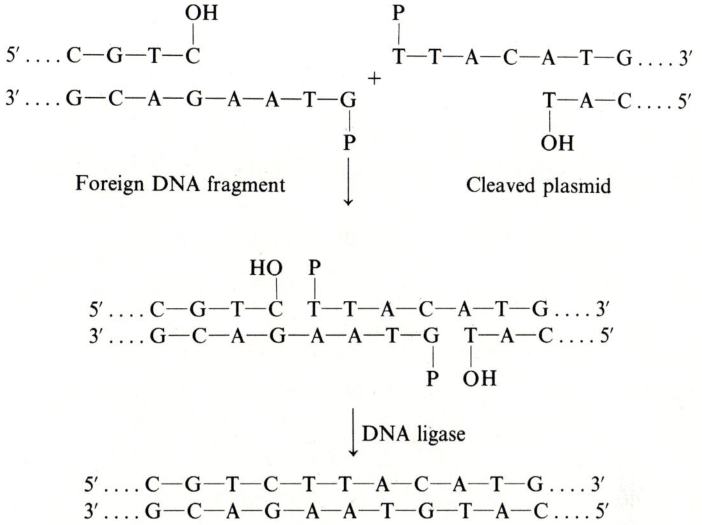

$$\begin{array}{ccc}
& \overset{\text{OH}}{\underset{|}{}} & \overset{\text{P}}{\underset{|}{}} \\
5'\ldots\text{C—G—T—C} & & \text{T—T—A—C—A—T—G}\ldots3' \\
3'\ldots\text{G—C—A—G—A—A—T—G} & + & \text{T—A—C}\ldots5' \\
& \overset{|}{\text{P}} & \overset{|}{\text{OH}}
\end{array}$$

Foreign DNA fragment ⟶ Cleaved plasmid

$$\begin{array}{cc}
& \overset{\text{HO}}{\underset{|}{}}\ \overset{\text{P}}{\underset{|}{}} \\
5'\ldots\text{C—G—T—C} & \text{T—T—A—C—A—T—G}\ldots3' \\
3'\ldots\text{G—C—A—G—A—A—T—G} & \text{T—A—C}\ldots5' \\
& \overset{|}{\text{P}}\ \overset{|}{\text{OH}}
\end{array}$$

⟶ DNA ligase

$$5'\ldots\text{C—G—T—C—T—T—A—C—A—T—G}\ldots3'$$
$$3'\ldots\text{G—C—A—G—A—A—T—G—T—A—C}\ldots5'$$

FIGURE 9.3
Insertion of a DNA fragment into a linearized plasmid using cohesive ends and DNA ligase.

Blunt ends are also joined by DNA ligase. These ends can be joined provided that the 5′ ends have phosphate groups and that a free hydroxy group is present at the 3′ end, as shown in Figure 9.3.

Joining cohesive or blunt ends is time-consuming and not especially versatile. A more widely used procedure for joining DNA fragments is the use of **homopolymer tails**, as shown in Figure 9.4. Here, a segment of 50 to 150 poly A (or G) residues is added to one fragment (usually the cleaved plasmid vector) and an equal length of poly T (or C) residues is added to the DNA fragment to be inserted. The tails are added by the action of deoxynucleotidyl transferase, an enzyme that catalyzes the addition of deoxynucleotides to the unblocked 3′ hydroxy ends of single- or double-stranded DNA. When the two products with homopolymer tails are incubated together, the insertion is completed by hydrogen bond formation between the complementary homopolymer tails. The hybrid DNA formed by the insertion then takes the form of circular, double-stranded DNA. Not all of the phosphodiester bonds are intact at the joints, but the long segment of complementary base pairing at the joints is sufficient to hold the insert intact. It has been shown that the optimal length of the dA · dT joint is about 100 base pairs.

The length of the homopolymer tail can be shortened if dC · dG homopolymer tails are used. Because of more favorable stacking interactions, only about 20 overlapping dG · dC pairs are necessary to achieve the same stability as 100 dA · dT pairs.

Finally, blunt-ended DNA fragments may be joined to vehicular DNA by using **synthetic linkers**. The ends of DNA fragments are elongated by the addition of short, synthetic DNA molecules (linkers), using DNA ligase.

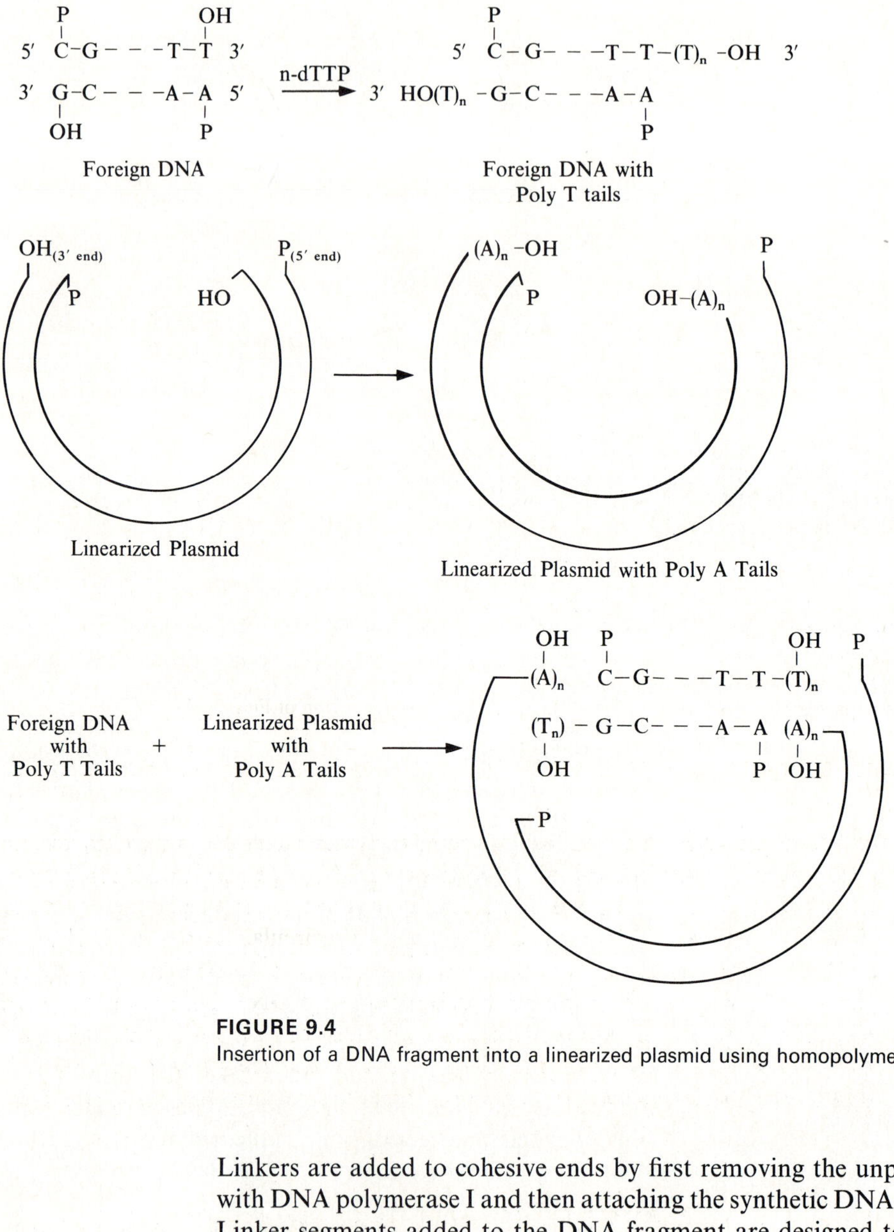

$$
\begin{array}{cc}
\text{P} & \text{OH} \\
| & | \\
5'\ \ \text{C–G–} - -\text{T–T}\ \ 3' \\
3'\ \ \text{G–C–} - -\text{A–A}\ \ 5' \\
| & | \\
\text{OH} & \text{P}
\end{array}
\qquad \xrightarrow{\ \text{n-dTTP}\ } \qquad
\begin{array}{cc}
\text{P} & \\
| & \\
5'\ \ \text{C–G–} - -\text{T–T–(T)}_n\ \text{–OH}\ \ \ 3' \\
3'\ \ \text{HO(T)}_n\ \text{–G–C–} - -\text{A–A} \\
& | \\
& \text{P}
\end{array}
$$

Foreign DNA   Foreign DNA with
                           Poly T tails

Linearized Plasmid   Linearized Plasmid with Poly A Tails

Foreign DNA   Linearized Plasmid
with   +   with
Poly T Tails   Poly A Tails

FIGURE 9.4
Insertion of a DNA fragment into a linearized plasmid using homopolymer tails.

Linkers are added to cohesive ends by first removing the unpaired bases with DNA polymerase I and then attaching the synthetic DNA with ligase. Linker segments added to the DNA fragment are designed to contain a restriction site. Action of the appropriate restriction endonuclease on the linker-modified DNA then leads to a newly formed cohesive end. The cohesive end may then be joined by ligase to the plasmid vector, which has been linearized by the same restriction enzyme. Although the use of synthetic linkers has great versatility as a joining method, it has disadvantages. One problem is that when the cohesive ends are produced by

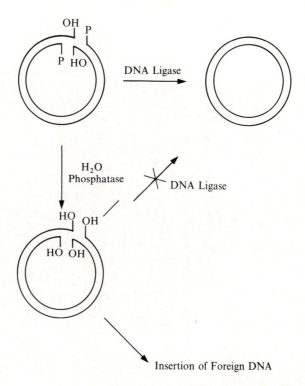

FIGURE 9.5
Removal of phosphate groups to minimize recyclization of linearized plasmid prior to insertion of foreign DNA fragment.

the action of a nuclease on the DNA fragment-linker molecule, cleavage sites for the nuclease may also be present in the DNA fragment.

The major difficulty encountered in the linking process is the tendency of the linearized plasmid to recircularize during the ligase reaction without insertion of the passenger DNA. (See Figure 9.5.) This leads to a lowered concentration of hybrid DNA and, therefore, a low efficiency of transformation. This problem can be minimized by removal of the 5′ terminal phosphates of the linearized plasmid. Treatment of the linear plasmid with a phosphatase will remove the terminal phosphates, as shown in Figure 9.5.

F. INTRODUCTION OF RECOMBINANT DNA INTO THE HOST CELLS

Now that the hybrid DNA has been selected and prepared, it must be introduced into a host cell where it can be replicated. Current methods for placement of the recombinant DNA into the host cell are still rather inefficient, with only one DNA molecule in 10,000 being successfully replicated. This low efficiency, however, leads to a sufficient quantity of hybrid DNA and protein products for analytical purposes.

The cloning of plasmid vectors is accomplished by the **transformation** of *E. coli* cells. It has been demonstrated that the introduction of hybrid plasmid and phage DNA into *E. coli* cells is promoted by calcium chloride (Mandel and Higa, 1970; Cohen, Chang, and Hsu, 1973). The general technique for bacterial transformation consists of washing the recipient *E. coli* cells with calcium chloride solution and incubating the washed cells with a solution of recombinant DNA in the presence of calcium chloride.

The suspension of cells and plasmid DNA is first chilled to ice temperatures and then incubated at 37°C for plasmid establishment. Upon introduction into the bacterial cell, the hybrid DNA is replicated and the selective markers are expressed. If the markers express drug resistance, the successfully transformed cells have the ability to grow in the presence of a particular antibiotic.

Bacteriophage DNA vehicles can be encapsulated or packaged into infectious viral particles. Host bacterial cells are grown to a midlog phase and infected with the virus containing the recombinant DNA. The viral particles enter the host cell, where the recombinant DNA is replicated with expression of the selective markers. Using current methods, the transfer efficiency for bacteriophage DNA is slightly better than for plasmid DNA.

Since the efficiency of successful hybrid DNA transplant and replication is so low, it is essential to be able to select and isolate those cells replicating the hybrid DNA. A method for selection will be illustrated using the plasmid vehicle pBR322 (Figure 9.6). Plasmid pBR322 contains two identifiable markers; both are antibiotic resistance genes. The location of these resistance genes is shown in Figure 9.6. The section of the plasmid labeled Amp^r confers resistance to ampicillin or penicillin, and that labeled Tet^r gives resistance to tetracycline. Assume that a DNA fragment, obtained by the restriction enzyme BamHI, is inserted into the BamHI restriction site of the plasmid. Cohesive ends are produced on the fragment and plasmid DNA. The final circularization is completed by DNA ligase. Since the insertion process is not highly efficient, the product will be a mixture of recircularized pBR322 with no inserted DNA and hybrid pBR322 containing the foreign DNA. When *E. coli* cells are incubated with the vehicle mixture in the presence of calcium chloride, three types of host cells are obtained: (1) *E. coli* cells containing neither pBR322 nor the hybrid pBR322, (2) cells containing pBR322, and (3) transformed cells containing hybrid plasmid.

The three types of cells will show different drug resistances. The normal *E. coli* cells will not survive in the presence of ampicillin, the cells containing unmodified pBR322 will survive in the presence of ampicillin *and* tetracycline, and the transformed cells will survive in the presence of

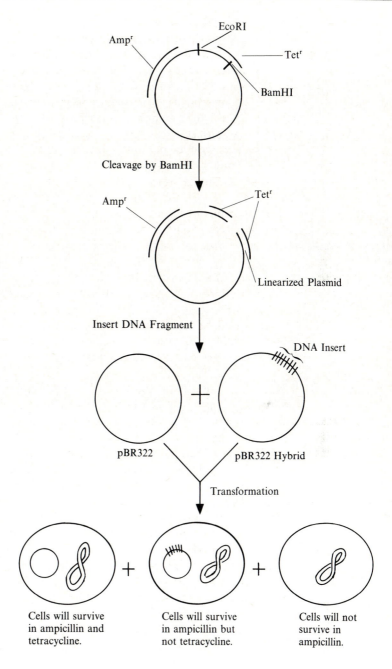

FIGURE 9.6
Procedure for selecting bacterial cells that contain hybrid DNA.

ampicillin but not tetracycline. Since the passenger DNA is inserted within the tetracycline-resistance gene, the hybrid cells are not able to produce proteins that protect them against this antibiotic. The cells are screened for antibiotic resistance using a technique called **replica plating** (Figure 9.7). Cells are streaked and allowed to grow on agar plates containing ampicillin, where only host cells containing pBR322 and hybrid pBR322 can survive. Some of the viable colonies are transferred to agar plates containing ampicillin or tetracycline. The transfer must be to identical locations on the two plates. These two plates are incubated for several hours at 37°C and searched for colonies that are growing on the ampicillin

FIGURE 9.7
Example of replica plating to screen bacterial cells for antibiotic resistance.

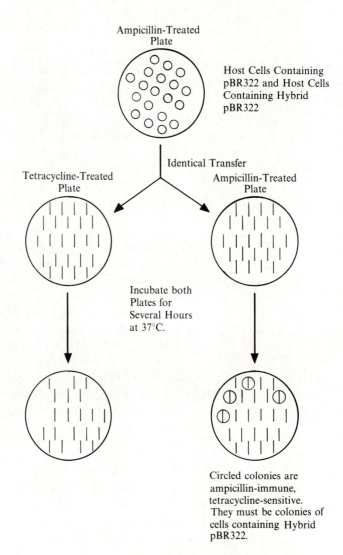

Ampicillin-Treated Plate

Host Cells Containing pBR322 and Host Cells Containing Hybrid pBR322

Identical Transfer

Tetracycline-Treated Plate

Ampicillin-Treated Plate

Incubate both Plates for Several Hours at 37°C.

Circled colonies are ampicillin-immune, tetracycline-sensitive. They must be colonies of cells containing Hybrid pBR322.

plate, but not on the tetracycline plate. Those colonies are most likely made up of cells containing the hybrid pBR322 plasmid. The ampicillin-resistant, tetracycline-sensitive *E. coli* colonies are removed from the plate and grown in liquid culture to replicate many copies of the hybrid plasmid. The plasmid can be amplified by adding chloramphenicol to the culture (Experiment 17). The plasmid DNA is then isolated for further analysis and characterization (Experiments 18–20).

Occasionally it is necessary to remove the inserted foreign DNA from the hybrid plasmid. This may be desirable in order to determine whether the passenger DNA underwent base modification and/or deletion during replication. In the example described here with hybrid pBR322, the inserted DNA may be removed by treatment with the same restriction nuclease (BamHI) used to cleave the original plasmid and foreign DNA.

G. ANALYSIS AND CHARACTERIZATION OF RECOMBINANT DNA

The analysis of recombinant DNA has, in the past, been a time-consuming and tedious process. Several methods are now available for the rapid isolation and analysis of small amounts of recombinant DNA. In a typical procedure, 1.5 ml of a culture of transformed *E. coli* cells is sufficient for plasmid analysis by agarose gel electrophoresis after digestion by a restriction enzyme (Birnboim and Doly, 1979; Holmes and Quigley, 1981).

One ultimate goal of plasmid isolation and analysis is to construct a **restriction enzyme map** for the plasmid. Such a map, shown in Figure 9.8, displays the sites of cleavage by several restriction endonucleases and the number of fragments obtained after digestion with each enzyme. At the center of the circle is shown the position of the two resistance genes, Amp^r and Tet^r. The innermost circle shows the site of attack by restriction enzymes that cleave at one location. The action of each of these on pBR322 will linearize the plasmid for insertion of a DNA fragment. A zero point is defined by the site cleaved by EcoR1. Each concentric circle represents the action of an individual restriction enzyme that produces several breaks in the plasmid. For example, the restriction enzyme TaqI acts at seven sites, producing fragments labeled A, B, C, D, E_1, E_2 and F. The map is of value in the selection of plasmids for cloning and for characterization of recombinant DNA molecules by fragment analysis.

A map is constructed by first digesting plasmid DNA with restriction endonucleases that yield only a few fragments. Each digest of DNA obtained with a single nuclease is analyzed by agarose gel electrophoresis, using standards for molecular weight determination. These are referred to as the primary digests. Second, each of the primary digests is treated with a series of additional restriction enzymes, and the digests are analyzed

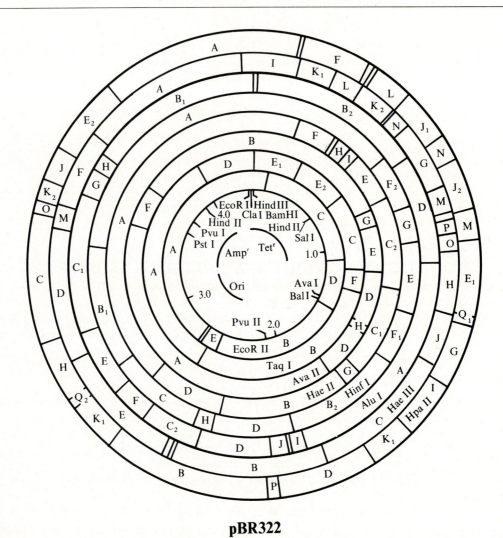

pBR322

FIGURE 9.8
A restriction enzyme map for the plasmid pBR322. From *Molecular Cloning, A Laboratory Manual*, by T. Maniatis, E. Fritsch, and J. Sambrook. Cold Spring Harbor Laboratory, 1982.

by agarose gel electrophoresis. The restriction map is then constructed by combining the various fragments by trial and error, and logic, much like the sequencing of a protein by proteolytic digestion by several enzymes and searching for overlap regions in the fragments. In the case of the restriction endonuclease digests, two characteristics are known for the DNA fragments, the approximate molecular weights (from electrophoresis) and the nature of the fragment ends (determined by the known selectivity of the individual restriction enzymes). The fragments may also be sequenced by the Sanger or Maxam-Gilbert methods.

It is probably not feasible for students in their first biochemistry laboratory experience to complete a typical DNA recombinant research problem. Such an undertaking could easily require one semester or more and may also demand several hours of work each day. However, many of the principles and techniques can be introduced in the undergraduate laboratory and applied to representative problems in modern nucleic acid biochemistry. Experiment 17 describes the growth of *E. coli* cells and amplification of ColE1 plasmids with chloramphenicol. Experiment 18 outlines the procedure for the isolation of plasmid DNA from the cell culture and purification by ultracentrifugation on a cesium chloride gradient. Experiment 20 describes the specific cleavage of ColE1 plasmids or other DNA molecules by a restriction enzyme and analysis of the fragments by agarose gel electrophoresis. If a single experiment in molecular cloning is desired, Experiment 21 may be completed. This experiment describes the rapid, small-scale isolation of plasmids from a cell culture or agar plate, cleavage of the plasmids by a restriction enzyme, and analysis of the fragments by electrophoresis.

REFERENCES

General

R.C. Bohinski, *Modern Concepts in Biochemistry*, Fourth Edition (1983), Allyn and Bacon (Boston), pp. 192–196. An introduction to recombinant DNA.

F. Bolivar and K. Backman in *Methods in Enzymology*, R. Wu, Ed., Vol. 68, (1979), Academic Press (New York), pp. 245–267. "Plasmids of *Escherichia coli* as Cloning Vectors."

M. Kahn, R. Kolter, C. Thomas, D. Figurski, R. Meyer, E. Remaut, and D. Helinski, *ibid.*, pp. 268–280. "Plasmid Cloning Vehicles Derived from Plasmids ColE1, F. R6K and RK2."

T. Maniatis, E. Fritsch, and J. Sambrook, *Molecular Cloning, A Laboratory Manual* (1982), Cold Spring Harbor Laboratory (Cold Spring Harbor, NY). This manual is highly recommended for all laboratories involved in recombinant DNA research. It contains details for the most effective procedures in all stages of molecular cloning.

J.D. Rawn, *Biochemistry* (1983), Harper and Row (New York), pp. 1018–1022. A brief introduction to genetic recombination.

R. Sinsheimer, *Ann. Rev. Biochem.*, *46*, 415–438 (1977). "Recombinant DNA."

E. Smith, R. Hill, I. Lehman, R. Lefkowitz, P. Handler, and H. White, *Principles of Biochemistry: General Aspects*, Seventh Edition (1983), McGraw-Hill (New York), pp. 724–725. A brief introduction to molecular genetics.

L. Stryer, *Biochemistry*, Second Edition (1981), Freeman (San Francisco), pp. 751–769. A complete chapter on DNA recombination.

W.B. Wood, J.H. Wilson, R.M. Benbow, and L.E. Hood, *Biochemistry, A Problems Approach*, Second Edition (1981). Benjamin/Cummings (Menlo Park, CA), pp. 371–402. A complete chapter with problems on recombinant DNA and genetic engineering.

G. Zubay, *Biochemistry* (1983), Addison-Wesley (Reading, MA), pp. 771–783. An introduction to the manipulation of DNA.

The entire volumes of the following references are dedicated to recombinant DNA.

Methods in Enzymology, Vol. 68, 100, 101.

Science, Vol. 209, Sept. 19, 1980.

Specific

H. Birnboim and J. Doly, *Nucleic Acids Res.*, 7, 1513 (1979). "A Rapid Alkaline Extraction Procedure for Screening Recombinant Plasmid DNA."

D. Clewell, *J. Bacteriol.*, 110, 667–676 (1972). "Nature of ColE1 Plasmid Replication in *Escherichia coli* in the Presence of Chloramphenicol."

S. Cohen, A. Chang, H. Boyer, and R. Helling, *Proc. Natl. Acad. Sci.*, 70, 3240 (1973). "Construction of Biologically Functional Bacterial Plasmids *in vitro*."

S. Cohen, A. Chang, and L. Hsu, *Proc. Natl. Acad. Sci.*, 69, 2110 (1973). "Nonchromosomal Antibiotic Resistance in Bacteria: Genetic Transformation of *Escherichia coli* by R. factor DNA."

V. Hershfield, H. Boyer, C. Yanofsky, M. Lovett, and D. Helinski, *Proc. Natl. Acad. Sci.*, 71, 3455 (1974). "Plasmid ColE1 as a Molecular Vehicle for Cloning and Amplification of DNA."

D. Holmes and M. Quigley, *Anal. Biochem.*, 114, 193 (1981). "A Rapid Boiling Method for the Preparation of Bacterial Plasmids."

M. Mandel and A. Higa, *J. Mol. Biol.*, 53, 154 (1970). "Calcium Dependent Bacteriophage DNA Infection."

J. Morrow, S. Cohen, A. Chang, H. Boyer, H. Goodman, and R. Helling, *Proc. Natl. Acad. Sci.*, 71, 1743 (1974). "Replication and Transcription of Eukaryotic DNA in *Escherichia coli*."

II

Experiments

Experiment **1**

Using the Biochemical Literature

RECOMMENDED READING:
Chapter 1, Section F

SYNOPSIS
The objective of this exercise is to introduce students to the biochemistry books and journals found in most libraries. Several problems typically encountered in biochemical research are posed, and students are asked to solve the problems or answer questions.

I. INTRODUCTION

The first step toward the solution of a problem in the biochemistry laboratory is often a visit to the library. Laboratory students should learn the value of the library early in the term. During almost every experiment, you will need information that can be found only in the library. You may require the pK for an amino acid, the molecular weight of a protein, or methods for the analysis of a biomolecule. Therefore, you should have some introductory experience in the library so that the use of available resources is familiar.

II. PROCEDURE—LIBRARY PROJECTS

Below are several problems or questions that might be encountered by research workers in biochemistry. Different types of references will be necessary to solve each of these problems. The references that will be of most value are research journals, biochemistry textbooks, biochemistry dictionaries, biochemistry handbooks, *Chemical Abstracts*, and methods books.

The exercise will begin with some examples with hints to help begin the library search.

1. Briefly define the medical-biochemical term *alkaptonuria*.

Solution: For basic definitions, a biochemistry dictionary or your biochemistry textbook are your best sources. For example, alkaptonuria is defined in *Dictionary of Biochemistry* (1975) by J. Stenesh, page 12: "A genetically inherited metabolic defect in man that is characterized by the urinary excretion of black melanin pigments formed from homogentisic acid; the defect is due to a deficiency of the enzyme homogentisic acid oxidase, which functions in the metabolism of phenylalanine and tyrosine."

2. What is the molecular weight of the protein insulin?

Solution: Physical constants such as molecular weights, pK values, $[\alpha]_D$, and extinction coefficients are found in textbooks or handbooks such as *Handbook of Biochemistry and Molecular Biology*. From *Biochemistry* by L. Stryer, the molecular weight of insulin is 5.8 kilodaltons or 5800.

3. List one journal article published in the past two years on the topic of cytochrome P-450.

Solution: When journal references are required, *Chemical Abstracts* should be the starting point. Since cytochrome P-450 is a chemical substance, the *Chemical Abstracts Chemical Substance Index* is used. Cytochrome P-450 will be listed in alphabetical order; for example, in Volume 100, 1984, the listing occurs on pages 2388CS–2391CS. Approximately seven columns of abstracts are listed. The first listing gives the abstract number R 2508d. The R indicates that this reference is a review article. The 2508d is the actual abstract number. Find this number in the Abstracts, volume 100, Number 1, p. 218. This abstract lists the title of the article ("Cytochrome P-450: a multifaceted catalyst"), the author (Coon, Minor J.), and the address of the author. It is published in *Transactions of the New York Academy of Sciences*, vol. 41, 1983, pages 41–48. The abstract also contains a brief description of the article.

To illustrate another use of *Chemical Abstracts*, find the author's name in the author index (Volume 100): Under the listing "Coon, Minor J." is the same abstract, R 2508d.

4. Briefly define each of the following biochemical terms. Give the reference used for each in the following order: Name of book, author, publisher, and page number.

(a) serotonin
(b) adipose tissue
(c) codons
(d) wobble hypothesis
(e) galactosemia
(f) sedimentation coefficient
(g) chemiosmotic coupling hypothesis
(h) immunoelectrophoresis
(i) restriction endonucleases
(j) isotachaphoresis

5. What is the molecular weight of each of the following substances?
   (a) maltose
   (b) ferritin
   (c) chymotrypsin, bovine pancreas
   (d) phenylalanine

6. What are the pK_a values for phosphoric acid, aspartic acid, and histidine?

7. What is the isoelectric pH (pH_I) of human ceruloplasmin?

8. What is the molar extinction coefficient (ε) at 280 nm for rabbit muscle aldolase?

9. What are the melting temperature and the % GC content of DNA from the λ virus?

10. What is the specific rotation ($[\alpha]_D$) for the carbohydrate D-ribose?

11. List five journal articles published in the past two years that are related to the restriction endonuclease EcoRI. List by title of article, author(s), journal name, volume, pages, and year.

12. Professor William N. Lipscomb of Harvard wrote a review article, published in 1983, on the structure and catalysis of enzymes. List the book title, article title, volume, and page numbers for the reference.

13. Find a reference that gives complete details for the purification of the calcium binding protein, calmodulin, from animal tissue.

14. Find a reference that gives complete details for the synthesis of sodium pyruvate. Hint: *Biochemical Preparations* or *Chemical Abstracts*.

15. List a reference that gives complete details for performance of SDS-polyacrylamide gel electrophoresis. Hint: *Methods in Enzymology* or other methods book.

16. Choose a research journal article related to biochemistry published within the last month and write a brief summary (5 or 6 sentences). Emphasize the analytical techniques used in the article.

Experiment **2**

Analytical Methods for Amino Acid Separation and Identification

RECOMMENDED READING:
Chapter 2, Section A; Chapter 3, Sections A, B, E; Chapter 4, Sections A, B

SYNOPSIS
Separation and identification of amino acids are operations that must be performed frequently by biochemists. The 20 amino acids present in proteins have similar structures but are unique in polarity and ionic characteristics. In this experiment, the student will use a combination of ion-exchange chromatography and paper chromatography to separate and identify the components of an unknown amino acid mixture.

I. INTRODUCTION AND THEORY

Twenty common amino acids are the fundamental building blocks of proteins. All proteins found in nature are constructed by amide bond linkages between α-amino acids. The amino acids isolated from protein material all have common structural characteristics. They possess at least one carboxyl group and at least one amino group (Figure E2.1). The distinctive physical, chemical, and biological properties associated with an amino acid are the result of the R group, which is unique for each amino acid.

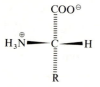

FIGURE E2.1
The general structure of the zwitterionic form of an amino acid. R represents the side chain.

The structure and biological function of a protein depend upon its amino acid content. It is a matter of basic importance to understand practical methods used for the separation and identification of the twenty common amino acids.

Acid-Base Chemistry of the Amino Acids

When proteins are heated in aqueous acid or base, the amide bonds are hydrolyzed, releasing the constituent amino acids in a free form. Separating and identifying the amino acids present in this complex mixture, which may contain up to twenty different molecules, requires a thorough knowledge of the physical and chemical properties of amino acid molecules. Recall from your biochemistry lectures that amino acids are amphoteric; that is, they can act as Brønsted acids (proton donating) or Brønsted bases (proton accepting) (Reactions 1 and 2). The zwitterionic form (II) is produced by proton dissociation of I (Reaction 1) and it can also participate in proton dissociation (Reaction 2), which forms anion III.

$$\overset{\oplus}{H_3}NCH(R)COOH \rightleftharpoons \overset{\oplus}{H_3}NCH(R)COO^\ominus + H^\oplus \qquad \text{(Reaction E2.1)}$$
$$\qquad\quad I \qquad\qquad\qquad\qquad\qquad II$$

$$\overset{\oplus}{H_3}NCH(R)COO^\ominus \rightleftharpoons H_2NCH(R)COO^\ominus + H^\oplus \qquad \text{(Reaction E2.2)}$$
$$\qquad\quad II \qquad\qquad\qquad\qquad\quad III$$

Each ionic reaction is defined by an acid ionization constant, K_a, and a pK_a ($-\log K_a$). All of the α-carboxyl groups of the 20 amino acids have similar but different pK_a values ($pK_a \approx 2$ to 3), as is the case also for the α-amino groups ($pK_a \approx 9$ to 10). (See Table III in the Appendix.) The pK_a value of an amino acid represents the pH at which two ionic forms are present in equal concentrations. Some amino acids also have ionizable groups on the R side chain. The extra acidic or basic groups, of course, add to the complexity of acid-base reactions of amino acids; but, as we shall see, this allows for greater resolution in analyzing amino acid mixtures. If you are not completely familiar with the acid-base properties of amino acids, refer to your biochemistry textbook.

Separation of Amino Acid Mixtures

Let us consider a typical problem that one often encounters in protein chemistry. When a protein is isolated in purified form, it is first characterized as to its amino acid composition. The protein is hydrolyzed, usually under acidic conditions, and the resulting mixture of free amino acids is subjected to qualitative and quantitative analysis. This determines what amino acids are present and how many molecules of each amino acid are present in the protein.

At first glance, this seems to be an insurmountable problem. The mixture of amino acids may contain up to 20 different, but similar, molecular structures. How does one resolve such a mixture of similar molecules? Several techniques ranging from chemical methods to chromatographic methods have been developed. Chapters 3, 4, and 7 describe three general methods useful for separation of biomolecules: chromatography, electrophoresis, and centrifugation. Centrifugation, which is used to separate molecules on the basis of size and specific gravity, would not be helpful since amino acids show little difference in these properties. Analysis of mixtures by electrophoresis is primarily dependent upon the sign and size of the charge on the individual molecules. Since amino acids are charged molecules at most pH values, this technique should have promise. Electrophoresis of the 20 amino acids at pH 1.0 might lead to separation into only two groups; one group of three amino acids nearer the cathode (lys, his, arg) and one group of seventeen amino acids between the origin and the group of three. In theory, this would not lead to satisfactory separation of the amino acids. In practice, however, electrophoresis is a valuable technique because, in addition to charge, molecules are also separated on the basis of polarity. This method is most appropriate when a solution of only a few amino acids or peptides is to be analyzed. In addition, two-dimensional electrophoresis and use of more effective stationary media are improvements that make electrophoresis a viable technique in amino acid and protein analysis.

We should also examine the third option mentioned above—chromatography. Does it have any potential in separating amino acid mixtures? Acknowledging that the major differences among the amino acid structures are polarity and charge, suitable chromatographic methods to exploit these differences are paper, thin-layer, ion-exchange, gas, and high pressure liquid chromatography. Differences in molecular weight of the amino acids are too small for effective separation by gel filtration, and affinity chromatography does not lend itself well to separation of small molecules. As discussed in Chapter 3, paper, thin-layer, ion-exchange, high pressure liquid, and gas chromatography are especially suitable for the separation of molecules that have, primarily, polarity differences. High pressure liquid and gas chromatography are excellent choices for the separation of amino acids, but they are limited for your use since derivatives of the amino acids must be prepared and expensive instrumentation is required. However, it should be noted that these two methods have certain advantages and are widely used. A technician who is familiar with the recent advances in HPLC and GC and who has the necessary instrumentation can analyze complex amino acid mixtures more effectively and more rapidly than with any other technique. A distinct advantage is the small sample size required for analysis. Unfortunately, GC and HPLC instrumentation is not yet readily available in the undergraduate biochemistry laboratory. Interested students should review some of the advances in the area (Pisano *et al.*, 1972; Downing and Mann, 1976; Zimmerman *et al.*, 1976, 1977; Black *et al.*, 1982).

A widely used and simple method for separating and identifying amino acids in a mixture is ion-exchange chromatography. As a first approximation, separation by ion-exchange chromatography is based on charge differences between the amino acid molecules. At a specific pH value, all amino acids may have similar charge signs. For example, most amino acids are cationic below pH 4, and most are anionic above pH 8. However, since each has a different series of ionization constants, the size of the charge at a specific pH will be different. The exact size of the charge can be calculated using the Henderson-Hasselbalch equation. The charge differences will be small for the majority of amino acids, and complete separation by ion-exchange chromatography is not predicted. This means that under certain conditions, the three groups of acidic, neutral, and basic amino acids can be separated as described for electrophoresis. However, we find that in the actual practice of cation-exchange chromatography, much better resolution than expected is attained.

Figure E2.2 illustrates the separation of amino acids by ion-exchange chromatography. Careful study of the order of elution of 17 amino acids reveals that within a charge group, they are separated on the basis of polarity. For example, both glutamic acid and aspartic acid have the same approximate net charge at pH 3.25, yet they are well resolved by cation-exchange chromatography. Perhaps the separation of these two amino acids can be explained by a net charge difference. The pH_I, the pH at which the amino acid has no net charge, is $(2.09 + 3.86)/2 = 2.98$ for aspartic acid and $(2.19 + 4.25)/2 = 3.22$ for glutamic acid. This means that at pH 3.25, aspartic acid has a net charge of approximately 0 and glutamic acid has a net charge of between 0 and $+1$. Indeed, there is a small difference between the net charges. A net charge of approximately 0 on aspartic acid would elicit little attraction with the anionic sites on the resin; however, the partial positive charge on glutamic acid would cause delay of elution but probably not enough to explain the wide separation. Perhaps some additional factor is affecting the rate of elution of the amino acid.

To examine this possibility, let's study another pair of amino acids, alanine and serine. At pH 3.0 both are slightly cationic, yet serine elutes much earlier than alanine from a cation-exchange column. In fact, alanine does not elute until after a buffer change from pH 3.25 to pH 4.25. In both cases, glutamic acid vs. aspartic acid and alanine vs. serine, note that the more polar amino acid elutes first. Aspartic acid, with one less methylene group than glutamic acid, would be expected to be more polar ($-CH_2-$ vs. $-CH_2-CH_2-$). The hydroxyl group of serine makes it more polar than alanine ($-CH_2OH$ vs. $-CH_3$). We believe that the slower elution of glutamic acid and alanine is due to more favorable hydrophobic interactions between the side chain of the amino acid and the nonpolar regions of the ion-exchange resin. Further confirmation

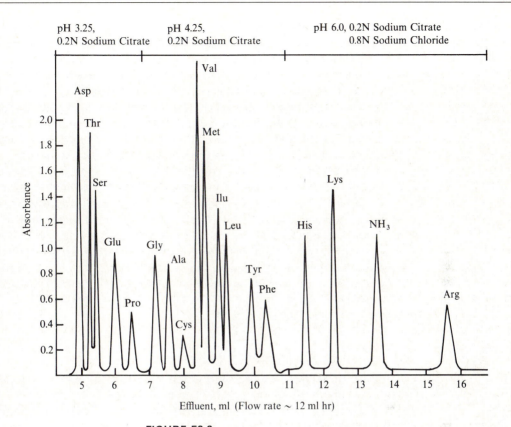

FIGURE E2.2

Separation of amino acids by cation-exchange chromatography. Three buffers, listed at the top, were used for elution. From *Biochemistry, A Problems Approach*, Second Edition by William B. Wood, John H. Wilson, Robert M. Benbow, and Leroy E. Hood. © 1981 The Benjamin/Cummings Publishing Company, Inc.

that polarity aids greatly in resolving the amino acids comes from the position of phenylalanine on the elution chart. Phenylalanine, which has a net charge similar to alanine and serine, is exceedingly late in column elution. Why? In summary, the separation of amino acids by cation-exchange chromatography is based upon two primary factors: (1) *charge*—the greater the positive charge on the amino acid at a specific pH, the slower the elution time, and (2) *polarity*—the more polar the amino acid side chain, the greater the rate of elution.

Before we proceed with a discussion of the experiment to be performed, some peculiarities of Figure E2.2 should be considered. Why does the chromatographic analysis shown in the figure record 17 amino acids and not 20? Which amino acids are missing and why? The amino acid mixture represented by the chromatogram is the product of acid-catalyzed hydrolysis of a typical protein. Under these conditions, trypto-phan is destroyed, and the amide side chains of asparagine and glutamine are hydrolyzed to free carboxyl groups (asn → asp; gln → glu). It should also be noted that serine and threonine are damaged by acid hydrolysis, so their quantitative recovery from protein hydrolysis is not possible.

The last four peaks in Figure E2.2, which elute with pH 6 buffer, represent basic molecules—histidine, lysine, ammonia, and arginine. What is the source of the ammonia? Note that the three amino acids elute with the pH 6 buffer even though they are cationic at this pH. (Draw the structures at pH 6 to prove this.) The ionic interactions between the amino acids (+ charged) and the resin (− charged) are disrupted by the addition of a salt (in this case, 0.8 N NaCl) to the buffer. This results in an increase in ionic strength of the eluting buffer. The large excess of Na^+ ions causes the displacement of the cationic amino acids, which are present in lower concentration. Consequently, we now recognize two methods for elution of bound amino acids from an ion-exchange resin. First, the pH of the eluting buffer can be changed, which alters the net charge on the bound ions. Second, the ionic strength of the buffer may be increased, thus disfavoring the interaction between the amino acid and the charged resin.

Practical Aspects of Ion-Exchange Chromatography

In practice, ion-exchange chromatography begins with the preparation of a buffered ion-exchange column as described in Chapter 3. Application of an amino acid mixture to the top of the resin and elution with buffers result in differential migration of the amino acids. Fractions that are collected from the column may be analyzed qualitatively and quantitatively. Quantitative analysis of the individual amino acids is best done by reaction of each fraction with ninhydrin, followed by absorbance or fluorescence measurement of the intensity of the blue-purple color. Qualitative identification of eluted amino acids is accomplished by comparison with a profile of standard amino acids, as in Figure E2.2. Amino acid analyzers automatically elute the ion-exchange column, treat the eluent with ninhydrin, and provide a recorded profile of elution as shown in Figure E2.2 (Duggan, 1957; Moore *et al.*, 1963).

Successful completion of this experiment requires careful preparation of the ion-exchange medium. This process is time-consuming, so it has been completed for you. Your instructor obtained a cation-exchange resin such as AG50W (Bio-Rad) or Dowex 50 (Dow) in the H^+ form (see Figure E2.3). The resin was first washed with 4 N HCl to remove contaminating metal ions. The resin was then washed with 2 N NaOH to convert the H^+ form to the Na^+ form. Now the resin can function as a cation exchanger.

Now that the ionic state of the resin has been defined, how does this form of the resin function in the separation of amino acids? Assume that

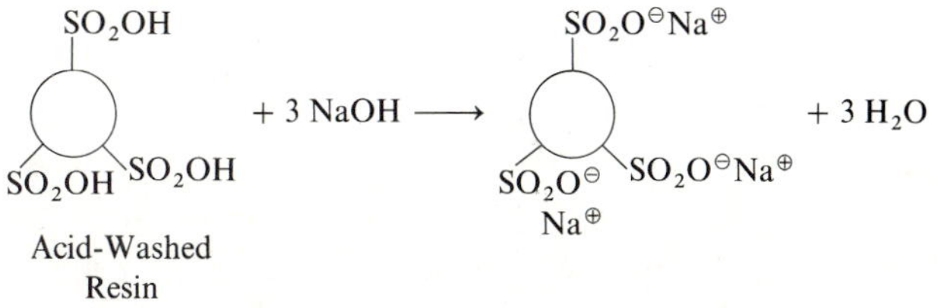

FIGURE E2.3
Conversion of the protonated ion-exchange resin to the cation-exchange form.

your amino acid mixture contains three components, lysine, alanine, and aspartic acid. They are dissolved in a pH 3.0 buffer. The predominant ionic form of each is shown below:

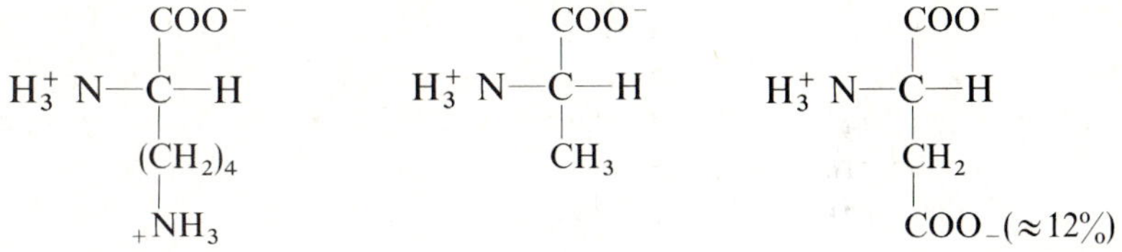

$$
\begin{array}{c}
COO^- \\
| \\
H_3^+N-C-H \\
| \\
(CH_2)_4 \\
| \\
{}_+NH_3
\end{array}
\qquad
\begin{array}{c}
COO^- \\
| \\
H_3^+N-C-H \\
| \\
CH_3
\end{array}
\qquad
\begin{array}{c}
COO^- \\
| \\
H_3^+N-C-H \\
| \\
CH_2 \\
| \\
COO_-(\approx12\%)
\end{array}
$$

lysine	alanine	aspartic acid
at	at	at
pH 3.0	pH 3.0	pH 3.0
net charge = +	net charge ≈ 0	net charge = slightly −

Since the stationary resin is negatively charged, only cations will favorably interact. Aspartic acid, with a negative charge, will therefore percolate through the column with buffer solution and elute first. Both lysine and alanine are cationic and will interact with the resin by exchanging with bound sodium ions. The rate of elution of lysine and alanine from the column will depend upon the strength of their interaction with resin. Lysine has a greater positive charge and will bind more tightly to the anionic resin at pH 3. Alanine, with a slightly positive charge, also will remain bound to the resin. At this point, one amino acid (aspartic acid) has eluted and, therefore, separated from the mixture. How is it possible to achieve selective elution of each of the two amino acids remaining on the column? Increasing the pH of the buffer should decrease the positive charge on the amino acids and weaken the ionic interactions between the bound amino acids and the resin. What pH should be chosen? At pH values greater than 6.0, alanine becomes negative and hence would repel the resin and be eluted. At pH values greater than the pH_I of lysine (9.74) it, too, becomes negative. Since stepwise elution of the amino acids is preferred, small pH changes are more effective than large pH changes. Changing from a buffer of pH 3.0 to pH 6.0 would elute the alanine, and then elution with a buffer of pH 9 would elute lysine. By a stepwise increase in pH, the separation of the three amino acids can be achieved. What would happen if the pH 6.0 buffer would be skipped and the column eluted with pH 3 buffer, then pH 9 buffer? If you desired to maintain a constant buffer pH throughout the experiment, how would you elute the more basic amino acids?

How do we know that the amino acids have been eluted from the column? They are not colored, so their progress through the column cannot be directly monitored. Ninhydrin reacts with all of the amino acids except proline to yield a purple pigment. Proline reacts with ninhydrin to form a yellow reaction mixture. The intensity of the color is proportional to the amount of amino acid present, so an estimate of the concentration of the amino acid can be made. You should refer to your textbook for a further discussion of the ninhydrin reaction.

It is more difficult to confirm the identity of each amino acid isolated from the ion-exchange column. Standard amino acids can be applied to

FIGURE E2.4
Flow chart for analysis of an amino acid mixture.

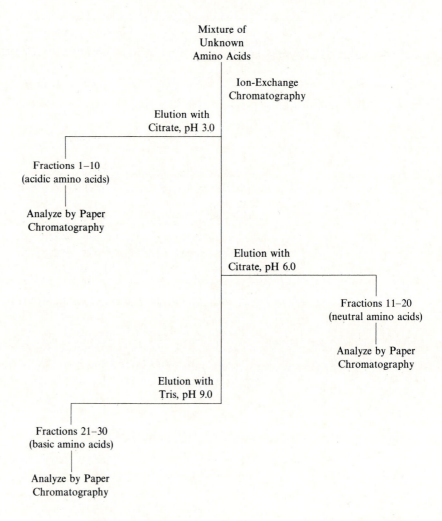

the column, and the volume of buffer needed to elute each amino acid can be measured, but it is better to confirm the identity of the amino acid by some other physical or chemical method such as paper or thin-layer chromatography. Paper chromatography will be used in this experiment because of its versatility, convenience, and low cost.

Overview of the Experiment

In this experiment you will conduct an analysis on a solution containing one to three unknown amino acids. The individual amino acids in the mixture will first be separated by cation-exchange chromatography. Each amino acid will then be identified by paper chromatographic analysis using known amino acids as standards. Study Figure E2.4 for an outline of the experiment.

The experiment can be completed in three hours. It is recommended that students work in pairs. While one student begins the preparation of the ion-exchange column (Part A), the other should begin spotting amino acid standards on the paper chromatogram (Part B).

II. EXPERIMENTAL

Materials and Supplies

Dowex 50-x8 or Amberlite IR-120 cation-exchange resin in 0.05 M citrate buffer, pH 3.0

Citrate buffer, 0.05 M, pH 3.0 and 6.0

Glass column, 1 × 25 cm, or glass buret

Tris buffer, 0.05 M, pH 9.0

Amino acid unknown mixture in citrate buffer, 2% w/v, pH 3.0

Ninhydrin solution in aerosol bottle

Small plug of glass wool

Chromatography paper, Whatman 3MM, 3 × 10 cm and 20 × 20 cm

Small glass funnel

Test tubes (30, 10 × 100 mm)

Solvent system for chromatography, acetonitrile and 0.1 M ammonium acetate (60:40), pH 4.0

Chromatography chamber

Capillary tubes or microcapillary pipets, 10 μl

Standard amino acid solutions, 2% w/v in H_2O

Oven at 110°C

Procedure

A. Separation of Amino Acid Mixture by Ion-Exchange Column Chromatography

Preparing the Column. Clamp a glass column or buret to a ring stand with two clamps, making sure the column is exactly vertical. With the stopcock closed, pour 2 to 3 ml of 0.05 M citrate buffer, pH 3.0, into the column. With a long glass rod, put a small plug of glass wool into the bottom of the column. Press the air bubbles out of the glass wool. This glass wool barrier prevents the ion-exchange resin from running out of the column. Use just enough glass wool to fill the hole in the column; otherwise the flow-rate will be greatly diminished. Obtain about 15 to 20 ml of cation-exchange resin in a small graduated beaker. Transferring the resin from the beaker to the column is not an easy task. Place a small glass funnel into the top of the column, mix the resin well by swirling or stirring with a glass rod, and pour the slurry into the column. Allow the resin to settle into a column 15 to 20 cm in height. When all the resin beads have settled, open the stopcock and allow the buffer to drain from the column until the buffer meniscus is just above the top of the resin. **DO NOT ALLOW THE COLUMN TO RUN DRY.**

Application of Sample and Collection of Fractions. Prepare 30 test tubes for fraction collection by numbering the tubes and marking the 1-ml level on each tube with a wax pencil. To do this, pour 1 ml of water into a test tube and mark the top level of water with the pencil. Use the height of the water to mark the other test tubes.

Using a graduated pipet, carefully apply 0.5 ml of an unknown amino acid mixture to the top of the resin. Apply the mixture dropwise so the resin is not disturbed. Open the stopcock and begin collecting fractions in test tube #1. When the top level of the buffer has just entered the resin, turn off the stopcock and carefully add 1 ml of citrate buffer (pH 3.0) to the column to wash any amino acid from the inside wall of the column. Allow the buffer to enter the resin and continue to collect fractions in tube #1 until 1 ml is collected; then begin collecting in tube #2.

When the buffer has all entered the resin, turn off the stopcock and carefully add 10 ml of buffer to the top of the column. Continue to collect 15 1-ml fractions. If an acidic amino acid is present in the unknown mixture, it will be eluted during this procedure. To test fractions for the presence of an amino acid, place 2 drops from each fraction onto a strip of chromatography paper (3 × 10 cm) as shown in Figure E2.5. Allow the spots to air-dry, and spray them in a hood with ninhydrin. Place in an oven at 110°C for 10 minutes. Red-purple spots indicate the presence of an amino acid.

If an amino acid has been completely removed from the column (no colored spot for fraction #10), change the eluting buffer by draining all

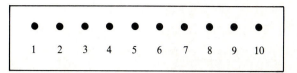

FIGURE E2.5
Analysis of fractions from the cation-exchange column.

the citrate buffer from the column (continue to collect fractions) and add 10 ml of 0.05 M citrate buffer, pH 6.0, to the column. Collect fractions until the top of the buffer has reached the top of the resin. Test each fraction as before for amino acid or check for pH change with pH paper. Neutral amino acids in the unknown mixture should have been eluted during this wash. Now wash the column with 10 ml of Tris, pH 9.0. Collect fractions until all the buffer has entered the resin. Spot two drops of each fraction onto chromatography paper and neutralize each spot by applying two drops of glacial acetic acid. Allow the paper to dry, spray with ninhydrin, and place in the oven as before. Determine which fractions contain amino acids, and save them for Part B.

B. Separation and Identification of Amino Acids by Paper Chromatography

Obtain a sheet of 20 × 20 cm chromatography paper. Handle the paper only with gloves. The chromatogram is prepared by applying spots of each amino acid solution along one edge of the paper. The spots should be no closer than 2 cm from the bottom or side edges. It is best to lay the sheet on a piece of paper and mark the 2-cm line on the underlying paper on both sides of the plate. On the same sheet of paper, mark 2-cm intervals to use as guides for spot application. You should have room for nine different spots—five standard amino acids, the original unknown mixture, and up to three unknown fractions from the chromatography column. Be sure to prepare a "key" for the identity of each application (see Figure E2.6). Using small capillary tubes drawn to a fine point or microcapillary pipets, apply each standard amino acid provided. Each spot on the chromatogram should be prepared with four to six applications of an amino acid solution. This will insure that there is enough

FIGURE E2.6
Preparation of a sheet for paper chromatography.

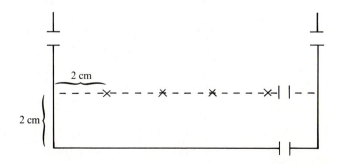

amino acid to be detected after solvent development of the paper. The diameter of each application should be kept as small as possible (0.2 to 0.3 cm) so that the developed spots do not overlap. When amino acid fractions are available from the column, spot these on the paper in the same manner. You may have one, two, or three unknown fractions, depending on how many amino acids were present in the unknown mixture. Use the unknown fractions that contain the most amino acid, as judged by the deepest red-purple color after ninhydrin spray (Part A).

After all of the spots have dried, roll the paper into a cylinder and staple along the edges. Do not overlap the paper at the stapled edges. Place the sheet in the chromatography jar containing acetonitrile and 0.1 M ammonium acetate (60:40), pH 4.0. Place the edge with the applied spots in the solvent, but be sure the solvent level is below the row of spots. Allow the solvent to rise to within 1 cm of the top of the paper (about 1 hour), remove it from the chromatography chamber, and dry it in a hood. Spray lightly with ninhydrin (HOOD!) and develop the color in a 110°C oven for 10 minutes.

III. ANALYSIS OF RESULTS

A. Ion-Exchange Chromatography

Determine which fractions contain amino acids by studying the papers sprayed with ninhydrin. How many different amino acids are present in your unknown, and to what class (acidic, basic, neutral) does each belong? Explain the order of elution of each amino acid. From this information, is it possible to positively identify each amino acid that was eluted from your column?

B. Paper Chromatography

Prepare a table of R_f values of the standard amino acids. Review Chapter 3 concerning the calculation of R_f. Describe in your notebook the color of each ninhydrin spot. Since the color depends on the amino acid, this helps in the identification of an unknown amino acid. Calculate the R_f values for the unknown amino acids. You should now be able to identify what amino acids are present in your unknown mixture. Can you identify the components of your unknown mixture using only the paper chromatography data?

IV. QUESTIONS AND PROBLEMS

1. Write the predominant ionic structure of lysine that would exist in aqueous solutions at the following pH values: pH 2, pH 5, and pH 11.

2. Draw all of the possible ionization states of glutamic acid.

3. Predict the relative order of R_f values for the amino acids in the following mixture: ser, lys, leu, val, and ala. Assume that the separation was carried out on paper with the solvent system described in this experiment.

4. Calculate the net charge on each of the following amino acids at the given pH.

   alanine, pH 3.0
   aspartic acid, pH 4.0
   lysine, pH 8.0
   serine, pH 6.0
   histidine, pH 9.0

5. A mixture of amino acids is subjected to paper electrophoresis at pH 6.0. The amino acids present in the mixture are val, glu, phe, ser, and lys. Predict which amino acids will migrate toward the cathode. Toward the anode?

6. Compare the effectiveness of the two techniques you used to separate the unknown amino acid mixture. Which one leads to better resolution of the mixture? Which technique is better for identification of the individual amino acids?

7. Predict the order of elution of the following mixture of amino acids from a cation-exchange column: glu, ala, ser, pro, and arg. Assume the procedure was the same as in this experiment. Explain your answer.

8. Predict the order of elution from a cation exchange column of a mixture containing the following amino acids and peptides: asp, ala, phe-ala, asp-asp, and phe-lys. Assume the same conditions and procedure detailed in this experiment.

9. For each pair of amino acids listed below, predict which one would be first to elute from a cation exchange column. Assume the buffer pH begins at 3.25 and is increased in a stepwise fashion to pH 6.
   (a) asp, glu   (d) lys, phe
   (b) ser, val   (e) leu, val
   (c) phe, ala   (f) gly, ala

V. REFERENCES

General

R. Bohinski, *Modern Concepts in Biochemistry*, Fourth Edition (1983), Allyn and Bacon (Boston), pp. 64–71. An introduction to biochemistry laboratory methods.

J. Brewer, A. Pesce, and R. Ashworth, *Experimental Techniques in Biochemistry* (1974), Prentice-Hall (Englewood Cliffs, NJ), pp. 74–79. An introduction to ion-exchange chromatography.

T. Cooper, *The Tools of Biochemistry*, (1977), John Wiley, Inc. (New York), pp. 136–168. An entire chapter on the theory and practice of ion-exchange chromatography.

A. Lehninger, *Principles of Biochemistry* (1982), Worth Publishers (New York), pp. 100–110. A brief introduction to amino acid analysis.

J.D. Rawn, *Biochemistry* (1983), Harper and Row (New York), pp. 33–58. A discussion of amino acid analysis.

L. Stryer, *Biochemistry*, Second Edition (1981), Freeman (San Francisco), pp. 19–22. A brief discussion of ion-exchange chromatography.

Specific

S.D. Black, M.J. Coon, *et al.*, *Anal. Biochem.*, *121*, 281 (1982). "HPLC of Phenylthiohydantoin-Amino Acids."

E. Duggan in *Methods in Enzymology*, S. Colowick and N. Kaplan, Editors (1957), Academic Press (New York), Vol. III, pp. 492–504. "Measurement of Amino Acids by Column Chromatography."

M.R. Downing and K.G. Mann, *Anal. Biochem.*, *74*, 298–319 (1976). "High-Pressure Liquid Chromatographic Analysis of Amino Acid Phenylthiohydantoins: Comparison with Other Techniques."

S. Moore and W. Stein in *Methods in Enzymology*, S. Colowick and N. Kaplan, Editors (1963). Academic Press (New York), Vol. VI, pp. 819–831. "Chromatographic Determination of Amino Acids by the Use of Automatic Recording Equipment."

J.J. Pisano, T.J. Bronzert, and H.B. Brewer, *Anal. Biochem.*, *45*, 43–59 (1972). "Advances in the Gas Chromatographic Analysis of Amino Acid Phenyl- and Methylthiohydantoins."

C.L. Zimmerman, E. Appella, and J.J. Pisano, *Anal. Biochem.*, *75*, 77–85 (1976). "Advances in the Analysis of Amino Acid Phenylthiohydantoins by High Performance Liquid Chromatography."

C.L. Zimmerman, E. Appella, and J.J. Pisano, *Anal. Biochem.*, *77*, 569–573 (1977). "Rapid Analysis of Amino Acid Phenylthiohydantoins by High-Performance Liquid Chromatography."

Experiment **3**

Sequential Analysis of a Dipeptide or Tripeptide

RECOMMENDED READING:
Chapter 3

SYNOPSIS

The complete primary structure of a protein is known only after the amino acid sequence has been determined. In this experiment the procedures that are in common use to determine protein primary structure are applied to an unknown di- or tripeptide. Amino acid composition of the peptide will be determined by acid hydrolysis followed by paper chromatography. The identity of the N-terminal amino acid will be achieved by the dansyl or Edman method followed by thin layer chromatography. The C-terminal amino acid is identified by paper chromatography after release by carboxypeptidase.

I. INTRODUCTION AND THEORY

Structural elucidation of natural macromolecules is an important step in understanding the relationships between the chemical properties of a biomolecule and its biological function. The techniques used in organic structure determination (NMR, IR, UV, and MS) are somewhat useful when applied to biomolecules, but the unique nature of natural molecules also requires the application of specialized chemical techniques. Proteins, polysaccharides, and nucleic acids are polymeric materials, each composed of hundreds or sometimes thousands of monomeric units (amino acids, monosaccharides, and nucleotides, respectively). But there is only a limited number of these units from which the biomolecules are synthesized. For example, only 20 different amino acids are found in proteins, but these different amino acids may appear several times in the same protein molecule. Therefore, the structure of a peptide or protein can be recognized only after the amino acid composition *and* sequence have been determined.

Amino Acid Composition by Hydrolysis of Peptide Bonds

The amide bonds in peptides and proteins can be hydrolyzed in strong acid or base. Treatment of a peptide or protein under either of these conditions yields a mixture of the constituent amino acids. Neither acid- nor base-catalyzed hydrolysis of a protein leads to ideal results, as both tend to destroy some constituent amino acids. Acid-catalyzed hydrolysis destroys tryptophan, causes some loss of serine and threonine, and converts asparagine and glutamine to aspartic acid and glutamic acid, respectively. Base-catalyzed hydrolysis leads to destruction of serine, threonine, cysteine, and cystine, and also results in racemization of the free amino acids. Since acid-catalyzed hydrolysis is less destructive, it is often the method of choice. The hydrolysis procedure consists of dissolving the protein sample in aqueous acid, usually 6 N HCl, and heating the solution in a sealed, evacuated vial at 110°C for 12 to 24 hours. The time interval necessary for hydrolysis depends upon the nature of the amino acid residues. Because of steric interactions, hydrolysis at valine and isoleucine is slower than at other amino acid residues.

Modification and Removal of the N-Terminal Amino Acid

The sequential analysis of amino acids in purified peptides and proteins is best initiated by analysis of the terminal amino acids. In a peptide there will be one amino acid with a free α-amino group (N-terminus end) and one amino acid with a free α-carboxyl group (C-terminus end). Many chemical methods have been developed to selectively tag and identify these terminal amino acids.

N-Terminal amino acid analysis is achieved by the use of (1) 2,4-dinitrofluorobenzene (Sanger reagent), (2) 1-dimethylaminonaphthalene-5-sulfonyl chloride (dansyl chloride), or (3) phenylisothiocyanate (Edman reagent). Figure E3.1 shows the structures of these reagents. Although the chemistry is different for each of these reagents, the same general concept

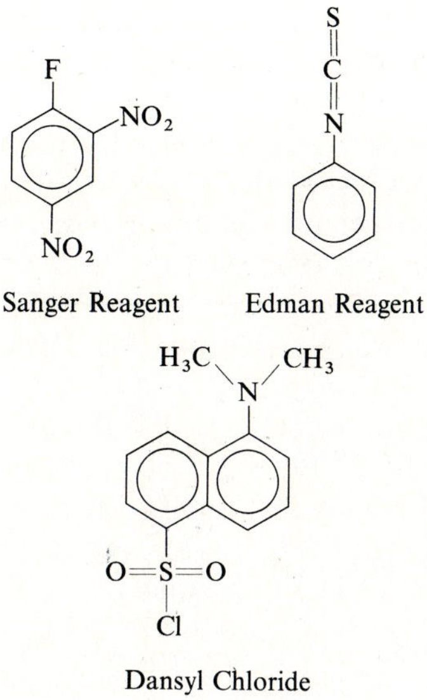

FIGURE E3.1
Three reagents useful in N-terminal analysis.

is used. These compounds react with the N-terminal amino acid of a peptide or protein to produce covalent derivatives that are stable to acid-catalyzed hydrolysis. The process is illustrated in Figure E3.2. The mixture of modified amino acid and free amino acids is separated, and the N-terminal amino acid is identified by thin-layer, paper, gas, or high pressure liquid chromatography. The first N-terminal method to be widely used was based on the Sanger reagent, which produces yellow-colored

FIGURE E3.2
Analysis of N-terminal amino acids.

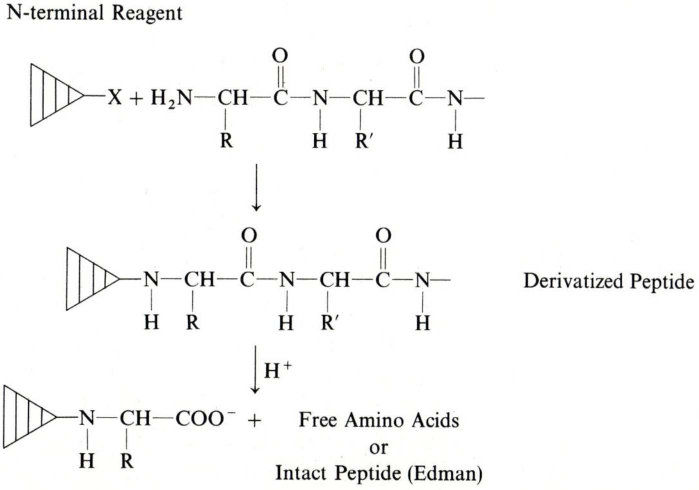

2,4-dinitrophenyl (DNP) derivatives of N-terminal amino acids. The Sanger method has several disadvantages, including poor yield of the DNP-peptide derivative, low sensitivity of analysis, and instability of some DNP-amino acids during acid hydrolysis. The dansyl chloride method has largely replaced the Sanger method, because very sensitive fluorescence techniques may be used for detection and analysis of the dansyl amino acid derivatives and the derivatives are more stable during acid hydrolysis.

Even more versatile than the dansyl method is the Edman method (Figure E3.3) (Konigsberg, 1972). The N-terminal amino acid is removed as its phenylthiohydantoin (PTH) derivative under anhydrous acid conditions, while all other amide bonds in the peptide remain intact. The derivatized amino acid is then extracted from the reaction mixture and identified by paper, thin-layer, gas, or high pressure liquid chromatography. The intact peptide (minus the original N-terminal amino acid) may be isolated and recycled by reaction with phenylisothiocyanate. Since this method is nondestructive to the remaining peptide (aqueous acid hydrolysis is not required) and it results in good yield, it can be used for stepwise sequential analysis of peptides. The method has now been automated.

Under all N-terminal labeling conditions, internal amino acid residues in a peptide may be modified, but in ways different from the N-terminal amino acid. For example, if the N-terminal amino acid is alanine, only the free α-amino group would be reactive to the selective

FIGURE E3.3
The Edman method of N-terminal amino analysis.

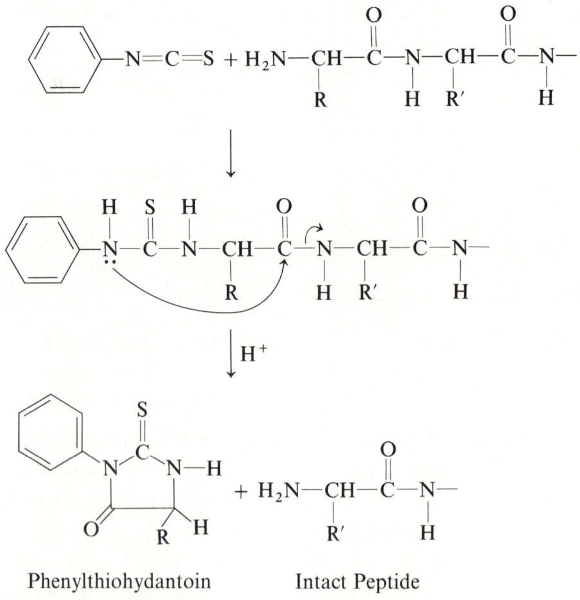

Phenylthiohydantoin   Intact Peptide

reagent. However, if lysine, serine, or other amino acids with nucleophilic side chains are present at the N-terminus, both nucleophilic groups (α-amino and ε-amino for lysine, α-amino and β-hydroxyl for serine) have the potential for reaction with the labeling reagent. If lysine or serine residues are located internally, only the side chain nucleophilic group (ε-amino for lysine, β-hydroxyl for serine) has the potential for reaction. Chromatographic techniques may be used to separate and identify each of these three types of amino acid modification (α-amino, side chain, and both).

Removal of the C-Terminal Amino Acid from a Peptide

The analysis of the C-terminal amino acid of a peptide or protein may be achieved by either chemical or enzymatic methods. The chemical methods are similar to the procedures for N-terminal analysis. C-terminal amino acids are converted to hydrazides by hydrazine or are reduced to amino alcohols by lithium borohydride. The modified amino acids are released by acid hydrolysis and identified by chromatography. Both of these chemical methods are difficult, and clear-cut results are not readily obtained. The method of choice is peptide hydrolysis catalyzed by carboxypeptidases A and B (Ambler, 1972). These two enzymes catalyze the hydrolysis of amide bonds at the C-terminal end of a peptide (Reaction 1), since carboxypeptidase action requires the presence of a free α-carboxyl group in the substrate.

$$-\overset{\overset{\displaystyle O}{\|}}{C}-\underset{\underset{\displaystyle H}{|}}{N}-\underset{\underset{\displaystyle R}{|}}{CH}-\overset{\overset{\displaystyle O}{\|}}{C}-\underset{\underset{\displaystyle H}{|}}{N}-\underset{\underset{\displaystyle R'}{|}}{CH}-COO^- + H_2O \xrightarrow{\text{carboxypeptidase}}$$

$$-\overset{\overset{\displaystyle O}{\|}}{C}-\underset{\underset{\displaystyle H}{|}}{N}-\underset{\underset{\displaystyle R}{|}}{CH}-COO^- + H_2N-\underset{\underset{\displaystyle R'}{|}}{CH}-COO^- \quad \text{(Reaction E3.1)}$$

Carboxypeptidase A catalyzes the hydrolysis of carboxyl-terminal acidic or neutral amino acids; however, the rate of hydrolysis depends upon the structure of the side chain R'. Amino acids with nonpolar aryl or alkyl side chains are cleaved more rapidly. Carboxypeptidase B is specific for the hydrolysis of basic carboxyterminal amino acids (lysine and arginine). Neither peptidase will function if proline occupies the C-terminal position or is the next to last amino acid.

It should be recognized that the release of the C-terminal amino acid by carboxypeptidase A or B uncovers yet another C-terminal amino acid

TABLE E3.1
Approximate Relative Rates of Release by Carboxypeptidase A

Rapid Release:	Tyr, Phe, Trp, Leu, Ile, Met, Thr, Gln, His, Ala, Val
Slow Release:	Asn, Ser, Lys
Very Slow Release:	Gly, Asp, Glu
Not Released:	Pro, Arg

From R. Ambler in *Methods in Enzymology*. C.H.W. Hirs and S.N. Timasheff, Eds., Vol. XXVB (1972), Academic Press (New York), pp. 436–445.

that may be cleaved by the enzyme. If the second amino acid is cleaved at a rate greater than the first amino acid, this may lead to ambiguity in C-terminal analysis, since both free amino acids may be present in approximately equal concentrations in the reaction mixture. To avoid this difficulty it is often possible to incubate the protein with a carboxypeptidase, remove aliquots from the reaction mixture at various times, analyze the free amino acids, and graph the release of individual amino acids as a function of time. The relative rates of removal of C-terminal amino acids by carboxypeptidase A are shown in Table E3.1. Figure E3.4 demonstrates results from C-terminal analysis of two different peptides. In order to obtain unambiguous results from identification of C-terminal amino acids, it is essential to consider the relative rates of hydrolysis of individual amino acids by carboxypeptidase.

Separation and Identification of Amino Acid Mixtures

Sequence determination of a dipeptide, tripeptide, or larger protein depends upon the availability of techniques for the separation and identification of free and derivatized amino acids. The most versatile, economical, and convenient techniques are based on chromatographic methods. Earlier workers relied on paper chromatography and thin-layer chromatography (TLC); however, more sensitive techniques are now available. Automated ion-exchange chromatography (amino acid analyzers), gas chromatography, electrophoresis, and high pressure liquid chromatography are now powerful tools for the qualitative and quantitative analysis of amino acids and derivatives. All of these techniques are discussed in Chapters 3 and 4. Because there is still a demand for rapid, qualitative, routine analysis of amino acids, thin-layer and paper chromatographic methods are still in the process of development and improvement. Amino acids and derivatives may be analyzed directly by TLC without further derivatization as is required for GC. Several support materials are available, but most analyses are carried out on silica gel or cellulose. Extensive tables of R_f values

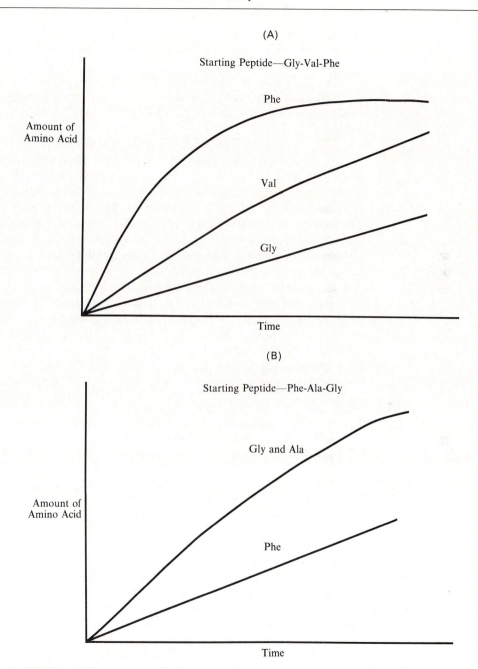

FIGURE E3.4
C-terminal analysis by carboxypeptidase hydrolysis; see text for details.

in different solvent systems are available (Niederwieser, 1972). The free amino acids can be detected on the developed chromatographic plates by reaction with ninhydrin. A pink-purple color is obtained for all amino acids except proline. A yellow color develops with proline. Since phenylthiohydantoin and dansyl derivatives of amino acids contain aromatic moieties, they can be detected on TLC plates by fluorescence under an ultraviolet lamp.

Gas chromatography is used to analyze volatile derivatives of amino acids. Phenylthiohydantoins (products from Edman degradation) may be analyzed directly by GC, but are better resolved if converted to their trimethylsilyl derivatives with N,O-bis(trimethylsilyl) acetamide. Free amino acids are generally converted to their N-trifluoroacetyl *n*-butyl esters or trimethylsilyl derivatives before GC analysis. For best results, all gas chromatography of amino acid derivatives should be done with a glass column and injection port, as contact with metals causes extensive decomposition of the derivatives.

High pressure liquid chromatographic techniques have been applied with success to the analysis of phenylthiohydantoin, 2,4-dinitrophenyl, and dansyl amino acid derivatives (Black *et al.*, 1982).

Overview of the Experiment

In this experiment, the sequence of amino acids in a dipeptide or tripeptide will be determined by using some of the techniques just described. The amino acid composition of the unknown peptide will be found by acid-catalyzed hydrolysis followed by chromatographic analysis of the amino acids. Sequence analysis will then be carried out by identification of the N- and/or C-terminus. See Figure E3.5 for a summary of the procedure. The general experimental procedure will be outlined, but the speci-

FIGURE E3.5
Flowchart for the analysis of a dipeptide or tripeptide.

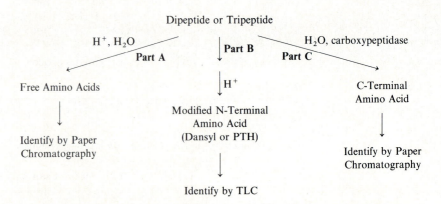

fic analyses you will perform will depend upon the unknown peptide. The following schedule is appropriate for this experiment.

Dipeptide

Method A: Total Hydrolysis and Dansyl Analysis
Period 1: (a) Begin acid hydrolysis of peptide. (b) Complete dansyl chloride reaction and begin acid hydrolysis of dansyl peptide. (c) Prepare paper chromatogram by applying standard amino acids.
Period 2: (a) Work up the dansyl hydrolysate and spot on the TLC plate along with standard dansyl amino acids. Develop the plate in solvent. (b) Work up the peptide acid-hydrolysate, spot on the paper chromatogram, and develop.
Method B: Total Hydrolysis and Edman Analysis
Period 1: (a) Same as 1(a, c) above. (b) Begin Edman reaction and proceed through toluene evaporation.
Period 2: (a) Complete Edman reaction and spot on TLC plate along with PTH amino acid standards. Develop in solvent. (b) Work up the peptide acid-hydrolysate. Spot on the paper chromatogram and develop.

Tripeptide

A combination of methods can be used for this analysis, but the following is recommended.
Periods 1 and 2: Same as Method A or B above.
Period 3: Carboxypeptidase reaction and paper chromatographic analysis.

Many variables must be considered as you plan the execution of this experiment. For example, there is no single paper or thin-layer chromatographic method that will successfully separate and identify every combination of two or three amino acids or derivatives. Successful results from this experiment depend partly upon the proper choice of unknown peptides by your instructor. Since each peptide unknown may require a specific set of analyses and reaction conditions, your instructor should give you some hints on how you should proceed.

Optional Experiments

(a) If your laboratory is equipped for GC or HPLC analysis of amino acid derivatives and you have sufficient time, ask your instructor for help in the use of these instruments.

(b) If time is limited, obtain a sample of a peptide or purified protein and identify the N- or C-terminal amino acid only.

II. EXPERIMENTAL

Materials and Supplies

Part A—Total Hydrolysis of Peptide
  Unknown dipeptide or tripeptide, 2 mg
  Melting point capillary tube, 1.5 × 100 mm
  6 N HCl
  Heat lamp
  Whatman 3MM paper, 20 × 20 cm
  Amino acid standard solutions, 1% in water
  Syringe, 50–100 μl
  Chromatography chambers (jars or sandwich-type by Kodak)
  Chromatography solvent, acetonitrile and 0.1 M ammonium acetate, 60:40, pH 4.0 or 5.0
  Oven at 100°C
  Ninhydrin spray
  Rotary evaporator or tank of N_2 gas

Part B—N-Terminal Analysis
  1. Dansyl chloride method
  Unknown dipeptide or tripeptide, 1 mg
  Dansyl chloride solution, 5 mg/ml in acetone
  Conical centrifuge tube, 12 ml
  Sodium bicarbonate, 0.2 M
  Hydrocarbon foil
  Constant temperature bath at 37°C
  Hydrolysis vial ($\frac{1}{2}$ dram with polyvinyl liner, Sargent-Welch)
  6 N HCl
  Acetone and 6 N HCl (1:1)
  Oven at 100°C
  Silica gel thin-layer plates, 20 × 20 cm
  Chromatography solvent in jar, chloroform: methanol: acetic acid (75:25:5)
  Dansyl amino acid standards in acetone/ethanol, 1 mg/ml
  UV detection lamp
  2. Edman method
  Unknown dipeptide or tripeptide, 3 mg
  Conical centrifuge tube, 12 ml
  Pyridine, 70%
  Phenylisothiocyanate
  Tank of purified N_2 gas
  Constant temperature water bath at 37°C
  Toluene
  Silica gel thin-layer plates, 20 × 20 cm
  Phenylthiohydantoin amino acid standards in dichloroethane

Chromatography solvent in jar, chloroform: isopropanol: water (28:8:1, v/v)

UV detection lamp

Anhydrous trifluoroacetic acid

Hydrocarbon foil

Ethyl acetate

Part C—C-Terminal Analysis

Unknown peptide, 2 mg

N-Ethyl morpholine acetate buffer, 0.2 M, pH 8.5

Solution of carboxypeptidase A in water, 1 mg/ml (50 EC units per ml)

Constant temperature bath at 37°C

Acetic acid, 0.1 M

Clinical centrifuge

Heat lamp

Solvent for paper chromatography (see Part A)

Ninhydrin spray

Rotary evaporator

Procedure

Part A—Total Hydrolysis of Unknown Peptide

Place approximately 1 mg of your peptide sample on a small watch glass and dissolve it in 0.05 ml of 6 N HCl delivered with a syringe. With a syringe, transfer the peptide solution into a melting point capillary, and seal the open end with a bunsen burner. Put the sealed tube in a small labeled beaker and place in a 110°C oven for 12 to 15 hours, or at room temperature for one week. After hydrolysis, open the capillary by scratching one end with a sharp file, and transfer the hydrolysate to a watch glass. Evaporate the liquid by *gentle* heating under a heat lamp (HOOD!), add 0.1 ml of distilled water and repeat the evaporation step. Finally, prepare the residue for chromatography by dissolving it in 0.1 ml of distilled water.

For paper chromatography, obtain a 20 × 20 cm sheet of Whatman 3MM. Wear disposable plastic gloves when you handle the paper in order to avoid transfer of amino acids from your fingers to the paper sheet. Prepare the chromatogram by applying spots of the unknown hydrolysis mixture and amino acid standards along one edge of the paper. The amino acid standards can be applied during Period 1 and the hydrolysis mixture at the beginning of Period 2. The spot should be no closer than 2 cm from the bottom edge or side edges. You may wish to mark the paper *lightly* 2 cm from one edge with a pencil. Prepare a "key" in your notebook to keep track of the identity of each spot (Figure E3.6). Using small capillary tubes drawn to a fine point or microcapillary pipets, make two separate

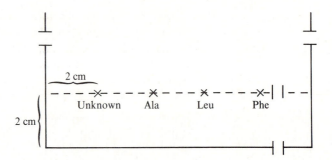

FIGURE E3.6
Preparation of a thin-layer or paper chromatogram.

applications of your unknown solution, one containing approximately 5 μl and another containing two or three times as much sample. It is best to apply the larger amount in small increments. Make one application and allow it to dry before more is applied to the spot. Apply the amino acid standards at selected spots along the 2-cm line; 5 to 10 μl of each standard should be sufficient. Applications of amino acids should be kept as small as possible (0.2 to 0.3 cm) so that the developed spots do not overlap. To develop the paper chromatogram, staple the sheet into a cylinder and place it in the chromatography jar containing a 1-cm layer of acetonitrile and 0.1 M ammonium acetate (60:40, pH 4.0 or 5.0) (Heimer, 1972). When preparing the cylinder, handle the paper only with gloves, and do not overlap the edges while stapling. Solvent development time is 40 to 60 minutes.

When the developing solvent reaches a level approximately 1 cm from the top of the chromatogram, remove the paper from the jar and allow it to dry in a hood. Remove the staples and spray the paper lightly with ninhydrin solution. Allow to dry in a hood for 10 minutes. The color may be developed by placing the chromatogram in an oven (110°C for 10 min) or at room temperature for 1 to 2 hours. All the spots should be circled with a pencil, as they will fade with time. Describe in your notebook the color of each spot.

Part B. Method 1—Dansyl Chloride

CAUTION:

Dansyl chloride is a corrosive organic acid halide, which should be used only in a hood. Wear disposable gloves while using the reagent. If it is spilled on the skin, sprinkle with sodium bicarbonate and wash with copious amounts of warm soapy water. Rinse well. For floor or bench spills, cover with sodium bicarbonate and transfer mixture to a beaker of water. Pour the aqueous solution down the drain with excess water.

Do not pipet solutions of dansyl chloride by mouth!

FIGURE E3.7
Evaporation of solvent by a stream
of N_2 gas.

Weigh about 1 mg of the unknown peptide and transfer into a conical centrifuge tube. Dissolve the peptide in 0.5 ml of 0.2 M sodium bicarbonate. Add 0.2 ml of the solution of dansyl chloride in acetone and mix well. Cover with hydrocarbon foil and incubate for 1 hour at 37°C or 2 hours at room temperature. After completion of the reaction, evaporate the solution to dryness under vacuum (rotary evaporator) or by a gentle stream of N_2 gas, with the reaction tube held in a beaker of warm water (Figure E3.7). Dissolve the residue in 0.5 ml acetone/6 N HCl solution and transfer it into the hydrolysis vial. Evaporate the acetone from the vial with a gentle stream of nitrogen gas. Seal the hydrolysis vial and heat in a 110°C oven for 10 to 16 hours.

Remove the hydrolysate from the vial and evaporate it on a watch glass under a heat lamp. Dissolve the residue in 0.1 ml of water and evaporate again to remove HCl. Redissolve the residue in 0.3 ml of water, and chromatograph it on a thin-layer silica gel plate using dansyl amino acid standards. Obtain a 20×20 cm silica gel sheet. These may be handled only by the edges or back. If the humidity is relatively high in your laboratory, heat the sheet in a 110°C oven for 10 minutes. After cooling, begin the application of the unknown hydrolysate and amino acid standard solutions. These should be applied in the same manner as for paper chromatography, except that no mark or line should be made on the plate. It is best to lay the sheet on a piece of paper and mark the 2-cm line on the underlying paper on both edges of the plate. On the same underlying sheet of paper, mark 2-cm intervals to use as guides for spot application. Be sure to prepare a "key" for the identity of each application. The same techniques and amounts should be used as described for paper chromatography. Apply two spots of the unknown dansyl derivative (5 μl and 10 μl). Apply the dansyl amino acid standards in 5-μl spots. Use chloroform:methanol:acetic acid (75:25:5) to develop the plates. Remove the plates after a development period of $1\frac{1}{2}$ hours even if the solvent hasn't reached the top. It will probably have developed only about 50%. Detection of the spots is accomplished with a UV lamp.

Part B. Method 2—Edman Method

CAUTION:
Phenylisothiocyanate is a lachrymator and should be handled only with gloves in a hood. If it is spilled on the skin, wash immediately with strong soap solution and rinse well. Small spills on the lab bench or floor should be absorbed on paper towels or vermiculite. Sweep the contaminated solid onto paper and allow the phenylisothiocyanate to evaporate in a hood. Check with your instructor for disposal of the paper. For large spills, absorb or mix with vermiculite, sodium bicarbonate, or sand. Package this in a paper carton and have your instructor dispose of it.

Trifluoroacetic acid is a corrosive liquid. Handle it only with disposable gloves in a hood. Wash skin spills immediately with soap and water and rinse well. Cover floor or bench spills with excess sodium bicarbonate and vermiculite. Mix well and transfer to a beaker of water. After reaction, pour solution down the drain with excess water.

Pyridine is a toxic, flammable liquid (flash point is 20°C) with a very irritating smell. Use only in a hood with no flames. Gloves should be worn while handling.

Toluene is a flammable liquid (flash point is 4°C). Use only in a hood with no flames.

Do not pipet these four reagents—pyridine, trifluoroacetic acid, toluene, or phenylisothiocyanate—by mouth!

To a conical centrifuge tube, add about 2 mg of the unknown peptide and dissolve in 1.0 ml of 70% aqueous pyridine (HOOD!). Add 0.1 ml of phenylisothiocyanate and mix well. Flush the tube with purified N_2 gas to exclude oxygen, and stopper it with a cork or cover with Parafilm. If O_2 is present, oxygen atoms will replace sulfur in the thiocarbonyl group of the phenylthiocarbamyl peptide. Incubate the reaction mixture in a constant temperature bath (37°C) for 2 hours. Evaporate the solvent using a rotary evaporator. An alternate method of solvent evaporation is to place the test tube in a beaker of warm water and flush the inside of the tube with N_2 gas as in Figure E3.7. Wash the residue remaining in the tube with two 1-ml portions of toluene to remove unreacted phenylisothiocyanate. Remove residual toluene on the rotary evaporator or in a warm water bath under a stream of N_2.

The cyclization of the phenylthiocarbamyl peptide and cleavage of the peptide bond are brought about by reaction with anhydrous trifluoroacetic acid. Add 0.5 ml of trifluoroacetic acid to the unknown peptide residue, cover with hydrocarbon foil, and heat for 10 minutes at 60°C or allow the reaction to proceed for 24 hours at room temperature. Remove the excess trifluoroacetic acid by placing the tube in a beaker of warm water with evaporation under a stream of N_2 in a hood. Extract the phenylthiohydantoin amino acid derivative from the residue with ethyl acetate (3 × 1 ml). Combine the ethyl acetate extracts in a small beaker and concentrate to 0.5 ml over a steam bath in a hood. Apply the concentrated ethyl acetate solution on two separate spots (5 μl and 10 μl) on a thin-layer plate as described in Part B.1. On the same plate, apply spots of phenylthiohydantoin amino acid standards (10 μl). Develop the plate in chloroform:isopropanol:water (28:8:1, v/v). Phenylthiohydantoins are detected on the plate by the use of a UV lamp.

Part C—C-Terminal Analysis using Carboxypeptidase A

Weigh about 2 mg of the unknown peptide into a test tube and dissolve in 1.0 ml of 0.2 N N-ethylmorpholine acetate buffer, pH 8.5. Add

0.1 ml (5 EC units) of the carboxypeptidase solution and mix well, but do not shake. Incubate at 37°C and withdraw small aliquots (0.1 ml) at timed intervals (0, 5, 10, 20, 30 and 40 minutes). Immediately after withdrawal, transfer each aliquot to a test tube containing 0.5 ml of 0.1 M acetic acid. This stops the enzyme-catalyzed hydrolysis and causes precipitation of the enzyme. Remove the precipitate from each acidified aliquot by centrifugation at 1000 rpm for 10 min. Evaporate 0.2 ml of each supernatant on separate watch glasses. Dissolve the residue in 0.1 ml of water and chromatograph against amino acid standards on Whatman 3MM paper as in Part A. Since a some-what "quantitative" analysis will be made of the C-terminal amino acid, be sure the same amount of solution is applied to each chromatogram. For detection, spray with ninhydrin solution.

III. ANALYSIS OF RESULTS

Part A—Total Hydrolysis of Unknown Peptide

Prepare a table of R_f values of the standard amino acids. You should also describe the color of the original ninhydrin spot. Since the colors vary slightly with the amino acids, this aids in the identification of an unknown amino acid. Calculate the R_f values for the constituent amino acids of the unknown peptide. From these data you should be able to identify what amino acids are present in your unknown.

Part B. Method 1—Dansyl Chloride

Calculate R_f values for the standard dansyl amino acids and compare with the unknowns. What is the N-terminal amino acid? What is the structure of your unknown dipeptide? Did your TLC plate have two or more fluorescent spots for the dansyl reaction mixture? Which one is the dansyl amino acid? Can you speculate on the origin of the other(s)?

Part B. Method 2—Edman Method

Determine the R_f values for the standard and unknown amino acid phenylthiohydantoins. Compare the R_f value for the unknown phenylthiohydantoin with the standards and try to identify the N-terminal amino acid in the unknown.

Part C—C-Terminal Analysis of Peptide using Carboxypeptidase A

Study the chromatograms of the released C-terminal amino acids and prepare a graph of intensity of each purple ninhydrin spot vs. time. For early reaction times you will probably see only one amino acid spot, but later aliquots may yield two or even three amino acid spots. If this is the case, make a separate line on the graph for each appearing amino acid. Estimate the intensity of each amino acid spot using such terms as light, medium, and dark or use a scale of 1 to 10 on your graph. Using data from all three parts of the experiment, draw the structure of your unknown tripeptide.

IV. QUESTIONS

1. Compare the advantages and disadvantages of the Edman and dansyl chloride methods of N-terminal analysis.

2. Why must the layer of solvent in a chromatography jar be below the origin line containing the applied samples on the chromatogram?

3. Why must the Edman reaction be carried out in a basic solvent such as aqueous pyridine?

4. An unknown tripeptide was analyzed according to the procedures described in this experiment and the following results were obtained.
   Part A: Total hydrolysis and paper chromatography—phe, leu, ala
   Part B: Dansyl chloride and thin layer chromatography—dansyl alanine
   Part C: Carboxypeptidase A—Relative order of release: phe > leu ≈ ala
   What is the sequence of the tripeptide?

5. Could you determine the amino acid sequence of a dipeptide using two steps, (1) total acid hydrolysis and (2) C-terminal analysis with carboxypeptidase?

V. REFERENCES

General

R.C. Bohinski, *Modern Concepts in Biochemistry*, Fourth Edition (1983), Allyn and Bacon (Boston), pp. 88–102. Determination of protein primary structure.

R. Haschemeyer and A. Haschemeyer, *Proteins* (1973), Wiley-Interscience (New York), pp. 68–89. This reference contains a chapter on primary structure analysis.

A. Lehninger, *Principles of Biochemistry* (1982), Worth Publishers, Inc. (New York), pp. 129–134. An introductory section on primary structure analysis.

J.D. Rawn, *Biochemistry* (1983), Harper and Row (New York), pp. 65–77. A section on determining amino acid composition and sequence of proteins.

L. Stryer, *Biochemistry*, Second Edition (1981), Freeman (San Francisco), pp. 21–27. A section on determining amino acid sequence.

E. Smith, R. Hill, I. Lehman, R. Lefkowitz, P. Handler, and A. White, *Principles of Biochemistry: General Aspects*, Seventh Edition (1978), McGraw-Hill Book Co. (New York), pp. 58–69. Determination of amino acid sequence.

Specific

R. Ambler in *Methods in Enzymology*, C.H.W. Hirs and S.N. Timasheff, Eds., Vol. XXVB, (1972), Academic Press (New York), pp. 436–445. "Carboxypeptidase."

S.D. Black and M.J. Coon, *Anal. Biochem.*, *121*, 281 (1982). "HPLC of Amino Acid Phenylthiohydantoins."

E.P. Heimer, *J. Chem. Ed.*, *49*, 547 (1972). "Paper Chromatography of Amino Acids."

W. Konigsberg, *Methods in Enzymology*, C.H.W. Hirs and S.N. Timasheff, Eds., Vol. XXVB, (1972), Academic Press (New York), pp. 326–332. "Edman Degradation."

A. Niederwieser, *ibid.*, pp. 60–99. "Thin Layer Chromatography of Amino Acids and Derivatives."

Experiment 4

Isolation and Purification of α-Lactalbumin, a Milk Protein

RECOMMENDED READING:
Chapter 2, Chapter 3, Sections A,D,E,F,H; Chapter 4, Chapter 5, Sections A,B,C and Chapter 7, Sections A,B

SYNOPSIS

α-Lactalbumin, a milk protein, is a component necessary for lactose synthesis. It can be readily extracted and purified from raw milk by a series of steps including salt and pH precipitation. Gel filtration on Sephadex G-50 provides more extensive purification of the protein. This experiment introduces the student to a series of protein purification steps that will yield α-lactalbumin in sufficient amounts and purity for further physical and biological characterization in Experiments 5 and 8.

I. INTRODUCTION AND THEORY

Protein purification is an activity that has occupied the time of biochemists throughout the history of biochemistry. In fact, a large percentage of the biochemical literature is a description of how specific proteins have been separated from the thousands of other proteins and biomolecules in tissues,

TABLE E4.1

Typical Sequence for the Purification of a Protein

1. Develop an assay for the desired protein.
2. Select the biological source of the protein.
3. Release the protein from the source and solubilize it in an aqueous buffer system.
4. Fractionate the cell components by physical methods (centrifugation).
5. Fractionate the cell components by differential solubility.
6. Ion exchange chromatography.
7. Adsorption chromatography.
8. Gel filtration.
9. Affinity chromatography.
10. Isoelectric focusing.

cells, and biological fluids. Biochemical investigations of all biological processes require, at some time, the isolation, purification, and characterization of a protein. Of course, there is no single technique or sequence of techniques that can be followed to purify all proteins. Most investigators approach the problem by trial and error. Fortunately, the experiences and discoveries of hundreds of biochemists have been combined so that today, a general and somewhat systematic approach is used for protein purification. The best purification procedure is one that yields a maximum amount of the purified protein in a minimum amount of time. The following discussion outlines the basic steps that must be considered in order to develop a protein purification scheme (Table E4.1).

Development of Protein Assay

First and foremost in any protein purification scheme is the development of an assay for the protein. This procedure, which may have a physical, chemical, or biological basis, is necessary in order to determine quantitatively or qualitatively the presence of the specific protein. During the early stages of purification, the particular protein desired must be distinguished from thousands of other proteins present in crude cell homogenates. The desired protein may be less than 0.1% of the total protein content of the crude extract. If the desired protein is an enzyme, the obvious assay will be based on biological function, that is, a measurement of the enzymatic activity after each isolation-purification step. The development of enzymatic assays is considered in greater detail in Experiment 7. If the protein to be isolated is not an enzyme or if the biological activity of the protein is unknown, physical or chemical methods must be used. One of the most useful analytical methods is electrophoresis (Chapter 4 and Experiment 5).

Source of the Protein

The selection of a source from which the desired protein is to be isolated should be considered. If the objective is simply to obtain a certain quantity of a protein for further study, then you would choose a source that contains large amounts of the protein. The best choice is an organ from a large animal that can be obtained from a local slaughterhouse. Microorganisms also serve as good sources, since they can be harvested in large quantities. If, however, you desire a specific protein from a specific type of cell, tissue, cell organelle, or biological fluid, then the source is limited.

Preparation of Crude Extract

Once the protein source has been selected, the next step is to release the desired protein from its natural cellular environment and solubilize it in aqueous solution. This calls for disruption of the cell membrane without damage to the cell contents. Proteins are relatively fragile molecules, and only gentle procedures are allowed at this stage. The gentlest methods for cell breakage are osmotic lysis, hand-operated glass homogenizers, and ultrasonic waves. These methods are useful for "soft tissue" as found in green plants and animals. When dealing with bacterial cells, where rigid cell walls are present, the most effective methods are grinding in a mortar with an inert abrasive such as sand or alumina, treatment with lysozyme, or both. Lysozyme is an enzyme that catalyzes the hydrolysis of polysaccharide moieties present in cell walls.

Osmotic lysis consists of suspending cells in a solution of relatively high ionic strength. This causes water inside the cell to diffuse out through the membrane. The cells are then isolated by centrifugation and transferred to pure water. Water rapidly diffuses into the cell, bursting the membrane.

Another alternative, gentle grinding, is best accomplished with a glass or Teflon homogenizer. This consists of a glass tube with a close-fitting piston. Several varieties are shown in Figure E4.1. The cells are forced against the glass walls under the pressure of the piston, and the cell components are released into an aqueous solution.

Ultrasonic waves, produced by a Sonicator, are transmitted into a suspension of cells by a metal probe (Figure E4.1c). The vibration set up by the ultrasonic waves disrupts the cell membrane, which allows the cell components to be released into the surrounding aqueous solution.

A less gentle, but widely used, solubilization method is the electric blender (Figure E4.1d). This method may be used for plant or animal tissue, but it is not effective for disruption of bacterial cell walls.

Since so many cell disruption methods are available, the experimenter must, by trial and error, find a convenient method that yields the maximum quantity of the protein with minimal damage to the molecules.

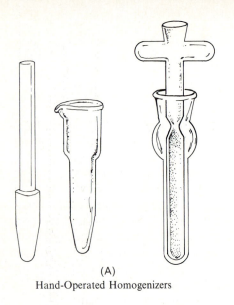

(A)

Hand-Operated Homogenizers

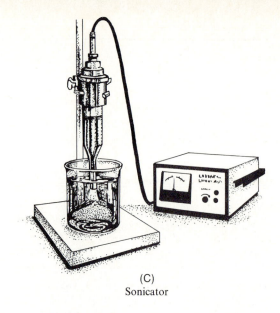

(C)

Sonicator

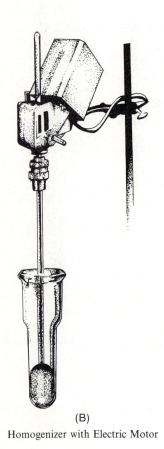

(B)

Homogenizer with Electric Motor

(D)

FIGURE E4.1

Tools for the preparation of a crude cell extract. (A) Hand-operated homogenizers, courtesy of Ace Glass, Incorporated, Vineland, NJ. (B) Homogenizer with electric motor, courtesy of VWR Scientific, Division of Univar. (C) Sonicator, Courtesy of Curtin Matheson Scientific, Inc. (D) Electric blender, photo courtesy of Mr. Mark Billadeau.

Stabilization of Proteins in a Crude Extract

Continued stabilization of the protein must always be considered in protein purification. While the protein is inside the cell, it is in a highly regulated environment. Cell components in these surroundings are protected against sudden changes in pH, temperature, or ionic strength, and against oxidation and enzymatic degradation. Once the cell wall barrier is destroyed, the protective processes are no longer regulated and degradation of the desired protein is likely to begin. An artificial environment that mimics the natural one must be maintained so that the protein retains its chemical integrity and biological function throughout the purification procedure. What factors are important in maintaining an environment in which proteins are stable? Although numerous factors must be considered, the most critical are (1) ionic strength and polarity, (2) pH, (3) the presence of metal ions, (4) oxidation, (5) endogeneous proteases, and (6) temperature.

The standard cellular environment is, of course, aqueous; because of the presence of inorganic salts, though, its ionic strength is relatively high. Addition of KCl, NaCl, or $MgCl_2$ to the cell extract may be necessary to maintain this condition. Proteins that are normally found in the hydrophobic regions of cells (i.e., membranes) are generally more stable in aqueous environments in which the polarity has been reduced by addition of 1 to 10% glycerol or sucrose.

The pH of a biological cell is controlled by the presence of natural buffers. Since protein structure is often irreversibly altered by extremes in pH, a buffer system must be maintained for protein stabilization (Johnson and Metzler, 1971). The importance of proper selection of a buffer system cannot be overemphasized. The criteria that must be considered in the selection of a buffer have been discussed in Chapter 2. Obviously a "perfect" buffer system is not available, but some are much better than others. Even at this advanced stage in our understanding of solution buffering, trial and error are still essential in selecting an effective, noninterfering buffer system. For most cell homogenates at physiological pH values, Tris and phosphate buffers are widely used.

The presence of metal ions in mixtures of biomolecules can be both beneficial and harmful. Metal ions such as Na^+, K^+, Ca^{2+}, Mg^{2+}, and Fe^{3+} may actually increase the stability of dissolved proteins. Heavy metal ions such as Ag^+, Cu^{2+}, Pb^{2+}, and Hg^{2+} are deleterious, particularly to those proteins that depend upon sulfhydryl groups for structural and functional integrity. The main sources of contaminating metals are buffer salts, water used to make buffer solutions, and metal containers and equipment. To avoid heavy metal contamination, you should use high-purity buffers, glass-distilled water, and glassware specifically cleaned to remove extraneous metal ions (Chapter 1). If metal contamination still persists, a chelating agent such as EDTA (1×10^{-4} M) may be added to the buffer.

This, of course, should not be used if the desired protein is an enzyme that is inhibited by EDTA.

Many proteins are susceptible to oxidation. This is especially a problem with proteins having free sulfhydryl groups, which are easily oxidized and converted to disulfide bonds. A reducing environment can be maintained by adding mercaptoethanol, cysteine, or dithiothreitol (1×10^{-3} M) to the buffer system.

Many biological cells contain degradative enzymes (proteases) that catalyze the hydrolysis of peptide linkages. In the intact cell, functional proteins are protected from these because the destructive enzymes are stored in cell organelles (lysosomes, etc.) and released only when needed. The proteases are freed upon cell disruption and immediately began to catalyze the degradation of protein material. This detrimental action can be slowed by the addition of specific protease inhibitors such as phenyl-methyl sulfonyl fluoride or certain peptides. It should be obvious that these inhibitors are to be used with extreme caution, since they are potentially toxic.

Many of the above conditions that affect the stability of proteins in solution are dependent upon chemical reactions. In particular, metal ions, oxidation processes, and proteases bring about chemical changes in proteins. It is a well-accepted tenet in chemistry that lower temperatures slow down chemical processes. We generally assume that proteins are more stable at cold temperatures. Although there are a few exceptions to this, it is fairly common practice to carry out all procedures of protein isolation under reduced temperature conditions (0 to 4°C).

After cell disruption, gross fractionation of the properly stabilized, crude cell homogenate may be achieved by physical methods, specifically centrifugation. Figure 7.9, Chapter 7, outlines the stepwise procedure commonly used to separate subcellular organelles such as nuclei, mitochondria, lysosomes, and microsomes. If the desired protein is known to be in one of these subcompartments of the cell, partial purification of the protein has already been achieved by isolating the organelle. If the protein is, instead, in the soluble cytoplasm of the cell, it will remain dissolved in the final supernatant obtained after centrifugation at $100,000 \times g$.

Separation of Proteins Based on Solubility Differences

Proteins are soluble in aqueous solutions primarily because their charged and polar amino acid residues are solvated by water. Any agent that disrupts these protein/water interactions will decrease the protein solubility because protein/protein interactions become more important. Protein/protein aggregates are no longer sufficiently solvated, and they precipitate from solution. Since each specific type of protein contains a unique amino acid composition and sequence, the degree and importance of water solvation will differ for each protein. Therefore, different proteins will precipitate at different concentrations of precipitating agent. The agents most often used for protein precipitation are (1) inorganic salts, (2) organic solvent, (3) polyethylene glycol (PEG), (4) pH, and (5) temperature. The most commonly used inorganic salt, ammonium sulfate, is highly solvated in water and actually reduces the water available for interaction with protein. As ammonium sulfate is added to a protein solution, a concentration of salt is reached at which there is no longer sufficient water present to maintain a particular type of protein in solution. The protein precipitates or is "salted out" of solution. The concentration of ammonium sulfate at which the desired protein precipitates from solution cannot be calculated, but must be established by trial and error. In practice, ammonium sulfate precipitation is carried out in stepwise intervals. For example, a crude cell extract is treated by slow addition of dry, solid, high-purity ammonium sulfate in order to achieve a change in salt concentration from 0 to 25%. The solution, now 25% in ammonium sulfate, is gently stirred for up to 60 minutes and is subjected to centrifugation at $20,000 \times g$. The precipitate that separates upon centrifugation and the supernatant are analyzed for the desired protein. If the protein is still predominantly present in the supernatant, the salt concentration is increased from 25 to 35%. This process of ammonium sulfate addition and centrifugation is continued until the desired protein is "salted out." Although other inorganic salts are used, ammonium sulfate has many advantages including high solubility in water and less chance of protein denaturation.

Organic solvents also decrease protein solubility, but they are not as widely used as ammonium sulfate. They are thought to function as precipitating agents in two ways: (1) by dehydrating proteins, much like ammonium sulfate, and (2) by decreasing the dielectric constant of the solution. The organic solvents used (which, of course, must be water soluble) include methanol, ethanol, and acetone. Organic solvents must be used with care because they are more likely than ammonium sulfate to cause protein denaturation.

A relatively new method of selective protein precipitation is the use of nonionic polymers. The most widely used agent in this category is polyethylene glycol (Honig and Kula, 1976). The polymer is available in a variety of molecular weights ranging from 400 to 7500. The biochemical literature reports successful use of different sizes, but lower molecular weight polymer has been shown to be the most specific. The principles behind the action of PEG as a protein precipitating agent are not completely understood. Possible modes of action include (1) complex formation between protein and polymer and (2) exclusion of the protein from part of the solvent (dehydration) followed by protein aggregation and precipitation. This method of protein precipitation should receive considerable attention in the future, since it has several advantages over other fractionation techniques. The use of nonionic polymers is simple, straightforward, and gentle. PEG-precipitated protein usually retains its native conformation and biological function. Possibly the most attractive advantage of PEG is its ability to fractionate proteins on the basis of size and shape as in gel filtration. Although it is too early to state a general rule, most fractionation experiments with PEG show that the larger the protein, the less soluble it is in highly concentrated polymer solution.

Finally, changes in pH and temperature have been used effectively to promote selective protein precipitation. A change in the pH of the solution alters the ionic state of a protein and may even bring some proteins to a state of charge neutrality. Charged protein molecules tend to repel each other and remain in solution; however, neutral protein molecules are not electrically repulsed from each other so they tend to aggregate and precipitate from solution. A protein is least soluble in aqueous solution when it has no net charge, that is, when it is isoelectric. This characteristic can be used in protein purification, since different proteins usually have different isoelectric pH values.

An increase in temperature generally causes an increase in the solubility of solutes. This general rule is followed by most proteins up to about 40°C. Above this temperature, many proteins aggregate and precipitate from solution. If the protein of interest is heat stable and still water soluble above 40°C, a major step in protein purification can be achieved since most other proteins precipitate below 40°C and can be removed by centrifugation.

With so many different protein fractionating methods available, it becomes difficult to choose the correct one. As a guideline, it is fair to state that all precipitation methods discussed will provide some purification for most proteins. The effectiveness of all of them is very nearly the same, so the selection should be based on the convenience of the method and the stability of the desired protein under the experimental conditions.

Selective Techniques in Protein Purification

After gross fractionation of proteins, as discussed above, more refined methods with greater resolution can now be attempted. These methods, in order of increasing resolution, are ion-exchange chromatography, adsorption chromatography, gel filtration, affinity chromatography, and isoelectric focusing. Since the basis of protein separation is different for each of these techniques, it is most effective and appropriate to use all of the techniques in the order given. At least three logical reasons can be cited for proceeding according to this order: (1) it is in the direction of increasing refinement and resolution, (2) the amount of protein material required decreases with each step, and (3) the time necessary to complete each step increases in the same direction. As a protein is purified, it generally becomes less abundant and more stable in solution, so techniques that take longer periods may be utilized in the final stages of protein purification. Since the final stages of purification are capable of the most resolution, the extra time required is often justified.

The chromatographic methods have been discussed in Chapter 3. However, we must consider an important topic, preparation of protein solutions for chromatography. Fractionation of heterogeneous protein mixtures by inorganic salts, organic solvents, or PEG usually precedes ion-exchange chromatography. The presence of the precipitating agents will interfere with the later chromatographic steps. In particular, the presence of ammonium sulfate increases the ionic strength of the protein solution and damps the ionic protein–ion exchange resin interactions. Three procedures are in current use to remove undesirable small molecules from protein solutions: (1) dialysis (Chapter 2), (2) gel filtration (Chapter 3, and Pharmacia Fine Chemicals, 1979), and (3) hollow fiber filtration (Chapter 2, and Cooper, 1977). All three methods effectively remove small molecules without serious damage to or loss of proteins. Dialysis requires extended periods of time, sometimes 1 or 2 days for completion, whereas gel filtration and hollow fiber filtration are accomplished in 30 to 60 minutes.

A wide variety of chromatographic methods and support media are available. The best known support for ion-exchange chromatography of proteins is DEAE cellulose, an anion exchanger. CM cellulose and phosphocellulose are both widely used as cation exchangers. The many uses of these chromatographic methods are discussed in Chapter 3.

To some individuals, especially those who recall their experiences in organic chemistry laboratory, the ultimate step in purification of a molecule is crystallization. The desire to obtain crystalline protein has long been powerful, and many proteins have been crystallized. However, there is a common misconception that the ability to form crystals of a protein ensures that the protein is homogeneous. For many reasons (entrapment of contaminants within crystals, aggregation of protein molecules, etc.), the ability to crystallize a protein should not be used as a criterion of purity. However, protein crystallization is a worthwhile endeavor in the final stages of protein purification since it would be expected to remove minor impurities. One of the most effective methods for crystallization utilizes the differential solubility of proteins in ammonium sulfate associated with temperature changes (Jakoby, 1971).

Purification of α-Lactalbumin

A crude preparation of α-lactalbumin was first obtained from milk in 1899. Since that time, several investigators have developed procedures for the isolation and purification of the protein. It has a molecular weight of 15,500 and contains 129 amino acid residues. α-Lactalbumin, which is present at concentrations averaging 1 mg/ml, was at first thought to serve a nutritional function. We now know that it is an essential component of the lactose synthetase system. A thorough discussion of the history and biological role of α-lactalbumin is available in the literature (Ebner, 1970). We will further consider the biological role of α-lactalbumin in Experiment 8, but here we are concerned with the isolation and purification of the protein.

The primary proteins in milk are the caseins. These may be removed from milk by acid precipitation at pH 4.6 or by heat and sodium sulfate treatment. α-Lactalbumin remains in solution after either of these treatments. Study the flow chart in Figure E4.2 for a summary of the procedure for α-lactalbumin purification. Step I provides a supernatant that contains

FIGURE E4.2

Flow chart for isolation of α-lactalbumin from raw milk.

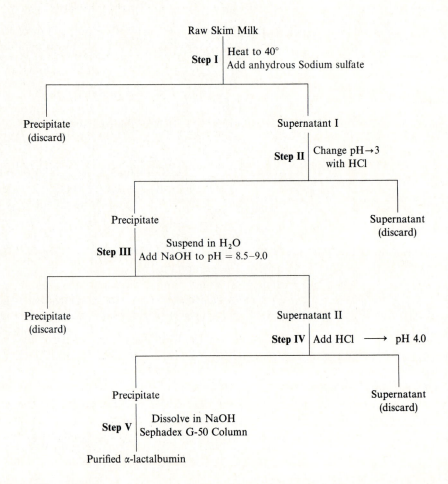

α-lactalbumin and other milk albumins but not the caseins. α-Lactalbumin is insoluble in solutions of less than pH 4, so addition of HCl to create a pH of 3 yields a precipitate that contains the desired protein (Step II). α-Lactalbumin is quite soluble in solutions above pH 6, and Step III takes advantage of this property to extract the desired protein from the precipitate. The base-soluble protein is further purified by a second acid precipitation (Step IV) and, finally, Sephadex gel filtration (Step V).

Recall from the previous discussion of protein purification that an important step is the development of a specific assay for the desired protein. The assay must be specific so that the desired protein can be detected both quantitatively and qualitatively in a solution containing thousands of other proteins. This is difficult for α-lactalbumin, because it has no unique physical, chemical, or biological property that can be rapidly measured. Measuring its biological activity as a modifier protein in the lactose synthetase system is the most specific way to analyze α-lactalbumin; however, the assay, which requires a coupled enzyme system, is too time-consuming to complete in this lab period. This biological assay for α-lactalbumin will be introduced in Experiment 8. Of course, it would be possible to detect the presence of α-lactalbumin in crude extracts using electrophoresis. However, this would not give quantitative information and it could not be completed during this lab period. Experiment 5 will detail the electrophoretic analysis and characterization of α-lactalbumin.

Proteins have a broad characteristic absorption spectrum that is centered at about 280 nm. The major absorption is due to the presence of aromatic moieties in the amino acids phenylalanine, tyrosine, and tryptophan. During α-lactalbumin purification described in this experiment, you will monitor the process by measuring the absorption at 280 nm (A_{280}) of column fractions to be sure the experiment is proceeding correctly. You must recognize that you are not measuring the concentration or presence of α-lactalbumin specifically, but the total amount of protein present. However, if you follow the written procedure carefully, you can be assured of a good yield of purified α-lactalbumin.

Overview of the Experiment

The complete isolation and purification of α-lactalbumin as described here requires about 6 hours. Steps I to IV can be completed in 3 hours, but step V requires another 3 hours. The first four steps may be completed during the first lab period, and the precipitate is dissolved in 0.05 M ammonium bicarbonate solution and frozen until the next lab period.

Alternatively, students may be given commercial α-lactalbumin that is further purified by step V, gel filtration. The experiment can then be completed in one lab period. Similar results are obtained with the Sephadex column whether the α-lactalbumin is commercially obtained or purified by steps I to IV.

II. EXPERIMENTAL

Materials and Supplies

Raw, unpasteurized skimmed milk, 100 ml

Anhydrous sodium sulfate, 20 g

Concentrated HCl

HCl, 1 N

Magnetic stirring motor and stir bar

NaOH, 1 N

Sephadex G-50 in 0.05 M ammonium bicarbonate, 100 ml of slurry; use fine or medium grade

Ammonium bicarbonate, 0.05 M, 1 liter

Cheese cloth

Glass column, 2.5 × 40 cm

2 matched quartz cuvets

Small wad of cotton

Test tubes for column fractions, 100, 10 × 100 mm

Centrifuge tubes, 50 ml or 250 ml

pH paper

Electric hot plate

Refrigerated centrifuge capable of 10,000 to 12,000 × g

UV-VIS spectrophotometer

Fraction collector

pH meter

Procedure

Step I

Pour 100 ml of raw milk into a 1-liter beaker and heat, on the electric plate with stirring, to 40°C. Weigh 20 g of anhydrous sodium sulfate and add, in small portions, to the stirred, warm milk. The salt should be added over a period of 10 to 15 minutes. Allow the solution to stir for 10 minutes after addition of the sodium sulfate. Filter the solution through several layers of cheese cloth with a Büchner funnel, with gentle suction from an aspirator. Discard the precipitate with the cheese cloth.

Step II

Pour the filtrate into a 1-liter beaker and gently stir with a magnetic stirring motor and bar. Adjust the pH to 3 ± 0.1 (pH meter) with dropwise addition of concentrated HCl. Centrifuge the suspension at $10,000 \times g$ for 15 minutes and discard the supernatant.

Step III

Suspend the precipitated protein in about 10 ml of distilled water and adjust the pH to 8.5 to 9.0 (pH paper) with dropwise addition of 1 N NaOH. Mix well, but do not shake. Most of the protein should dissolve. Centrifuge at $12,000 \times g$ for 10 minutes. Decant the supernatant into a 50 ml beaker.

Step IV

Adjust the pH of the supernatant to pH 4.0 (pH meter) with dropwise addition of 1 N HCl. Stir gently while adding HCl. Centrifuge the acidified solution at $12,000 \times g$ for 10 minutes. Decant and discard the supernatant.

Step V

Dissolve the precipitate from step IV in a minimal amount of 0.05 M ammonium bicarbonate. If necessary, this solution may be frozen and stored for the next lab period.

Packing the Sephadex Column. The laboratory assistants have prepared the Sephadex G-50 for your use. They have preswollen the dry beads in water for several hours, removed fine particles that may slow column flow, and equilibrated the gel in 0.05 M ammonium bicarbonate. If chromatography columns are not available, one can easily be constructed from a glass tube (2.5×40 cm), cork stoppers, Tygon tubing, a screw clamp, and small glass tubing as shown in Figure E4.3. With the screw clamp closed, add 10 to 15 ml of 0.05 M ammonium bicarbonate and insert a small piece of cotton into the bottom of the column. Force air bubbles out of the cotton with a long glass rod. Pour a well-mixed slurry of Sephadex G-50 into the column. Open the screw clamp to allow a slow column flow. As the gel settles into a bed, add more slurried G-50 until the settled bed reaches about 30 cm. During this packing process, never allow the column to run dry. If more bicarbonate solution is necessary, add carefully without disturbing the gel. After a bed of desired height has been prepared, continue eluting the column with bicarbonate. Cut a piece of filter paper into a circle with a diameter of 2 to 2.3 cm and put it on top of the Sephadex bed. Be careful not to disturb the gel.

Sample Application. Remove the α-lactalbumin extract (step IV) from the freezer and allow it to thaw in ice. If you are using commercial α-lactalbumin, prepare a solution of 10 mg of protein in 2 ml of 0.05 M ammonium bicarbonate. After the eluting solvent has just drained into the top of the G-50 column (excess bicarbonate may be removed from

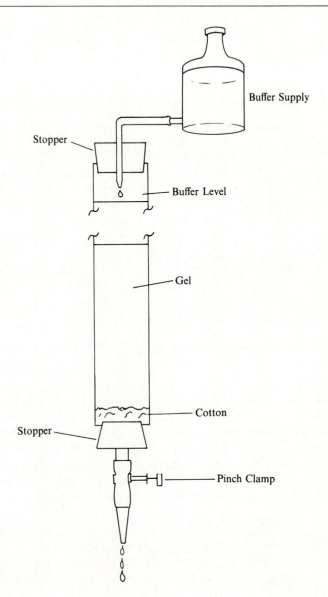

Buffer Supply

Stopper

Buffer Level

Gel

Cotton

Stopper

Pinch Clamp

FIGURE E4.3
A simple column for gel filtration.

the top of the column with a pipet), add the α-lactalbumin solution drop-wise to the top of the gel. Use a pipet and avoid disturbing the top sur-face of the gel. After the protein solution has just entered the gel, slowly and carefully add 10 ml of ammonium bicarbonate solution to the top of the column without disturbing the gel surface and allow this to penetrate the surface of the column. Then, immediately add bicarbonate solution to fill the column. Set up an eluting solvent supply as shown in Figure E4.3. This will allow solvent to enter the column at the same rate as solvent elution from the column.

   Column Development, Fraction Collection, and Analysis of Fractions. The stopper in the top of the column must be air-tight. Adjust the screw clamp so that the flow rate of the column is about 10 to 15 ml/hour. Collect 5-ml fractions in test tubes in a fraction collector. Transfer the contents of every fifth fraction to a quartz cuvet and measure the absorbance at 280 nm. For a blank, use 0.05 M ammonium bicarbonate. Record the A_{280} values in your notebook. When you begin to detect protein eluting from the column ($A_{280} > 0$), measure the A_{280} of every fraction. On a single sheet of graph paper, plot A_{280} vs. fraction number. Continue collecting fractions until no more protein is eluted (about 30 fractions). Save all fractions that have A_{280} over 0.2. All of the fractions should be frozen until use in Experiments 5 and 8.

   When you are finished with the Sephadex column, pour the gel into a container labeled "used Sephadex 50." The gel may be recycled and used later.

III. ANALYSIS OF RESULTS

Study your experimentally derived plot of A_{280} vs. Sephadex column fraction number. How many major protein peaks are present? Which peak has the greatest amount of protein? Do you believe that each peak contains a single type of protein, or does it contain a mixture? If each peak represents a mixture of proteins, what one thing does each protein have in common?

   The presence of α-lactalbumin in a fraction cannot be confirmed until completion of Experiments 5 and 8. However, if you know that α-lactalbumin is among the smallest proteins in milk (molecular weight is 15,500), could you determine which protein peak contains α-lactalbumin?

IV. QUESTIONS

1. What other purification techniques could be applied to the purification of α-lactalbumin? Study the following references for methods that have been applied to the isolation and purification of α-lactalbumin. W.G. Gordon and W.F. Semmett, *J. Amer. Chem. Soc.*, *75*, 328 (1953). W.G. Gordon and J. Ziegler, *Biochemical Preparations*, *4*, 16 (1955). U. Brodbeck, W.L. Denton, N. Tanahashi, and K.E. Ebner, *J. Biol. Chem.*, *242*, 1391 (1967). F.J. Castellino and R.L. Hill, *J. Biol Chem.*, *245*, 417 (1970).

2. What methods of analysis could you perform on the purified α-lactalbumin in order to determine the extent of purity?

3. Absorbance measurements of column fractions at 280 nm were used in this experiment to detect the presence of protein material. Do you think this method of analysis could lead to a quantitative determination of protein concentration? What other biomolecules might interfere with this measurement?

V. REFERENCES

General

R. Bohinski, *Modern Concepts in Biochemistry*, Fourth Edition (1983), Allyn and Bacon (Boston), pp. 90–102. A brief introduction to protein purification.

J.M. Brewer, A.J. Pesce, and R.B. Ashworth, *Experimental Techniques in Biochemistry* (1974), Prentice-Hall (Englewood Cliffs, NJ), pp. 32–96. Chromatographic techniques.

T.G. Cooper, *The Tools of Biochemistry* (1977), Wiley-Interscience (New York), pp. 355–405. A detailed chapter on protein purification.

R.H. Haschemeyer and A.E.V. Haschemeyer, *Proteins* (1973), Wiley-Interscience (New York), pp. 31–53. Purification and analysis of proteins.

J.D. Rawn, *Biochemistry* (1983), Harper and Row (New York), pp. 163–192. Isolation and purification of biomolecules.

G. Zubay, *Biochemistry* (1983), Addison-Wesley (Reading, MA), pp. 39–67. Analysis of protein structure.

Specific

T.G. Cooper, *The Tools of Biochemistry*, (1977), Wiley-Interscience (New York), pp. 384–386. Hollow-fiber filtration.

K.E. Ebner, *Acct. Chem. Res.*, *3*, 41–47 (1970). Biological role of α-lactalbumin.

W. Honig and M.R. Kula, *Anal. Biochem.*, *72*, 502–512 (1976). Selective protein precipitation with PEG.

W.B. Jakoby in *Methods in Enzymology*, W. Jakoby, Editor, Vol. XXII (1971), Academic Press (New York), pp. 248–252. Crystallization of proteins.

R.J. Johnson and D.E. Metzler in *Methods in Enzymology*, W. Jakoby, Editor, Vol. XXII (1971), Academic Press (New York), pp. 3–5. Preparation of buffers.

Pharmacia Fine Chemicals, *Gel Filtration, Theory and Practice* (1979). 800 Centennial Avenue, Piscataway, NJ 08854. A good review with discussion of experimental use.

Characterization of α-Lactalbumin

RECOMMENDED READING:
Chapter 2, Section E; Chapter 4, Sections A, B, D; Chapter 5, Sections A, B, C.

SYNOPSIS
In this experiment, α-lactalbumin isolated from milk in Experiment 4 is analyzed by polyacrylamide gel electrophoresis, the Lowry protein assay, and ultraviolet spectroscopy.

I. INTRODUCTION

In Experiment 4, α-lactalbumin was isolated from milk and purified by selective precipitation and gel filtration. Recall that the procedure provided you with a final product that had an ultraviolet absorption at 280 nm, thus indicating that protein material may be present in solution. However, no analysis was done to confirm the presence of protein material, to identify the protein, or to determine the amount isolated. The form of α-lactalbumin isolated in Experiment 4 was a solution in ammonium bicarbonate buffer. Several fractions containing protein were obtained from a Sephadex G-50 column. This is a form quite amenable to further analysis and characterization.

Quantitative Analysis of Protein

Several useful methods for the quantitative determination of protein solutions were discussed in Chapter 2. Two of those methods, a colorimetric protein assay and the spectrophotometric assay, will be applied to the α-lactalbumin solutions. Neither of these assays is specific for a certain type of protein; rather, they both estimate total protein content.

Analysis of Purity

It is unlikely that the protein fractions from Experiment 4 contain a single type of protein. How many different proteins are present? What is the relative abundance of each protein? Is α-lactalbumin the predominant protein in the Sephadex fractions? These questions may be answered by analysis of the pooled fraction by electrophoretic methods. The technique of polyacrylamide gel electrophoresis will be introduced and applied to isolated and standard α-lactalbumin.

Characterization by UV Spectroscopy

The absorption spectrum of a protein is a characteristic property of that specific protein and can be used to aid in the identification of an unknown. However, it is helpful to have available standard spectra of known proteins for comparison.

A UV spectrum can also be used for the measurement of physical constants characteristic of a specific protein. The α-lactalbumin fraction will first be characterized by measurement of the spectrum from 240 to 340 nm. From this, the extinction coefficient is calculated using the Beer-Lambert relationship. Finally, the A_{280}/A_{290} ratio is calculated; this may be used as an indicator of the purity of the isolated α-lactalbumin.

Overview of the Experiment

Part A, gel electrophoresis, may require two class periods for completion; however, much of the time can be used to complete Parts B and C. It is recommended that Part A be done in groups of four to six students. Preparation of the polyacrylamide gels requires approximately 1 to $1\frac{1}{2}$ hours. The electrophoresis operation will run 1 to 2 hours, but constant attention is not required during this time interval. Staining of the gels (Coomassie blue) requires several hours or even days, but again constant attention is not essential. Silver staining requires less time and is more sensitive. The specific conditions given here are for disc gel electrophoresis in the column form. The analysis may also be done by slab gel electrophoresis.

Part B, protein measurement by the Lowry method, is completed in 1 to $1\frac{1}{2}$ hours, including a 30-minute period for color development. The Bradford assay may be used for this analysis.

The UV spectrum, Part C, may be completed in about 30 minutes.

II. EXPERIMENTAL

Materials and Supplies

CAUTION:
Acrylamide in the unpolymerized form is a skin irritant and a potential neurotoxin. Wear gloves and a mask while weighing the dry powder. Do not breathe the dust. Prepare all acrylamide solutions in the hood. Do not mouth pipet any solutions used for gel formation or staining.

Part A. Polyacrylamide Disc Gel Electrophoresis (Modification based on the procedure of Davis, 1964)

Gel electrophoresis system with power supply

Glass tubes, 6 mm ID × 7.5 cm, cleaned in chromic acid solution (Chapter 1) and rinsed in distilled water

α-Lactalbumin samples (standards and samples isolated in Experiment 4)

Stock Solutions

    A. Tris-glycine buffer, pH 8.3

    B. Tris buffer containing TEMED, pH 8.9

    C. 28% acrylamide plus 0.74% bis-acrylamide

    D. Tris buffer containing TEMED, pH 6.9

    E. 10% acrylamide plus 2.5% bis-acrylamide

    F. Riboflavin, 4 mg/dl H_2O

    G. Ammonium persulfate, 0.14% in H_2O; prepare just before use

    H. Sucrose, 40 g/dl H_2O

    I. Bromphenol blue tracking dye, 0.02% in H_2O

Fluorescent lamp

Rubber septum caps

Parafilm or plastic wrap

Syringe and needle

Staining with Coomassie Blue:

    Dye solution, 0.25% Coomassie blue in methanol:acetic acid:water (5:1:5)

    Destaining solution, acetic acid:methanol:water (7:7:86)

Staining with Silver:

    Prefixing solution 1, methanol:water:acetic acid (50:40:10)

    Prefixing solution 2, water:acetic acid:methanol (88:7:5)

    Fixing solution, 10% glutaraldehyde in H_2O

    Silver nitrate, 0.1% in H_2O

Developing solution, 50 μl of 37% formaldehyde in 1 dl of 3% sodium carbonate

Dithiothreitol solution, 5 μg/ml H_2O

Citric acid, 2.3 M

Part B. The Protein Assay

CAUTION:

The Folin-Ciocalteu solution and Bradford dye reagent are poisonous. Do not mouth pipette!

Lowry Assay

α-Lactalbumin, isolated from Experiment 4

Bovine serum albumin standard, 0.1 mg/ml in H_2O

Solution A, $CuSO_4$, 1% in H_2O

Solution B, Na_2CO_3, 2% in 0.1 M NaOH

Solution C, sodium tartrate, 2% in H_2O

Solution D, Folin-Ciocalteau (phenol) reagent, 2 N

Spectrophotometer for reading 540 nm with glass cuvets (1 or 3 ml)

Test tubes, 10 × 100 mm

Bradford Assay

α-Lactalbumin, isolated from Experiment 4

Bovine gamma globulin standard, 0.1 mg/ml in H_2O

Bradford dye reagent. This is a commercially available mixture of Coomassie Brilliant Blue G-250 dye, phosphoric acid, and methanol. One volume of dye reagent should be diluted with 4 volumes of distilled water prior to use. Filter and store in a glass container at room temperature. The dilute solution is stable for 2 weeks.

Spectrophotometer for reading A_{595} with glass cuvets (1 or 3 ml)

Test tubes, 10 × 100 mm

Part C. Ultraviolet Spectrum

Recording spectrophometer with a matched pair of quartz cuvets

Ammonium bicarbonate, 0.05 M

α-Lactalbumin, isolated from Experiment 4 and standard

Procedure

A. Polyacrylamide Disc Gel Electrophoresis

Preparation of the Polyacrylamide Gels

Prepare the following working solutions. The letters refer to the stock solutions listed under Part A, above.

I. Separation gel
   1.0 ml of Solution B
   2.0 ml of Solution C
   1.0 ml of H_2O

II. Stacking gel
   1.0 ml of Solution D
   2.0 ml of Solution E
   1.0 ml of Solution F
   4.0 ml of Solution H

Obtain two clean glass tubes (6 mm ID × 7.5 cm) for each protein sample to be analyzed by electrophoresis. It is recommended that each student group run a total of four gels, two gels of different protein concentrations on isolated α-lactalbumin and two gels on standard α-lactalbumin. Cut 2 × 2 cm pieces of parafilm or plastic wrap, one for each glass tube. Wrap each film around an end of a glass tube and insert that end in a rubber septum. Prepare the separation gel solution by mixing 4 ml of separation gel working solution I with 4 ml of ammonium persulfate, Solution G. Mix well and *immediately* transfer, with a Pasteur pipet, 1 ml (20 drops) of the solution to each glass tube. Gently place a drop of water on top of the gel solution. Allow the gel solution to polymerize (about 30 minutes).

During the polymerization process, prepare the stacking gel working solution, II. After the polymerization period, carefully remove the water layer above the separation layer with a Pasteur pipet. Carefully transfer 0.2 ml (4 drops) of working solution II to the top of the separation gel, add a drop of water as before, and place the tubes under a fluorescent lamp for about 30 minutes in order to polymerize the stacking gel. Remove the layer of water after polymerization.

Remove the glass tubes from the rubber septa and unwrap the plastic film. Label the gel tube and place it in the gel electrophoresis system with the stacking gel up. The bottom end of each tube must be in contact with the lower chamber buffer (Tris-glycine buffer, A) as shown in Figure E5.1. When all positions in the electrophoresis system are fitted with gel-filled glass tubes (or rubber stoppers, if not enough glass tubes are available), pour Tris-glycine buffer A into the top unit until the tops of all the tubes are covered with buffer. Remove air bubbles from the glass tubes with a Pasteur pipet. For specific instructions on the use of your electrophoresis chamber, ask your instructor or read the directions accompanying the electrophoresis unit.

Application of the Protein Samples onto the Gels

For each sample to be analyzed, carry out the following procedure. Mix two portions of each protein solution (0.1 ml and 0.2 ml) with 0.1 ml of the sucrose solution (H). Add 0.1 ml of the bromphenol tracking dye

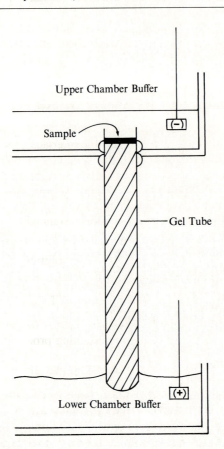

FIGURE E5.1
Placement of a disc gel column in
an electrophoresis chamber.

solution to each. With a long-tipped pipet, transfer a 0.1 ml protein sample to the top of a glass tube, one sample per tube. When applying, put the pipet tip through the upper buffer to the top of the gel in the glass tube. The sample solution, which is made more dense than the buffer by the sucrose solution, will settle onto the top of the gel.

Electrophoresis Procedure

CAUTION:
Do not touch the electrophoresis chamber or wires while the electrophoretic operation is in progress. Voltages may reach as high as 300 volts and shocks may be fatal.

Complete the construction of the electrophoresis unit and have your instructor check the apparatus before it is connected to the power supply.

Subject the tubes to electrophoresis at 5 mA/tube until the bromphenol tracking dye has moved to about one cm from the bottom end of the tubes (1 to 2 hours). *Turn off the power supply current*, open the unit, and remove the gel tubes.

Removing and Staining the Gels

To remove the gels from the glass tubes, fill a syringe with water, insert the needle carefully into the interface between the glass wall and gel, and slowly force water from the syringe while rotating the tube. Place the gel into a petri dish (or a screw-cap tube) and cut the gel just at the bottom of the tracking dye. Discard the small tip of the gel (about 1 cm).

Procedure for Coomassie Stain

Pour Coomassie dye solution into the petri dish or test tube until the gel is covered. Allow the dye solution to work on the gel for about 30 minutes. Remove the dye solution and cover the gel with the destaining solution. At intervals of a few hours, remove the colored destaining solution by decantation and replace with fresh destaining solution. This is repeated until most of the color is out of the gel except for the colored bands of protein. The destaining procedure may require several days.

Procedure for Silver Stain (Morrissey, 1981)

Prepare the gels for fixing by soaking in prefixing solution 1 for 30 minutes, then in prefixing solution 2 for 30 minutes. The gels are then fixed with 10% glutaraldehyde for 30 minutes to retard protein movement or diffusion. Rinse the gels with distilled water, making several changes during a 2-hour interval. Soak in dithiothreitol solution (5 μg/ml) for 30 minutes. Decant the solution and soak in 0.1% silver nitrate for 30 minutes. Rinse once with distilled water and twice with formaldehyde developing solution. Allow the gels to soak in the developing solution until the protein bands are readily visible. Stop the staining process with 5.0 ml of 2.3 M citric acid. Finally, rinse the stained gels several times with water.

Analysis of Stained Gels

Measure the total length of each gel and the distance migrated by each protein and the tracking dye from the top. In your notebook, draw a picture of each gel with protein bands.

B. The Protein Assay

Lowry Assay

Set up ten 10 × 100 mm test tubes (colorimetric tubes) and add reagents according to Table E5.1. Tube 1 is used as a blank and tubes 2 through 6 are for construction of a standard calibration curve. Tubes 7

TABLE E5.1

Procedure for the Lowry Protein Assay

| Reagents | Tube # | | | | | | | | | |
	1	2	3	4	5	6	7	8	9	10
water	1.0	0.9	0.8	0.6	0.4	0.2	0.7	0.7	0.4	0.4
standard BSA or gamma globulin	—	0.1	0.2	0.4	0.6	0.8	—	—	—	—
α-lactalbumin	—	—	—	—	—	—	0.3	0.3	0.6	0.6
alkaline CuSO₄ solution	5.0 ——————————————————————————————————→									
mix well, wait 10 min										
Folin-Ciocalteau	0.25 —————————————————————————————————→									
mix well, wait 30 min										
Read A_{540}										

to 10 are duplicates of two different concentrations of the isolated α-lactalbumin solution. Notice that water is added to give a final volume of 1.0 ml in each tube.

Prepare a fresh working solution of alkaline copper sulfate following these exact directions: Transfer 1.0 ml of solution C (sodium tartrate) into a 125 ml Erlenmeyer flask. Add, with stirring, 1.0 ml of solution A ($CuSO_4$) and then 98 ml of solution B (Na_2CO_3). Mix well while adding. This solution will remain stable for about one day. Add 5 ml of the alkaline copper sulfate solution to each tube and *immediately* mix well, using a vortex mixer if possible. Allow the tubes to stand for 10 minutes at room temperature. Add 0.25 ml of 2 N Folin-Ciocalteau solution and immediately mix well. After a 30-minute period for color development, read and record the absorbance at 540 nm. Use Tube 1 as a blank to zero the spectrophotometer.

Bradford Assay

Set up 10 test tubes (10 × 100 mm, colorimetric tubes) and add water and proteins according to the top three rows of Table E5.1. Tube 1 is used as a blank and tubes 2 through 6 are for construction of a standard calibration curve. Tubes 7 to 10 are duplicates of two different concentrations of the isolated α-lactalbumin solution. Water is added to give a final volume of 1.0 ml in each tube. Add 5.0 ml of dilute Bradford dye reagent to each tube and mix well by gentle inversion. After a period of at least 5 minutes, read A_{595} for each tube, using Tube 1 as a blank. The tubes should be read within an hour after adding the dye.

C. Ultraviolet Spectrum

Turn on the recording spectrophotometer and the hydrogen/deuterium lamp. Allow the instrument to warm up for about 15 minutes. Set the wavelength to 260 nm. During the warm-up period, prepare samples by transferring 1 or 3 ml of ammonium bicarbonate buffer into a quartz cuvet and an equal volume of α-lactalbumin solution into a matched cell. Place the cuvet with buffer only in the sample beam and adjust the absorbance to zero. Remove the blank cuvet and replace with the cuvet containing α-lactalbumin. Read and record the absorbance at 260 nm. If the A_{260} is greater than 1.0, dilute the sample with ammonium bicarbonate and repeat the reading.

Using the above procedure, read and record the absorbances at 280 and 290 nm.

Obtain the continuous spectrum of the α-lactalbumin sample from 240 to 340 nm. Do this by placing a cuvet containing ammonium bicarbonate in the reference beam and α-lactalbumin in the sample beam. Record the ultraviolet spectrum. Dilute the α-lactalbumin solution with ammonium bicarbonate if the recorder runs off-scale. Record the spectrum with standard α-lactalbumin.

III. ANALYSIS OF RESULTS

A. Disc Gel Electrophoresis

Study the destained gels or diagrams and estimate the number of protein components in each sample. Calculate the *relative mobility* of each protein band using Equation 1.

$$\text{relative mobility} = \frac{\text{distance moved by protein}}{\text{distance moved by dye}} \qquad \text{(Equation E5.1)}$$

Prepare a table of relative mobilities of all bands in each gel. Compare the gels for isolated α-lactalbumin to those for standard α-lactalbumin. Is α-lactalbumin the predominant protein in your preparation from milk? Try to estimate the percentage of the isolated sample that is α-lactalbumin. Assume that all proteins on the gel stain to the same extent with the dye, even though this is probably not true.

B. The Protein Assay

Prepare a standard calibration curve using the data obtained from bovine serum albumin or gamma globulin (Table E5.1, tubes 2 to 6). Plot the absorbance on the *y* axis and mg of standard protein per assay on the *x* axis. Use the standard curve to calculate the protein concentration in the isolated α-lactalbumin fractions (in units of mg/ml).

C. Ultraviolet Spectrum

Calculation of Protein Concentration. Calculate the ratio A_{280}/A_{260}. Use Table 2.4 in Chapter 2 to find the factor F that corresponds to the percentage of nucleic acid in the fraction and then calculate the protein concentration using both of the following relationships.

$$\text{protein concentration (mg/ml)} = A_{280} \cdot F \qquad \text{(Equation E5.2)}$$

or

$$\text{protein concentration (mg/ml)} = 1.55\, A_{280} - 0.76\, A_{260}$$

$$\text{(Equation E5.3)}$$

Compare these two results with the results from the protein assay.

Calculation of Extinction Coefficient. Use the Beer-Lambert relationship (Equation 4) to calculate the extinction coefficient in terms of $E^{1\%}$ at 280 nm.

$$A = Ebc \qquad \text{(Equation E5.4)}$$

where

A = absorbance at 280 nm

E = extinction coefficient, $E^{1\%}$

b = pathlength (usually 1 cm)

c = concentration of protein in mg/ml

The literature value for $E^{1\%}_{280}$ of purified bovine α-lactalbumin is 20.1. Calculate the ratio A_{280}/A_{290}, and compare to the standard value for purified α-lactalbumin of 1.30.

Ultraviolet Spectrum. Compare the spectrum of isolated α-lactalbumin with that of standard α-lactalbumin. Describe and explain any differences.

IV. QUESTIONS

1. What assumptions must be made about the relative mobility of bromphenol when used as a tracking dye?

2. What would be the effect of each of the following changes on the relative mobility of α-lactalbumin in disc gel electrophoresis?
   (a) lower the pH of all buffers
   (b) increase the ionic strength of the buffers

(c) change the temperature of the electrophoretic operation

(d) increase the concentration of α-lactalbumin

(e) increase the concentration of bis-acrylamide in the gels

(f) decrease the electrophoretic running time

3. Describe the chemistry involved in the Lowry assay (see Chapter 2).

4. Describe the procedure for obtaining a "difference spectrum" of isolated α-lactalbumin vs. standard α-lactalbumin.

5. Why do most protein solutions absorb light of 280 nm wavelength?

6. Comment on the validity of the following statement. "Ideally, the best Lowry (or Bradford) calibration curve for this experiment would use purified α-lactalbumin rather than bovine serum albumin or gamma globulin."

V. REFERENCES

Specific

B.J. Davis, *Ann. N.Y. Acad. Sci.*, *121*, 404 (1964). "Disc Electrophoresis II: Method and Application to Human Serum Proteins."

J.H. Morrissey, *Anal. Biochem.*, *117*, 307 (1981). "Silver Stain for Proteins in Polyacrylamide Gels."

Protein-Ligand Interactions: Binding of Fatty Acids to Serum Albumin

RECOMMENDED READING:
Chapter 2, Section E; Chapter 5, Sections A, D, E, and F.

SYNOPSIS
Serum albumin circulates in the blood stream, transporting essential nutrients to peripheral tissues. Nonpolar molecules have a special affinity for selected binding sites on the protein. In this experiment, the dynamics of fatty acid binding to albumin will be investigated by exploiting the fluorescent changes that occur when a dye, 1-anilino-naphthalene-8-sulfonate, binds to albumin.

I. INTRODUCTION AND THEORY

The majority of the processes that occur in an organism are the result of interactions between molecules. The action of hormones is familiar to all. A hormone response is the consequence of a weak, but specific, interaction between the hormone molecule and a receptor protein in the membrane of the target cell. Before a metabolic reaction can occur, a small substrate molecule must physically interact in a certain well-defined manner with a macromolecular catalyst, an enzyme. The biochemical action of a drug also depends upon molecular interactions. The drug is first distributed throughout the body via the blood stream. Drugs in the blood stream are often bound to plasma proteins, which act as carriers. When the drug molecules are transported to their site of action, a second molecular interaction is likely to occur. Many drugs elicit their effects by interfering with biochemical processes. This may take the form of enzyme inhibition, where the drug molecule binds to a specific enzyme and prohibits its catalytic action. Table E6.1 lists several other molecular interactions that lead to some dynamic biochemical action.

All of these molecular interactions have at least two common characteristics. (1) The forces that are the basis of these interactions are weak and noncovalent. We usually define the interactions in terms of hydrogen bonding, hydrophobic stabilization, van der Waals forces, and electrostatic interactions. (2) The binding between molecules is close, selective, and specific. Imagine that the interaction brings together two molecular surfaces. Where the surfaces are in contact, the forces must be complementary. If on one surface there is a nonpolar molecular group (phenyl ring, hydrophobic alkyl chain, etc.), the adjacent region on the other surface must also be hydrophobic and nonpolar. If a positive charge exists on one surface, there may be a neutralzing negative charge on the other surface. Very simply stated, the two molecules must be compatible.

Quantitative Treatment of Binding

A thorough understanding of the biochemical significance of ligand binding to macromolecules comes only from a quantitative analysis of the strength of binding. (In biochemistry, a small molecule that binds to a macromolecule is called a **ligand**.) Binding affinity between two molecules is often expressed as an equilibrium constant, the **formation constant**, K_f, which is derived from the law of mass action. Consider the specific interaction between a small molecule, L (for ligand), and a macromolecule, M

TABLE E6.1

Examples of Ligand: Macromolecule Interactions

Type of Interaction	Biochemical Significance
Lipids: Proteins	Cell membranes
Substrate: Enzyme	Metabolic reactions
Hormone: Receptor protein	Regulation
Inhibitor: Enzyme	Metabolic regulation
Ligand: Carrier protein	Membrane transport
Antigen: Antibody	Immune response
Coenzyme: Enzyme	Metabolic reactions

(Equation 1). These two species combine to form a complex, LM.

$$L + M \longrightarrow LM \qquad \text{(Equation E6.1)}$$

K_f, the formation constant for the complex, is defined by Equation 2.

$$K_f = \frac{[LM]}{[L][M]} \qquad \text{(Equation E6.2)}$$

Do not confuse K_f with K_d, the dissociation constant. The relationship between K_f and K_d is defined in Equation 3.

$$K_d = \frac{[L][M]}{[LM]} = \frac{1}{K_f} \qquad \text{(Equation E6.3)}$$

The larger the value of K_f, the greater the strength of binding between L and M. (Large K_f infers a high concentration of LM relative to L and M.) Return to Equation 2 and note that in order to determine K_f, a method must be developed to measure equilibrium concentrations of L, M, and the complex LM. In a later section, we will describe experimental techniques that are applied to these measurements of binding constants, but first we must reorganize Equation 2 into a form that contains more readily measurable terms. We will begin with the assumption that the macromolecule, M, has several binding sites for L and that these sites do not interact with each other. That is to say, K_f is identical for all binding sites. The following definitions are necessary for the reorganization of Equation 2.

$[L]$ = equilibrium concentration of free or unbound ligand

$[M]$ = equilibrium concentration of macromolecule with no bound L; or the concentration of unoccupied binding sites

$[LM]$ = equilibrium concentration of ligand:macromolecule complex; or the concentration of occupied sites

$[M]_0$ = total or initial concentration of macromolecule; or total concentration of available binding sites

$[L]_0$ = total concentration of bound and unbound ligand; or initial concentration of ligand

v = fraction of available sites on M that are occupied; or the fraction of M that has L in binding site:

$$v = \frac{[LM]}{[M]_0} \qquad \text{(Equation E6.4)}$$

The term v is particularly significant because it can be considered a ratio of the number of occupied sites to the total number of potential binding sites on M. It can be measured experimentally, but first it must be redefined in the following manner: Since $[M]_0 = [LM] + [M]$, then

$$v = \frac{[LM]}{[LM] + [M]}$$

From Equation 2, $[LM] = K_f[L][M]$. Therefore,

$$v = \frac{K_f[L][M]}{K_f[L][M] + [M]}$$

Simplifying,

$$v = \frac{K_f[L]}{K_f[L] + 1} \qquad \text{(Equation E6.5)}$$

You should recognize the similarity of Equation 5 to the Michaelis-Menten equation for enzyme catalysis. A graph of v vs. $[L]$ yields a hyperbolic curve (see Figure E6.1) which approaches a limiting value or saturation level. At this point, all binding sites on M are occupied. Because of the difficulty of measuring the exact point of saturation, this nonlinear curve is seldom used to determine K_f. Linear plots are more desirable, so Equation 5 is converted to an equation for a straight line. The equation will now be put into a more general form to account for

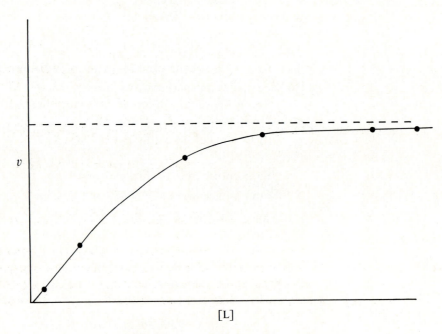

FIGURE E6.1
A plot illustrating saturating of binding sites with ligand.

any number of potential binding sites on M. The symbol $\bar{v}$ will be used to represent the average number of occupied sites per M, and n will represent the number of potential binding sites per M molecule. Assuming that all the binding sites on M are equivalent, Equation 5 becomes:

$$\bar{v} = \frac{nK_f[L]}{K_f[L] + 1} \qquad \text{(Equation E6.6)}$$

If $\bar{v}$ is the average number of occupied sites per M molecule, then $n - \bar{v}$ is the average number of unoccupied sites per M molecule.

$$n - \bar{v} = n - \frac{nK_f[L]}{K_f[L] + 1}$$

Simplifying,

$$(n - \bar{v})(K_f[L] + 1) = n(K_f[L] + 1) - nK_f[L]$$

$$(n - \bar{v}) = \frac{n}{K_f[L] + 1} \qquad \text{(Equation E6.7)}$$

To further simplify Equation 7, let the term $\bar{v}/(n - \bar{v})$ represent the ratio of occupied sites to nonoccupied sites on M; it can be mathematically represented as

$$\frac{\bar{v}}{n - \bar{v}} = \left(\frac{nK_f[L]}{K_f[L] + 1}\right)\left(\frac{K_f[L] + 1}{n}\right)$$

or

$$\frac{\bar{v}}{n - \bar{v}} = K_f[L] \qquad \text{(Equation E6.8)}$$

A more desirable form for graphical use is known as **Scatchard's equation:**

$$\frac{\bar{v}}{[L]} = K_f(n - \bar{v}) \qquad \text{(Equation E6.9)}$$

If a plot of $\bar{v}/[L]$ vs. $\bar{v}$ yields a straight line, shown by the solid line in Figure E6.2, then all the binding sites on M are identical and independent. K_f and n are estimated as shown in the figure.

The derivation of Equation 9 assumes that K_f is identical for all binding sites; that is, the binding of one molecule of L does not influence the binding of other L molecules to binding sites on M. However, it is common for ligand:macromolecule interactions to display such influences. The binding of one L molecule to M may encourage or inhibit the binding of a second L molecule to M. For example, the binding of dioxygen to one of the four subunits of hemoglobin increases the affinity of the other subunits for oxygen. There is said to be **cooperativity** of sequential binding. If the sites do show cooperative binding, the plot is nonlinear, as shown by the dotted line in Figure E6.2. The shape of the nonlinear curve may be used to determine the number of types of binding sites. The dotted line

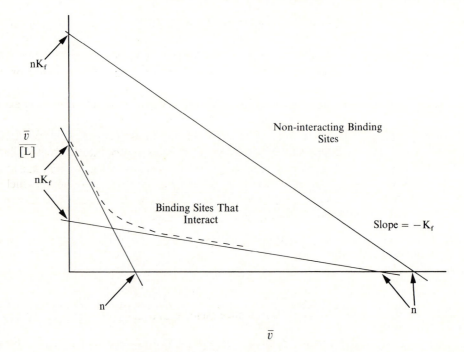

FIGURE E6.2
Scatchard plot. Two types of Scatchard curves are illustrated. The upper plot (solid line) represents binding to a macromolecule with noninteracting sites. The binding of a ligand molecule at one site is independent of the binding of a second ligand molecule at another site. The plot can be extrapolated to each axis and the n and k_f calculated. The lower, dotted line represents binding of ligand molecules to binding sites that interact. One ligand molecule bound to the macromolecule influences the rate of binding of other ligand molecules. The dotted line is evaluated as shown in the figure. The curved line has two distinct slopes. It can be resolved into two straight lines, each of which may be evaluated for n and K_f. This would indicate that there are two types of binding sites, each with a unique n and K_f.

in Figure E6.2 can be resolved into two lines, indicating that two types of binding sites are present on M. The n and K_f for each type of binding site may be estimated by resolving the smooth curve into straight lines as shown in Figure E6.2. K_f and n can be estimated by extrapolating the two straight lines to the axes.

Experimental Detection of Binding and Measurements of K_f and n

One of the simplest and most convenient methods for detecting ligand binding is the differential method, which detects and quantifies some measurable change that accompanies the binding of L and M. Most often, this is a change in spectral absorption or fluorescence. Fluorescence measurements are more sensitive, with detection at the level of 10^{-6} to 10^{-9} M.

The Scatchard equation for defining protein-ligand interactions is easily adaptable to the differential method. Fluorescence or absorption data can be expressed in terms of [bound]/[free] ligands or [occupied]/[unoccupied] sites on M. The fluorimetric method in this experiment is based on the following discussion. Our analysis assumes that L and LM, but not M, contribute to the measured fluorescence. Even though free L is capable of fluorescence, the fluorescence yields are usually enhanced when L is associated with M. This leads to a change in fluorescence that is proportional to the ratio of bound L to unbound L, which is exactly what is needed for the Scatchard equation. The fluorescence observed, F_{obs}, from a mixture of L and M is

$$F_{obs} = F_L + F_{LM} \qquad \text{(Equation E6.10)}$$

where

F_L = fluorescence yield of free L

F_{LM} = fluorescence yield of bound L

Fluorescence data may be used directly to calculate the fraction of ligand bound:

$$\text{fraction of ligand bound} = f = \frac{F_{obs} - F_L}{F_{total} - F_L} \qquad \text{(Equation E6.11)}$$

F_{total} is the total fluorescence of LM when all the L molecules in solution are bound to M. The term $F_{total} - F_L$ is proportional to the total number of binding sites on M, and $F_{obs} - F_L$ represents the average number of occupied sites on M caused by a certain concentration of L. For cases where the fluorescence of free ligand, F_L, is very small, Equation 11 becomes:

$$f = \frac{F_{obs}}{F_{total}} \qquad \text{(Equation E6.12)}$$

F_{total}, the fluorescence of LM when all of the L molecules in solution are bound, must be determined at several ligand concentrations. This requires the use of excessive amounts of the macromolecule, M, which may be expensive and scarce. Instead, the fluorescence of a standard compound at the same concentration as the ligand may be used to calculate F_{total}. The fraction of L bound to M at any concentration of L is calculated using Equation 13.

$$f = \frac{F_{obs}}{RF_{std}} \qquad \text{(Equation E6.13)}$$

where

$$R = \frac{F_{total}}{F_{std}}$$

F_{std} = fluorescence of the standard compound
at concentrations equal to that of the ligand,
L. Quinine is used as the standard
compound in this experiment.

For graphical analysis of the experimental data, $\bar{v}$ must be calculated as described in Equation 14.

$$\bar{v} = \frac{F_{obs}}{F_{total}} \cdot \frac{[L]_0}{[M]_0} = f\frac{[L]_0}{[M]_0} \qquad \text{(Equation E6.14)}$$

Recall that $\bar{v}$ is the average number of occupied sites per M divided by the number of potential binding sites on M.

Transport Properties of Serum Albumin

Albumin, with a concentration of about 50 mg/100 ml, is the major protein in human plasma. Up to 40% of the total albumin is in circulation, transporting essential nutrients (especially those that are sparingly soluble in plasma). The fatty acids, which are important fuel molecules for the peripheral tissues, are distributed by albumin. In addition, albumin is the plasma transport protein for other substances, including bilirubin, thyroxine, and possibly some steroid hormones. Also, many drugs (including aspirin, sulfonilamides, clofibrate, and digitalis) bind to albumin and are most likely carried to their sites of action by the protein.

Extensive research on albumin has led to a relatively clear picture of fatty acid binding (Spector, 1975). Each albumin molecule is capable of binding up to five molecules of fatty acid. Quantitative binding studies have shown that as the alkyl chain of a molecule increases in length, the strength of its binding to albumin increases. What does this indicate about the type of interactions stabilizing the fatty acid:albumin complexes?

Many years ago, Laurence (1952) demonstrated that when small fluorescent molecules are combined with proteins, the fluorescence spectrum undergoes changes in intensity. The spectroscopic changes that occur can be used to characterize the type of ligand binding. Most binding studies of albumin rely on the fluorescent probe, 1-anilino-naphthalene-8-sulfonate (ANS). This fluorescent dye binds to nonpolar regions of albumin to the extent of five moles of dye per mole of protein (Daniel and Weber, 1966; Santos and Spector, 1972). The fluorescence intensity due to albumin-bound ANS is at least two hundred times greater than the fluorescence of free ANS. Some nonpolar chemicals such as fatty acids have been shown to quench the fluorescence of albumin-bound ANS (Santos and Spector,

1972; 1974). This means that the added compound is able to displace bound ANS from albumin. Such a result is significant, because it implies that the two substances (fatty acid and ANS) compete for the same binding sites on albumin. The displacement of bound ANS from albumin, as measured fluorometrically, is now widely used to detect and quantify the binding of nutrients, drugs, and other physiologically important chemicals to albumin.

Overview of the Experiment

In this experiment, you will first evaluate the binding of the fluorescent probe, ANS, to serum albumin. Increasing amounts of ANS will be added to a fixed concentration of serum albumin. Fluorescence measurements will be obtained and used in Equation 14 to calculate $\bar{v}$. A Scatchard plot of $\bar{v}$ vs. $\bar{v}/[\text{ANS}]$ will be prepared and used for determination of n and K_f. The effectiveness of fatty acids in displacing ANS from albumin will be evaluated by adding various concentrations of a fatty acid to ANS:albumin complex.

The time requirements for this experiment are:

Part A: Concentration of bovine serum albumin—15 minutes.

Part B: Binding of ANS to albumin—1½ hours.

Part C: Binding of fatty acids to albumin—1 hour.

II. EXPERIMENTAL

Materials and Supplies

Bovine serum albumin in phosphate buffer, 5×10^{-6} M BSA
   This solution should be prepared and ready for use at the beginning of the laboratory period. Most commercial preparations of BSA contain bound fatty acids that must be removed before binding studies. The fatty acids can be removed by charcoal treatment (Chen, 1967).

Quinine, 2.5×10^{-4} M in 0.2 M H_2SO_4

Sodium phosphate buffer, 0.05 M, pH 7.4

1-Anilino-naphthalene-8-sulfonate solution, 2.5×10^{-4} M in sodium phosphate buffer, pH 7.4

Fatty Acids.
   Dissolve the sodium salt of each fatty acid in phosphate buffer listed above. Final concentration of fatty acid solutions should be 5×10^{-4} M. Any fatty acids may be used, but caproic and lauric acids are recommended.

Fluorescence cuvets, two 3-ml cuvets

Matched quartz cuvets, 1 or 3 ml

Microsyringes, 100 μl

Constant temperature bath at or near room temperature

UV-VIS spectrophotometer

Fluorescence spectrophotometer

Procedure

A. Concentration of Bovine Serum Albumin Solution

It is essential to know the exact concentration of the protein solution. This can be determined by measuring the absorbance of the solution at 280 nm. Turn on the spectrometer and UV lamp and set the wavelength to 280 nm. After a warm-up period of 15 to 20 minutes, pipet 1.0 or 3.0 ml of the standard phosphate buffer into a 1- or 3-ml quartz cuvet. Place the cuvet in the sample position of the spectrometer and adjust the absorbance to 0.0. Now, carefully pipet 1.0 or 3.0 ml of the BSA solution into the matched cuvet. Obtain the A_{280}. If the absorbance reading is off scale, dilute the BSA solution and again read A_{280}. This is used to calculate directly the concentration of the BSA as described in Analysis of Results.

B. Binding of ANS to Albumin

Before beginning this part of the experiment, read the section on fluorescence in Chapter 5. Also, you should be familiar with the use of the spectrofluorometer. Turn on the instrument; set the excitation wavelength to 380 nm and the emitted wavelength to 470 nm. Use Table E6.2 to prepare the various solutions for fluorescence measurements. All reagents should be in a constant temperature bath at or near room temperature. If your fluorometer has a circulating water bath for samples, this should be set at the same temperature as the constant temperature bath. Each cuvet is prepared separately as described in Table E6.2. After addition of all reagents to a fluorescence cuvet, mix well by holding a small square of

TABLE E6.2[1]

Preparation of Cuvets for Part B

Reagent	Cuvet Number					
	1	2	3	4	5	6
BSA, 5×10^{-6} M	2.0	2.0	2.0	2.0	2.0	2.0
Phosphate buffer	1.0	0.85	0.80	0.75	0.70	0.65
ANS, 2.5×10^{-4} M	0.0	0.15	0.20	0.25	0.30	0.35

[1] Units are milliliters.

TABLE E6.3

Preparation of Standard Cuvets for Part B

Reagent	1	Cuvet Number				
		2	3	4	5	6
Phosphate buffer	3.00	2.85	2.80	2.75	2.70	2.65
Quinine solution	0.0	0.15	0.20	0.25	0.30	0.35

hydrocarbon foil over the cuvet and inverting it several times. *Do not shake the cuvets!* Zero the fluorometer with a cuvet containing 3 ml buffer and obtain a fluorescence reading on the sample cuvet. Continue this procedure until you have obtained the fluorescence of all six cuvets.

Standardize the fluorescence readings using the quinine solution. Prepare six cuvets as described in Table E6.3. Prepare the cuvets, mix well, and obtain the fluorescence of each cuvet. Zero the instrument with phosphate buffer (cuvet 1). The excitation and emitted wavelength are the same as before.

C. Binding of Fatty Acids to ANS-Albumin

Prepare sample cuvets according to Table E6.4. Prepare each cuvet as described in the procedure for part B. Record the fluorescence intensity for each of the sample cuvets. Use a cuvet of phosphate buffer to zero the instrument. Repeat the procedure in Table E6.4 with another fatty acid if time allows.

III. ANALYSIS OF RESULTS

A. Concentration of BSA

The molar extinction coefficient, ε, for BSA is $4.36 \times 10^{-4} \text{ M}^{-1} \text{ cm}^{-1}$. Use Beer's law, $A = \varepsilon C l$, to calculate the concentration, C, of the BSA solution. (See Chapter 2.)

TABLE E6.4

Binding of Fatty Acids to ANS-Albumin

Reagent	Cuvet Number				
	1	2	3	4	5
BSA, 5×10^{-6} M	2.0	2.0	2.0	2.0	2.0
Phosphate buffer	0.85	0.80	0.75	0.70	0.60
ANS, 2.5×10^{-4} M	0.10	0.10	0.10	0.10	0.10
Caproic acid, 5×10^{-4} M	0.05	0.10	0.15	0.20	0.30

B. Binding of ANS to Albumin

For cuvets 1 to 6 in Table E6.2, you have obtained a fluorescence reading, F_{obs}. Table E6.3 yields F_{std} values for quinine at the same concentrations as ANS. Prepare a table as shown below and calculate the various terms. R for the conditions in this experiment is 9.85.

Cuvet #	F_{obs}	F_{std}	f	$[ANS]_0$	$\bar{v}$	$[ANS]$	$\dfrac{\bar{v}}{[ANS]}$
		F_{obs}/RF_{std}			$f\dfrac{[ANS]}{[BSA]_0}$	$(1-f)[ANS]_0$	

Now, plot $\bar{v}/[ANS]$ vs. $\bar{v}$ as in Figure E6.2 and evaluate the line for n and K_f. Any curvature in the graph implies cooperative effects in the binding of ANS. If this is the case, evaluate the constants as shown for the dotted line in Figure E6.2.

C. Binding of Fatty Acids to Albumin

Calculate the percentage change of the fluorescence for each concentration of acid added. Explain the decrease in fluorescence as the concentration of fatty acid is increased.

IV. QUESTIONS AND PROBLEMS

1. A drug, X, was studied for its affinity for albumin. When X was bound to albumin, an increase in fluorescence was noted. The $\bar{v}$ values were determined from these fluorescence measurements. Use the data below to determine K_f and n for the interaction between X and albumin. Prepare two types of graphs and compare the results. In one graph plot $\bar{v}$ vs. $[X]$, and in the second plot $\bar{v}/[X]$ vs. $\bar{v}$. Which is the better method? Why?

$[X]$	$\bar{v}$
0.36	0.43
0.60	0.68
1.2	1.08
2.4	1.63
3.6	1.83
4.8	1.95
6.0	1.98

2. You are attempting to develop a fluorescent probe for use in binding studies. List several requirements that the probe must meet in order to

be effective. For example, it must bind at specific locations of the macro-molecule. Can you think of other requirements?

3. A research project you are working on involves the study of sugar binding to human albumin. The sugars to be tested are not fluorescent, and you do not wish to use a secondary probe such as ANS. Human albumin has only one tryptophan residue, and you know that this amino acid is fluorescent. You find that the tryptophan-fluorescence spectrum of human albumin undergoes changes when various sugars are added. Can you explain the results of this experiment and discuss the significance of the finding?

4. Equation 5 in this experiment can be used to determine K_f values, but hyperbolic plots are obtained. Can you convert Equation 5 into an equation that will yield a linear plot without going through all the changes necessary for the Scatchard equation? Hint: Study the conversion of the Michaelis-Menten equation to the Lineweaver-Burk equation.

V. REFERENCES

General

J.M. Brewer, A.J. Pesce, and R.B. Ashworth, *Experimental Techniques in Biochemistry* (1974), Prentice-Hall (Englewood Cliffs, NJ), pp. 216–262. Fluorescence techniques.

F.W. Dahlquist in *Methods in Enzymology*, C.H.W. Hirs and S.N. Timasheff, Editors (1978), Vol. 48 F, Academic Press (New York), pp. 270–299. "The Meaning of Scatchard and Hill Plots."

D. Freifelder, *Physical Biochemistry*, 2nd Edition (1982), Freeman (San Francisco), pp. 654–684. Introduction to ligand binding.

D.E. Metzler, *Biochemistry* (1977), Academic Press (New York), pp. 182–189. Theory of ligand binding.

Specific

R.F. Chen, *J. Biol. Chem.*, *242*, 173 (1967). "Removal of Fatty Acids from Serum Albumin by Charcoal Treatment."

E. Daniel and G. Weber, *Biochemistry*, *5*, 1893 (1966). "Cooperative Effects in Binding by Bovine Serum Albumin. I. The Binding of 1-Anilino-8-naphthalenesulfonate. Fluorimetric Titrations."

D.J. Laurence, *Biochem. J.*, *51*, 168 (1952). "Adsorption of Dyes on Bovine Serum Albumin by the Method of Fluorescence."

E.C. Santos and A.A. Spector, *Biochemistry, 11*, 2299 (1972). "Effect of Fatty Acids on the Binding of 1-Anilino-8-naphthalenesulfonate to Bovine Serum Albumin."

E.C. Santos and A.A. Spector, *Mol. Pharmacol., 10*, 519 (1974). "Effects of Fatty Acids on the Interaction of 1-Anilino-8-naphthalenesulfonate with Human Plasma Albumin."

A.A. Spector, *J. Lipid Res., 16*, 165 (1975). "Fatty Acid Binding to Serum Albumin."

Kinetic Analysis of the Tyrosinase-Catalyzed Oxidation of 3,4-Dihydroxyphenylalanine

RECOMMENDED READING:
Chapter 5, Sections A, B, C.

SYNOPSIS

Tyrosinase, a copper-containing oxidoreductase, catalyzes the ortho-hydroxylation of monophenols and the aerobic oxidation of catechols. The enzyme activity can be assayed by monitoring the oxidation of 3,4-dihydroxyphenylalanine (DOPA) to the red-colored dopachrome. The kinetic parameters of K_M and V_{max} will be evaluated using Lineweaver-Burk or direct linear plots. Inhibition of tyrosinase by benzoate and thiourea will also be studied. Finally, two stereoisomers, L-DOPA and D-DOPA, will be tested and compared as substrates.

I. INTRODUCTION AND THEORY

Enzymes are biological catalysts. Without their presence in a cell, most biochemical reactions would not proceed at the required rate. The physico-chemical and biological properties of enzymes have been investigated since the early 1800s. The unrelenting interest in enzymes is due to several factors—their dynamic and essential role in the cell, their extraordinary catalytic power, and their selectivity. Two of these dynamic characteristics will be evaluated in this experiment, namely, a kinetic description of enzyme activity and molecular selectivity.

Enzyme-catalyzed reactions proceed through an ES complex as shown in Reaction 1. The individual rate constants, k_n, are placed with each arrow.

$$E + S \underset{k_2}{\overset{k_1}{\rightleftharpoons}} ES \underset{k_4}{\overset{k_3}{\rightleftharpoons}} E + P \qquad \text{(Reaction E7.1)}$$

E represents the enzyme, S the substrate or reactant, and P the product. For a specific enzyme, only one or a few different substrate molecules can bind in the proper manner and produce a functional ES complex. The substrate must have a size, shape, and polarity compatible with the active site of the enzyme. Some enzymes catalyze the transformation of many different molecules as long as there is a common type of chemical linkage in the substrate. Others have absolute specificity and can form reactive ES complexes with only one molecular structure. In fact, some enzymes are able to differentiate between D and L isomers of substrates.

Kinetic Properties of Enzymes

The initial reaction velocity, v_0, of an enzyme-catalyzed reaction varies with the substrate concentration, [S], as shown in Figure E7.1. The Michaelis-Menten equation has been derived to account for the kinetic properties of enzymes. (Consult a biochemistry textbook for a derivation of this equation and for a discussion of the conditions under which the equation is valid.) The common form of the equation is

$$v_0 = \frac{V_{max}[S]}{K_M + [S]} \qquad \text{(Equation E7.1)}$$

where

v_0 = initial reaction velocity

V_{max} = the maximal reaction velocity; attained when all enzyme active sites are filled with substrate molecules

[S] = substrate concentration

K_M = the Michaelis constant = $\dfrac{k_2 + k_3}{k_1}$

The important kinetic constants, V_{max} and K_M, can be graphically determined as shown in Figure E7.1. Equation 1 and Figure E7.1 have all of the disadvantages of nonlinear kinetic analysis. V_{max} can only be estimated because of the asymptotic nature of the line. The value of K_M, the substrate concentration that results in a reaction velocity of $V_{max}/2$, depends upon V_{max}, so both are in error. By taking the reciprocal of both sides of the Michaelis-Menten equation, however, it is converted into the Lineweaver-Burk relationship (Equation 2).

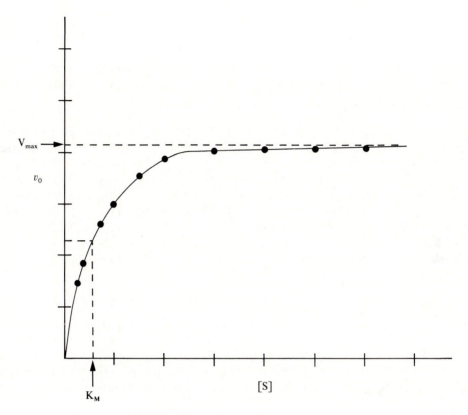

FIGURE E7.1
Michaelis-Menten plot for an enzyme-catalyzed reaction.

$$\frac{1}{v_0} = \frac{K_M}{V_{max}}\frac{1}{[S]} + \frac{1}{V_{max}} \qquad \text{(Equation E7.2)}$$

This equation, which is in the form $y = mx + b$, gives a straight line when $1/v_0$ is plotted against $1/[S]$ (Figure E7.2). The intercept on the $1/v_0$ axis is $1/V_{max}$ and the intercept on the $1/[S]$ axis is $-1/K_M$. A disadvantage of the Lineweaver-Burk plot is that the data points are compressed in the high substrate concentration region. A third type of graphical analysis results from the Eisenthal and Cornish-Bowden modification (Equation 3).

$$V_{max} = v_0 + \frac{v_0}{[S]}K_M \qquad \text{(Equation E7.3)}$$

This modification is a direct linear graph that results from plotting each pair of experimental values of v_0 and $[S]$. Figure E7.3 is an example of the Eisenthal and Cornish-Bowden analysis. K_M and V_{max} values are read directly from the graph.

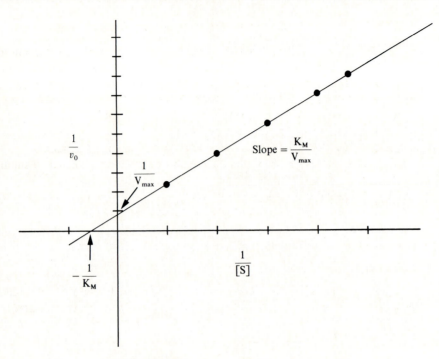

FIGURE E7.2
Lineweaver-Burk plot for an enzyme-catalyzed reaction.

FIGURE E7.3
Eisenthal and Cornish-Bowden
(direct linear) plot for an enzyme-
catalyzed reaction.

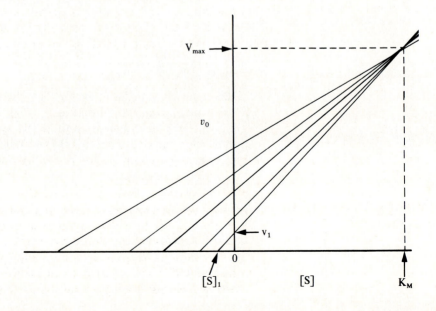

Significance of Kinetic Constants

The **Michaelis constant**, K_M, for an enzyme-substrate interaction has two meanings: (1) K_M is the substrate concentration that leads to an initial reaction velocity of $V_{max}/2$ or, in other words, the substrate concentration that results in the filling of one-half of the enzyme active sites, and (2) $K_M = (k_2 + k_3)/k_1$. The second definition of K_M has special significance in certain cases. When $k_2 \gg k_3$, then $K_M = k_2/k_1$ so K_M is equivalent to the dissociation constant of the ES complex. When $k_2 \gg k_3$, a large K_M implies weak interaction between E and S, whereas a small K_M indicates strong binding between E and S.

V_{max} is important because it leads to the analysis of another kinetic constant, k_3, the **turnover number**. The analysis of k_3 begins with the basic rate law for an enzyme-catalyzed process (Equation 5), which is derived from Equation 1:

$$v_0 = k_3[ES] \qquad \text{(Equation E7.5)}$$

If all of the enzyme active sites are saturated with substrate, then [ES] in Equation 5 is equivalent to $[E_T]$, the total concentration of enzyme, and v_0 becomes V_{max}; hence

$$V_{max} = k_3[E_T] \qquad \text{(Equation E7.6)}$$

or

$$k_3 = \frac{V_{max}}{[E_T]}$$

For an enzyme with one active site per molecule, the turnover number, k_3, is the number of substrate molecules transformed to product by one enzyme molecule per unit time, usually in minutes or seconds. The turnover number is a measure of the efficiency of an enzyme.

Inhibition of Enzyme Activity

Nonsubstrate molecules may interact with enzymes, leading to a decrease in enzymatic activity. The study of enzyme inhibition is of interest because it often reveals information about the mechanism of enzyme action. Also, many toxic substances, including drugs, express their action by enzyme inhibition.

The process of reversible inhibition is described by an equilibrium interaction between enzyme and inhibitor. Most inhibition processes can be classified as **competitive** or **noncompetitive**, depending on how the inhibitor impairs enzyme action. A competitive inhibitor is usually similar in structure to the substrate and is capable of reversible binding to the enzyme active site. In contrast to the substrate molecule, the inhibitor molecule cannot undergo chemical transformation to a product; however, it does interfere with substrate binding. A noncompetitive inhibitor does not bind in the active site of an enzyme, but binds at some other region of the enzyme molecule. Upon binding of the noncompetitive inhibitor, the enzyme is reversibly converted to a nonfunctional conformational state, and the substrate, which is fully capable of binding to the active site, is not converted to product.

Any complete inhibition study requires the experimental differentiation between competitive and noncompetitive inhibition. The two inhibitory processes are kinetically distinguishable by application of the Lineweaver-Burk or direct linear plot. For each inhibitor studied, at least two sets of rate experiments are completed. In all sets, the enzyme concentration is identical. In set 1, the substrate concentration is varied and no inhibitor is added. In set 2, the same variable substrate concentrations are used as in set 1, and a constant amount of inhibitor is added to each assay. If additional data are desired, more sets are prepared with variable substrate concentrations as in set 2; but a constant, and different, concentration of inhibitor is present. These data, when plotted on a Lineweaver-Burk graph, lead to three lines as shown in Figures E7.4 or E7.5. The Lineweaver-Burk plot of a competitively inhibited process is shown in Figure E7.4. All three lines intersect at the same point on the $1/v_0$ axis;

FIGURE E7.4
Lineweaver-Burk plot for competitive inhibition.

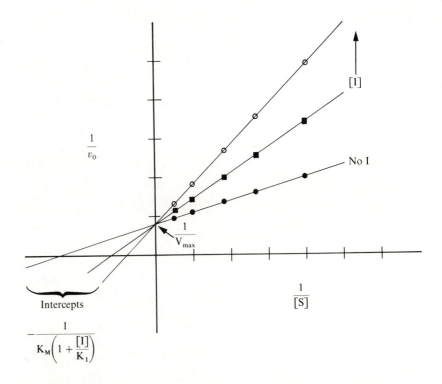

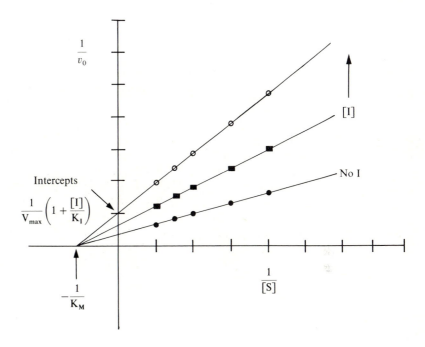

FIGURE E7.5
Lineweaver-Burk plot for non-competitive inhibition.

hence, V_{max} is not altered by the competitive inhibitor. If enough substrate is present, the competitive inhibition can be overcome. The apparent K_M value (measured on the $1/[S]$ axis) changes with each change in inhibitor concentration. Figure E7.4 reveals several kinetic characteristics of the inhibition process: K_M, V_{max}, and K_I. The quantity K_I is the dissociation constant for the EI complex, and its value indicates the strength of binding between enzyme and inhibitor. A large K_I implies relatively weak interaction between E and I.

A Lineweaver-Burk plot of enzyme kinetics in the presence and absence of a noncompetitive inhibitor is shown in Figure E7.5. V_{max} in the presence of a noncompetitive inhibitor is decreased, but K_M is unaffected. The effect of a competitive inhibitor on the direct linear plot is shown in Figure E7.5.

Units of Enzyme Activity

The actual molar concentration of an enzyme in a cell-free extract or purified preparation is seldom known. Only if the enzyme is available in a pure crystalline form, carefully weighed, and dissolved in a solvent, can the actual molar concentration be accurately known. It is, however possible to develop a precise and accurate assay for enzyme activity. Consequently, the amount of a specific enzyme present in solution is most often expressed in units of activity. Three units are in common use, the

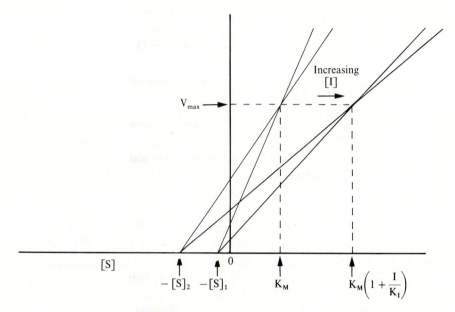

FIGURE E7.6
Direct linear plot for competitive inhibition.

International unit (IU), the **katal**, and **specific activity**. The International Union of Biochemistry Commission on Enzymes has recommended the use of a standard unit, the **International unit**, or just **unit**, of enzyme activity. One IU of enzyme corresponds to that amount that catalyzes the transformation of 1 μmole of substrate to product per minute under specified conditions of pH, temperature, ionic strength, and substrate concentration. If a solution containing enzyme converts 10 μmoles of substrate to product in 5 minutes, the solution contains 2 enzyme units. A new unit of activity, the katal, has been recommended. One **katal** of enzyme activity represents the conversion of 1 mole of substrate to product in one second. One International unit is equivalent to 1/60 μkatal or 0.0167 μkatal. One katal, therefore, is equivalent to 6×10^7 International units. It is convenient to represent small amounts of enzyme in millikatals (mkatals), microkatals (μkatals), or nanokatals (nkatals). The enzyme activity in the above example was 2 units, which can be converted to katals as follows: Since one katal is 6×10^7 units, 2 units are equivalent to $2/6 \times 10^7$ or 33 nkatals (0.033 μkatals). If the enzyme is an impure preparation in solution, the activity is most often expressed as units/ml or katals/ml.

Another useful quantitative definition of enzyme efficiency is **specific activity**. The specific activity of an enzyme is the number of enzyme units or katals per milligram of protein. This is a measure of the purity of an enzyme. If a solution contains 20 milligrams of protein that express 2 units of activity (33 nkatals), then the specific activity of the enzyme is 2 units/20 mg = 0.1 units/mg or 33 nkatals/20 mg = 1.65 nkatals/mg. As an enzyme is purified, its specific activity will increase. That is, during purification, the enzyme concentration will increase relative to the total protein concentration until a limit is reached. The maximum specific activity is attained when the enzyme is homogeneous or in a pure form.

The activity of an enzyme depends on several factors, including substrate concentration, cofactor concentration, pH, temperature, and ionic strength. The conditions under which enzyme activity is measured are critical and must be specified when activities are reported.

Design of an Enzyme Assay

Whether an enzyme is obtained commercially or prepared in a multi-step procedure, an experimental method must be developed to detect and quantify the specific enzyme activity. During isolation and purification of an enzyme, the assay is necessary to determine the amount and purity of the enzyme and to evaluate its kinetic properties. An assay is also essential for a further study of the mechanism of the catalyzed reaction.

The design of an assay requires certain knowledge of the reaction:

1. the complete stoichiometry,
2. what species are required (substrate, metal ions, cofactors, etc.) and their kinetic dependence,
3. optimal pH, temperature, and ionic strength.

In addition to these conditions, a method must be available to identify and monitor any physical, chemical, or biological change that occurs during enzyme-catalyzed conversion of substrate to product. The most straightforward approach is to monitor the change in substrate or product concentration during the reaction. Many techniques, including spectrophotometry, fluorimetry, acid-base titration, and radioactive counting are used to make these measurements. The choice depends primarily upon the structural characteristics of the substrate and/or product and the chemistry that occurs during the reaction.

If an enzyme assay involves continuous monitoring of substrate or product concentration, the assay is said to be **kinetic**. On the other hand, if a single measurement of substrate or product concentration is made after a specified reaction time, a **fixed-time assay** results. The kinetic assay is more desirable because the time course of the reaction is directly observed and any discrepancy from linearity can be immediately detected.

Figure E7.7 displays the kinetic progress curve of a typical enzyme-catalyzed reaction and illustrates the advantage of a kinetic assay. The

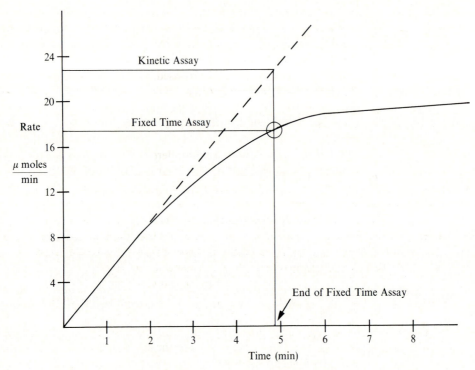

FIGURE E7.7
Kinetic progress of an enzyme-catalyzed reaction. See text for details.

rate of product formation decreases with time. This may be due to any combination of a decrease in substrate concentration, denaturation of the enzyme, and product inhibition of the reaction. The solid line in Figure E7.7 represents the continuously measured time course of a reaction (kinetic assay). The true rate of the reaction is determined from the slope of the dotted line drawn tangent to the experimental result. From the data given, the rate is 5 μmoles of product formed per minute. Data from a fixed-time assay are also shown on Figure E7.7. If it is assumed that no product is present at the start of the reaction, then only a single measurement after a fixed period is necessary. This is shown by a circle on the experimental rate curve. The measured rate is now 16 μmoles of product formed every 5 minutes or about 3 μmoles/minute, considerably lower than the rate derived from the continuous, kinetic assay. Which rate measurement is correct? Obviously, the kinetic assay gives the true rate because it corrects for the decline in rate with time. The fixed-time assay can be improved by changing the time of the measurement, in this example, to 2 minutes of reaction time, when the experimental rate is still linear. It is possible to obtain true rates with the fixed-time assay, but one must choose the time period very carefully. In the laboratory, this is done by removing aliquots of a reaction mixture at various times and measuring the concentration of product formed in each aliquot. Figure E7.7 reinforces an assumption used in the derivation of the Michaelis-Menten equation. Only measurements of initial velocities lead to true reaction rates. This avoids the complications of enzyme denaturation, decrease of [S], product inhibition, and reversion of the product to substrate.

Several factors must be considered when the experimental assay conditions are developed. The reaction rate depends upon the concentrations of substrate, enzyme, and necessary cofactors. In addition, the reaction rate is under the influence of environmental factors such as pH, temperature, and ionic strength. Enzyme activity increases with increasing temperature until the enzyme becomes denatured. The enzyme activity then decreases until all enzyme molecules are inactivated by denaturation. During kinetic measurement, it is essential that the temperature of all reaction mixtures be maintained constant.

Ionic strength and pH should also be monitored carefully. Although some enzymes show little or no change in activity over a broad pH range, most enzymes display maximum activity in a narrow pH range. Any assay developed to evaluate the kinetic characteristics of an enzyme must be performed in a continuously buffered solution, preferably at the optimal pH. Enzymes vary greatly in their response to inorganic salts. Some enzymes require the presence of anions or cations for activity, whereas others are inhibited. Trial and error using common salts (KCl, NaCl, etc.) is the best approach.

Applications of an Enzyme Assay

The conditions used in an enzyme assay depend upon what is to be accomplished by the assay. There are two primary applications of an enzyme assay procedure. First, it may be used to measure the concentration of active enzyme in a preparation. In this circumstance, the measured rate of the enzyme-catalyzed reaction must be proportional to the concentration of enzyme; stated in more kinetic terms, there must be a linear relationship between initial rate and enzyme concentration (the reaction is first order in enzyme concentration). To achieve this, certain conditions must be met: (1) the concentrations of substrate(s), cofactors, and other requirements must be in excess, (2) the reaction mixture must not contain inhibitors of the enzyme, and (3) all environmental factors such as pH, temperature, and ionic strength should be optimal. Under these conditions, a plot of enzyme activity (μmole/minute) vs. enzyme concentration is a straight line and can be used to estimate the concentration of active enzyme in solution.

Second, an enzyme assay may be used to measure the kinetic properties of an enzyme such as K_M, V_{max}, and inhibition characteristics. In this situation, different experimental conditions must be used. If K_M for a substrate is desired, the assay conditions must be such that the measured initial rate is first order in substrate. The concentration of enzyme and all other factors must be in excess. To determine K_M of a substrate, constant amounts of enzyme are incubated with varying amounts of substrate. A Lineweaver-Burk plot ($1/v_0$ vs. $1/[S]$) or direct linear plot may be used to determine K_M and V_{max}. If a reaction involves two or more substrates, each must be evaluated separately. The concentration of only one substrate is varied while the other is held constant at a saturating level. The same procedure holds for determining the kinetic dependence on cofactors. Substrate(s) and enzyme are held constant, but at excess levels, and the concentration of the cofactor is varied.

Inhibition kinetics are included in the second category of assay applications. An earlier discussion outlined the kinetic differentiation between competitive and noncompetitive inhibition. The same experimental conditions that pertain to evaluation of K_M and V_{max} hold for K_I estimation. Conditions must be optimal, with excess enzyme and cofactor concentration. A constant level of inhibitor is added to each assay, but the substrate concentration is varied as for K_M determination. In summary, a study of enzyme kinetics is approached by measuring initial reaction velocities under conditions where only one factor (substrate, enzyme, cofactor) is varied and all others are held constant and in excess.

Properties of Tyrosinase

Tyrosinase, also commonly called polyphenol oxidase, has two catalytic activities; *o*-hydroxylation of monophenols and aerobic oxidation of *o*-diphenols (Reactions 2 and 3).

monophenol + O_2 ⟶ catechol + H_2O (Reaction E7.2)

2 catechol + O_2 ⟶ 2 *o*-quinone + 2 H_2O (Reaction E7.3)

The official Enzyme Commission classification and number are monophenol, *o*-diphenol: O_2 oxidoreductase, EC 1.14.18.1. Tyrosinases are present in plant and animal cells. The biological function of tyrosinase in plant cells is unknown, but its presence is readily apparent. Upon injury, many plant products such as potatoes, apples, and bananas undergo a browning process. This is due to the formation and further oxidation of natural quinones as shown in Reaction 3. In mammalian cells, tyrosinase catalyzes two steps in the biosynthesis of melanin pigments from tyrosine. The reaction sequence, shown in Figure E7.8, is localized in melanocytes or pigment cells and produces the melanin substances that impart color to skin, hair, and eyes. Tyrosinase located in skin is activated by the ultraviolet rays of the sun, thus causing greater melanin production and, ultimately, suntan.

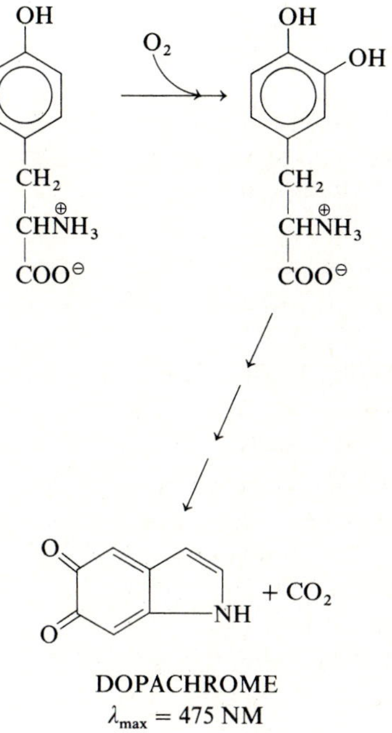

FIGURE E7.8
The oxidation of DOPA as catalyzed by tyrosinase.

DOPACHROME
$\lambda_{max} = 475\ NM$

The many tyrosinases in nature differ in cofactor requirement, metal ion content, substrate specificity, molecular weight, and oligomeric structure. The diverse properties of the tyrosinases will not be discussed; instead, the properties of a single enzyme, mushroom tyrosinase, will be outlined (Decker, 1977; Duckworth and Coleman, 1970; Nelson and Mason, 1970).

Mushroom tyrosinase is tetrameric, with a total molecular weight of 128,000. There are four atoms of Cu^+ associated with the active enzyme. Two types of substrate binding sites exist in the enzyme, one type for the phenolic substrate and one type for the dioxygen molecule. The copper atom is associated with the dioxygen binding site; hence, chemicals that form complexes with copper atoms are potent inhibitors of tyrosinase activity. The inhibitors azide, cyanide, phenylthiourea, and cysteine compete with oxygen binding. Nonphenolic, aromatic compounds compete with the binding and oxidation of phenols and catechols.

Overview of the Experiment

Several kinetic characteristics of mushroom tyrosinase will be examined in this experiment. A spectrophotometric assay of tyrosinase activity will be introduced and applied to the evaluation of substrate specificity, K_M of the natural substrate, 3,4-dihydroxyphenylalanine (L-DOPA), and inhibition characteristics.

Many assays for tyrosinase activity have been developed. Procedures in the literature include the oxygen electrode, oxidation of tyrosine followed at 280 nm, and oxidation of DOPA followed at 475 nm. The most convenient assay involves following the tyrosinase-catalyzed oxidation of DOPA by monitoring the initial rate of formation of dopachrome at 475 nm (Figure E7.8).

Two substrates are required in the tyrosinase-catalyzed reaction, phenolic substrate (DOPA) and dioxygen. The conditions described in the experiment are such that the reaction mixtures are saturated with dissolved dioxygen. Therefore, when measurements are made for K_M, only the concentration of DOPA is limiting, so the rate of the reaction depends on DOPA concentration.

The dopachrome assay is extremely flexible, as it can be applied to a variety of studies of tyrosinase. Projects not described here, but of potential interest, are pH optimum studies, inhibition studies, and formation of the apoenzyme by dialysis and reconstitution (Pomerantz, 1966).

Time requirements for this experiment are:

Part A. Tyrosinase concentration—15 minutes

Part B. Tyrosinase level for kinetic assay—60 minutes

Part C. K_M of DOPA—2 hours

Part D. Inhibition—2 hours

It is recommended that Part A be done as a group project so that enzyme is not wasted. Parts B and C can be completed in a 3-hour laboratory period. Part D can be completed during a second 3-hour period, because the determination of K_M of L-DOPA must be repeated if Part D is done later.

II. EXPERIMENTAL

Materials and Supplies

Sodium phosphate buffer, 0.1 M, pH 7.0

Mushroom tyrosinase, 100 units/ml.
> 1 unit represents the amount of enzyme that causes an A_{280} change of 0.001 using tyrosine as substrate. Prepare solution in phosphate buffer. A_{280} should be about 0.20. Store in icebath during laboratory period.

L-DOPA, 3 mg/ml in phosphate buffer

D-DOPA, 3 mg/ml in phosphate buffer

Thiourea, 1×10^{-3} M in phosphate buffer

Benzoic acid, sodium salt, 3×10^{-3} M in phosphate buffer

Cuvets, glass and quartz

Stopwatch

Constant temperature bath at 25–30°C

UV-VIS Spectrophotometer with recorder and constant temperature circulating bath or colorimeter

Procedure

A. Measurement of Tyrosinase Concentration

In order to determine the specific activity of the enzyme, the exact concentration of the enzyme must be known. The concentration of the solution of tyrosinase may be determined as a class project by the following procedure. Turn on the spectrophotometer and the UV lamp. Adjust the wavelength to 280 nm. Allow the instrument and lamp to warm up for 15 to 20 minutes. Transfer 1.0 or 3.0 ml of the phosphate buffer to a 1- or 3-ml quartz cuvet. Place it in the sample position of the spectrophotometer and adjust the balance to zero absorbance. Discard the buffer, and clean and dry the cuvet. Transfer 1.0 or 3.0 ml of the tyrosinase solution into the quartz cuvet. Place in the sample position and record the absorbance at 280 nm. If the A_{280} is greater than 1, dilute some tyrosinase solution with phosphate buffer and repeat the absorbance reading. Be sure the dilution is quantitative. That is, into a 3-ml cuvet, for example, add 1.5 ml of phosphate buffer with a pipet, and likewise, 1.5 ml of tyrosinase solution; mix well before reading the absorbance. Calculate the tyrosinase concentration as described in the Analysis of Results section.

B. Determination of the Tyrosinase Level for Kinetic Assays

All solutions listed under Materials and Supplies, except tyrosinase, should be stored in a constant temperature bath. The tyrosinase solution should be stored in an icebath. If there is no constant temperature circulating bath on the spectrophotometer, maintain all solutions except tyrosinase at room temperature. Otherwise, set both the constant temperature bath and the circulating bath at the same temperature. The recommended temperature range is 25° to 30°C. Turn on the spectrophotometer and recorder for warm-up. Adjust the wavelength to 475 nm.

Before kinetic constants can be evaluated, it is critical to find the correct concentration of enzyme to use for the assays. If too little enzyme is used, the overall absorbance change for a reaction time period will be so small that it is difficult to detect differences due to substrate concentration changes or inhibitor action. On the other hand, too much enzyme will allow the reaction to proceed too rapidly, and the leveling-off of the time course curve as shown in Figure E7.7 will occur very early because of the rapid disappearance of substrate. A rate that is intermediate between these two extremes is best. For the dopachrome assay, it is desirable to use the level of tyrosinase that gives a linear absorbance change at 475 nm for 3 minutes.

The general procedure for the assay is as follows. Pipet phosphate buffer and L-DOPA into a glass cuvet according to the directions in Table E7.1. Mix by inversion (cover the cuvet with hydrocarbon foil), and place the cuvet in the spectrometer sample position. If a recorder is available, adjust it to a full-scale absorbance of 1 and set the pen to 0 or the 0.10 absorbance line with the recorder zero adjust dial. Set the recorder speed to 1 or 2 cm/minute and turn on the recorder. Observe whether there is a "blank" rate (no enzyme). Obtain a recorder trace for 3 minutes. Now, pipet the appropriate amount of enzyme into the cuvet. Mix by inversion of the cuvet after covering it with hydrocarbon foil (DO NOT SHAKE!), and immediately place in the spectrophotometer. Turn on the recorder pen and observe the recorder trace for 3 minutes. Be sure you mark, on the recorder paper, the point of tyrosinase addition. With a straight-edge, draw a line tangent to the recorder trace as shown in Figure E7.7. Is the time course linear for the full 3 minutes? Determine the longest period of linearity and calculate the average ΔA/minute over this time period. If you observed a "blank" rate with no enzyme present, calculate its ΔA/minute and subtract it from the enzyme-catalyzed reaction rate.

If no recorder is available, you must take absorbance readings at 475 nm every 30 seconds. Proceed as follows. To the cuvet, add buffer and L-DOPA according to Table E7.1. Mix well and place in the photometer. Immediately adjust the absorbance to 0. Begin timing with a stopwatch and record in your notebook an absorbance reading every 30 seconds. Continue for 3 minutes. Now, add the appropriate level of tyrosinase, mix (DO NOT SHAKE), immediately place the cuvet in the photometer and set the absorbance to 0. Start the stopwatch and record an absorbance reading every 30 seconds for 3 minutes. On graph paper, prepare a plot of A_{475} vs. time. Calculate ΔA/minute over a period during which the assay

TABLE E7.1

General Procedure for the Tyrosinase Assay

Reagent	Asssay				
	1	2	3	4	5
Phosphate buffer	1.95	1.90	1.80	1.70	1.60
L-DOPA	1.0	1.0	1.0	1.0	1.0
Tyrosinase	0.05	0.10	0.20	0.30	0.40

is linear. On the same graph paper, plot A_{475} vs. time for the "blank." If there is a significant rate (greater than 5% of the enzyme rate), subtract it from the enzyme-catalyzed rate before preparing the graph.

Repeat the above procedures (with or without recorder) with a different concentration of tyrosinase. If the corrected rate (ΔA/min for enzyme minus ΔA/min for blank) was less than 0.03, double the amount of enzyme added. If the corrected rate was greater than 0.3/minute, reduce the enzyme by one-half. Determine the rate for a total of five different enzyme levels. The enzyme levels chosen should cover a ΔA/minute range from about 0.025/minute to 0.25/minute. The tyrosine levels listed in Table E7.1 are only recommended; you may have to try higher or lower amounts of tyrosinase.

C. K_M of L-DOPA and D-DOPA

Now that the appropriate enzyme level has been determined, the kinetic constants may be evaluated. The K_M for L-DOPA can be obtained by setting up the same assay as in Part B, except that the factor to vary will be the concentration of L-DOPA. The concentration of L-DOPA in Part B was sufficient to saturate all the tyrosinase active sites, so the rate depended only on the enzyme concentration. In Part C, L-DOPA levels will be varied over a range that is nonsaturating.

Choose the level of enzyme that will give a convenient rate, at saturating levels of L-DOPA, of 0.10 to 0.15 ΔA/minute. Prepare a table, for five assays, similar to Table E7.1. The total volume for each assay must be constant at 3.0 ml. Since the enzyme added to each assay remains constant, only the volume of buffer and L-DOPA is varied. The recommended levels of L-DOPA are 0.05, 0.10, 0.20, 0.40, 0.80, and 1.0 ml. The procedure to follow is identical to Part B. The enzyme should be added last, after a "blank" rate is determined for each assay. If the recommended levels of L-DOPA do not result in a saturating rate, increase the amount of L-DOPA. From the slope of each recorder trace or graph of absorbance vs. time, calculate the ΔA/minute for each level of L-DOPA. In order to determine the K_M of D-DOPA, repeat the above procedure using D-DOPA rather than L-DOPA. Calculate the ΔA/minute for each concentration level of D-DOPA.

D. Inhibition of Tyrosinase Activity

Whether an inhibitor acts in a competitive or noncompetitive manner is deduced from a Lineweaver-Burk or direct linear plot using varying concentrations of inhibitor and substrate. In separate assays, two substances will be added to the DOPA-tyrosinase reaction mixture, and the effect on enzyme activity will be quantified. The structures of the two chemicals, sodium benzoate and thiourea, are shown in Figure E7.9. The inhibition assays must be done immediately following the K_M studies. To

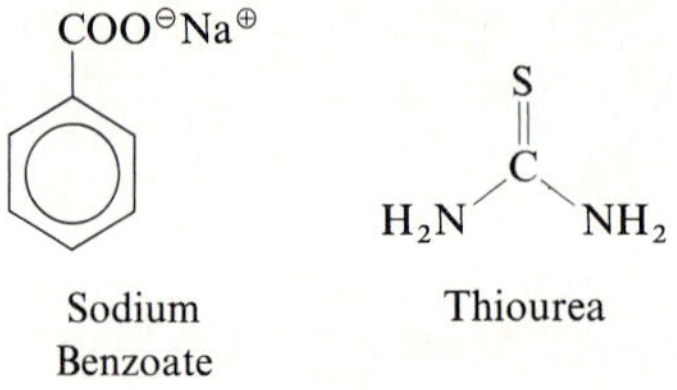

COO$^\ominus$Na$^\oplus$

Sodium
Benzoate

S
‖
C
H$_2$N   NH$_2$

Thiourea

FIGURE E7.9
Two inhibitors of tyrosinase.

measure inhibition, reaction rates both with and without inhibitor must be used and the tyrosinase activity must not be significantly different. If it is necessary to do the inhibition studies later, the K_M assay for L-DOPA must be repeated with freshly prepared tyrosinase solution.

To set up the inhibition assay, prepare a table similar to Table E7.1. Inhibitor should appear in the list of reagents before tyrosinase. Choose a level of tyrosinase that results in a relatively high reaction rate (a ΔA/min of approximately 0.2). Vary the amount of L-DOPA as in Part C. A constant amount of inhibitor (benzoate or thiourea) should be added to each cuvet. You will have to determine this level of inhibitor by trial and error. The desired inhibition rate with saturating substrate is 10 to 20% of the uninhibited rate. Add all reagents except tyrosinase, mix well, and determine the blank rate, if any. Add tyrosinase, mix, and immediately record A_{475} for 3 minutes. From recorder traces or graphs of A_{475} vs. time, calculate ΔA/minute for each assay.

III. ANALYSIS OF RESULTS

A. Concentration of Tyrosinase Solution

If the tyrosinase preparation is homogeneous, the concentration of enzyme may be determined by absorbance measurement at 280 nm. The extinction coefficient, $E_{280}^{1\%}$, for tyrosinase is 24.9. That is, the absorption of a 1% (w/v) solution of pure tyrosinase in a 1-cm cell at 280 nm is 24.9. Beer's law states that

$$A = \varepsilon bc$$

where

A = absorbance

ε = extinction coefficient

c = concentration of solution

b = cell path length (usually 1 cm)

For a 1% solution of tyrosinase, A is equal to ε, if b is 1 cm,

$$A = E_{280}^{1\%}$$

Therefore, the concentration of tyrosinase in solution is calculated from a simple ratio,

$$\frac{C_1}{A_1} = \frac{C_2}{A_2}$$

where

C_1 = concentration of a standard solution of tyrosinase = 1%

A_1 = absorption of the standard solution of tyrosinase at 280 nm
= 24.9

C_2 = concentration of unknown solution of tyrosinase in % (w/v)

A_2 = absorption of unknown solution of tyrosinase at 280 nm

Convert your concentration (now in w/v%) to mg of tyrosinase/ml.

B. Determination of the Tyrosinase Level for Kinetic Assays

Prepare a table of rate (ΔA/minute) vs. enzyme concentration (mg per assay). ΔA/minute units are converted to a more useful form, amount of product formed/time, usually μmole/minute. The conversion is straightforward when Beer's law is used in the form of Equation 7,

$$c = \frac{A}{\varepsilon b}$$

(Equation E7.7)

where A refers to the absorbance change that occurs per minute, b is the light path length through the cuvet (1 cm), ε is the extinction coefficient for the product dopachrome (3600 M^{-1} cm^{-1}), and c is the concentration of product. According to this relationship, the absorbance change during 1 minute for production of 1 mole of dopachrome is 3600:

$$1 \text{ mole} = \frac{A}{3600 \text{ M}^{-1} \text{ cm}^{-1} \times 1 \text{ cm}}$$

$$A = 3600$$

In order to convert the raw data to moles of product formed per minute, each ΔA/minute is divided by 3600. For example, a ΔA/minute of 0.1 becomes $0.1/3600 = 2.78 \times 10^{-5}$ mole/minute or 27.8 μmole. Add another column to your table, label it μmoles product formed per minute, and calculate the appropriate rate for each enzyme concentration. Prepare a graph of rate (μmole/minute on the ordinate, y axis) vs. enzyme concentration in each assay (mg, on the abscissa, x axis). Connect as many of the points as possible with a straight line passing through the origin. If most of the points are on this line, the assay and the standard curve can be used to quantify an unknown level of tyrosinase. The standard curve also provides the experimenter with a choice of enzyme levels to use for further kinetic studies.

C. K_M of L-DOPA and D-DOPA

The treatment of results will be described for L-DOPA. The procedure for D-DOPA is identical. Prepare a table of L-DOPA concentration per assay (mmolar) vs. ΔA/minute. Convert all ΔA/minute units to μmoles/minute as described in Part B. Prepare a Michaelis-Menten curve (μmoles/min vs. [S]) as in Figure E7.1, and a Lineweaver-Burk plot (1/μmole/minute vs. 1/[S]) as in Figure E7.2. Alternatively, you may wish to use the direct linear plot. Estimate K_M and V_{max} from each graph. The intercept on the rate axis of the Lineweaver-Burk plot is equal to $1/V_{max}$. For example, if the line intersects the axis at 0.02, then $V_{max} = 1/0.02$ or 50 μmoles of product formed per minute. The line intersects the 1/[S] axis at a point equivalent to $-1/K_M$. If the intersection point on the 1/[S] axis is -0.67, then $K_M = -1/-0.067 = 15$ μmolar. Repeat this procedure for the data obtained for D-DOPA. Compare the K_M and V_{max} values and explain any differences.

Some enzymes are able to differentiate between D and L isomers of substrates. DOPA, with a chiral center, exists in enantiomeric pairs, D and L. The naturally occurring DOPA molecule has the L-configuration, so it might be expected that the enzyme-catalyzed oxidation of L-DOPA is more facile than that of D-DOPA. The relative reactivity can be determined in a quantitative way by comparing K_M values. Does tyrosinase exhibit stereoselectivity?

D. Inhibition of Tyrosinase Activity

Two Lineweaver-Burk plots are prepared, one for each chemical inhibitor tested. Each plot of $1/v_0$ vs. 1/[S] consists of two lines, one representing the absence of inhibitor, the other representing a specific concentration of inhibitor.

Convert all ΔA/minute data into μmole/minute and calculate the reciprocal of each data point. Plot against the reciprocal of the corresponding substrate concentration (1/mmolar). One line on each graph is obtained from the set of data acquired from the L-DOPA K_M experiment in Part C. Draw the two separate lines with a straight-edge, making sure the appropriate data points are used. Compare the two graphs with Figures E7.4 and E7.5 to determine the type of inhibition for sodium benzoate and thiourea. Explain the inhibitory effect of each substance.

IV. QUESTIONS AND PROBLEMS

1. One of the many straight-line modifications of the Michaelis-Menten equation is the Eadie-Hofstee equation:

$$v_0 = -K_M \frac{v_0}{[S]} + V_{max}$$

Beginning with the Lineweaver-Burk equation, derive the Eadie-Hofstee equation. Explain how it may be used to plot a straight line from experimental rate data. How are V_{max} and K_M calculated?

2. The table below gives initial rates of an enzyme-catalyzed reaction along with the corresponding substrate concentration. Use any graphical method to determine K_M and V_{max}.

v_0 μmole/minute	[S] mole/liter
130	65×10^{-5}
116	23×10^{-5}
87	7.9×10^{-5}
63	3.9×10^{-5}
30	1.3×10^{-5}
10	0.37×10^{-5}

3. The enzyme concentration used in each assay in Problem 2 is 5×10^{-7} mole/liter. Calculate k_3, the turnover number, in units of sec^{-1}.

4. Using the data in Problems 2 and 3, calculate the specific activity of the enzyme in units/mg and katal/mg. Assume the enzyme has a molecular weight of 55,000 and the reaction mixture for each assay is contained in a total volume of 1.00 ml.

5. Each of the compounds listed below is known to inhibit tyrosinase activity.

sodium azide, NaN_3 sodium cyanide, NaCH

L-phenylalanine   8-hydroxyquinoline

tryptophan   diethyldithiocarbamate

cysteine   thiourea

(a) Study the structure of each and try to predict whether the substance is a competitive or noncompetitive inhibitor of tyrosinase activity as measured by the dopachrome assay. Assume that the inhibition data were evaluated using a Lineweaver-Burk plot of $1/v_0$ vs. $1/[DOPA]$.

(b) An alternate method for measuring tyrosinase activity is the use of an oxygen electrode to measure the rate of dioxygen utilization during phenol oxidation (see Reactions 1 and 2). Predict whether each substance is a competitive or noncompetitive inhibitor of tyrosinase as measured by an oxygen electrode. Assume that a Lineweaver-Burk plot of $1/v_0$ vs. $1/[O_2]$ was used to evaluate the rate data.

6. Compound X was tested as an inhibitor of the enzyme in Problem 2. Use the rate data in Problem 2 and the following inhibition data to evaluate compound X. Is it a competitive or noncompetitive inhibitor? Calculate K_I for the inhibitor.

| $[X] = 3.7 \times 10^{-4} \, M$ | | $[X] = 1.58 \times 10^{-3} \, M$ |
v_0 μmole/min	$[S]$ mole/liter	v_0 μmole/min
125	6.5×10^{-4}	110
102	2.3×10^{-4}	80
70	7.9×10^{-5}	40
45	3.9×10^{-5}	20
20	1.3×10^{-5}	—
5	0.37×10^{-5}	—

V. REFERENCES

General

R.C. Bohinski, *Modern Concepts in Biochemistry*, Fourth Edition (1983), Allyn and Bacon (Boston), pp. 132–143. An introduction to enzyme kinetics.

A. Cornish-Bowden, *Fundamentals of Enzyme Kinetics* (1979), Butterworths (London). An entire book devoted to enzyme kinetics.

J.D. Rawn, *Biochemistry* (1983), Harper and Row (New York), pp. 227–267. A full chapter on enzyme kinetics.

E. Smith, R. Hill, I. Lehman, R. Lefkowitz, P. Handler, and A. White, *Principles of Biochemistry: General Aspects*, Seventh Edition (1983), McGraw-Hill (New York), pp. 179–209. A chapter on enzyme kinetics.

L. Stryer, *Biochemistry*, Second Edition (1981), Freeman (San Francisco), pp. 103–120. Introduction to enzymes and kinetic analysis.

W.B. Wood, J.H. Wilson, R.M. Benbow, and L.E. Hood, *Biochemistry, A Problems Approach*, Second Edition (1981), Benjamin/Cummings (Menlo Park, CA), pp. 144–172. Enzyme kinetics with an introduction to the Eisenthal and Cornish-Bowden direct linear plot.

G. Zubay, *Biochemistry* (1983), Addison-Wesley (Reading, MA), pp. 137–159. Detailed discussion of enzyme kinetics.

Specific

L. Decker, *Worthington Enzyme Manual* (1977), Worthington Biochemical Corporation (Freehold, NJ). pp. 74–76.

H.W. Duckworth and J.E. Coleman, *J. Biol. Chem.*, *245*, 1613 (1970). "Physicochemical and Kinetic Properties of Mushroom Tyrosinase."

R.M. Nelson and H.S. Mason in *Methods in Enzymology*, S.P. Colowick and N.O. Kaplan, Editors, Vol. 17A, Academic Press (New York), pp. 626–32. "Mushroom Tyrosinase."

S.H. Pomerantz, *J. Biol. Chem.*, *241*, 161 (1966). "The Tyrosine Hydroxylase Activity of Mammalian Tyrosinase."

Experiment **8**

Evaluation of a Regulatory Enzyme System: Lactose Synthetase

RECOMMENDED READING:
Chapter 5, Sections A, B, C

SYNOPSIS
Lactose synthetase is an enzyme system in milk and mammary gland that catalyzes the transfer of galactose from UDP-galactose to glucose. The enzyme system is unique in that it consists of two protein components, a modifier subunit (α-lactalbumin) and a catalytic subunit (galactosyl transferase). In this experiment, the regulatory properties of lactose synthetase will be studied using a coupled enzyme assay.

I. INTRODUCTION

Regulation of Enzymatic Activity

Students often have the misconception that enzymes are fully active under all cellular conditions and that metabolic products are produced at maximum rates at all times. However, the metabolic reactions occurring in cells are under strict regulatory control. Metabolic intermediates are synthesized and degraded according to the particular needs of the cell at that time. Throughout the process of evolutionary change, biological cells have developed intricate processes by which they balance the concentrations of metabolites and, therefore, conservatively use scarce nutrients.

One of the most important ways for cells to control metabolic activity is by regulating the activity of enzymes. Nature has developed several ways to control enzyme activity. The simplest, most readily apparent method of regulation is dependence upon cellular conditions. The rates of many enzyme-catalyzed reactions are under the influence of pH and intracellular concentration of substrates, products, and cofactors. Secondly, some of the digestive enzymes such as trypsin and chymotrypsin are initially synthesized in an inactive or **zymogen** form. When these particular enzymes are needed, the zymogen form is converted to an active form. A third type of enzyme regulation occurs at the gene level, where the rate of enzyme synthesis is controlled. Some enzymes are not even synthesized until there is a specific need for their action. By genetic control, cells are able to maintain enzyme activities at levels that are necessary at any particular time.

Metabolic regulation is also under the control of chemical messengers (hormonal substances) that are secreted by the endocrine system. The control of carbohydrate metabolism by insulin and glucagon is a well-known example.

Finally, metabolic activity may be controlled by the action of non-substrate molecules on **regulatory** or **allosteric** enzymes. The small regulatory molecules affecting enzyme activity are called **modulators** or **effectors,** and they may act to inhibit or stimulate an enzyme-catalyzed reaction.

Allosteric enzymes have two types of binding sites, primary binding sites for substrate binding and secondary binding sites for modulator binding. When a secondary binding site on an allosteric enzyme holds a modulator, the conformation of the enzyme is changed so that it functions at a slower ($-$ modulator) or faster ($+$ modulator) rate than in the absence of the modulator.

Modulators are usually small, nonsubstrate molecules, but they often have some relationship to the substrate. For example, the end product of a sequence of metabolic reactions is often a negative modulator for the enzyme catalyzing the first step in the sequence.

Characterization of Allosteric Enzymes

Allosteric enzymes can easily be identified because they do not obey classical Michaelis-Menten kinetics. Many allosteric enzymes display a sigmoidal rather than a hyperbolic relationship, when v_0 is plotted against [S]. This indicates cooperative binding of substrate; binding of the first substrate molecule influences the binding of subsequent substrate molecules. The influence can be an enhancement of substrate binding ($+$ cooperativity) or a decrease ($-$ cooperativity). Most allosteric enzymes are oligomeric and have more than one of each type of binding site. Allosteric enzymes may respond in different ways when under the influence of a modulator. Some allosteric enzymes display a change in substrate K_M when the modulator is bound, whereas others show a change in V_{max}. However, not all allosteric enzymes can be categorized in this manner. As more and more allosteric enzymes are identified and characterized, it becomes increasingly difficult to categorize the enzymes into groups with similar properties. It seems that each newly discovered regulatory enzyme system represents a unique type of molecular control of a biological reaction.

A case in point is the allosteric enzyme system of lactose synthetase. The biosynthesis of lactose is outlined in Reactions 1 to 3, where UTP represent uridine 5′-triphosphate and UDP is uridine diphosphate.

$$UTP + glucose\ 1\text{-}P \longrightarrow UDP\text{-glucose} + PP_i \qquad \text{(Reaction E8.1)}$$

$$UDP\text{-glucose} \longrightarrow UDP\text{-galactose} \qquad \text{(Reaction E8.2)}$$

$$UDP\text{-galactose} + glucose \longrightarrow lactose + UDP \qquad \text{(Reaction E8.3)}$$

Reaction 1 is catalyzed by glucose 1-phosphate uridyl transferase (EC 2.7.7.9), Reaction 2 by UDP-galactose 4-epimerase (EC 5.1.3.2), and Reaction 3 by lactose synthetase (EC 2.4.1.22).

The lactose synthetase system has been isolated from bovine mammary glands and bovine skim milk. Purification of the enzyme by gel filtration led to the discovery that two separate proteins are required for lactose synthetase activity (Reaction 3). The separate proteins, called A and B, are not able to catalyze Reaction 3, but when they are combined lactose synthetase activity is observed (Brew, Vanaman, and Hill, 1968; Ebner, 1970). Protein B has since been obtained in crystalline form from bovine milk and found to be identical to α-lactalbumin, a protein first isolated from milk in 1899. Until 1965, the biological function of α-lactalbumin was not known; in fact, it was assumed only to be a nutritional factor in milk.

Protein A in purified form has enzyme activity in the absence of α-lactalbumin (Protein B). Protein A alone catalyzes the transfer of galactose to N-acetylglucosamine (Reaction 4).

$$UDP\text{-galactose} + N\text{-acetyl-D-glucosamine} \longrightarrow$$
$$N\text{-acetyllactosamine} + UDP \quad \text{(Reaction E8.4)}$$

α-Lactalbumin is an inhibitor of Reaction 4. Glucose may be substituted for N-acetylglucosamine, but lactose synthesis is very slow unless α-lactalbumin is added. These observations indicate that the actual enzymatic activity of lactose synthetase is associated with the A protein. If α-lactalbumin is not present, then the preferred substrate is N-acetylglucosamine or other amino sugars. However, in the presence of α-lactalbumin, the galactosyl transferase prefers glucose as a substrate, and the binding of N-acetylglucosamine is inhibited. α-Lactalbumin changes the acceptor specificity of protein A from N-acetylglucosamine to glucose. Thus, α-lactalbumin is called a "specifier protein." One explanation for the action of α-lactalbumin is that when it is bound to a site on Protein A, the conformation of the active site is changed to accept glucose more readily.

Several researchers have demonstrated that, under appropriate conditions, α-lactalbumin interacts with galactosyl transferase to form a detectable complex (Powell and Brew, 1975; Bell, Beyer, and Hill, 1976).

The control of galactosyl transferase by α-lactalbumin has physiological and clinical significance. The activity of the A protein is high during pregnancy, but lactose synthetase activity is very low. The amino sugars produced in Reaction 4 are important for antibody production during pregnancy. At the time of birth, α-lactalbumin levels in the mammary gland increase, inhibiting Reaction 4 and stimulating the synthesis of lactose.

Assay for Lactose Synthetase Activity

In this experiment, the properties of an allosteric enzyme are evaluated using a kinetic assay to monitor enzymatic activity. The design of an assay for enzyme activity was outlined in Experiment 7; however, only one protein, tyrosinase, and a substrate, DOPA, were required. An assay for lactose synthetase requires finding optimum concentrations for both proteins A and B.

The reaction to be assayed is shown again below (Reaction 3). It does not lend itself directly to a spectrophotometric assay because it is not accompanied by significant changes in the visible or ultraviolet wavelength region.

$$\text{UDP-galactose + glucose} \longrightarrow \text{lactose + UDP} \qquad \text{(Reaction E8.3)}$$

However, it is possible to "couple" Reaction 3 to one or more reactions that offer a measurable absorbance change. The formation of UDP in Reaction 3 is coupled to three reactions (Reactions 5, 6, and 7), the last of which produces an absorbance change.

$$\text{UDP + ATP} \longrightarrow \text{UTP + ADP} \qquad \text{(Reaction E8.5)}$$

$$\text{PEP + ADP} \longrightarrow \text{pyruvate + ATP} \qquad \text{(Reaction E8.6)}$$

$$\text{pyruvate + NADH + H}^+ \longrightarrow \text{lactate + NAD}^+ \qquad \text{(Reaction E8.7)}$$

The phosphate exchange of Reaction 5 is catalyzed by nucleoside diphosphokinase. Reaction 6, the transfer of phosphate from phosphoenolpyruvate to ADP catalyzed by pyruvate kinase, is a familiar one from the glycolytic pathway. Finally, the pyruvate generated by Reaction 6 is reduced at the expense of NADH. Lactate formation is catalyzed by lactate dehydrogenase. In order for the four reactions (3, 5, 6, and 7) to be a properly coupled system, Reactions 5, 6, and 7 must occur at rates equivalent to but not slower than Reaction 3. Stated in other terms, reaction conditions must be adjusted so that Reaction 3 is the rate-limiting step. This is accomplished by using excess amounts of ATP, PEP, and NADH. The experimental conditions, therefore, are fixed so that absorbance change occurring in Reaction 7 will be directly proportional to the concentration of UDP produced in Reaction 3. NADH absorbs strongly at 340 nm, whereas, NAD^+ is relatively transparent. Thus, the rate of disappearance of NADH is a measure of the rate of appearance of UDP.

The standard assay for lactose synthetase requires careful attention because many different components must be added. One simplifying and economical feature of the assay is that a crude preparation of pyruvate kinase serves as a substitute for several required enzymes. Crude pyruvate kinase samples also contain nucleoside diphosphokinase and lactate dehydrogenase.

Overview of the Experiment

Kinetic characteristics of lactose synthetase will be evaluated by assaying the enzyme system under various conditions. The following procedures are outlined in the experimental section.

A. Find the appropriate concentration level of galactosyl transferase with glucose or N-acetylglucosamine as substrate.

B. Demonstrate the requirement for α-lactalbumin with glucose as substrate.

C. Determine the effect of α-lactalbumin on galactosyl transferase with N-acetylglucosamine as substrate.

This experiment may be completed as a single exercise or it may be combined with Experiment 4, "Isolation and Purification of α-Lactalbumin" or Experiment 5, "Characterization of α-Lactalbumin." The α-lactalbumin isolated in Experiment 4 may be used, or the protein modifier may be purchased from a commercial supplier. All three parts of this experiment can be completed in a 3 to 4 hour laboratory period.

II. EXPERIMENTAL

Materials and Supplies

Tris buffer, 0.05 M, pH 7.5, containing 0.004 M $MnCl_2$. This buffer is used for all solutions below.

Glucose solution, 0.8 M

α-Lactalbumin solution, 1.5 mg/ml or use α-lactalbumin from Sephadex 50 column in Experiment 4

Phosphoenolpyruvate (PEP), 0.01 M

ATP, 0.001 M

N-acetylglucosamine (NAG), 0.03 M

Galactosyl transferase, 0.2 units/ml

Crude pyruvate kinase, 26 mg/ml

UDP-galactose, 0.002 M

NADH, 0.002 M

Working solution for coupled assay (good for 25 assays). Prepare just before use.

PEP	3.0 ml
ATP	3.0 ml
Pyruvate kinase	0.6 ml
NADH	3.0 ml

Spectrophotometer with recorder and temperature control (set at 25–30°C). Full absorbance range should be 0–1.0 absorbance unit.

Quartz cuvets, 1 ml

Constant temperature water bath, set at same temperature as temperature control on spectrometer

Procedure

Part A. Finding the Appropriate Concentration Level of Galactosyl Transferase

Turn on the spectrophotometer and UV lamp for warm-up. Set the wavelength at 340 nm. Adjust the temperature control for the cuvet holder to the desired temperature. Incubate all solutions except galactosyl transferase in a constant temperature water bath. Store the galactosyl transferase on ice. Prepare assay mixtures one at a time as outlined in Table E8.1. For assay 1, transfer the appropriate amount of each reagent into

TABLE E8.1

Preparation of Assay Mixtures for Part A

	Assay				
Reagents	1	2	3	4	5
Tris/MnCl$_2$ buffer	0.3	0.25	0.20	0.15	0.10
Working solution	0.3	$\longrightarrow$			
Glucose or NAG	0.1	$\longrightarrow$			
α-Lactalbumin	0.2	$\longrightarrow$			
UDP-galactose	0.1	$\longrightarrow$			
Galactosyl transferase	0	0.05	0.10	0.15	0.2

a 1 ml cuvet. For a 3-ml cuvet, triple all quantities. Place the cuvet in the spectrophotometer cell holder and allow one minute for temperature-equilibration. Then record the absorbance change at 340 nm for 3 to 5 minutes. Assay 1 represents a blank rate (minus enzyme). Assays 2 through 5 with varying amounts of galactosyl transferase may be done in one of two ways. If four cuvets are available, assays can be prepared all at once. Transfer the appropriate amount of each reagent, *except galactosyl transferase*, into cuvets and incubate in the water bath. Then add the proper amount of galactosyl transferase to cuvet 2, place it in the spectrophotometer, and measure the absorbance change at 340 nm for 3 to 5 minutes. Repeat this procedure for cuvets 3 through 5. Record ΔA for each assay.

Repeat the assays in Table E8.1, but use N-acetylglucosamine instead of glucose as substrate. Use the same amounts of reagents and the same procedure as for glucose. Record ΔA/min for each assay.

Part B. The Requirement for α-Lactalbumin

The influence of α-lactalbumin on the standard assay in Part A is evaluated by varying the concentration of α-lactalbumin in a series of assays. Prepare a table similar to Table E8.1, but vary the amount of α-lactalbumin in each cuvet. Recommended amounts of 1.5 mg/ml α-lactalbumin solution are 0.01, 0.02, 0.04, 0.08, 0.15, and 0.2 ml. Vary the amount of Tris buffer as needed to keep the volume constant. Maintain all other reagents at constant levels. Use a galactosyl transferase level from Part A that resulted in a ΔA/min of about 0.10.

Part C. Effect of α-Lactalbumin on Galactosyl Transferase with NAG as Substrate

Prepare a table with all required reagents. Use N-acetylglucosamine as substrate rather than glucose. Vary the amount of α-lactalbumin in each assay. Recommended levels are 0.005, 0.01, 0.02, and 0.05 ml of 1.5 mg/ml α-lactalbumin solution. Determine ΔA/min for each level of α-lactalbumin.

III. ANALYSIS OF RESULTS

Part A. Finding the Appropriate Concentration of Galactosyl Transferase

Prepare a table of rate (ΔA/min) vs. galactosyl transferase concentration (units per assay). Since rate units of ΔA/min are meaningless in terms of amount of product produced, they should be changed to units of μmole product formed per minute. Refer to Experiment 7 for a discussion of this conversion. The extinction coefficient, ε, for NADH is 6.22×10^3 M^{-1} cm^{-1}. Assay 1 is a blank and the rate obtained, if any, should be subtracted from assays 2 through 5. Prepare a graph of rate (μmole/min on the ordinate, y axis) vs. enzyme concentration in each assay (units of enzyme activity on the abscissa, x axis). Connect the points with the best straight line passing through the origin. Repeat this analysis for NAG as substrate. Describe how these graphs could be used to determine the activity of a solution of galactosyl transferase of unknown catalytic activity.

Part B. The Requirement for α-Lactalbumin

Convert rate measurements, originally in ΔA/min, to μmole product formed per minute. Prepare a table with all data, amount of α-lactalbumin in each assay, ΔA/min, and μmole product formed per minute. Finally, prepare a graph of rate (μmole product/min, y axis) vs. α-lactalbumin concentration (mg/assay). What is the shape of the graph? Is this what you would expect? Explain. Describe how this graph could be used to measure the concentration of α-lactalbumin in an unknown solution. Prepare a Lineweaver-Burk plot or an Eisenthal and Cornish-Bowden linear plot and determine the K_M for α-lactalbumin.

Part C. Effect of α-Lactalbumin on Galactosyl Transferase with NAG as Substrate

Convert all ΔA/min rates to μmole product formed/min and prepare a graph of μmole product formed/min vs. α-lactalbumin per assay. Explain the shape of the curve. What is the effect of α-lactalbumin in the assay?

IV. QUESTIONS

1. Suggest two experimental techniques that could be used to search for a complex between α-lactalbumin and galactosyl transferase. Describe possible experimental conditions.

2. Study the results obtained in Part C of this experiment and describe experimental methods characterizing α-lactalbumin as an inhibitor, (competitive, noncompetitive, etc.)

3. Describe how you would use the coupled assay to study each of the following problems:
    (a) substrate specificity of lactose synthetase.
    (b) K_M for glucose in the presence and absence of α-lactalbumin.
    (c) The concentration of α-lactalbumin in an unknown solution.
    (d) The concentration of NAG in an unknown solution.

V. REFERENCES

General

R.C. Bohinski, *Modern Concepts in Biochemistry*, Fourth Edition (1983), Allyn and Bacon (Boston), pp. 151–160. An introduction to allosteric enzymes.

J.D. Rawn, *Biochemistry* (1983), Harper & Row (New York), pp. 528–532. Regulation of metabolic activity.

E. Smith, R. Hill, I. Lehman, R. Lefkowitz, P. Handler, and A. White, *Principles of Biochemistry: General Aspects*, Seventh Edition (1983), McGraw-Hill (New York), pp. 199—209, 823. Discussion of allosteric enzymes and lactose synthetase.

L. Stryer, *Biochemistry*, Second Edition (1981), Freeman (San Francisco), pp. 120–123, 522–523. Examples of allosteric enzymes.

G. Zubay, *Biochemistry* (1983) Addison-Wesley (Reading, MA), pp. 185–193, 451. A discussion on regulation of enzyme activity and lactose synthesis.

Specific

J.E. Bell, T.A. Beyer, and R.L. Hill, *J. Biol. Chem.*, *251*, 3003 (1976). "The Kinetic Mechanism of Bovine Milk Galactosyl Transferase."

K. Brew, T.C. Vanaman, and R.L. Hill, *Proc. Natl. Acad. Sci. U.S.*, *59*, 491 (1968). "The Role of α-Lactalbumin and the A Protein in Lactose Synthetase."

K.E. Ebner, *Acct. Chem. Res. 3*, 41 (1970). "A Biological Role for α-Lactalbumin as a Component of an Enzyme Requiring Two Proteins."

J.T. Powell and K. Brew, *J. Biol. Chem.*, *250*, 6337 (1975). "On the Interaction of α-Lactalbumin and Galactosyl Transferase during Lactose Synthesis."

Experiment **9**

Isolation, Purification, and Partial Characterization of a Lipid Extract from Nutmeg

RECOMMENDED READING:
Chapter 3, Sections A, B, D.

SYNOPSIS
Lipids, substances found in both plant and animal cells, are among the most ubiquitous of biomolecules. A lipid extract of ground nutmeg will be purified by adsorption chromatography on a silica gel column. Analysis of the lipid extract by thin-layer chromatography will provide the classification of the unknown lipid. The molecular structures of the major components present in the extract will be elucidated in Experiment 10.

I. INTRODUCTION AND THEORY

Lipids are low-molecular-weight biomolecules that can be extracted from plant and animal tissues by organic solvents. The major classes of compounds represented in the lipid group are triacylglycerols, glycerophosphatides, sphingolipids, glycolipids, and sterols (Figure E9.1). Lipids are

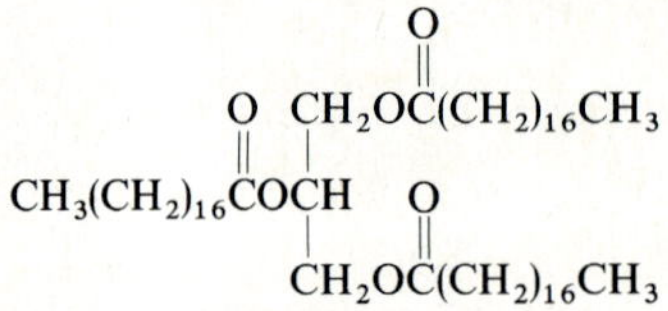

$$O$$
$$\parallel$$
$$O \quad CH_2OC(CH_2)_{16}CH_3$$
$$\parallel \quad |$$
$$CH_3(CH_2)_{16}COCH \quad O$$
$$| \quad \parallel$$
$$CH_2OC(CH_2)_{16}CH_3$$

A Triacylglycerol
(tristearin)

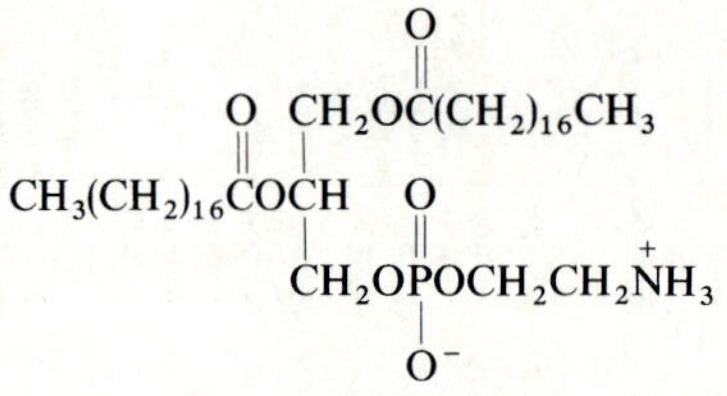

$$O$$
$$\parallel$$
$$O \quad CH_2OC(CH_2)_{16}CH_3$$
$$\parallel \quad |$$
$$CH_3(CH_2)_{16}COCH \quad O$$
$$| \quad \parallel \quad +$$
$$CH_2OPOCH_2CH_2NH_3$$
$$|$$
$$O^-$$

A Glycerophosphatide
(phosphatidyl ethanolamine)

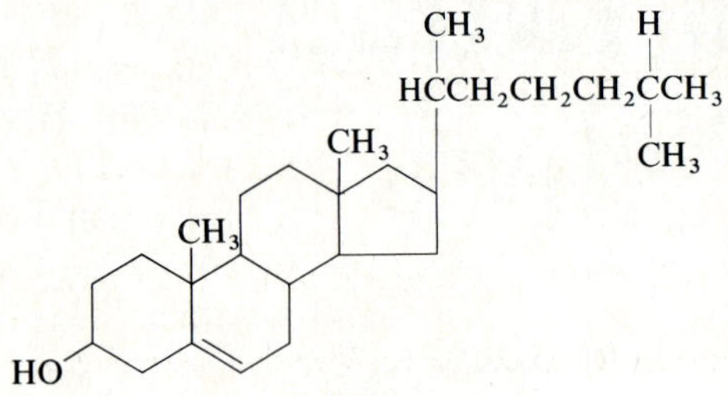

$$CH_3 \qquad H$$
$$HCCH_2CH_2CH_2CCH_3$$
$$CH_3 \qquad\qquad CH_3$$

$$CH_3$$

HO

A Steroid
(cholesterol)

$$H$$
$$CH_3(CH_2)_{12}C=CCHCHCH_2OR$$
$$H \quad HO \quad NH$$
$$C=O$$
$$(CH_2)_n$$
$$CH_3$$

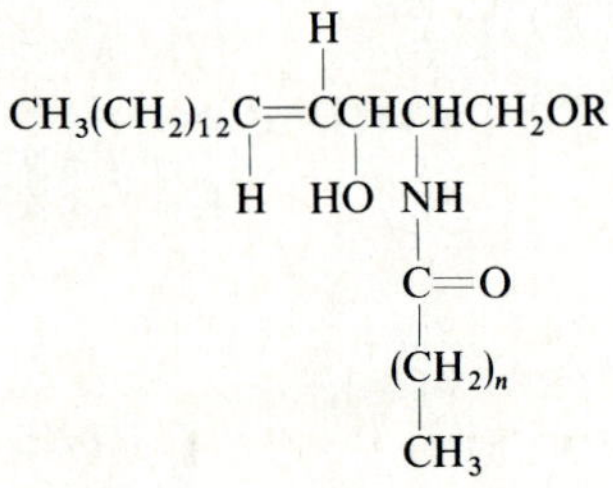

A Sphingolipid or Glycolipid

R = H, Ceramide
R = Phosphocholine, Sphingomyelin
R = galactose, Cerebroside

FIGURE E9.1
Structures of the major lipid classes.

found in most cells and tissue, but rarely are they present in a "free" or "uncombined" state. They are usually complexed with proteins or carbohydrates, which complicates their extraction and structural identification. Lipids have not only a wide variety of molecular structures, but also are involved with a whole array of biological functions, including membrane structural integrity, vitamin activity, cell recognition, and cellular energy.

In this chapter we will approach the world of lipids with an introduction to analytical procedures currently applied to lipid extraction, purification, and characterization.

Lipid Extraction

The extraction procedure used to isolate lipids from biological tissue depends upon the class of lipid desired and the nature of the biological source (animal tissue, plant leaf, plant seed, cell membranes, etc.). Because lipids are generally less polar than other cell constituents, they may be selectively extracted with the use of organic solvents. Early studies of lipids used ether, acetone, hexane, and other organic solvents for extraction; however, these solvents extract only lipids bound in a nonpolar or hydrophobic manner. In the 1950's, Folch's group reported the use of chloroform and methanol (2:1) in lipid extractions (Folch, 1951, 1957). This became the solvent system of choice for total lipid extraction because it was especially useful in dissociating lipid-protein complexes of plasma membranes (Radin, 1969). The only apparent disadvantage of this solvent system was its tendency to dissolve some nonlipids, including proteins. Increasing awareness of safety in the laboratory (both methanol and chloroform are toxic) led to a search for other solvent systems. A more desirable lipid solvent is hexane and isopropanol (3:2) (Radin, 1981). In addition to its lower toxicity, it has the advantage over chloroform-methanol that it dissolves very little nonlipid material. Whatever solvent system is used, chloroform-methanol or hexane-isopropanol, a total lipid extraction process yields a complex mixture of organic compounds in which all components have one common characteristic, a relatively nonpolar structure.

Solvent extraction of lipids from biological tissues is best carried out in a homogenizer as described in Experiment 4. More drastic extraction conditions, such as ultrasonic vibration or grinding with sand, may be necessary if the biomaterial contains excessive amounts of connective tissue. After the solvent extraction, several "work-up" steps should be completed before isolation of the crude lipid residue. If chloroform-methanol has been used, it is essential to wash the extracted lipid solution with an aqueous salt solution (0.5 to 1% KCl, NaCl or $CaCl_2$) in order to remove water and nonlipid materials. It is generally not necessary to wash hexane-isopropanol extractions, but if desired, they may be extracted with 7 to 10% aqueous sodium sulfate. This treatment removes any nonlipid materials that may be present and also reduces the volume of the extract because it removes water-soluble isopropanol without removing desired lipid material. Reducing the volume of the extraction solvent is particularly beneficial if further purification of the lipids by chromatography is under consideration.

The lipid extract, now washed to remove nonlipid materials, must be dried with anhydrous sodium sulfate or magnesium sulfate to remove the last traces of water. Evaporation of the water-free solvent extract is accomplished by rotary evaporation under vacuum. The crude, dry lipid residue may be susceptible to chemical modification by air oxidation, so it should be stored in the cold ($-20°C$) and dark, under an atmosphere of N_2 gas. Lipids containing polyunsaturated fatty acids are especially sensitive to air oxidation, so protective measures should be taken if these substances are of interest. For a more detailed discussion of lipid extraction, consult a recent monograph published by The American Oil Chemist's Society (Nelson, 1975).

Lipid Purification

Separation of crude lipid extracts into individual lipid classes is a difficult and time-consuming activity. In some cases a crude separation of lipids can be attained by selective solvent extraction. The more nonpolar neutral lipids (triacylglycerols and steroids) may be efficiently extracted from biological material with chloroform, ether, or acetone. The more polar lipids (phospholipids and glycolipids) may be selectively isolated by initial extraction of the tissue with cold acetone to remove neutral nonpolar lipids, followed by extraction of more polar lipids with hexane-isopropanol. For more extensive purification of lipids, the researcher must turn to chromatography. Chromatographic methods are capable of both resolving a complex lipid mixture into the various classes of lipids and separating the individual components of a single class of lipids. Silica gel chromatography (column and thin-layer) is now the most frequently used method for the resolution of lipid extracts. For large scale separation of lipid classes, column chromatography using various adsorbents (silica gel, Florisil, ion-exchange celluloses, and lipophilic Sephadexes) is most effective.

Adsorption chromatography (Chapter 3) on columns of silica gel is the best preparative method for the separation and purification of neutral lipids (Sweeley, 1969). Silica gel (sometimes called silicic acid) is chemically represented as $SiO_2 \cdot xH_2O$ and is a relatively polar adsorbent. Neutral or nonpolar lipids are not readily adsorbed by the support and can be separated from the more polar lipids, which are preferentially adsorbed. In practice, a glass column is packed with a slurry of silica gel in a hydrocarbon solvent (petroleum ether, pentane, or hexane). The lipid fraction to be purified is dissolved in the same hydrocarbon solvent and applied to the top of the column. To elute the various lipid classes, solvents of increasing polarity are allowed to percolate through the silica gel column and are collected at the bottom of the column. Table E9.1 outlines the order of elution of lipids from silica gel columns. The polarity of the eluting solvent is increased by adding a gradually increasing proportion of ethyl ether. For example, 1% ether in petroleum ether (Fraction I) will elute only the least polar of the lipids—paraffins and fatty acid esters; 4% ethyl ether in petroleum ether has sufficient polarity to elute triacylglycerols, and 25% to 75% ethyl ether in petroleum ether will elute diacylglycerols and monoacylglycerols. For elution of phospholipids and

TABLE E9.1

Order of Elution of Lipids from Silica Gel Columns (Sweeley, 1969)

Fraction Number	Solvent	Lipid Component
Neutral		
I	1% Ethyl ether/petroleum ether	Paraffins
II	1% Ethyl ether/petroleum ether	Squalene, waxes, fatty acid esters
III	1% Ethyl ether/petroleum ether	Cholesterol esters
IV	4% Ethyl ether/petroleum ether	Triacylglycerols, fatty acids
V	8% Ethyl ether/petroleum ether	Steroids (cholesterol)
VI	25% Ethyl ether/petroleum ether	Diacylglycerols
VII	100% Ethyl ether	Monoacylglycerols
Polar		
VIII	5% Methanol/Chloroform	Ceramides, phosphatidic acid, cardiolipins
IX	20% Methanol/Chloroform	Phosphatidylethanolamine, phosphatidylserine, phosphatidylinositol
X	30% Methanol/Chloroform	Phosphatidylcholine
XI	50% Methanol/Chloroform	Sphingomyelin
XII	70% Methanol/Chloroform	Lysophosphatidylcholine

glycolipids, solvent systems of increasing polarity such as chloroform-methanol must be used; however, silica gel columns are most effective for resolution of the neutral lipids.

Once several silica gel column fractions have been collected, the eluted fractions must be analyzed for the presence of lipids. Thin-layer chromatography on silica gel plates will be introduced in this experiment for the detection and identification of neutral lipids.

As previously discussed in Chapter 3, thin-layer chromatography utilizes a thin coating of silica gel on a glass or plastic plate to separate components of a mixture (Skipski and Barclay, 1969). The mixture of chemical substances to be analyzed is applied near one edge of the plate, and development of the plate with a solvent system allows the applied samples to move a distance that is related to the polarity of the sample. A solvent system consisting of hexane, diethyl ether, and acetic acid will separate triacylglycerols, cholesterol esters, fatty acid esters, and fatty acids (see Figure E9.2). More polar lipids, such as glycerophosphatides, sphingolipids, and glycolipids, remain at the origin. Pure standard samples of lipids are evaluated on the same TLC plate in order to help identify unknowns. The lipids on the solvent-developed plate are not colored, but

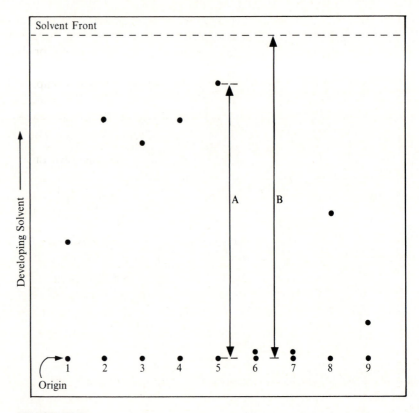

FIGURE E9.2
Typical thin-layer chromatogram of lipids. The solvent was hexane: diethyl ether: acetic acid (90:10:1). (1) Cholesterol, (2) fatty acid, (3) triacylglycerol, (4) fatty acid methyl ester, (5) cholesterol ester, (6) glycerophosphatide, (7) sphingolipid, (8) 1,3- or 1,2-diacylglycerol, (9) 1- or 2-monoacylglycerol. The R_f for (5), cholesterol ester, is calculated as follows:

$$R_f = \frac{\text{distance of migration of cholesterol ester}}{\text{distance of migration of solvent}} = \frac{A}{B} = \frac{14.3 \text{ cm}}{17 \text{ cm}} = 0.84$$

will react with iodine vapor to produce dark red-brown spots on a yellow background.

Overview of the Experiment

In this experiment you will isolate an unknown lipid mixture from its natural source by solvent extraction. The source of the unknown lipid is *Myristica fragrans* seed (nutmeg). This particular seed is unusual in that it contains an appreciable quantity of a single type of lipid. In fact, a single compound makes up 70 to 80% of the total lipid extract. The lipid is of sufficient homogeniety to allow its isolation as a solid with a relatively sharp melting point. The crude isolated material will be purified by silica gel column chromatography or recrystallization, and classified by thin-layer chromatography with standard lipids. For a flow chart of the procedure, see Figure E9.3. Complete identification of the molecular structure of the lipid will be the objective of Experiment 10. Approximate time requirements for each experimental section are:

A. Isolation of the Unknown Lipid—1 hour
B. Purification of the Unknown Lipid
   1. Recrystallization—$\frac{1}{2}$ hour
   2. Silica gel column—1 hour
C. Characterization by thin layer chromatography—$1\frac{1}{2}$ hours

II. EXPERIMENTAL

Materials and Supplies

Nutmeg, ground or whole

Hexane:isopropanol, 3:2

FIGURE E9.3
Flow chart for isolation of lipid from nutmeg.

Flow Chart of Procedure

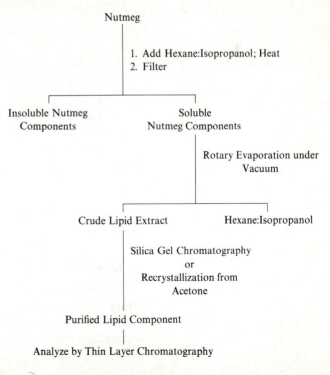

Nutmeg

1. Add Hexane:Isopropanol; Heat
2. Filter

Insoluble Nutmeg Components   Soluble Nutmeg Components

Rotary Evaporation under Vacuum

Crude Lipid Extract   Hexane:Isopropanol

Silica Gel Chromatography
or
Recrystallization from Acetone

Purified Lipid Component

Analyze by Thin Layer Chromatography

Acetone

Diethyl ether

Hexane

Silica gel, 100–200 mesh

Glass column (1 × 25 cm)

Glass wool or cotton

Small funnel and qualitative grade filter paper

Silica gel thin layer plates (20 × 20 cm and 5 × 10 cm)

Chromatography solvent system, hexane:diethyl ether:acetic acid (80:20:1) in a chromatography jar

Lipid standards in chloroform: 2% solutions of a triacylglycerol (triolein), cholesterol ester, fatty acid, fatty acid methyl ester, and glycerophosphatide (phosphatidylcholine, phosphatidylethanolamine, etc.)

Microcapillaries or tapered capillaries

Iodine chamber. Place a few crystals of iodine in an empty chromatography jar and cover.

Rotary evaporator

Procedure

A. Isolation of the Unknown Lipid

CAUTION:

Hexane, isopropanol, acetone, and ether are volatile and extremely flammable. Always measure out quantities in a hood. Be sure no lighted bunsen burners or flames are in the vicinity of your work!

Place 5 grams of finely crushed or ground nutmeg into a 125 ml Erlenmeyer flask. Add 50 ml of hexane:isopropanol (3:2) and warm on a steam bath or hot plate in a hood for 15 minutes. Mix well while heating. Filter the extraction mixture rapidly through fluted qualitative grade filter paper and pour an additional 20 ml of warm hexane:isopropanol through the solid residue in the filter. Remove the solvent from the extract under vacuum on a rotary evaporator to obtain a yellow oil or off-white solid. Alternatively, the solvent may be removed on a steam bath (HOOD!) while flushing the inside of the flask with N_2 gas. Weigh the crude product and retain a 1- to 2-mg portion for chromatographic analysis (Part C).

B. Purification of the Unknown Lipid

Partial and rapid purification of the lipid is possible by recrystallization from acetone. If more time is available, and a purer sample of the lipid is desired, a silica gel column should be run.

1. Recrystallization. To recrystallize the unknown, dissolve the crude residue in a minimum amount (about 20 ml) of warm acetone on a steam bath (HOOD!). After slow cooling of the acetone solution to room temperature and then in ice, filter by suction and wash the solid product with ice-cold acetone. A typical yield is about 0.5 gram of purified lipid. Save the purified lipid for Experiment 10.

2. Silica Gel Chromatography. To prepare the silica gel column, clamp the glass column to a ring stand. Prepare a slurry of 6 grams of silica gel in 20 ml of hexane. Put a small piece of glass wool or cotton into the bottom of the column and pour about 5 ml of hexane into the column. Be sure the stop-cock is closed! With a long glass rod, tap the glass wool or cotton to drive out air bubbles and push it tightly into the column. Place a small funnel on top of the column and begin to pour the well-stirred slurry of silica gel into the column. As the silica gel begins to pack into a column, open the stopcock to allow release of solvent. Continue to add the slurry of silica gel until a tightly packed column of 10 to 12 cm is attained. *Do not allow the column to run dry during packing.* Allow 1 to 2 cm of solvent to remain above the level of the silica gel.

Prepare the lipid for chromatography by dissolving 50 to 75 mg of crude lipid in a minimum of hexane (5 to 10 ml). Open the stopcock and allow excess solvent to drain from the column until the level of solvent just reaches the top of the silica gel column. Very carefully add the solution of crude lipid to the top of the column. This should be done by using a Pasteur pipette and allowing drops of the solution to run down the inside of the glass column to the top of the silica gel bed. It is important not to disturb the top of the silica gel column. After addition of the lipid solution, begin to collect a fraction from the bottom of the column into a 25 ml Erlenmeyer flask. Set the flow rate to about 2 drops per second. When the level of solution in the column reaches the top of the silica gel, turn off the stopcock and carefully rinse the inside of the glass column with 1 ml of hexane. Open the stopcock until the solvent just enters the column. This solvent washes the lipid material that may have adhered to the inside glass walls. Now fill the column with hexane and continue to collect the eluting solvent at 2 drops per second. Collect a total of 20 ml of solvent in the first fraction. During this collection, continue to add solvent to the top of the column. Now change the eluting solvent to 90% hexane and 10% diethyl ether. Pour this into the column and collect four 20-ml fractions. Be sure to number the collection flasks. Add solvent to the column as necessary. Finally, elute the column with 80% hexane and 20% diethyl ether and collect two more 20-ml fractions.

Each column fraction is analyzed for lipid material by spotting on a 5 × 10 cm silica gel thin-layer plate and exposing it to iodine vapor as follows. Prepare seven tapered capillary tubes and use these to place a spot of each solution on the TLC plate. Put at least 10 capillary applications from a single fraction on a spot. Your final plate should then have seven spots, one for each fraction. Set the plate in an iodine chamber and allow it to remain for about 15 minutes or until some spots are yellow or red-brown. The presence of lipid in a fraction is indicated by the red-brown color. Retain all the column fractions that appear to have lipid. Each of these fractions will be analyzed in Part C by thin-layer chromatography.

C. Partial Characterization of the Unknown Lipid

CAUTION:
Chloroform is suspected as a cancer-causing agent. Avoid breathing the vapors, and confine its use to a hood. Waste chloroform should be collected in a container and repurified by distillation.

Analysis of the purity of each silica gel column fraction and classification of the unknown lipid can be accomplished by thin-layer chromatography on silica gel plates. On a single plate will be spotted (1) a solution of the crude lipid extract from part A, (2) aliquots from each lipid fraction of the column (or recrystallized lipid), and (3) solutions of standard lipids (listed in the Materials section). On a 20 × 20 cm silica gel plate there is room for nine different analyses. Prepare a 1% solution of the crude lipid from part A in chloroform (10 mg/1 ml). If recrystallized lipid is to be analyzed rather than column-purified lipid, prepare a 1% solution in chloroform as for the crude lipid. Following the procedure you used in Experiment 2 (amino acid analysis), apply each sample in a row across one side of the silica gel plate, 2 cm from the edge (see Figure E9.2). Use small, tapered capillary tubes or microcapillaries. To apply each solution, dip the capillary into the solution, and gently touch the end of the filled capillary on the TLC plate where you desire the spot. Allow the solution from the capillary to spread onto the TLC plate to form a circle no wider than 0.5 cm diameter. Lift the capillary from the plate and allow the solvent on the plate to dry. Again touch the end of the capillary to the same spot and reapply solution. Repeat this application process about 20 times for each sample. Of course, each sample to be analyzed is placed on a different location along the edge of the TLC plate.

Develop the plate in hexane:diethylether:acetic acid (80:20:1) by placing the TLC plate in a chamber containing the solvent system, making sure the edge with the applied samples is down. The solvent level in the chamber must not be above the application spots on the plate. (Why?) Leave the chromatogram in the chromatography jar until the solvent front rises to about 1 cm from the top of the plate (45 to 60 min). Remove the plate and make a small scratch at the solvent level. Allow the chromatogram to dry (*HOOD!*) and then place it in an iodine chamber for several minutes. Remove the plate and lightly trace, with a pencil or other sharp object, around each red-brown spot. This should be done promptly, as the colors will fade with time. Calculate the mobility of each standard and unknown lipid relative to the solvent front (R_f):

$$R_f = \frac{\text{distance from the origin migrated by a compound}}{\text{distance from origin migrated by solvent}}$$

III. ANALYSIS OF RESULTS

Calculate the approximate percentage lipid composition of the nutmeg (assume the lipid recovery is quantitative). Prepare a table listing the R_f values obtained for each standard lipid and for the crude and purified unknown. By comparing R_f values, identify the general class of lipid to which your purified extract belongs. How effective was your purification step? If impurities are present in the crude or purified lipid, can you identify them? If you recrystallized the crude lipid, calculate the recovery yield.

$$\text{recovery yield} = \frac{\text{g purified lipid}}{\text{g crude lipid}} \times 100$$

IV. QUESTIONS

1. If you wished to determine the complete structure of the isolated lipid, what experimental approach would you follow?

2. Why is acetic acid added as a minor component (1%) of the chromatography solvent system?

3. Try to explain the relative order of R_f values you obtained for the standard lipids.

4. How does iodine react to produce the red-brown spots on a developed chromatogram? Why do some lipids give darker spots than others?

5. Calculate the R_f for standard lipid number 2 (a fatty acid) on Figure E9.2.

6. Why should whole nutmeg be ground before the solvent extraction?

V. REFERENCES

General

R.C. Bohinski, *Modern Concepts in Biochemistry*, Fourth Edition (1983), Allyn and Bacon (Boston), pp. 224–233. An introduction to lipid structure-analysis.

M.I. Gurr and A.T. James, *Lipid Biochemistry*, 2nd Edition (1975), Halsted Press (New York). An excellent book for the undergraduate.

M. Kates, "Methods in Lipidology" in *Laboratory Techniques in Biochemistry and Molecular Biology*, T.S. Work and E. Work, Editors (1971), North Holland/American Elsevier. An excellent reference-techniques book.

J.D. Rawn, *Biochemistry* (1983), Harper and Row (New York), pp. 449–458. Introduction to lipid analysis, structure and function.

L. Stryer, *Biochemistry*, 2nd Edition (1981), Freeman (San Francisco), pp. 205–229. Introductory material on lipid structure-function.

G. Zubay, *Biochemistry* (1983), Addison-Wesley (Reading, MA), pp. 471–476. Structure and function of lipids.

Specific

J. Folch et al., *J. Biol. Chem.*, *191*, 833 (1951) and *226*, 497 (1957). Use of chloroform-methanol as a lipid solvent.

G.J. Nelson in *Analysis of Lipids and Lipoproteins*, E.G. Perkins, Editor (1975), The American Oil Chemist's Society (Champaign, IL), pp. 1–19. General extraction methods.

N.S. Radin in *Methods in Enzymology*, J.M. Lowenstein, Editor, Vol. XIV (1969), Academic Press (New York), pp. 245–254, 268–272. "Preparation of Lipid Extracts."

N.S. Radin, *ibid.*, Vol. 72 (1981), pp. 5–7. Extraction of lipids with hexane-isopropanol.

V.P. Skipski and M. Barclay, *ibid.*, Vol. XIV (1969), pp. 530–598. "Thin Layer Chromatography of Lipids."

C.C. Sweeley, *ibid.*, Vol. XIV (1969), pp. 255–267. Silica gel column chromatography.

Characterization and Positional Distribution of Fatty Acids in Triacylglycerols

RECOMMENDED READING:
Chapter 3, Section C.

SYNOPSIS
Fatty acids that are used for energy metabolism are stored in triacylglycerols. Three fatty acids are linked through ester bonds to the glycerol skeleton. In this experiment, a triacylglycerol sample will be characterized by identifying the fatty acids at carbon 2 and comparing them to the fatty acids at carbons 1 and 3 combined. The methods used include release of the fatty acids by saponification and lipase-catalyzed hydrolysis followed by analysis of the corresponding fatty acid methyl esters with gas chromatography.

I. INTRODUCTION AND THEORY

Triacylglycerols, the neutral, saponifiable lipids found in most organisms, serve as biochemical energy reserves in the cell. Neutral lipids may be isolated from natural sources by extraction with nonpolar solvents (ethyl ether, chloroform, hexane-isopropanol) as in the isolation of the unknown

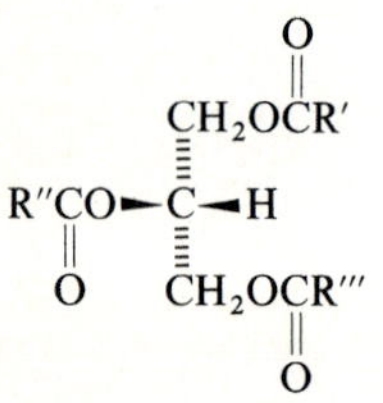

$$\begin{array}{c}
\qquad\qquad O \\
\qquad\qquad \| \\
\qquad CH_2OCR' \\
\qquad \vdots \\
R''CO\!\!-\!\!C\!\!-\!\!H \\
\| \qquad \vdots \\
O \quad CH_2OCR''' \\
\qquad\qquad \| \\
\qquad\qquad O
\end{array}$$

FIGURE E10.1
General structure for a triacylglycerol. R represents the alkyl chains
of fatty acids.

lipid from nutmeg (Experiment 9). Chemically, the triacylglycerols are fatty acid esters of the trihydroxy alcohol, glycerol (Figure E10.1). The figure also illustrates the recommended stereospecific numbering (sn) scheme for glycerol derivatives. According to IUPAC (1967), "The carbon atom that appears on top in that Fischer projection that shows a vertical carbon chain with the secondary hydroxyl group to the left is designated as C-1." To differentiate such numbering from conventional numbering conveying no steric information, the prefix "sn" (stereospecific numbering) is used; therefore, the structure in Figure E10.1 is designated 1, 2, 3-triacyl-*sn*-glycerol.

A variety of saturated and unsaturated fatty acids is present in triacylglycerols. Among the most abundant fatty acids are myristic (14:0), palmitic (16:0), stearic (18:0), and oleic (18:1$^{\Delta 9}$). To completely characterize the neutral lipids, one must know the identity of the fatty acids *and* the position that each fatty acid occupies on the glycerol skeleton (carbon number 1, 2, or 3). The biosynthetic distribution of fatty acids in positions 1, 2, or 3 of glycerol was once considered to be random; however, recent studies indicate some selectivity. For example, the proportion of unsaturated to saturated fatty acids in position 2 is quite often greater than at positions 1 and 3. Analytical techniques are available to determine the identity and position of each fatty acid in a triacylglycerol sample. Since glycerol is a **prochiral** molecule, it should be recognized that the 1 and 3 positions of glycerol are *not* equivalent. Methods have been devised for distinguishing between the fatty acids at positions 1 and 3 (stereospecific analysis of triacylglycerols) but a discussion is beyond the scope of this experiment (Brockerhoff, 1975). Because rather extensive experimentation is required to distinguish between fatty acids at positions 1 and 3, in this experiment you will consider these fatty acids as a group and instead distinguish between the fatty acids at position 2 and those at 1 and 3 combined. This will allow you to compare the amount of unsaturated fatty acids at position 2 relative to 1 and 3 combined.

Characterization of Triacylglycerols

The development of procedures in lipid enzymology, gas chromatography, and thin-layer chromatography now makes it possible to characterize triacylglycerols. Fatty acid content (identity and percentage composition) of triacylglycerols is most easily determined by complete saponification (NaOH/methanol) followed by esterification of the released fatty acids (Reaction 1). Gas chromatographic analysis of the fatty acid methyl esters (FAMEs), in conjunction with the analysis of the appropriate standard FAMEs, conveniently provides both qualitative and quantitative fatty acid analysis (Ackman, 1969; Patton et al., 1981; and Pelick, 1975).

$$\text{(Reaction E10.1)}$$

Positional distribution of fatty acids in triacylglycerols is elucidated by exploiting the selectivity of the lipolytic enzymes called lipases. Lipases (EC 3.1.1.3) are found in animals, microorganisms, and higher plants, and they catalyze ester hydrolysis, with preference for the acyl linkages at positions 1 and 3 of triacylglycerols (Decker, 1977). The cellular products of lipase-catalyzed hydrolysis of triacylglycerols are 2-monoacylglycerols and the free fatty acids from positions 1 and 3 (Reaction 2).

$$\text{(Reaction E10.2)}$$

Lipases act most efficiently in a heterogeneous medium, when the lipid substrate is dispersed as an emulsion in the buffered, aqueous phase containing the enzyme. The free fatty acids retrieved from the 1 and 3 positions of the triacylglycerol are characterized by esterification and gas chromatographic analysis. Information from this analysis leads directly to identification of the fatty acids at the combined positions 1 and 3. The 2-monoacylglycerol can be isolated from the reaction mixture by thin-layer or silica gel column chromatography. Fatty acid composition of the 2-monoacylglycerol is then determined by saponification, esterification, and gas chromatography of the fatty acid methyl esters.

Overview of the Experiment

In this experiment you will characterize an oil or the purified lipid isolated from nutmeg in Experiment 9. You will release the fatty acids by complete saponification, and identify and determine the mole percentage

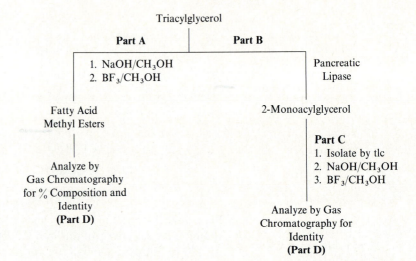

FIGURE E10.2
Flow chart for characterization of
a triacylglycerol.

composition of each fatty acid by gas chromatography (see Parts A and
D in Figure E10.2). The positional distribution of the acids is elucidated
by lipase action followed by gas chromatography (Parts B and C in Figure E10.2). Become familiar with Figure E10.2 before you begin the experiment. The time required for each part of the experiment is as follows:

A. Saponification of Triacylglycerols—1 hour

B. Lipase-Catalyzed Hydrolysis of Triacylglycerols—2 hours

C. Isolation and Saponification of 2-Monoacylglycerol—3 hours

D. Gas Chromatography of Fatty Acid Methyl Esters—2 hours

II. EXPERIMENTAL

Materials and Supplies

- Several fat and oil samples (lard, butter, vegetable oil, etc.)
  Unknown lipid from Experiment 9
- Methanolic sodium hydroxide, 0.5 N
- BF$_3$/methanol, 14%
- Hexane
- Saturated NaCl
  Anhydrous MgSO$_4$
  Tris buffer, 1 M, pH 8
  H$_2$SO$_4$, 1 M
  Aqueous CaCl$_2$, 40%

[handwritten annotations: "500 ml · Ⓐ", "Ⓑ", "1L", "dissolve in methanol store in plastic"]

Porcine pancreatic lipase (Steapsin), 50 triacetin units/mg

Fatty acid methyl ester standards (12:0, 14:0, 16:0, 18:0, 16:1$^{\Delta 9}$, 18:1$^{\Delta 9}$, 18:2$^{\Delta 9,12}$, 18:3$^{\Delta 9,12,15}$, solutions in hexane; 2 mg/ml)

Ethyl ether, peroxide free

Methanol containing 1% acetic acid

Silica gel thin-layer plates (2.5 × 10 cm)

Silica gel plates, preparative (20 × 20 cm, 1–2 mm thickness)

Chromatography solvent system, hexane:ethyl ether:acetic acid (80:20:1) in chromatography jar

Centrifuge tubes, 10-ml conical

Iodine chamber. Place a few crystals of iodine in an empty chromatography jar and cover.

Capillary tubes

Constant temperature water bath at 37°C

Steam bath

Gas chromatograph
  Conditions:
    gas flow: 50–70 ml/min
    $T_{injector}$: 200°C
    T_{column}: 185°C
    $T_{detector}$: 200°C
    Attenuation: 2 or 4
    Injection sample size: 5–10 μl
  Column: 6 ft, 10% Diethylene glycol succinate (DEGS) on Chrom W, 60/80 mesh

---

CAUTION:

All procedures in Parts A, B, and C requiring the heating or evaporating of hexane, ethyl ether, and methanol should be carried out in a well-ventilated hood. If hood space is not available, a vent may be constructed according to Figure E10.3. There should be no flames of any kind in the laboratory.

Procedure

A. Saponification of Triacylglycerol for Fatty Acid Composition

Select a triacylglycerol sample for analysis. This may be the unknown lipid that was isolated and purified in Experiment 9 or a fat or oil supplied by your instructor. Weigh 25 to 50 mg of the sample into a small conical

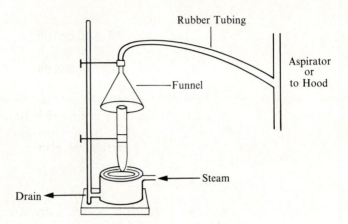

FIGURE E10.3
Construction of a vent for removal of noxious vapors from laboratory.

test tube. Add 3 ml of 0.5 N methanolic sodium hydroxide. Heat the mixture over a steam bath (HOOD!) until a homogeneous solution is obtained. Add 5 ml of BF_3/methanol to the saponification reaction mixture and boil 2 to 3 minutes. Cool and transfer the solution into a separatory funnel containing 25 ml of hexane and 20 ml of saturated NaCl solution. Shake well, but gently, and allow the layers to separate. Vigorous shaking will give rise to an emulsion. The hexane layer, containing the fatty acid methyl esters, is dried with about 1 g of anhydrous $MgSO_4$ and filtered or decanted into a small vial. Concentrate the solution on the steam bath to a volume of about 0.5 ml (HOOD!). This solution of fatty acid methyl esters should be capped to prevent air oxidation and stored in a ~~refrigerator~~ *freezer* until it is to be analyzed by gas chromatography.

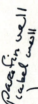

B. Lipase-Catalyzed Hydrolysis of Triacylglycerol (Brockerhoff, 1975)

In a conical test tube, dissolve or suspend 25 to 50 mg of another sample of the triacylglycerol used in Part A in 0.2 ml of hexane. Add the following reagents to the triacylglycerol:

1. 1.5 ml of 1 M Tris buffer, pH 8.0
2. 0.005 ml of 40% aqueous $CaCl_2$
3. 50 mg of pancreatic lipase suspended in 0.3 ml of 1 M Tris buffer, pH 8.0

After a brief (5 second) homogenization with a vortex mixer, or vigorous mixing, place the enzymatic reaction mixture in a constant temperature bath at 37°C. Shake the mixture every few minutes. The extent of reaction can be monitored by TLC analysis of aliquots removed at various times (1, 30, and 60 minutes). Aliquots of 1 drop are removed and clarified with a few drops of methanol containing 1% acetic acid. The clarified solution is spotted on a small (2.5 × 10 cm) silica gel plate, developed in the hexane: ethyl ether:acetic acid solvent, and placed in the iodine chamber for detection. When the triacylglycerol spot (fastest running) has nearly disappeared (the 60 min aliquot), the maximum yield of 2-monoacylglycerol is obtained. Stop the lipase reaction at this time by adding 1 ml of 1 M H_2SO_4. The spot on the TLC plate near the origin represents the 2-monoacylglycerol, and it should have increased in intensity with time of incubation. The end point of the lipase reaction can also be monitored by noting a turbidity or grainy appearance due to the precipitation of calcium salts of the released free fatty acids.

C. Isolation and Saponification of 2-Monoacylglycerols

To isolate the 2-monoacylglycerol from the reaction mixture, extract it three times with 3 ml of ethyl ether. Combine and dry the ether layers with 0.5 g of anhydrous $MgSO_4$ and concentrate on a steam bath (HOOD!) to about 1 ml. If necessary, the ether extract may be stored until the next lab period. The ether solution is streaked on a preparative silica gel plate (20 × 20 cm) in a 15 cm long band about 2.5 cm from the bottom edge. Develop the plate in hexane:ethyl ether:acetic acid (80:20:1). The components of the reaction mixture will separate into about four rows of spots. The components may be visualized by placing the plate in an iodine chamber for a few minutes. Mark the monoacylglycerol area (bands closest to the origin; lowest R_f), scrape the silica gel from this area with a razor blade, and suspend in 5 ml of ethyl ether. Mix the suspension for 5 minutes and filter by gravity through fluted paper into a 10-ml test tube. Dry with anhydrous $MgSO_4$ and carefully remove the ether on a steam bath or warm water bath (HOOD!). Add 1 ml of 0.5 N methanolic NaOH to the residue and heat on the steam bath for 5 minutes or until all the residue is dissolved. Add 2 ml of BF_3/MeOH to the saponification mixture and boil for 2 to 3 minutes (HOOD!). Cool and transfer the solution into a separatory funnel containing 20 ml of hexane and 15 ml of saturated NaCl solution. Shake gently and allow the layers to separate. The hexane layer, containing the methyl esters of the fatty acids originally at position 2 of the monoacylglycerol, is dried with anhydrous $MgSO_4$ and filtered into a small vial. Concentrate the solution on the steam bath to a volume of about 0.5 ml (HOOD!). The sample is now ready for gas chromatographic analysis.

D. Gas Chromatography of FAMEs

Study the section on gas chromatography in Chapter 3. If you are not familiar with the use of the instrument, have your instructor assist you. Be sure to record all instrumental conditions in your notebook (temperatures of the column, oven, and detector; rate of gas flow (ml/min); attenuation setting; sample size and recorder chart speed). It is recommended that you begin with the FAME standards, using 5 to 10 μl injection samples with an attenuator setting of 2. Two peaks should appear, one for the hexane solvent and a second representing the FAME. The hexane peak will be off scale; if the FAME peak is also off the recorder paper, reduce the injection sample or increase the attenuation setting to 8 or 16. If the recorder peaks for the FAMEs are too small, decrease attenuation setting (1 or 2). When you are analyzing the standard FAMEs, you may inject a second sample as soon as the FAME from the first sample has been eluted from the column. This occurs when the recorder pen returns to the original baseline.

Use the same instrumental conditions to analyze the samples from Parts A and C. The samples will contain a mixture of fatty acid methyl esters, so several recorder peaks will be obtained. Do not inject a second sample until you are sure all the fatty acids have been eluted from the first sample. The sample from Part C is sometimes contaminated with an organic binder that is present in the silica gel. Fortunately, this substance has a retention time greater than the C_{18} fatty acid methyl esters. Determine the retention time for each standard FAME and for each FAME in the unknown samples. Retention time is the time interval between injection of sample and maximum response of the recorder (see Figure 3.7 in Chapter 3).

III. ANALYSIS OF RESULTS

From the data that you have collected, it is possible not only to identify the fatty acids, but to calculate the composition in mole percentage of each fatty acid present in the triacylglycerol. Your experimental results also allow you to calculate the positional distribution of fatty acids, that is, the mole percentage of each fatty acid in the secondary position (carbon 2) and primary positions (carbons 1 and 3).

Prepare a table listing the retention time for each standard FAME. Use this table to identify the fatty acids present in each triacylglycerol you analyzed. Alternatively, plot the log of the retention time against the chain length of each saturated FAME (see Figure 3.8, Chapter 3). Unknown saturated fatty acids can be identified from experimental retention times using this plot. Unsaturated fatty acids cannot be identified from the plot of saturated FAMEs. A separate plot of log retention time vs. the chain length must be prepared for each level of saturation (saturated, monounsaturation, diunsaturation, etc.).

The amount of each FAME present is proportional to the area under its peak on the recorder chart. If the peaks are symmetrical, the peak area may be calculated by triangulation:

peak area = peak height × peak width at half height

or

$$\text{peak area} = HW_{1/2}$$

Thermal conductivity detectors used in gas chromatographs do not respond equally to all FAMEs. To correct for varying detector sensitivity, peak area for each FAME should be multiplied by the proper **response correction factor** (RCF) (Table E10.1). If extra time is available, you may want to calculate your own response correction factors for the fatty acid methyl esters. The factors are experimentally determined on a gas chromatograph by comparing the area under a GC peak due to a known amount of compound to the area under a GC peak represented by a reference compound.

$$RCF_1 = \frac{\text{amount of compound 1}}{\text{peak area of 1}} \times \frac{\text{peak area of reference}}{\text{amount of reference}}$$

The response correction factors for compound 1 and other compounds are calculated relative to the response correction factor for the reference, which is assumed to be 1.00.

The mole percentage of each FAME is calculated according to the following sequence of equations.

1. Correction of peak area of $FAME_1$ for varying detector sensitivity.

$$\text{peak area}_1 \times RCF_1 = \text{corrected peak area}_1$$

2. Conversion of corrected peak area (now in weight %) to moles.

$$\frac{\text{corrected peak area}_1}{\text{molecular weight}_1} = \text{mole-corrected peak area}_1$$

3. Conversion of mole-corrected peak area$_1$ to mole % of $FAME_1$.

$$\text{mole \%}_1 = \frac{\text{mole-corrected peak area}_1}{\text{total of mole-corrected peak areas}} \times 100$$
$$\text{for all FAMEs} \qquad \text{(Equation E10.1)}$$

TABLE E10.1

Fatty Acid Methyl Ester Response Factors Measured with a Thermal Conductivity Detector, DEGS Column at 190°C

FAME	Response Correction Factor (RCF)
Myristic (14:0)	0.908
Palmitic (16:0)	0.954
Stearic (18:0)	1.010
Linoleic $(18:2^{\Delta 9,12})$	1.071
Linolenic $(18:3^{\Delta 9,12,15})$	1.172

The denominator for Equation 1 is calculated by performing steps 1 and 2 for all of the FAMEs present in a single triacylglycerol sample. The mole-corrected peak areas are added together to obtain the total of all mole-corrected areas. If we assume that the extent of conversion of free fatty acids to FAMEs is essentially quantitative, or at least equal for all fatty acids in our experiment, the above calculation leads directly to the combined mole percent composition of fatty acids in all positions of the triacylglycerol.

The fatty acid composition (mole %) in positions 1 and 3 is obtained by subtracting the mole % of each fatty acid in position 2 (from analysis of 2-monoacylglycerol) from the mole % of the same fatty acid in the triacylglycerol:

$$\text{mole \% of } FA_A \text{ in positions 1 and 3}$$
$$= \frac{[3(\text{mole \% } FA_A \text{ in TG}) - (\text{mole \% } FA_A \text{ in MG})]}{2}$$

where

FA_A = a fatty acid present in position 2

TG = triacylglycerol

MG = monoacylglycerol

For the lipid sample that you analyzed, report the fatty acid composition in mole %, and the positional distribution of each fatty acid in mole %. Compare your results with those of other students.

IV. QUESTIONS

1. Positional distribution analysis of hundreds of naturally occurring triacylglycerols has led to the discovery of general trends in fatty acid composition. Position 2 tends to contain unsaturated fatty acids, while positions 1 and 3 contain saturated fatty acids. Did you observe this general trend in your analysis?

2. Why are FAMEs rather than free fatty acids used for gas chromatographic analysis?

3. Why was $CaCl_2$ added to the lipase reaction mixture?

4. Your TLC plate of the ether extract of the lipase reaction mixture should have four spots. Identify each spot, and explain the order of spots.

5. What is the identity of the lipid that was extracted from nutmeg seed (Experiment 9)?

V. REFERENCES

General

M.I. Gurr and A.T. James, *Lipid Biochemistry*, 2nd Edition (1975), Halsted Press (New York). An excellent introduction to lipid structure, function, and analysis.

M. Kates and M.K. Wassef in *Ann. Rev. Biochem.*, E.S. Snell, Editor, Vol. *39* (1970), Annual Reviews, Inc. (Palo Alto, CA), pp. 323–358. "Lipid Chemistry."

G. Zubay, *Biochemistry* (1983), Addison-Wesley (Reading, MA), pp. 474–476. An introduction to lipid structure and analysis.

Specific

R.G. Ackman in *Methods in Enzymology*, J.M. Lowenstein, Editor, Vol. XIV (1969), Academic Press (New York), pp. 329–381. "Gas-Liquid Chromatography of Fatty Acids and Esters."

H. Brockerhoff in *Methods in Enzymology*, J.M. Lowenstein, Editor, Vol. XXXVB (1975), Academic Press (New York), pp. 315–325. "Determination of the Positional Distribution of Fatty Acids in Glycerolipids."

L.A. Decker, Editor, *Worthington Enzyme Manual* (1977), Worthington Biochemical Corporation, (Freehold, NJ), pp. 122–124. Review of Hog Pancreas Lipase.

IUPAC-International Union of Biochemistry, *J. Biol. Chem.*, *242*, 4845–4849 (1967).

G.M. Patton, S. Cann, H. Brunengraber, and J.M. Lowenstein in *Methods in Enzymology*, Vol. 72 (1981), pp. 8–20. "Separation of Methyl Esters of Fatty Acids by Gas Chromatography on Capillary Columns."

N. Pelick and V. Mahadevan in *Analysis of Lipids and Lipoproteins*, E.G. Perkins, Editor (1975), The American Oil Chemist's Society (Champaign, IL), pp. 23–35. "Gas-Liquid Chromatography of Lipid Derivatives."

Experiment **11**

Polarimetric Analysis
of Carbohydrates

RECOMMENDED READING:
Chapter 5, Section G

SYNOPSIS
Carbohydrates are found in all forms of life. They are unique among biologically significant molecules in that they can exist in many stereochemical forms. The naturally occurring carbohydrates are asymmetric molecules; therefore, they can be detected and identified by measurements of optical rotation. An unknown monosaccharide and disaccharide will be identified by chemical and polarimetric characterization.

I. INTRODUCTION AND THEORY

Carbohydrate Structure and Function

Carbohydrates are naturally occurring compounds that have the basic formula $C_x(H_2O)_x$. We are all familiar with the central role of carbohydrates in energy metabolism, with glucose being the molecule of primary significance. But, in addition to their importance in metabolism, carbohydrates are essential components of cell membranes and bacterial cell walls. Here they play a role in structural integrity or a more dynamic role in cellular recognition and communication.

In chemical terms, carbohydrates are polyhydroxy aldehydes or ketones. Two common carbohydrates, glucose and fructose, are shown in Figure E11.1. Glucose is a polyhydroxy aldehyde with six carbons or an aldohexose, whereas fructose is a polyhydroxy ketone or a ketohexose.

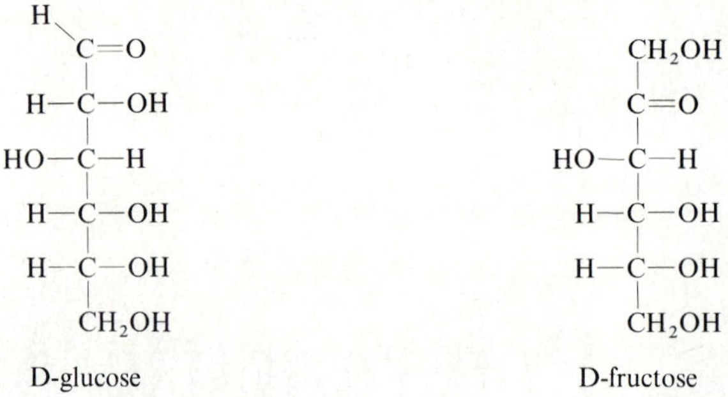

FIGURE E11.1
Two common carbohydrates, D-glucose and D-fructose.

$$
\begin{array}{c}
\text{H} \\
\quad\diagdown \\
\text{C}=\text{O} \\
\text{H}-\text{C}-\text{OH} \\
\text{HO}-\text{C}-\text{H} \\
\text{H}-\text{C}-\text{OH} \\
\text{H}-\text{C}-\text{OH} \\
\text{CH}_2\text{OH}
\end{array}
$$

D-glucose

$$
\begin{array}{c}
\text{CH}_2\text{OH} \\
\text{C}=\text{O} \\
\text{HO}-\text{C}-\text{H} \\
\text{H}-\text{C}-\text{OH} \\
\text{H}-\text{C}-\text{OH} \\
\text{CH}_2\text{OH}
\end{array}
$$

D-fructose

Glucose and fructose are **monosaccharides**. Larger carbohydrates formed by the combination of two or more monosaccharides are **disaccharides**, **trisaccharides**, and so on; in general, they are called **oligosaccharides**. Carbohydrates with 10 or more monosaccharides are called **polysaccharides**.

Carbohydrates may contain one or many **asymmetric carbon atoms** or **chiral centers**. Glucose has four chiral centers, and fructose has three. The stereochemistry of glucose, fructose, and other carbohydrates is defined by comparison to one of the simplest carbohydrates, glyceraldehyde. Glyceraldehyde, as shown in Figure E11.2, has one chiral center and thus can exist in two possible arrangements or enantiomers, D-glyceraldehyde and L-glyceraldehyde. Solutions of each of the isomers will rotate the plane of polarized light to the same extent, but in opposite directions. Carbohydrates are classified as D or L depending on how their stereochemical structures compare to D- and L-glyceraldehyde. (In the R,S system, D-glyceraldehyde is R. The stereochemistry of other carbohydrates may also be assigned by the R,S format; however, they are traditionally denoted by D or L.) Since most carbohydrates contain more than one chiral center, convention dictates that the stereochemistry of the asymmetric center most remote (highest in number) from the aldehyde end determines the classification. If the stereochemical configuration around that carbon atom is identical to D-glyceraldehyde, the carbohydrate is said to be in the D-family; if the configuration is opposite to D-glyceraldehyde (or identical to L-

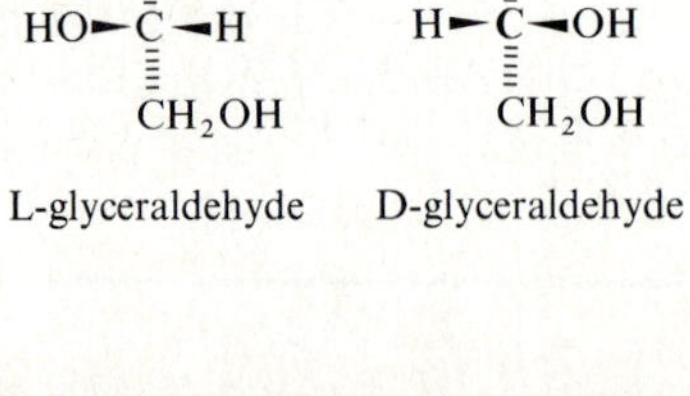

FIGURE E11.2
The two stereoisomers of glyceraldehyde.

L-glyceraldehyde   D-glyceraldehyde

glyceraldehyde), the carbohydrate is L. Both of the carbohydrates shown in Figure E11.1 are of the D-family. Note that the classification of the carbohydrates is not dependent on the direction of the rotation of plane polarized light, but is done strictly on a configurational basis. In fact, D-glucose has a specific rotation, $[\alpha]_D^{20}$, of $+52.2°$, whereas D-fructose has an $[\alpha]_D^{20}$ of $-92.0°$.

When glucose, fructose, and other carbohydrates are dissolved in water, the open chain form in Figure E11.1 (the Fischer projection structure) is transformed to an aldehyde hydrate, which rearranges to a cyclic structure as shown in Reactions 1 and 2.

α-D-Glucose   β-D-Glucose   (Reaction E11.1)

α-D-Fructose   β-D-Fructose   (Reaction E11.2)

Formation of the intramolecular hemiacetals results in two forms of the compound that differ in the stereochemistry at the hemiacetal (C-1 of glucose) or hemiketal (C-2 of fructose) carbon. For glucose, for example, these two forms are called α-D-glucose and β-D-glucose.

An aqueous solution of pure β-D-glucose has an initial $[\alpha]_D^{20}$ of $+19°$. However, if the solution is allowed to sit for several hours, the optical rotation slowly changes, resulting in a final $[\alpha]_D^{20}$ of $+52.2°$. Likewise, a solution of α-D-glucose has an initial $[\alpha]_D^{20}$ of $+113.4°$, but after several hours the specific rotation is $+52.2°$. In each case, the same equilibrium mixture of α-D- and β-D-glucose has been produced. This process of interconversion, called mutarotation, has been observed for many carbohydrate solutions when the sugars can easily form a five- or six-membered ring. Fructose, in Reaction 2, also undergoes mutarotation, resulting in an equilibrium mixture of β-D-fructose and α-D-fructose. In this case, a five-membered hemiketal ring is formed. The two predominant forms of the sugars (α and β) are referred to as **anomers**. The mutarotation process is catalyzed by acid or base.

Identification and Characterization of Carbohydrates by Polarimetry

Detection of mutarotation by optical rotation can aid in the identification of carbohydrates. Unknown carbohydrates are often identified by an initial measurement of optical rotation under controlled conditions. Standard values for $[\alpha]_D^{20}$ are readily available for most carbohydrates. Table E11.1 gives several of these values for common monosaccharides. Identification of an unknown carbohydrate by optical rotation measurements is similar to identifying an organic compound by melting point. However, two or more carbohydrates may have similar $[\alpha]_D^{20}$ values, and the absolute identity of the unknown cannot be determined by this measurement alone. Measurement of the specific rotation after mutarotation has occurred to produce an equilibrium mixture will yield a second optical rotation value. The equilibrium mixture attained by mutarotation of a sugar is reproducible; therefore, standard $[\alpha]_D^{20}$ values can be determined for each sugar. The specific optical rotations of several equilibrium mixtures of α and β anomers are given in Table E11.1. Now, two optical rotation values can be experimentally measured and used for the identification of unknown sugars.

The method for measuring optical rotations of pure sugars is a relatively straightforward process. A standard solution of the sugar is prepared and an initial optical rotation measurement is rapidly made. If this mea-

TABLE E11.1

Specific Optical Rotation, $[\alpha]_D^{20}$, in Degrees, for Several Carbohydrates.

	α	β	Equilibrium Mixture After Mutarotation
Monosaccharides			
L-arabinose	+55.4	+190.6	+104.5
D-fructose	—	−133.5	−92
D-galactose	+150.7	+52.8	+80.2
D-glucose	+112	+18.7	+52.7
D-mannose	+29.3	−16.3	+14.5
D-ribose	−23.7	—	−23.7
L-sorbose	—	—	−43.1
D-xylose	+93.6	—	+18.8
Disaccharides			
cellobiose	—	+14.2	+34.6
lactose	+90	+34.2	+53.6
maltose	+173	+112	+130
trehalose	—	—	+199
Trisaccharides			
raffinose	—	—	+101, +123

surement is made within a few minutes, the result will be the observed optical rotation for the pure anomer. A drop of dilute acid or base is then added to the carbohydrate sample and the optical rotation is monitored until a constant value is obtained. The final, constant reading corresponds to the equilibrium mixture of α and β anomers of the unknown carbohydrate.

Characterization of Glycosides by Polarimetry

Disaccharides and polysaccharides are formed when two or more monosaccharides are chemically combined. The bond linking the monosaccharides is formed between the hemiacetal or hemiketal functional group of one carbohydrate and a hydroxy functional group of a second carbohydrate (Reaction 3).

(Reaction E11.3)

The new linkage is called a **glycosidic bond** and the product has the general name **glycoside**. In the example, two glucose units are joined by an $\alpha(1 \rightarrow 4)$ glycosidic linkage to produce a disaccharide, maltose. Several variations of glycosidic bonds are observed in nature, and they are shown in Figure E11.3. Milk sugar, lactose, is a disaccharide consisting of galactose and glucose linked by a $\beta(1 \rightarrow 4)$ glycosidic bond. Common table sugar, sucrose, consists of glucose and fructose linked by an $\alpha(1 \rightarrow 2)$ glycosidic bond. Many identical monosaccharide molecules may combine by glycosidic bonds to form polysaccharides. Starch consists of glucose units linked by $\alpha(1 \rightarrow 4)$ glycosidic bonds, whereas cellulose is made up of glucose units linked by $\beta(1 \rightarrow 4)$ glycosidic bonds.

Glycosidic bonds are subject to hydrolysis under certain conditions. They hydrolyze readily in the presence of dilute aqueous acid or enzymes called **glycosidases**. Using enzymes for the specific cleavage of these bonds is a valuable technique for structure determination of di- and polysaccharides.

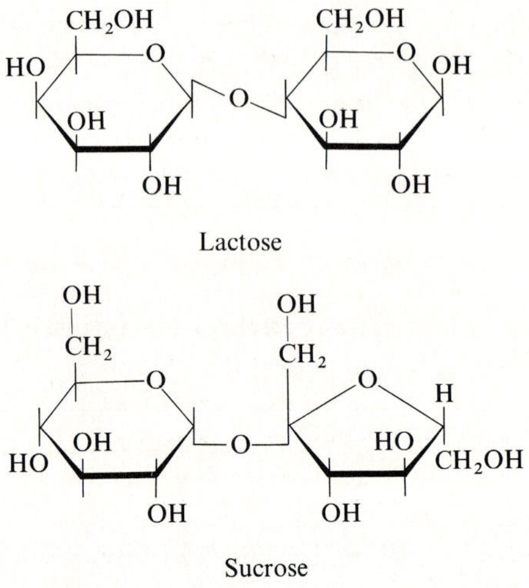

Lactose

Sucrose

FIGURE E11.3
Two common disaccharides, lactose and sucrose.

An interesting enzyme that catalyzes the hydrolysis of specific glycosidic bonds is invertase. The overall reaction catalyzed by invertase is shown in Reaction 4.

$$\text{sucrose} \rightleftharpoons \text{glucose} + \text{fructose} \qquad \text{(Reaction E11.4)}$$

The final outcome is an equimolar mixture of glucose and fructose. The name of the enzyme is derived from the observation that sucrose is dextrorotatory (plane polarized light is rotated to the right) and an equimolar mixture of glucose and fructose is slightly levorotatory (plane polarized light is rotated to the left). Therefore, the enzyme-catalyzed reaction results in an inversion of the sign of the optical rotation. Invertase is specific in its action, preferring sucrose and raffinose as substrates.

In this experiment, one sample of an oligosaccharide will be treated with acid and a second sample with invertase. After a brief incubation period under controlled conditions, an optical rotation measurement will be made on each reaction mixture. The experimental data are then compared to the optical rotation of an aqueous solution of the oligosaccharide. By using proper data analysis, it can be determined whether the oligosaccharide is susceptible to acid and/or invertase-catalyzed hydrolysis. In addition, if the oligosaccharide is cleaved, the percentage of substrate remaining in the sample can be calculated. Hence, polarimetry can be used to monitor the rate of glycoside bond hydrolysis.

There is one difficulty in measuring invertase-catalyzed sucrose inversion by the polarimetric method. The rate of the enzyme-catalyzed reaction is rapid, but the mutarotation of the monosaccharide products is slower. Therefore, the optical rotation of the reaction mixture will slowly change due to mutarotation even after the enzyme is inactivated. The undesirable effect of slow mutarotation is avoided by adding base to the reaction mixture after a suitable reaction period. The added base inactivates the enzyme and catalyzes the mutarotation. If the inversion of sucrose catalyzed by acid is studied, there is no difficulty due to mutarotation because acid acts as a catalyst for both the cleavage of the glycoside and the mutarotation of the products.

Overview of the Experiment

The two major exercises in this experiment are outlined in Figure E11.4. Part A involves the identification of an unknown carbohydrate by polarimetry measurements before and after mutarotation, and Part B consists of the application of polarimetry to characterizing glycosidic bonds

FIGURE E11.4
Flowchart for analysis of an unknown carbohydrate.

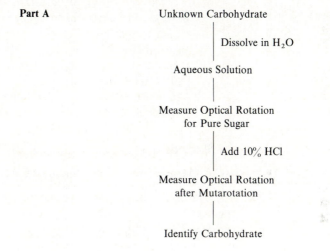

Part A

Unknown Carbohydrate

Dissolve in H_2O

Aqueous Solution

Measure Optical Rotation
for Pure Sugar

Add 10% HCl

Measure Optical Rotation
after Mutarotation

Identify Carbohydrate

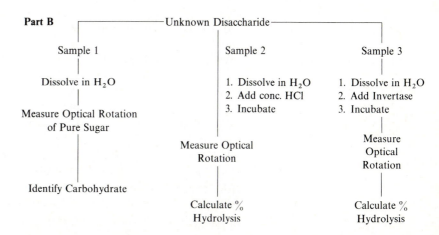

Part B Unknown Disaccharide

Sample 1	Sample 2	Sample 3
Dissolve in H_2O	1. Dissolve in H_2O 2. Add conc. HCl 3. Incubate	1. Dissolve in H_2O 2. Add Invertase 3. Incubate
Measure Optical Rotation of Pure Sugar	Measure Optical Rotation	Measure Optical Rotation
Identify Carbohydrate	Calculate % Hydrolysis	Calculate % Hydrolysis

and carbohydrate structure. Each part will require approximately $1\frac{1}{2}$ hours. Since most biochemistry laboratories have only one polarimeter available, a schedule of users must be prepared.

II. EXPERIMENTAL

Materials and Supplies

A. Identification of an Unknown Carbohydrate
   Unknown carbohydrates, 10 g samples
   10% HCl
   Polarimeter and 1 dm cells
   Volumetric flasks, 50 ml

B. Characterization of Glycosidic Bonds
   Several oligosaccharides: sucrose, maltose, lactose, and raffinose
   Concentrated HCl
   Invertase solution in acetate buffer, pH 5.0, 200 units/ml
   Constant temperature water bath at 37°C
   Polarimeter and 1 dm cells
   Volumetric flasks, 50 and 100 ml
   Saturated Na_2CO_3 solution

Procedure

A. Identification of an Unknown Carbohydrate

Before you begin the experimental procedure, become familiar with the use of the polarimeter. Read the instructions or have your instructor assist you. You may be fortunate enough to have the use of an automatic polarimeter with meter or digital readout of degrees. If not, you will have to operate the polarimeter manually.

Obtain a polarimeter cell and rinse it several times with distilled water. Be careful while handling the cell. Do not scratch or chip the glass lenses at the ends of the cell. Fill the cell to overflowing with distilled water and slide the glass lens over the end. There should be no air bubbles trapped under the lens in the cell. If air bubbles are present, remove the lens and add water to the cell with a disposable pipet. Close the cell and wipe the glass end pieces with lens paper. Place the filled cell into the polarimeter and zero the instrument in the following way. An automatic polarimeter has a "zero adjust" button. Push this button and 00.00 degrees should appear on the readout. A manual polarimeter is adjusted by rotating the analyzer. Look through the eyepiece and note the two halves of the circular field. Rotate the analyzer until the semicircles of the field are equal in illumination, as shown in Figure E11.5. Read the degree setting on the analyzer (sign and number) and record it in your notebook as the zero point. Turn the analyzer a few degrees off the zero point and again adjust

FIGURE E11.5
Adjustment of the analyzer for polarimetry measurements.

to the zero point. Record the second reading. Now, turn the analyzer a few degrees in the opposite direction as before and again adjust to the zero point. Record the third reading. Average the three zero point readings and use this as your zero adjust setting. Pour the water out of the cell and allow it to air dry.

The next step should be completed rapidly to avoid mutarotation of the sugar. Weigh 5.0 ± 0.01 g of your unknown carbohydrate and transfer it to a 50 ml volumetric flask. Add 40 ml of distilled water to dissolve the sugar. Fill to the mark with distilled water. Mix well and fill the polarimeter cell with the sugar solution. Check for air bubbles in the cell and remove them as before. Place the filled cell in the polarimeter and record three optical rotation readings as described above, readjusting the analyzer each time. Average the readings.

Transfer the contents of the polarimeter cell to the original volumetric flask and then add 1 drop of 10% HCl to the flask. Mix well and again fill the polarimeter cell. Record the optical rotation every 10 minutes for 30 minutes or until a constant reading is obtained. The final constant reading is that of the equilibrium mixture after mutarotation. Refer to the Analysis of Results section for instructions to calculate specific rotation, $[\alpha]_D^{20}$.

B. Characterization of Glycosidic Bonds

Obtain three 50-ml volumetric flasks. Label them with numbers from 1 to 3. Into each, weigh 2.5 ± 0.01 g of the unknown oligosaccharide. Each sample of unknown will be treated separately. The procedure is completed more rapidly if the samples are started at different times. Begin the preparation of Sample 2 immediately after beginning the incubation of Sample 1, and so forth.

Sample 1: Dissolve the sugar in about 40 ml of distilled water and then add water to the mark. Mix well, and cool to 20°C in a water bath. Transfer to a polarimeter cell and obtain three optical rotation measurements as described in Part A. Use water as a blank for zero point measurement.

Sample 2: Dissolve the sugar in about 40 ml of distilled water and then add 5.0 ml of concentrated HCl. Add water to the mark, mix well, transfer to a 125 ml Erlenmeyer flask, and incubate the mixture at 37°C for 15 minutes. Cool to 20°C, transfer to a polarimeter cell, and obtain three optical rotation measurements. Use water as a blank.

Sample 3: Dissolve the sugar in about 40 ml of distilled water and add 5.0 ml of invertase solution. Add water to the mark, mix well, transfer to a 125-ml Erlenmeyer flask, and incubate the mixture at 37°C for 15 minutes. Add 2.0 ml of saturated Na_2CO_3 to inactivate the invertase and accelerate mutarotation. If a precipitate forms, centrifuge for 20 minutes at 3000 rpm. Incubate an additional 15 minutes and obtain three optical rotation measurements. Use water as a blank.

III. ANALYSIS OF RESULTS

A. Identification of an Unknown Carbohydrate

Calculate average optical rotation measurements for the water blank and the unknown sample. Correct for the water blank by subtracting its rotation from that of the carbohydrate sample. Calculate the specific rotation, $[\alpha]_D^{20}$, for the unknown according to Equation 1.

$$[\alpha]_D^{20} = \frac{[\alpha]_{obs}}{\ell c} \qquad \text{(Equation E11.1)}$$

where

$[\alpha]_{obs}$ = observed optical rotation in degrees

ℓ = length of the polarimeter cell in decimeters

c = concentration of the unknown sugar in g/ml

Calculate the $[\alpha]_D^T$ for the equilibrium mixture after mutarotation in the same fashion. Compute the statistical significance of the data. Compare the two values with those in Table E11.1 and identify the unknown carbohydrate by name and type of anomer.

B. Characterization of Glycosidic Bonds

Calculate the optical rotation of each of the three samples. Remember to correct the reading for Sample 3, which was diluted by addition of 2 ml of Na_2CO_3. [Multiply the observed degrees of rotation (after water blank correction) by 52/50.] The result from Sample 1 gives the optical rotation of the pure oligosaccharide before inversion and mutarotation. Samples 2 and 3 give the rotation of mixtures of oligosaccharide and monosaccharide(s) caused by hydrolysis of the glycosidic bonds.

Identify the unknown disaccharide from the specific rotation of Sample 1. Using the data from Samples 1, 2, and 3, calculate the percentage of disaccharide remaining after treatment with acid or enzyme. Use the following relationships. In each case X = the amount of disaccharide hydrolyzed in gram %. The polarimeter cell pathlength is assumed to be 1 dm.

Sucrose

$$X = \frac{\alpha_{obs,\,Sample\,1} - \alpha_{obs,\,after\,inversion}}{7.7} \times 100$$

Maltose

$$X = \frac{\alpha_{obs,\,Sample\,1} - \alpha_{obs,\,after\,inversion}}{8.1} \times 100$$

Lactose

$$X = \frac{\alpha_{obs,\,Sample\,1} - \alpha_{obs,\,after\,inversion}}{1.5} \times 100$$

Describe the effectiveness of base and invertase at cleaving your unknown disaccharide.

IV. QUESTIONS

1. Illustrate the structure for the aldopentose D-ribose in the straight chain (Fischer projection) and the cyclic form (Haworth structure).

2. Why are there no optical rotation values listed in Table E11.1 for α and β forms of sucrose?

3. Predict the action of invertase on raffinose, shown below.

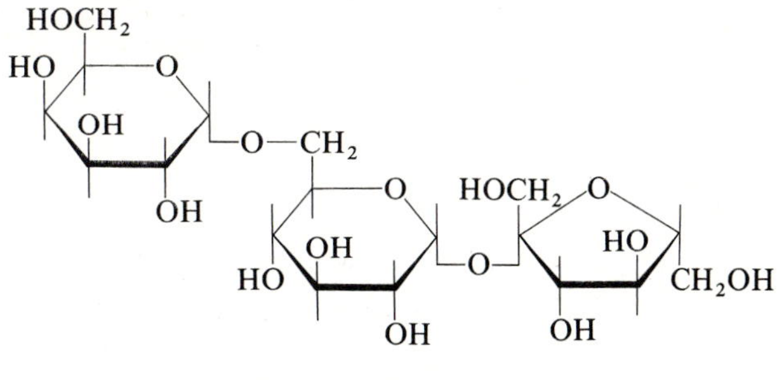

4. An unknown carbohydrate was dissolved in water (5 g/100 ml) and the optical rotation was immediately measured as $-6.75°$ at 20°C. The carbohydrate solution was allowed to remain in the polarimeter cell for 24 hours, and a second measurement was found to be $-4.55°$ at 20°C. After 24 hours more, the optical rotation was still $-4.55°$. Study Table E11.1 and identify the unknown. Assume that the "zero point" reading on the polarimeter was 0.01°. The pathlength of the cell is 1 dm.

5. The following optical rotation readings were measured for a water blank and an unknown carbohydrate solution.

α_{obs}, Blank	α_{obs}, Carbohydrate
+0.01	+3.24
0.00	+3.15
+0.02	+3.30
−0.01	+3.20
0.00	+3.21
+0.01	+3.19
−0.02	+3.17
0.00	+3.23
+0.01	+3.20
−0.01	+3.25

(a) Calculate the sample mean.
(b) Calculate the standard deviation.
(c) Calculate the 95% confidence levels for the measurements.

V. REFERENCES

General

R. Bentley, *Ann. Rev. Biochem., 41*, 953–996 (1972). "Configurational and Conformational Aspects of Carbohydrate Biochemistry."

R.C. Bohinski, *Modern Concepts in Biochemistry*, Fourth Edition (1983), Allyn and Bacon (Boston), pp. 198–222. Introduction to the structure and function of carbohydrates.

A. Lehninger, *Principles of Biochemistry*, (1982) Worth Publishers, Inc. (New York), pp. 277–294. Introduction to the structure of carbohydrates.

J.D. Rawn, *Biochemistry* (1983), Harper & Row (New York), pp. 268–319. Introduction to carbohydrates.

E. Smith, R. Hill, I. Lehman, R. Lefkowitz, P. Handler, and A. White, *Principles of Biochemistry: General Aspects*, Seventh Edition (1983), McGraw-Hill (New York), pp. 83–106. Structure and function of carbohydrates.

L. Stryer, *Biochemistry*, Second Edition (1981), Freeman (San Francisco), pp. 257–258. A brief introduction to carbohydrate structure.

G. Zubay, *Biochemistry* (1983), Addison-Wesley (Reading, MA), pp. 437–446. Introduction to the structure and function of carbohydrates.

Experiment **12**

Characterization of Erythrocyte Membrane Using Lipolytic Enzymes

RECOMMENDED READING:
Chapter 5, Sections A, B, C; Experiments 9, 10.

SYNOPSIS
The plasma membrane is a physical structure made up of lipids and proteins. It forms the boundary that maintains the intracellular components separate from the cellular environment. In this experiment, the arrangement of membrane lipids will be evaluated by studying the action of phospholipases on intact erythrocyte membranes.

I. INTRODUCTION AND THEORY

Structure and Function of the Membrane

The plasma membrane is the physical barrier that separates the interior of the cell from its exterior environment. The membrane provides structural integrity to the cell as well as a boundary to regulate the entry of essential nutrients and exit of metabolic waste products.

The biological significance of the plasma membrane has made it the target of numerous biochemical investigations. The plasma membrane consists of lipids, proteins, and carbohydrates. The ratio of these constituents varies and depends on the source of the membrane. Two major classes of lipids are present, the phospholipids, represented by the glycerophospholipids and sphingomyelin, and the steroids, represented by cholesterol. Structures for these membrane components are illustrated in Figure E12.1. The lipids are brought together primarily by hydrophobic interactions to form the basic structural unit of a membrane, a **lipid bilayer**.

The proteins of biological membranes are of two types, **peripheral proteins**, those on the surface of the membrane, and **integral proteins**, those embedded in the lipid bilayer. The primary function of the proteins is to carry out dynamic processes such as molecular transport and enzymatic action.

FIGURE 12.1

Typical lipids found in biological membranes. The arrows point to bonds that are susceptible to attack by various enzymes. PLA_2 = phospholipase A_2, PLC = phospholipase C, and SM = sphingomyelinase.

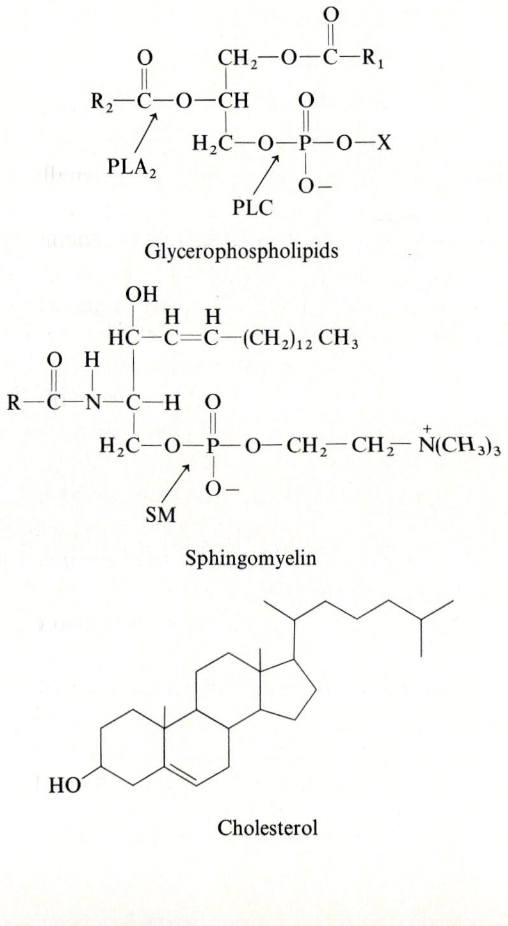

Glycerophospholipids

Sphingomyelin

Cholesterol

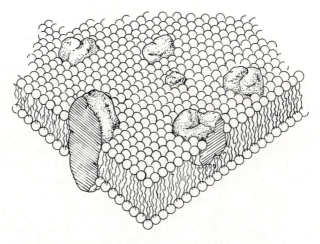

FIGURE E12.2
The fluid-mosaic model for the biological membrane. See text for explanation. From
S.J. Singer and G.L. Nicolson. *Science, 175,* 723 (1972). Copyright 1972 by the
American Association for the Advancement of Science.

The carbohydrate in the membrane is covalently linked to the lipid
and protein and, therefore, present as glycolipids and glycoproteins. Because of this, the membrane is usually considered to be a complex of lipid
and protein arranged as shown in Figure E12.2. This **fluid-mosaic model**,
proposed by S. Singer and G. Nicolson, consists of a lipid bilayer (phospholipids and cholesterol) embedded with proteins, with some on the surface and others passing through the entire bilayer. At the present time,
this model best illustrates the known physical, chemical, and biochemical
properties of plasma membrane. The model is supported by x-ray diffraction data, which reveal two areas of high electron density represented by
the two sides of the bilayer, and a central area of low electron density,
represented by the hydrophobic "tails" of the lipids. The dynamic transport properties can also be explained by the model. Nonpolar molecules
are able to diffuse through the membrane because of the hydrophobic
nature of the central region of the membrane. Polar, hydrophilic molecules
must be assisted in transport by specific proteins located in the membrane.

The fluid mosaic model is also consistent with the experimental observation of asymmetric organization of membrane components. That is,
some peripheral proteins are located on the exterior surface of the membrane, whereas others are on the interior surface. Biochemically, this results
in unidirectional or vectorial membrane processes such as the transport
of some nutrients from the outside to the inside of the cell, but not the
reverse.

Isolation and Characterization of Membranes

Before membrane structure and the molecular components can be studied, membrane material must be isolated in a more or less intact form. If the transport function of membranes is to be studied, then intact membrane systems must be prepared. Intact cytoplasmic membrane (**vesicles**) can be obtained from bacterial cells by removing cell wall material and intracellular components. **Erythrocyte ghosts** are prepared by removing hemoglobin and other intracellular components from red blood cells.

In order to isolate and characterize the constituents of cell membranes, the membrane material must be extracted under various conditions. Integral proteins are solubilized by homogenizing membrane preparations with detergents to dissociate the proteins from the lipid-protein complexes. Peripheral proteins are extracted with salt solutions. Further purification and characterization of proteins are carried out by chromatography and electrophoresis, as described in Chapters 3 and 4 and Experiment 4. Lipids are extracted with organic solvent systems such as isopropanol-hexane or methanol-chloroform, and are purified by chromatography.

One of the best sources of plasma membrane for experimental study is the mammalian erythrocyte (red blood cell). Not only is there a relatively abundant supply of such cells, but they do not contain the usual subcellular organelles; thus, there are no intracellular membrane systems. Therefore, red blood cells are a source of pure plasma membrane.

There is much recent interest in studying the asymmetric character and localization of lipids and proteins in biological membranes. In this experiment, the positioning of phospholipids in membrane will be investigated. The action of lipolytic enzymes on intact erythrocyte cell membranes provides information about the content and arrangement of phospholipids on the exterior surface of the intact erythrocyte membrane (Roelofsen, 1971; Zwaal, 1973).

The composition of the human erythrocyte membrane is approximately 50% protein, 40% lipid, and 10% carbohydrate. The ratio of phospholipids to cholesterol is about 1:1; and the relative amounts of individual phospholipids are phosphatidyl choline > phosphatidyl ethanolamine ≥ sphingomyelin ≫ phosphatidyl serine.

Three lipolytic enzymes that act in a selective way on phospholipids are phospholipase A_2 (PLA$_2$), phospholipase C (PLC), and sphingomyelinase (SM). Each enzyme catalyzes the hydrolysis of a specific bond in a phospholipid, as shown in Figure E12.1. When each of these phospholipases is incubated with intact erythrocytes, only exterior surface phospholipids are available for interaction with the lipolytic enzymes. There are three possible results from such incubation studies:

1. Phospholipid hydrolysis and hemolysis (erythrocyte breakdown and release of hemoglobin).

2. Phospholipid hydrolysis, but no hemolysis.

3. No phospholipid hydrolysis or hemolysis.

When the phospholipases act on the surface lipids, breaks may be produced in the membrane that allow release of hemoglobin into the solution. The extent of hemolysis is determined by spectrophotometric measurement of the released hemoglobin. Knowing the type of enzyme that causes hemolysis is of interest because it indicates that substrates of that enzyme are present in the membrane and are susceptible to hydrolytic breakdown.

Phospholipid breakdown in the membrane is determined by measuring the phospholipid content of the intact membrane before and after incubation with an enzyme. The phospholipids are extracted by treatment of the membrane with methanol-chloroform. Each lipid extract is analyzed for the presence of phosphate by a modification of the Fiske and Subba Row method (Bartlett, 1959). To perform this assay, all organophosphate materials in the extract are degraded, leaving inorganic phosphate. The inorganic phosphate is then treated with ammonium molybdate to form phosphomolybdate. The phosphomolybdate is reduced with a mixture of sodium bisulfite, sodium sulfite, and 1-amino-2-naphthol-4-sulfonic acid to form a phosphomolybdenum blue complex. The intensity of the blue color, measured at 660 ± 40 nm, is proportional to the concentration of inorganic phosphate. A standard curve with known phosphate concentrations is prepared for comparison. If phospholipid breakdown has occurred upon treatment of the membrane with the lipolytic enzyme, less inorganic phosphate will be detected in the lipid extract.

Mixtures of lipolytic enzymes may be able to act in sequence on membrane lipids. One enzyme may degrade a specific type of lipid on the surface, but a break in the membrane does not occur until a second enzyme degrades the new surface uncovered by the first. Information from these experiments leads to conclusions about the localization of specific lipids in membranes.

Chemical agents that disrupt membrane structure may also assist lipolytic enzymes in causing hemolysis. It would be expected that the presence of a detergent such as deoxycholate, SDS, or Triton X-100 in the incubation mixture would facilitate the lipolytic breakdown of the membrane. A detergent decreases the strength of the hydrophobic interactions that are responsible for holding the lipid-protein complexes together. This may result in alteration of the molecular organization of the membrane components, making the phospholipids more susceptible to hydrolytic breakdown.

The protein bovine serum albumin (BSA) readily binds fatty acids and other small hydrophobic molecules. The effect of this protein on membrane structure will also be evaluated by incubating membrane with a phospholipase in the presence of BSA. Removal of free fatty acids and nonpolar lipolytic products may increase the hemolytic effect of the phospholipase.

The phospholipases used in this experiment are available from a variety of sources. Phospholipase A_2 (phosphatide acylhydrolase, EC 3.1.1.4) may be isolated from porcine pancreas or the venom of the Indian cobra (*Naja naja*). Phospholipase C (phosphatidylcholine choline phosphohydrolase, EC 3.1.4.3) has been isolated from *Bacillus cereus* and *Clostridium welchii*. Sphingomyelinase (sphingomyelin choline phosphohydrolase, EC 3.1.4.12) may be isolated from human placenta and *Staphylococcus aureus*.

Overview of the Experiment

The action of lipolytic enzymes on intact human or bovine red blood cells will be investigated in this experiment. Erythrocytes, isolated from whole blood, will be incubated under various conditions with phospholipases and the extent of induced hemolysis will be monitored by spectrophotometric methods. The influence of surface active agents (detergents) and phospholipase mixtures will also be evaluated.

An observation of no hemolysis does not necessarily mean that no phospholipid breakdown has occurred. Membrane lipid degradation may occur under some conditions, but it may not be extensive enough to lead to hemolysis. In order to determine the extent of lipid breakdown, the erythrocyte membrane will be analyzed for phospholipid content before

FIGURE E12.3

Flowchart for characterization of erythrocyte membranes.

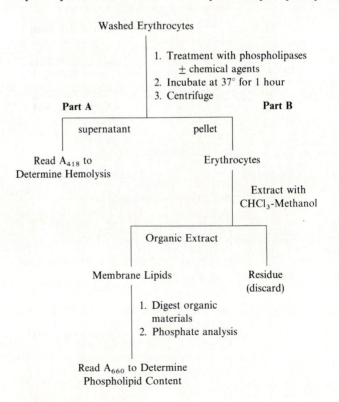

Washed Erythrocytes

1. Treatment with phospholipases
   $\pm$ chemical agents
2. Incubate at 37° for 1 hour
3. Centrifuge

Part A   **Part B**

supernatant   pellet

Read A_{418} to
Determine Hemolysis

Erythrocytes

Extract with
$CHCl_3$-Methanol

Organic Extract

Membrane Lipids   Residue
                         (discard)

1. Digest organic
   materials
2. Phosphate analysis

Read A_{660} to Determine
Phospholipid Content

and after lipase action. The amount of phospholipid present will be quantified by measuring inorganic phosphate. A flowchart of the procedures is outlined in Figure E12.3.

Part A, measurement of hemolysis, will require 3 hours of laboratory time. If phospholipid measurements are to be made (Part B), two additional 3-hour periods are required, not counting a 5-hour heating period for degradation of organic matter before phosphate analysis. If all parts of the experiment are to be completed, it is recommended that Part A be done during the first period. Extraction of the membrane lipids may be completed during the second period. Between the second and third periods, teaching assistants or selected students can carry out the long-term heating. Phosphate analysis is completed during the third period.

II. EXPERIMENTAL

Materials and Supplies

A. Phospholipase Action on Erythrocytes

Human or bovine blood, collected in heparinized or citrate-dextrose tubes

Isotonic saline, 0.87% NaCl, pH 7.4 with Tris, plus 0.25 mM $CaCl_2$

Refrigerated centrifuge with 12 or 15 ml tubes

Phospholipase A_2 in H_2O, 800 units/ml

Phospholipase C in buffer, 50 mM Tris, 5 mM $CaCl_2$, 50% glycerol, pH 7.2, 160 units/ml

Sphingomyelinase in buffer, 5 mM Tris, 5 mM $CaCl_2$, 50% glycerol, 2 mM $MgCl_2$, pH 7.6, 50 units/ml

Constant temperature bath at 37°C

Spectrophotometer

Conical centrifuge tubes, 12 ml

Triton X-100, 10^{-3} M in H_2O

Bovine serum albumin, 100 mg/ml in H_2O

EDTA solution, 2 mg/ml in H_2O

B. Extraction and Analysis of Phospholipids

Centrifuged red blood cells from Part A

Methanol

Chloroform

Clinical centrifuge

Steam bath

Absolute ethanol

Chloroform-methanol (1:1)

30% H_2O_2

10 N H_2SO_4

Oven or dry sand bath at 150–160°C

Test tubes, 12 × 100 mm, cleaned with phosphate-free detergent; rinse well with distilled H_2O

Molybdate reagent

Fiske and Subba Row reducer, 1-amino-2-naphthol-4-sulfonic acid, sodium sulfite, and sodium bisulfite

Standard phosphate solution, 0.001 M

Spectrophotometer for reading at 660 ± 40 mm

Procedure

A. Phospholipase Action on Intact Erythrocytes (Roelofsen, Zwaal, and Woodward, 1974)

Isolate intact erythrocytes by adding 2 ml of isotonic saline to 4 ml of fresh whole blood and centrifuging the mixture at 2000 × g for 10 minutes at 4°C (Hanahan, 1974). Use screw cap centrifuge tubes. Carefully remove the plasma and buffy coat (supernatant) with a disposable Pasteur pipet. Wash the pelleted cells by resuspending them in 4 ml of isotonic saline with very gentle mixing and centrifuge again at 2000 × g for 10 minutes. Remove the supernatant with a Pasteur pipet and repeat the washing step twice more with fresh 4-ml portions of isotonic saline. Remove the final supernatant and store the washed erythrocytes on ice. These will also be used in Part B.

Set up 12 conical centrifuge tubes and prepare each as described in Table E12.1. When complete, gently mix each tube and incubate for 1 hour at 37°C. Gently mix each tube every 15 minutes. At the end of a one-hour period, add 1 ml of EDTA solution to each tube to inhibit the phospholipase action. Centrifuge the tubes for 10 minutes at 2000 × g. Dilute 0.1 ml of each supernatant with 5 ml of distilled water, and read

TABLE E12.1
Preparation of Tubes for Incubation in Part A

Reagent	Tube 1	2	3	4	5	6	7	8	9	10	11	12
Distilled water	5.0	—	—	—	—	—	—	—	—	—	—	—
Isotonic saline	—	4.95	4.9	4.9	4.9	4.9	4.95	4.9	4.9	4.9	4.85	4.85
Washed erythrocytes	0.25	→										
Phospholipase A_2	—	0.05	—	—	—	0.05	—	—	0.05	—	—	—
Phospholipase C	—	—	0.1	—	0.05	—	—	0.05	0.05	—	—	0.05
Sphingomyelinase	—	—	—	0.1	0.05	0.05	—	—	—	0.05	0.05	—
Triton X-100	—	—	—	—	—	—	0.05	0.05	—	0.05	—	—
BSA	—	—	—	—	—	—	—	—	—	—	0.10	0.10

TABLE E12.2

Preparation of Standard Phosphate Curve

Reagent	Tube					
	1	2	3	4	5	6
Distilled water	1.0	0.90	0.80	0.60	0.30	—
Standard phosphate	—	0.10	0.20	0.40	0.70	1.0
Acid molybdate	1.0					→
Fiske-Subba Row reducer	0.25					→

and record the absorbance at 418 nm using water as a blank. Tube 1, containing water rather than saline buffer, represents 100% hemolysis.

B. Extraction and Analysis of Phospholipids

Estimate the volume of erythrocytes remaining from the Part A isolation of blood cells (should be approximately 1 ml) and add to each tube an equal volume of isotonic saline. Add 5 ml of methanol to each sample, cover with a screw cap, and shake well for at least 2 minutes. To each tube add 5 ml of chloroform and shake well for about 5 minutes. Centrifuge the tubes for 5 minutes at $2000 \times g$. Transfer each chloroform-methanol supernatant to a 25 ml Erlenmeyer flask and repeat the above extraction on the pellet twice more. After each extraction and centrifugation, transfer each supernatant to the proper Erlenmeyer. In other words, pool the proper organic extracts. Evaporate each extract to dryness with a steambath under a flow of air or under reduced pressure on a rotary evaporator. If water remains in the residue, add 1 ml of absolute ethanol and again evaporate to dryness. Dissolve the final residue in 1 ml of 1:1 chloroform-methanol.

Pipet duplicate samples (0.1 ml) of the above lipid extracts into labeled test tubes (12 × 100 mm). Evaporate the solvent and treat the phospholipid residue in the following way to decompose all organic material and convert organic phosphate to inorganic phosphate. Add 0.1 ml of 10 N H_2SO_4 and heat the tubes in a dry bath or oven at 150–160°C for at least 3 hours. Allow the tubes to cool and add 2 drops of 30% H_2O_2 to each tube. Heat at 150–160°C for another $1\frac{1}{2}$ hours. Allow the tubes to cool to room temperature and add 1.0 ml of acid molybdate and 0.25 ml of Subba Row reducer. Mix well after each addition. After 10 minutes at room temperature, read and record the absorbance at 660 ± 40 nm. Use distilled water as a blank to adjust the spectrophotometer.

A standard inorganic phosphate calibration curve is prepared as described in Table E12.2. Pipet the reagents (H_2O and phosphate standard) into *clean* test tubes and add 1.0 ml of the acid molybdate solution. Mix well and add 0.25 ml of the Fiske–Subba Row reducing solution. After 10 minutes, read and record the absorbance at 660 ± 40 nm, using distilled water for blank.

III. ANALYSIS OF RESULTS

A. Phospholipase Action on Erythrocytes

Prepare a table with the results of phospholipase treatment. Use these headings: test number (1–12), treatment (phospholipase $\pm$ chemical agent), and % hemolysis. Use the results from tube 1 for 100% hemolysis of the erythrocytes. Explain the results of each experiment and relate them to the possible arrangement of the various phospholipids in the erythrocyte membrane. What is the effect of BSA and the detergents on the lipolytic action? Try to explain the action of these agents. Explain the combined action of sphingomyelinase and phospholipase C on erythrocyte membrane.

B. Extraction and Analysis of Phospholipids

Prepare a standard calibration curve for phosphate, using the data from Table E12.2. Correct the absorbance for tubes 2 through 6 by subtracting the control, tube 1. Plot absorbance on the y axis and inorganic phosphate concentration (μmoles present in assay) on the x axis. Draw the best straight line connecting the points.

Prepare a table with the results of the phospholipid analysis. List sample number, actual absorbance reading, corrected absorbance reading (actual absorbance reading minus control, tube 1 in Table E12.2), and μmoles of phosphate present (from standard curve).

1. What samples displayed phospholipid breakdown? Use the results from tube 1, Table E12.1, to represent no phospholipid breakdown.

2. Was there phospholipid breakdown in any sample that did not display hemolysis?

3. Did you observe hemolysis under conditions where phospholipids remained intact? Explain.

4. Write a brief summary paragraph on the significance of these results in terms of the Singer-Nicolson fluid-mosaic model for the membrane.

IV. QUESTIONS

1. Using your results, what phospholipids are likely to be present at the exterior surface of the erythrocyte membrane?

2. Is it valid to use tube 1 from Table E12.1 for the control representing no phospholipid breakdown? Explain.

3. Why can tube 1 from Table E12.1 be used also to represent 100% hemolysis? Explain.

4. Membrane peripheral proteins are usually extracted and "dissolved" by a solution of salt or urea, whereas isolation of integral proteins requires the action of a detergent such as sodium dodecyl sulfate. Explain.

5. How would you identify the specific lipids isolated by extraction of the membrane with $CHCl_3$-MeOH?

V. REFERENCES

General

R.C. Bohinski, *Modern Concepts in Biochemistry*, Fourth Edition (1983), Allyn and Bacon (Boston), pp. 243–249. A brief introduction to plasma membranes.

M. Kates, "Techniques of Lipidology" in *Laboratory Techniques in Biochemistry and Molecular Biology*, T.S. Work and E. Work, Editors, Vol. 3, Part II (1972), Elsevier (Amsterdam). An excellent reference book for methods in lipid chemistry.

A. Lehninger, *Principles of Biochemistry* (1982), Worth Publishers (New York), pp. 322–326. A brief discussion of membrane structure and function.

J.D. Rawn, *Biochemistry* (1983), Harper & Row (New York), pp. 461–475. A discussion of membrane structure-function.

E. Smith, R. Hill, I. Lehman, R. Lefkowitz, P. Handler, and A. White, *Principles of Biochemistry: General Aspects*, Seventh Edition (1983), McGraw-Hill (New York), pp. 268–285. A chapter on membrane structure and function.

L. Stryer, *Biochemistry*, Second Edition (1981), Freeman (San Francisco), pp. 205–231. A chapter on membrane components and structure.

G. Zubay, *Biochemistry* (1983), Addison-Wesley (Reading, MA), pp. 573–658. Two chapters on membrane structure and function.

Specific

G. Bartlett, *J. Biol. Chem.*, *234*, 466 (1959). "Phosphorus Assay in Column Chromatography."

D. Hanahan and J. Ekholm in *Methods in Enzymology*, S. Fleischer and L. Packer, Editors, Vol. 31A (1974), Academic Press (New York), pp. 168–172. "The Preparation of Red Cell Ghosts (Membranes)."

B. Roelofsen, R. Zwaal, P. Comfurius, C. Woodward, and L. Van Deenen, *Biochim. Biophys. Acta, 241*, 925 (1971). "Action of Pure Phospholipase A_2 and Phospholipase C on Human Erythrocytes."

B. Roelofsen, R. Zwaal, and C. Woodward in *Methods in Enzymology*, S. Fleischer and L. Packer, Editors, Vol. 32B (1974). Academic Press (New York), pp. 131–140. "The Action of Pure Phospholipases on Native and Ghost Red Cell Membranes."

Sigma Chemical Company, "The Colorimetric Determination of Inorganic Phosphorus," (1979) Technical Bulletin No. 670. P.O. Box 14508, St. Louis, Missouri 63178.

R. Zwaal, B. Roelofsen, and C. Colley, *Biochim. Biophys. Acta, 300*, 159 (1973), "Localization of Red Cell Membrane Constituents."

Experiment **13**

Characterization of a Glycoprotein, Hemoglobin A_{Ic}

RECOMMENDED READING:
Chapter 3, Sections D, E and H; Chapter 5, Sections A, B, C.

SYNOPSIS
Hemoglobin A_{Ic}, a glycoprotein, comprises about 5% of the hemoglobin in normal adult red blood cells. In individuals with diabetes mellitus, its concentration may double. In this experiment, HbA_{Ic} and other glycohemoproteins will be isolated from human hemolysate by cation-exchange chromatography or affinity chromatography. Characterization of the glycoproteins is accomplished by visible spectrophotometry and carbohydrate analysis.

I. INTRODUCTION

Glycoproteins—Structure, Function, and Analysis

Proteins are classified into two broad categories: **simple**, those that consist only of amino acids, and **conjugated**, those proteins containing other moieties, including lipids, metal ions, nucleic acids, coenzymes, and carbohydrates. We are just beginning to realize the universal occurrence and multifaceted biological functions of the conjugated proteins. Perhaps the most widely diverse group is the **glycoproteins**, those that contain co-valently bound carbohydrate groups. A glycoprotein contains one, a few, or several carbohydrate units. The most common sites of carbohydrate attachment are the side chains of asparagine, serine, and threonine, although 5-hydroxyproline and lysine are also important. Figure E13.1 illustrates the manner in which carbohydrates are linked to the common amino acid residues. In Figure E13.1A is shown a typical N-glycosidic bond between the amide side chain of asparagine and C_1 of N-acetyl-glucosamine. The attachment of a sugar to threonine, shown in Figure E13.1B, consists of an O-glycosidic bond. The most common sugars found in glycoproteins are glucose, mannose, galactose, fucose, xylose, N-acetylglucosamine, N-acetylgalactosamine, and sialic acid. These sugars are found individually or, more commonly, linked together in oligosaccharide units that are attached to the protein.

The diverse structures of the glycoproteins would be expected to lead to great diversity in biological function. Table E13.1 lists common glyco-proteins classified according to function or source. The specific biological functions of many of the glycoproteins are still not completely understood. They are important components of cell membranes, where they coat the

FIGURE E13.1

Linkage of carbohydrates to amino acids in glycoproteins. (A) N-Glycosine bond be-tween asparagine and N-acetyl glucosamine. (B) O-Glycoside bond between threonine and glucose.

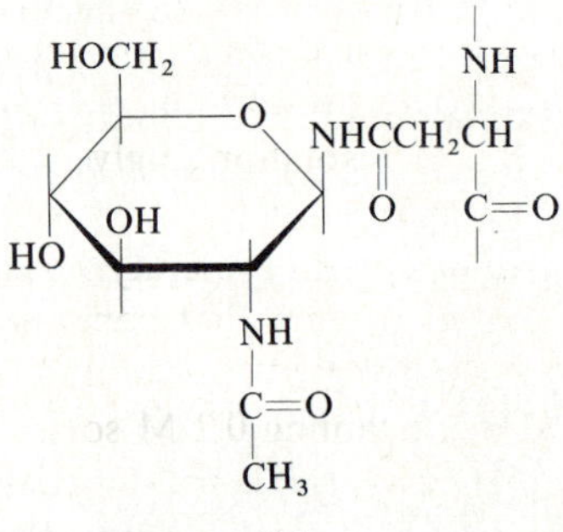

(A)

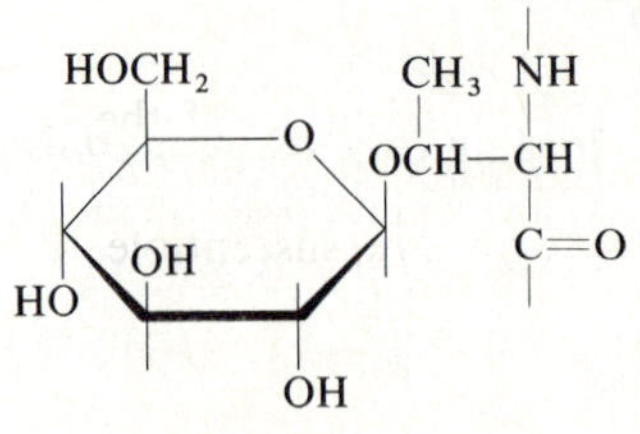

(B)

TABLE E13.1
Representative Examples of Glycoproteins.

Blood

Fibrinogen
Immunoglobulins
Blood-group proteins
Glycosylated hemoglobins
"Antifreeze proteins"

Hormones

Chorionic gonadotrophin
Follicle-stimulating hormone

Mucus Secretions

Submaxillary mucins
Gastric mucin

Cell Membranes

Glycophorin
Fibronectin

extracellular membrane surface (see Experiment 12). In membranes they may play a role in cellular recognition by immunoglobulins or provide sites for cell-cell interactions. Many plasma proteins, including the blood-group substances, are glycoproteins, but their functions are unclear.

Isolation, separation, and structural analysis of glycoproteins follow the same general procedures common for simple proteins. Ion-exchange chromatography, gel filtration, and electrophoresis are routinely used to isolate and purify glycoproteins. One new and very promising technique for the specific isolation of glycoproteins is based on the principles of affinity chromatography. A solid support, Glyco-Gel B[1], has been developed that contains covalently linked boronic acid on cross-linked agarose. Any glycoprotein that contains a carbohydrate unit with two neighboring hydroxyl groups is retained by the Glyco-Gel B (see Figure E13.2). A buffer containing 0.2 M sorbitol is used to elute the glycoprotein from the affinity gel. Sorbitol, a sugar alcohol with vicinal diols, ($\vdash$———$\dashv$), displaces the bound glycoprotein.
OH OH

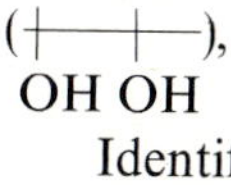

Identification of the amino acid residues involved in the protein-sugar linkage and characterization of the carbohydrate units require specific removal of the carbohydrate. O-Glycosidic linkages between the seryl or threonyl and sugar are susceptible to attack by a mild basic solution

---

[1] Trademark of Pierce Chemical Company.

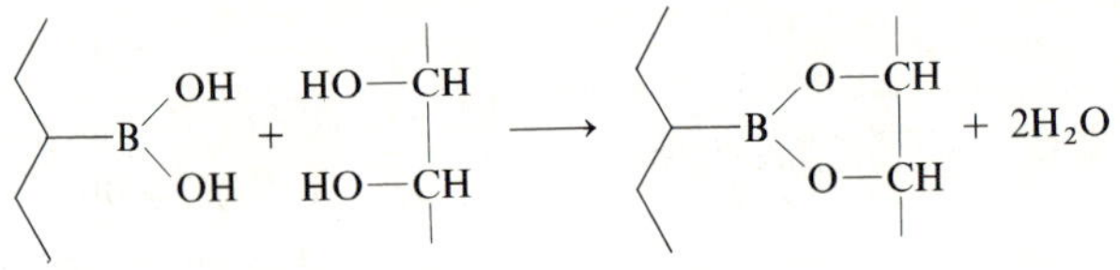

FIGURE E13.2
Affinity gel, showing boronic acid covalently linked to agarose. Attachment of a glycoprotein via two hydroxy groups is shown on the right side of the equation.

containing sodium borohydride. The reducing conditions lead to the
production of a sugar alcohol. N-glycosidic linkages between asparagine
residues and the carbohydrate unit are stable under these conditions, so
the two major types of sugar-protein linkage can be distinguished by this
analysis. Carbohydrate units linked to amino acids by N-glycosidic bonds
are cleaved by acid-catalyzed hydrolysis. The released carbohydrate unit,
which is separated from the protein by ion-exchange chromatography or
gel filtration, is analyzed for sugar composition. Sugar analysis is achieved
by complete hydrolysis of all internal glycosidic linkages, conversion of
the monosaccharides to suitable derivatives, and identification by gas
chromatography or high-pressure liquid chromatography.

Completion of the above procedures identifies the nature of the pep-
tide-sugar linkage and the total sugar composition, but does not completely characterize the carbohydrate unit unless the total glycoside unit
on the protein is a monosaccharide. Still to be defined are the sequence of
sugars in the carbohydrate chain and the type of sugar linkages. The types
of sugar linkages must be described in two ways, by determining the anomeric configuration (α or β) and by noting what hydroxyl groups are
used in the linkages ($1 \rightarrow 4$, $1 \rightarrow 6$, etc.). These problems are solved by degrading the intact carbohydrate unit with exoglycosidases or endoglyco-
sidases that catalyze the hydrolysis of specific sugar linkages. The released
carbohydrates are analyzed by GC or HPLC. The results of such studies
are combined, and the original structure is deciphered, using the overlap
method in much the same way as for protein sequence determination.

Properties of Hemoglobin A$_{1c}$

The major hemoglobin protein in most adult humans is HbA, which
has the subunit structure $\alpha_2\beta_2$. HbA comprises 90 to 95% of adult hemoglobin. Minor components that make up the remaining 5 to 10% are
HbA$_2$ (2.5%), HbF (0.5%), HbA$_{1a}$ (0.2%), HbA$_{1b}$ (0.4%), and HbA$_{1c}$ (3–5%)
(Bunn, Gabbay and Gallop, 1978). The most abundant minor component,
HbA$_{1c}$, has received considerable attention in the past decade because its
concentration is found to be approximately doubled in individuals with

diabetes mellitus (Rahbar, Blumenfeld, and Ranney, 1969). Figure E13.3 shows the most likely route of biosynthesis and the partial structure of HbA_{Ic}. The subunit structure of the hemoglobin is the normal $\alpha_2\beta_2$, but the N-terminal amino group (valine) of each of the two β chains is covalently linked to a ketohexose. The synthesis of HbA_{Ic} is unusual since it is thought to be nonenzymatic. Under physiological conditions, glucose or glucose-6-phosphate may be the source of the ketohexose. The rate of formation of HbA_{Ic} in the red blood cell depends upon the glucose concentration; hence, untreated diabetics will produce greater amounts of the minor hemoglobin. In the red blood cell, this glucose-modified hemoglobin is able to bind oxygen more strongly than HbA. This is probably due to a decreased reactivity of HbA_{Ic} with 2,3-diphosphoglycerate. More research is necessary in order to clarify the specific functional role, if any, of HbA_{Ic}.

The link between HbA_{Ic} and diabetes mellitus has sparked a great interest in the development of methods for separation, characterization, and analysis of variant hemoglobins. Cation exchange chromatography on Bio-Rex 70 is one of the most effective methods for separation of the several hemoglobins in human red blood cells (McDonald, Shapiro, Bleichman, Solway, and Bunn, 1978). All of the minor hemoglobin components are well-resolved and widely separated from normal HbA.

The characteristic absorption spectrum of oxyhemoglobin A, with λ_{max} at 415, 542, 577 nm, which is due to the heme prosthetic group, is well known. What may not be so well understood by students are the spectral changes that occur when hemoglobin is treated with inorganic or organic phosphates. Inorganic phosphate, 2,3-diphosphoglycerate (2,3-DPG), and inositol hexaphosphate (IHP) bind to hemoglobin in a cleft between the two β chains. When one of these molecules (for example, 2,3-DPG) occupies the small space between the β chains, the hemoglobin undergoes a conformational change. This change, which is primarily quaternary in nature, causes a slight movement of the heme prosthetic groups. This places the heme groups in a different environment and leads to the observed spectral changes of hemoglobin in the presence of phosphates. Recall from your biochemistry lectures that 2,3-DPG and other

FIGURE E13.3
Possible route of synthesis of HbA_{Ic} and partial structure of the protein. The sugar is covalently attached to the N-terminal end of each β chain in hemoglobin.

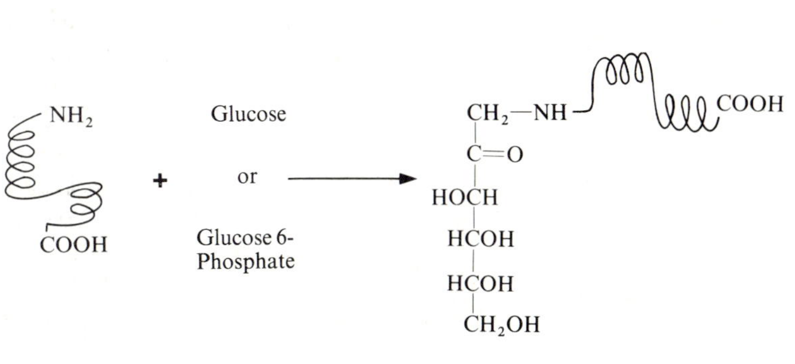

phosphates also decrease the oxygen affinity of deoxyhemoglobin. When glucose or other small molecules are covalently linked to the N-terminal amino acid of the β-chains, as in HbA_{Ic}, the phosphate binding site is blocked. This interferes with the noncovalent binding of 2,3-DPG or inositol hexaphosphate. Therefore, smaller spectral changes should occur when phosphate is added to HbA_{Ic} than when added to normal HbA. Also, the oxygen affinity of HbA_{Ic} should be little affected by the addition of 2,3-DPG.

One of the most effective and convenient methods for detecting and characterizing ligand-hemeprotein interactions is **difference spectroscopy**. This technique was introduced in Chapter 5.

A procedure for partial characterization of glycohemoglobins involves a colorimetric determination of the carbohydrate (Winterhalter, 1981). When some glycoproteins are treated with oxalic acid at 100°C, the bound carbohydrate is released in the form of 5-hydroxymethyl furfural (5-HMF). 5-HMF produces a colored adduct with thiobarbituric acid (TBA) that has a characteristic absorbance spectrum with $\lambda_{max} = 443$ nm. This test is relatively specific, since 5-HMA is formed only from carbohydrates bound to proteins via ketoamine linkages. If time permits, each of the hemoglobins that you isolate in this experiment will be characterized by the TBA test. A positive test indicates the presence of a ketohexose bound to an amino group as in Figure E13.3.

Glycosylated Hemoglobins and Diabetes Mellitus

Quantitative measurements of glycosylated hemoglobins in hemolysates provide a sensitive and accurate diagnosis in mild cases of diabetes, where glucose tolerance tests are sometimes variable. Clinical tests for glycosylated hemoglobins fall into three categories:

1. Direct Analysis of Erythrocytes
   (a) Hemolysate is prepared and treated with oxalic acid at 100°C. The 5-HMA released is mixed with thiobarbituric acid and the colored adduct is determined at 443 nm. This method is simple, fast, and reproducible, but it measures the blood content of all glycosylated hemoglobins.
   (b) The differential binding of inositol hexaphosphate by glycosylated hemoglobins compared to normal hemoglobin can be used to measure the concentration of glycohemoglobins. The higher the content of N-terminal glycosylated hemoglobin in a blood sample, the smaller the spectral difference in the visible spectrum. This test has been commercially developed by Abbott Laboratories (Moore and Stroupe, 1980).

(c) An antibody against HbA$_{1c}$ has been prepared and used directly on hemolysate; however, the sensitivity is acceptable only at levels of HbA$_{1c}$ of 6% or greater (Javid, 1977). An advantage is that only HbA$_{1c}$ is detected and measured.

2. Measurements of Separated Glycohemoglobins

Most tests based on this method separate the glycosylated hemoglobins as a group, and measurement is of total glycohemoglobins. The glycoproteins are separated by affinity chromatography, ion-exchange chromatography, and high-pressure liquid chromatography. This measurement is more accurate than those in group 1; however, it measures a mixture of glycohemoglobins rather than HbA$_{1c}$.

3. Measurement of Separated HbA$_{1c}$

Cation-exchange chromatography or gel-electrofocusing are used to separate HbA$_{1c}$ from the other minor hemoglobins and HbA. The purified HbA$_{1c}$ may be quantified by colorimetry or spectrophotometry.

Ideally, any clinical test developed for diagnosis of diabetes should be based on the measurement of HbA$_{1c}$, not total glycohemoglobins. However, the separation of the glycoproteins by ion-exchange chromatography is not readily applicable to rapid analysis of many blood samples, as would be required in a clinical laboratory.

Overview of the Experiment

In this experiment, human hemolysate will be subjected to cation-exchange or affinity chromatography. Most of the minor glycohemoglobins can be separated on a Bio-Rex 70 cation exchange column. Each column fraction containing the separated minor hemoglobins will be characterized by visible spectrophotometry and carbohydrate analysis. This procedure will require two laboratory periods of 3 to 4 hours each.

If time is limited, an alternate procedure may be used in this experiment. Rapid separation of glycoproteins from hemolysate can be accomplished with an affinity column. As was discussed in the introductory material, hemolysate is applied to the column, which is then eluted with buffer. Nonglycosylated proteins such as HbA will pass directly through the column, and glycoproteins will be retained by the solid support. The glycoproteins, which are eluted with sorbitol buffer, may be characterized by spectrophotometry and colorimetric analysis of carbohydrate content. The glycoprotein fraction from the affinity column will contain HbA$_{1c}$, but also other glycoproteins that are present in hemolysate. Therefore, your characterization tests will actually be carried out on a mixture of glycoproteins. HbA$_{1c}$ will make up a least 80% of this glycoprotein fraction, so it will have the greatest influence over the results of your analysis.

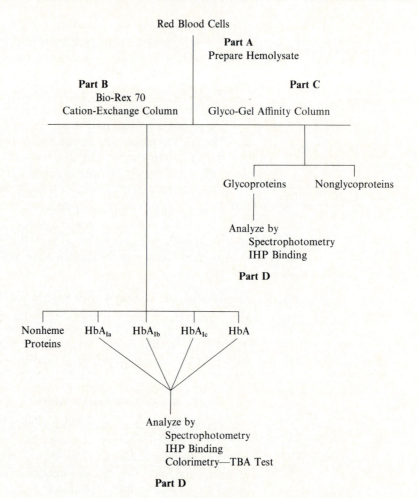

FIGURE E13.4
Flowchart for characterization of glycohemoglobins.

The alternative experiment with the affinity column will require about 3 hours, not including the thiobarbituric acid test for carbohydrates.

Figure E13.4 shows the experimental options available.

II. EXPERIMENTAL

Materials and Supplies

A. Preparation of Hemolysate

Whole Blood Samples. The samples can often be obtained from a hospital laboratory. They should be collected in heparinized or EDTA-treated tubes. These reagents prevent coagulation of the blood. The samples may be stored for up to one week at $0-5°C$. If students desire to use their own blood for this experiment, be sure the student health center or some other authorized agency is present to draw blood.

NaCl solution, 1% in water

Centrifuge, capable of 12,000 rpm

B. Cation-Exchange Chromatography

Bio-Rex 70 cation-exchange resin, 100-200 mesh, sodium form. Obtained from Bio-Rad Laboratories, Richmond, CA and prepared as described by the supplier. The final buffer for the slurry should be buffer A below.

Glass column, 2.5 × 30 cm

Buffer A, 0.05 M potassium phosphate, pH 6.6

Buffer B, 0.05 M potassium phosphate, 0.05 M NaCl, pH 6.6

Buffer C, 0.05 M potassium phosphate, 0.1 M NaCl, pH 6.6

Buffer D, 0.05 M potassium phosphate, 1.0 M NaCl, pH 6.6

Test tubes 50, 10 × 100 mm

Glass cuvets, 2, 3.0 ml

Fraction collector

Spectrophotometer

C. Affinity Chromatography

Glass column, 0.5 × 5 cm. A disposable pipet will work well.

Glyco-Gel B, obtained from Pierce Chemical Co., Rockford, IL. Prepare as described by the company. Final buffer for slurry should be the wash buffer below. Prepacked columns are also available. Each column can be regenerated twice.

Wash buffer, 0.25 M ammonium acetate, 0.05 M $MgCl_2$, 200 mg/liter NaN_3, pH 8.5

Elution buffer, 0.1 M Tris, pH 8.5, containing NaN_3 (200 mg/liter) and sorbitol (0.2 M)

Test tubes, 10, 10 × 100 mm

Hydrocarbon foil

Glass cuvets, 2, 3.0 ml

Spectrophotometer

D. Analysis of Glycoproteins

Double-beam spectrophotometer for difference spectroscopy

Glass cuvets

0.01 M inositol hexaphosphate (phytic acid)

0.1 M imidazole buffer, pH 6.8, plus 0.2 mM $K_4Fe(CN)_6$

Heating block or constant temperature bath at 100°C

1.0 M oxalic acid

Trichloroacetic acid, 40% in water

Thiobarbituric acid, 0.025 M in water

Fructose standard solution, 0.001 M in water

Glucose standard solution, 0.001 M in water

Test tubes, 10, 10 × 100 mm

Procedure

A. Preparation of Hemolysate

CAUTION:
Working with human blood presents a potential biohazard. Do not pipet any solutions by mouth. Wash hands with hot water and soap before eating, drinking, or smoking.

The red blood cells must be lysed so the hemoglobin can be released and dissolved in solution. Whole blood (2 ml) is centrifuged for ten minutes at 4000 × g to isolate erythrocytes. Decant and discard the supernatant and wash the erythrocytes by adding 3 ml of 1% NaCl. Centrifuge at 4000 × g for 10 minutes. Discard the supernatant and repeat the wash step two more times. To the isolated and washed red blood cells, add 6 ml of distilled water in order to disrupt the cell membrane by osmotic pressure. Stir and allow the mixture to stand for 20–30 minutes for cell lysis. Centrifuge cell debris for 20 minutes at 30,000 × g. The supernatant contains hemoglobin and other red blood cell constituents. The sediment contains fragments of cell membranes and other cell debris.

B. Bio-Rex 70 Column Chromatography (McDonald et al., 1978)

If possible, this column should be run in a cold room or in a refrigerated chamber. Clamp the glass column to a ring stand. Pour a slurry of Bio-Rex 70 into the column and allow the resin to pack to a height of 20 cm. Use the same general procedure that you followed in Experiments 2 and 4. Never allow the column to run dry. When the resin has settled to a final height of 20 cm, cut a small, round piece of filter paper and allow it to settle on top of the column. Place 5 ml of the hemolysate prepared in Part A onto the column by adding 1 to 2 ml dropwise with a disposable pipet. Pour the remaining 3 to 4 ml of hemolysate carefully down the inside wall of the column.

Begin to collect 3-ml fractions from the column into test tubes in a fraction collector. The column flow rate should be approximately 2 to 3 ml/min. When all the hemolysate has entered the resin, add dropwise 2 ml of buffer A. Allow the buffer to enter the resin and then elute the column with 200 ml of buffer A, continuing to collect 3 ml fractions. Allow buffer A to completely enter the column and then add buffer B. Elute the column with B, collecting 3-ml fractions, until no more red protein is eluted (approximately 100 ml). Change the eluting buffer to C and collect 3 ml fractions until the column eluent is colorless. Finally, elute the column with buffer D to wash off hemoglobin A. While the column is running, you should be taking absorbance measurements on each fraction. Read A_{415} on each fraction collected with buffers A and B. Read A_{540} for each fraction collected with buffer C and A_{670} for buffer D. The different wavelengths are used because later fractions may be off the absorbance scale at 415 nm. In each case, zero the spectrometer with the proper buffer. Prepare a graph of absorbance (A_{415}, A_{540} or A_{670}) vs. fraction number. Combine the fractions represented by each peak. Do this by combining 3 or 4 of the most intensely colored fractions from each peak. The first major red peak is $HbA_{Ia_1} + HbA_{Ia_2}$. The second is HbA_{Ib} and the third is HbA_{Ic}. The largest and last peak is normal hemoglobin A. Label the four fractions I, II, III, and IV. Skip the next section on affinity chromatography and begin Section D, Analysis of Glycoprotein.

C. Affinity Chromatography using Boronic Acid Gel

CAUTION:

The buffers used in this part of the experiment contain sodium azide, a deadly poison. **Do not pipet any solutions by mouth**. Sodium azide may react with copper or lead plumbing to form explosive azides. For disposal, flush waste buffer down the drain with large volumes of water.

Prepare the small glass column or disposable pipet by placing a very small piece of glass wool or cotton into the constriction. Then plug the constricted end with hydrocarbon foil. Clamp the column to a ring stand and pour 1 ml of the Glyco-Gel B slurry into the column. Allow some buffer to pass out of the column by removing the parafilm, but the column should not run dry. Close off the column with hydrocarbon foil when most of the buffer is gone. Obtain 0.1 ml of the hemolysate. Allow all the buffer to enter the gel and immediately and carefully add the sample of hemolysate to the top of the column. Allow it to enter the gel and immediately add 0.5 ml of the ammonium acetate buffer to wash the remaining hemoglobin completely into the gel. Begin to collect a fraction from the column and apply 19.5 ml of the wash buffer (ammonium acetate) in a dropwise fashion. Collect 3-ml fractions in test tubes. When the wash buffer has just entered the gel, add 5 ml of eluting buffer containing sorbitol and collect all the eluting buffer in a single test tube. This last fraction contains glycoproteins. Read the A_{415} of all fractions and plot A_{415} vs. fraction number. Save all fractions for carbohydrate analysis in Section D.

D. Analysis of Glycoproteins

Spectrophotometry. Record a visible spectrum (340 nm to 700 nm) for each pooled column fraction. To do this, dispense 3 ml of a fraction into a glass cuvet and run the spectrum, using as reference the buffer that eluted that particular fraction. For example, for fraction I of the Bio-Rex column, use Buffer A in the reference beam. If the spectrometer pen runs off the absorbance scale, dilute the sample with the appropriate buffer and repeat the scan. If you performed the separation using the Bio-Rex column, you should have four fractions (I, II, III, IV) on which to record spectra. For fractions from the affinity column, run spectra on the final fraction (eluted with sorbitol buffer) and on only one of the earlier fractions.

Thiobarbituric Acid Test. Each hemoglobin fraction may be analyzed for carbohydrate content by using the thiobarbituric acid test as described in the introduction. The test is carried out on column fractions and on standard fructose and glucose solutions. The procedure is outlined in Table E13.2.

Tube 1 is a control containing only water and the required chemical reagents. Tubes 2 and 3 contain standard carbohydrates, fructose and glucose. Additional tubes contain column fractions with glyco- or non-glycohemoglobins. After addition of oxalic acid to each tube, mix well and heat at 100°C for 4 hours. Cover each test tube with a marble to avoid evaporation. Add 1.0 ml of 40% trichloroacetic acid to each tube to precipitate the protein. Centrifuge each reaction mixture for ten minutes at $2000 \times g$. Transfer 2.0 ml of each supernatant to a separate tube and mix with 2.0 ml of 0.025 M thiobarbituric acid. Incubate the reaction mixtures at 40° for 1 hour. Allow the tubes to cool to room temperature and record the spectrum of each reaction mixture from 400 to 500 nm.

TABLE E13.2

Thiobarbituric Acid Test, Part D

Reagent-Procedure	1	2	Tube 3	4	5
Distilled water	2.0	1.8	1.8	—	—
Standard fructose	—	0.2	—	—	—
Standard glucose	—	—	0.2	—	—
Hemoglobin column fractions	—	—	—	—	—
I	—	—	—	2.0	—
II, etc.	—	—	—	—	2.0
Oxalic acid, 1.0 M	1.0 ——————————→				
Mix well and heat					
Add TCA, 40%	1.0 ——————————→				
Centrifuge					
Transfer 2.0 ml of each supernatant to separate tubes					
Add TBA, 0.025 M	2.0 ——————————→				
Heat and record spectrum					

Differential Binding of Inositol Hexaphosphate (IHP). Obtain two matched glass cuvets for spectral measurements. To check whether the cuvets are well matched, fill both with water or a buffer and scan from 400 to 600 nm. A straight line should be obtained. Into each cuvet, transfer 0.5 ml of a hemoglobin fraction. This may be one of the two fractions from affinity chromatography or Fraction I, II, III or IV from the Bio-Rex column. Add 2.5 ml of the imidazole buffer containing $K_4Fe(CN)_6$ to each cuvet. To the reference cuvet, add 0.1 ml of inositol hexaphosphate solution. To the sample cuvet, add 0.1 ml of distilled water. Mix each cuvet by holding a small square of hydrocarbon foil over the opening and inverting several times. Do not shake. Place the cuvets in the proper compartments of a double-beam spectrometer. With the 0 setting or balance knob, set the recorder pen to midscale (0.5 A) at 400 nm. The recorder pen could move in either direction, so it must have sufficient chart space for movement. Using the above procedure, obtain an IHP difference spectrum for as many of your hemoglobin fractions as time allows.

III. ANALYSIS OF RESULTS

B and C. Separation of Hemoglobins

If you separated the hemoglobins by ion-exchange chromatography, compare the elution profile with Figure 1 in McDonald et al. (1978). Note, and record in your notebook, any similarities and differences. Identify the hemoglobin fraction represented by each peak in the profile. Can you estimate the quantity of each fraction and calculate the percentage of each type of hemoglobin?

If you used affinity chromatography, define the type of hemoglobin in each of the two fractions. What is the ratio of glycohemoglobins to nonglycohemoglobins? Calculate the percent of glycohemoglobins using Equation 1.

$$\% \text{ Glycohemoglobin} = \frac{A_{414} \text{ (Bound Hb)} \times 100}{A_{414} \text{ (Bound Hb)} + 5A_{414} \text{ (Unbound Hb)}}$$

A_{414} (Bound Hb) = Absorbance at 414 nm of the fraction eluted with sorbitol buffer

A_{414} (Unbound Hb) = Absorbance at 414 nm of the unbound hemoglobins

(Equation E13.1)

D. Analysis of Glycoproteins

Spectrophotometry

Do you note any differences between the absorption spectra of hemoglobin A and the glycohemoglobins? Should there be spectral differences?

Thiobarbituric Acid Test

Explain the results of your TBA test with standard fructose and glucose. Did you obtain a color for each? Which hemoglobin fraction shows the greatest formation of 5-HMA? Are these the expected results?

Differential Binding of IHP

Describe the difference spectrum obtained for each hemoglobin. What does your result imply about the interaction of IHP with hemoglobin? Compare the IHP difference spectrum of hemoglobin A with that of hemoglobin A_{Ic} (or the mixture of glycohemoglobins). Explain any differences. Which hemoglobin, A or A_{Ic}, binds IHP more tightly? Why? Compare your difference spectra with those in Chapter 5, Figure 5.10.

IV. QUESTIONS

1. Describe the principles behind the separation of hemoglobins by Bio-Rex chromatography.

2. Many of the proteins in the hemolysate are "nonheme" proteins. How would you detect their elution from a Bio-Rex or Glyco-Gel column?

3. In the early literature describing ion-exchange chromatographic separation of the hemoglobins, glycohemoglobins were called "fast hemoglobins." Why?

4. Did both standard fructose and glucose give an intense color for the TBA test? Explain the results.

5. Could you substitute Sephadex G-100 for Bio-Rex 70 and expect an effective resolution of the hemoglobins?

6. Why was it necessary in the difference spectrum experiment to have identical concentrations of protein in both cuvets?

7. Why must a "matched pair" of cuvets be used for difference spectroscopy?

8. When you monitor the Bio-Rex ion exchange column at 415 nm, what chromophore are you measuring?

V. REFERENCES

General

H.F. Bunn, K.H. Gabbay, and P.M. Gallop, *Science, 200,* 21 (1978). "The Glycosylation of Hemoglobin: Relevance to Diabetes Mellitus."

D. Freifelder, *Physical Biochemistry* (1976), Freeman (San Francisco), pp. 395–399. Applications of difference spectroscopy.

R. Kornfeld and S. Kornfeld, *Ann. Rev. Biochem.*, *45*, 217 (1976). "Comparative Aspects of Glycoprotein Structure."

E. Smith, R. Hill, I. Lehman, R. Lefkowitz, P. Handler, and A. White, *Principles of Biochemistry: General Aspects*, Seventh Edition (1983), McGraw-Hill (New York), pp. 460–469. An introduction to glycoprotein structure, analysis, and synthesis.

L. Stryer, *Biochemistry*, Second Edition (1981), Freeman (San Francisco), pp. 221–22, 714–719. Synthesis of glycoproteins.

Specific

H.F. Bunn, K.H. Gabbay, and P.M. Gallop, *Science*, *200*, 21 (1978). "The Glycosylation of Hemoglobin: Relevance to Diabetes Mellitus."

J. Javid, P.K. Pettis, R.J. Koenig, and A. Cerami, *Blood*, *50*, 110 (1977). Using antibodies to measure HbA_{1c}.

M.J. McDonald, R. Shapiro, M. Bleichman, J. Solway, and H.F. Bunn, *J. Biol. Chem.*, *253*, 2327 (1978). "Glycosylated Minor Components of Human Adult Hemoglobin."

E.G. Moore and S.D. Stroupe, U.S. Patent No. 4,200,435, April 29, 1980. Diagnostics Division of Abbott Laboratories, *Chem. Ab.*, *93*, 110212y (1980). Using differential spectroscopy to measure glycosylated hemoglobin in blood.

S. Rahbar, O. Blumenfeld, and H.M. Ranney, *Biochem. Biophys. Res. Comm.*, *36*, 838 (1969). "Studies of an Unusual Hemoglobin in Patients with Diabetes Mellitus."

K.H. Winterhalter in *Methods in Enzymology*, E. Antonini, L. Rossi-Bernardi, and E. Chiancone, Editors, Vol. 76 (1981), Academic Press (New York), p. 732. "Determination of Glycosylated Hemoglobin."

Experiment **14**

Photoinduced Proton Transport and Phosphorylation by Spinach Chloroplasts

RECOMMENDED READING:
Chapter 2, Section A; Chapter 7, Sections, A, B, C.

SYNOPSIS

Photosynthesis occurs in the plant cell organelle called the chloroplast. During the process of photosynthesis, electrons are transferred from H_2O to $NADP^+$ via an electron carrier system. The energy released by electron transport is converted into the form of a proton gradient and coupled to ADP phosphorylation. In this experiment a method is introduced to demonstrate the formation of the proton gradient across the chloroplast membranes.

I. INTRODUCTION

The Photosynthetic Process

Photosynthesis occurs only in plants, algae, and some bacteria, but all forms of life are dependent upon its product. In photosynthesis, electromagnetic energy from the sun is used as the driving force for a thermodynamically unfavorable chemical reaction, the synthesis of carbohydrates from CO_2 and H_2O (Reaction 1).

$$CO_2 + H_2O \xrightarrow{hv} (CH_2O) + O_2 \qquad \text{(Reaction E14.1)}$$

Here, (CH_2O) represents a carbohydrate molecule. The simplicity of Reaction 1 is deceiving, because the process of photosynthesis is a complex one; the details are still under investigation. The fundamental photosynthetic event is initiated by the interaction of a photon with a chlorophyll molecule, causing excitation of an electron in chlorophyll. The activated electron is passed through a chain of electron carriers aligned in order of decreasing reduction potential. The electron transport chain is similar to that in mitochondrial respiration. Energy released during electron transport is coupled to the synthesis of ATP from ADP and phosphate, P_i. This process is outlined in Figure E14.1. Two photosystems, I and II, are at work during photosynthesis in plants. Each photosystem contains

FIGURE E14.1
The photosynthetic process, showing coupling of electron transport and ADP phosphorylation. See text for details. The dashed line shows electron flow in cyclic photophosphorylation.

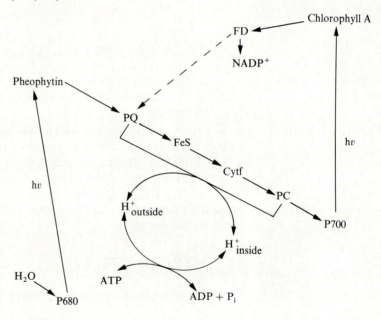

PQ = Plastoquinone
FeS = Iron Sulfur Protein
Cytf = Cytochrome f
PC = Plastocyanin
FD = Ferredoxin

reaction centers (chlorophylls P680 and P700) that collect the light. The two systems are linked by the electron transport chain, allowing for electron flow from H_2O (to produce O_2) ultimately to $NADP^+$. In other words, H_2O is oxidized, and the electrons are carried to $NADP^+$, forming the reduced cofactor, NADPH. The overall reaction is shown in Reaction 2. Note that this is the reverse of the respiratory process occuring in mitochondria, where NADPH is oxidized.

$$H_2O + NADP^+ + nADP + nP_i \xrightarrow{hv} NADPH + H^+ + nATP + \tfrac{1}{2}O_2$$

(Reaction E14.2)

When electrons flow from photosystem I to photosystem II, protons are transported across the chloroplast membranes as indicated in Figure E14.1. This aspect of photosynthesis will be discussed in a later section.

The light reactions illustrated in Figure E14.1 are accompanied by a sequence of "dark reactions" leading to the formation of carbon intermediates and sugars. This sequence of reactions, called the Calvin cycle, incorporates CO_2 into carbohydrate structures.

Figure E14.1 illustrates the photosynthetic process as it occurs in higher plants. This is called **noncyclic photophosphorylation** to distinguish it from **cyclic photophosphorylation** in photosynthetic bacteria. Cyclic photophosphorylation requires only photosystem I and a second series of electron carriers to return electrons to the electron-deficient chlorophyll. The dashed line in Figure E14.1 indicates the flow of electrons in cyclic photophosphorylation. ATP is produced during the cyclic process just as in the noncyclic process, but NADPH is not.

The photosynthetic process in green plants occurs in subcellular organelles called **chloroplasts**. These organelles resemble mitochondria; they have two outer membranes and a folded inner membrane called the **thylakoid**. The apparatus for photosynthesis, including the chlorophyll reaction centers and electron carriers, is in the thylakoid membrane. The chemical reactions of the Calvin cycle take place in the **stroma**, the region around the thylakoid membrane.

Mechanism of Photophosphorylation

One very active area of biochemical investigation is the study of energy transfer in mitochondria and chloroplasts in both oxidative phosphorylation and photophosphorylation, electron flow through a carrier chain is linked in some way to the generation of ATP. Most research is directed toward determining how the energy provided by electron transport is coupled to the phosphorylation of ADP. In recent years, three theories of energy coupling have received attention: chemical coupling, conformational coupling, and chemiosmotic coupling. The **chemical coupling hypothesis** proposes the formation of high-energy chemical intermediates; the **conformational coupling hypothesis** proposes a high-energy molecular conformation; and **chemiosmotic coupling** involves a charge separation or concentration gradient across a membrane.

The chemiosmotic coupling hypothesis, proposed by P. Mitchell in 1961 is the most attractive explanation, and many experimental observations now support this idea. Simply stated, Mitchell's hypothesis suggests that electron transfer is accompanied by a transport of protons across the membrane, resulting in a proton concentration gradient. When the proton gradient (a high-energy condition) collapses, the released energy is coupled to the phosphorylation of ADP. In mitochondria, the proton transport is through the inner membrane, from the inside (matrix side) to the outside (cytoplasmic side), whereas in chloroplasts protons are transported across the thylakoid membrane from the outside to the inside.

Experimental results supporting the Mitchell hypothesis were first reported from studies using chloroplasts. Two critical experiments providing unequivocal evidence for linkage between proton transport and ADP phosphorylation were reported. Jagendorf (1964) generated a high-energy proton gradient by illuminating chloroplasts in the absence of ADP and phosphate. As one would predict from the Mitchell hypothesis, Jagendorf observed a rise in the pH of the medium, indicating proton uptake by the chloroplasts. In later experiments, Jagendorf (1966) induced an artificial proton gradient across the thylakoid membrane by incubating chloroplasts under acidic conditions (pH 4) and then rapidly transferring them to pH 8 buffer. When ADP and P_i were added to the chloroplast suspension, the synthesis of ATP was observed, and the previously induced pH gradient disappeared. Similar experiments with mitochondria also confirm the transport of protons in oxidative phosphorylation, but in the opposite direction from photophosphorylation.

Continued study of light-induced proton uptake by chloroplasts has led to the observation that proton uptake is accompanied by Mg^{2+} and K^+ extrusion from the chloroplasts. The specific function of these exchanges, other than ionic balance, is not understood.

Practical Aspects of the Experiment

Chloroplasts will be isolated by careful extraction of spinach leaves, using Tris buffer containing sucrose. The crude extract contains both whole and fragmented chloroplasts, but both contain all the necessary photosynthetic components and are capable of photophosphorylation. The preparation described in this experiment retains almost all of the chlorophyll in the chloroplasts. The total chlorophyll content of the chloroplasts will be determined by extracting the pigment with aqueous acetone and measuring the absorption at $\lambda = 652$ nm. The chlorophyll concentration is calculated according to Equation 1 (Arnon, 1949).

$$C = \frac{A}{\varepsilon \ell} = \frac{A_{652}}{34.5 \times \ell}$$

(Equation E14.1)

where

C = concentration of chlorophyll in mg/ml

A_{652} = absorption of chlorophyll at 652 nm

ε = extinction coefficient for chlorophyll, 34.5

ℓ = pathlength of cuvet

The average range of chlorophyll concentration for the described chloroplast preparation is 0.3 to 1.0 mg/ml.

The light-induced pH changes are measured by monitoring the pH of a suspension of illuminated chloroplasts. If protons are taken up into the chloroplasts, the pH of the suspension medium should increase. Figure E14.2 shows the typical results from such an experiment. The proton transport process is reversed by eliminating the light source (indicated by the dotted line in Figure E14.2). Figure E14.2 shows the result of illuminated chloroplasts alone. It has been found that the presence of a light-sensitive redox dye stimulates the proton transport process. In this

FIGURE E14.2
Light-induced pH changes in chloroplasts. Light is turned on and off at indicated times.

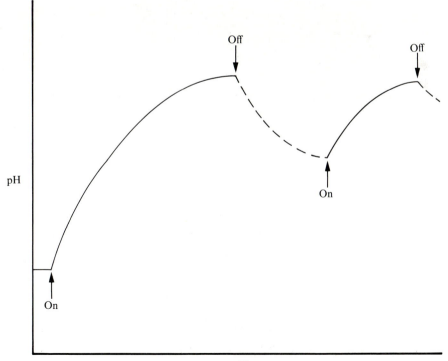

experiment, phenazine methosulfate (PMS) will be used, although others, including methyl viologen and trimethylhydroquinone, may be used as an electron transport cofactor.

Overview of the Experiment

Chloroplasts will be isolated from spinach leaves and used to explore some intricacies of proton transport and photophosphorylation. The chloroplast preparation should be used as soon as possible (within 2 to 3 hours) since biological activity will decrease with time. The experimental apparatus for measurement of proton uptake should be assembled and all reagents prepared before the chloroplast extract is begun. The time required for the various parts of the experiment is as follows.

A. Preparation of Spinach Chloroplasts—30 minutes

B. Determination of Chlorophyll Content—15 minutes

C. Measurement of Proton Uptake—1 to 2 hours, depending on the choice of experiments.

Other optional experiments may be completed if time allows. For example, the effectiveness of various redox dyes may be analyzed. In addition to those listed in the text, FMN, ferricyanide, and dichlorophenolindophenol may be tested (Neuman and Jagendorf, 1964). It has been shown that NH_4Cl and amines stimulate proton uptake (Crofts, 1967). If a potassium ion-specific electrode is available, the light-induced efflux of K^+ from spinach chloroplasts may be studied (Dilley, 1964). Other methods for measuring the rate of ATP formation have been reported (Jagendorf and Uribe, 1966).

II. EXPERIMENTAL

Materials and Supplies

A. Preparation of Spinach Chloroplasts

Fresh spinach leaves, about 50 grams. These should be washed well with cold water, placed in plastic bags, and kept cold for several hours before use.

Homogenizing buffer, 0.02 M Tris, pH 8, containing 0.01 M NaCl and 0.4 M sucrose. **Keep cold**.

Sharp knife

Blender

Cheesecloth

Refrigerated centrifuge

Chloroplast suspension solution, 0.4 M sucrose containing 0.01 M NaCl

B. Determination of Chlorophyll Content

    Chloroplast preparation from Part A

    80% acetone solution in water

    Conical centrifuge tube

    Glass cuvet, one pair

    Spectrophotometer for measurement at 652 nm

    Clinical or table-top centrifuge

C. Measurement of Proton Uptake

    pH meter, with recorder if possible. This should have a full-scale pH change of 0.5 pH unit.

    500-watt tungsten lamp, with a Corning 1-69 filter or a solution containing a trace of $CuSO_4$ to remove infrared radiation

    Magnetic stirrer and small stir bar

    Phenazine methosulfate (PMS)

    Timer

    Large test tube, 2×20 cm

    Phosphate solution, 0.01 M K_2HPO_4

    Adenosine diphosphate, 0.2 M

Procedure

A. Preparation of Spinach Chloroplasts (Walker, 1971, 1980)

Chloroplasts are fragile and unstable; therefore, this part of the experiment should be done as rapidly as possible and in subdued lighting. Maintain all solutions at 4°C or below.

Cut and discard the midribs from 50 g of spinach leaves and tear the leaves into small pieces. Immediately place the leaves into a precooled blender containing 100 ml of ice-cold homogenizing buffer. Grind the leaves at top speed no longer than 5 seconds. Strain the homogenate through eight layers of cheesecloth. Squeeze all the liquid from the cheesecloth. Centrifuge the filtrate in precooled tubes at $1000 \times g$ for 1 minute to remove whole cells. Transfer the supernatant to precooled centrifuge tubes and spin at $6000 \times g$ for about 15 seconds. Decant the supernatant and suspend the chloroplasts (pellet) in 50 ml of Tris homogenizing buffer. Centrifuge the suspension at $6000 \times g$ for about 10 seconds. (Note the color of the supernatant; the deeper the green, the more chlorophyll was lost from the chloroplasts.) Pour off and discard the supernatant. Resuspend the isolated chloroplasts by stirring into 100 ml of suspension solution (sucrose containing NaCl). Store the chloroplast preparation in ice until use. It will remain stable for 2 to 4 hours.

B. Determination of Chlorophyll Content

Extract the chlorophyll from the chloroplasts by mixing, in a conical centrifuge tube, 0.1 ml of well mixed chloroplast suspension with 9.9 ml of 80% acetone in water. Spin in a table-top centrifuge for 10 minutes. Transfer the supernatant to a glass cuvet and read the absorbance at 652 nm using 80% acetone in water as blank. Calculate the concentration of chlorophyll in the chloroplast suspension using Equation 1.

C. Measurement of Proton Uptake (Dilley, 1972)

CAUTION:
Phenazine methosulfate is a skin irritant, so it should be handled with care. Avoid contact with skin and do not breathe dust.

The experimental arrangement is shown in Figure E14.3. Maintain the temperature of the water bath at 10°C by adding ice. Add a trace of $CuSO_4$ solution to the water bath to eliminate infrared radiation from the lamp. Obtain a test tube that will hold the pH electrode and 20 to 30 ml of solution. Dilute the chloroplast solution with suspension buffer to a chlorophyll concentration of about 150 μg/ml. Adjust the pH meter with standard pH 7 buffer. The following three experimental conditions are recommended. Each condition should be tested individually and done in duplicate.

1. Chloroplasts with no redox cofactor

   20 ml of chloroplast suspension (about 2.7 mg chlorophyll)

   1 ml of suspension buffer

2. Chloroplasts with redox cofactor

   20 ml of chloroplast suspension

   5 mg of phenazine methosulfate or other redox dye

   1 ml of suspension buffer

3. Chloroplasts with redox cofactor, ADP, and phosphate

   20 ml of chloroplast suspension

   5 mg of phenazine methosulfate or other redox dye

   0.5 ml of ADP solution

   0.5 ml of phosphate buffer

To complete each experiment, prepare the reaction mixture as described above. Mix well and transfer to the reaction tube. Insert a small magnetic stir bar and place the pH electrode into the reaction mixture. Turn on the magnetic stirrer to a slow rate and check to see whether the

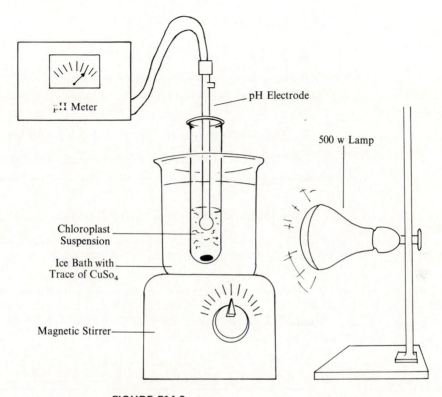

FIGURE E14.3
Experimental arrangement for measuring proton uptake by illuminated chloroplasts.
See text for details.

stirrer affects the pH meter. Adjust the pH of the reaction mixture to a pH reading in the range of 6 to 6.2 with acid or base. Turn on the pH meter and recorder, if available, and then turn on the lamp to illuminate the chloroplasts. Mark the recorder sheet at the time the lamp is turned on. Allow the recorder to trace the pH reading with time for 60 seconds, and then turn off the light but continue to record the pH. Turn the lamp on after a dark interval of 30 seconds. Continue to record the pH for 60 seconds, again turn the lamp off, and continuously record the pH. Repeat the experiment with a fresh portion of chloroplasts.

If no pH recorder is available, pH readings can be taken from the meter or digital readout and a graph of pH vs. time is prepared. Record a pH reading every 10 seconds.

Experimental conditions 2 and 3, with their various additions, are completed in an identical fashion. If chloroplasts still remain stable and time permits, other experiments outlined in the overview may be completed.

III. ANALYSIS OF RESULTS

A. Preparation of Spinach Chloroplasts

Write all observations regarding the preparation in your notebook. Record the color of each supernatant and pellet.

B. Determination of Chlorophyll Content

Use Equation 1 to calculate the concentration of chlorophyll in the chloroplast suspension. What is the total yield of chlorophyll in mg?

C. Measurement of Proton Uptake

If no recorder tracing is available, prepare a graph of pH vs. time for each experiment. Prepare and complete a table with the following headings:

Addition	Rate of pH Change (sec/0.5 pH)	Relative Rate
none	_____	_____
+ PMS	_____	_____
+ PMS, ADP, P_i	_____	_____

Compare the relative rates of proton uptake for each experimental condition and explain any differences. Did the addition of light-sensitive redox cofactor affect the rate of pH shift? What is the effect of the addition of ADP and phosphate on the rate of pH shift? Explain the effect of alternating light and dark intervals.

IV. QUESTIONS

1. Write a flow chart outlining the chloroplast preparation procedure and describe the purpose of each step.

2. Explain the chlorophyll extraction procedure in chemical terms. Why is acetone effective in the extraction? How could the experiment be modified to increase the yield of chlorophyll extraction?

3. Describe other methods to measure the rate of ATP generation during photophosphorylation.

4. Why is the amount of chlorophyll kept constant in all experiments described here?

5. Describe methods that could be used to measure extrusion of K^+, Mg^{2+}, or Ca^{2+} from chloroplasts.

6. Explain the effect, if any, of the redox cofactor on the rate of proton uptake.

7. Why does the pH of the suspension solution decrease during dark intervals?

V. REFERENCES

General

D. Arnon, *Trends Biochem. Sci, 9*, 258–262 (1984). "The Discovery of Photosynthetic Phosphorylation."

M. Avron, *Ann. Rev. Biochem., 46*, 143 (1977). "Energy Transduction in Chloroplasts."

R.C. Bohinski, *Modern Concepts in Biochemistry*, Fourth Edition (1983) Allyn and Bacon (Boston), pp. 363–383. Introduction to photosynthesis.

J.D. Rawn, *Biochemistry* (1983), Harper & Row (New York), pp. 703–741. Basic concepts of photosynthesis.

E. Smith, R. Hill, I. Lehman, R. Lefkowitz, P. Handler, and A. White, *Principles of Biochemistry: General Aspects*, Seventh Edition (1983), McGraw-Hill (New York), pp. 478–506. An introduction to photosynthesis.

L. Stryer, *Biochemistry*, Second Edition (1981), Freeman (San Francisco), pp. 431–453. A chapter on photosynthesis.

G. Zubay, *Biochemistry* (1983), Addison-Wesley (Reading, MA), pp. 409–435. An introduction to photosynthesis.

Specific

D. Arnon, *Plant Physiol., 24*, 1 (1949). Determination of chlorophyll content of chloroplasts.

A. Crofts, *Biochem. Biophys. Res. Comm., 24*, 127 (1966). Uptake of NH_4^+ by illuminated chloroplasts.

R. Dilley in *Methods of Enzymology*, A. San Pietro, Editor, Vol 24B, (1972), Academic Press (New York), pp. 68–74. Transport of H^+, K^+, and Mg^{2+} in isolated chloroplasts.

A. Jagendorf and E. Uribe, *Proc. Nat. Acad. Sci. U.S., 55*, 170 (1966). "ATP Formation Caused by Acid-Base Transition of Spinach Chloroplasts."

J. Neumann and A. Jagendorf, *Arch. Biochem. Biophys.*, *107*, 109 (1964). "Light-Induced pH Changes Related to Phosphorylation by Chloroplasts."

D. Walker in *Methods in Enzymology*, A. San Pietro, Editor, Vol. 23A, (1971), Academic Press (New York), pp. 211–220. Isolation of chloroplasts and grana using aqueous procedures.

D. Walker, *ibid.*, Vol. 69c, (1980), Academic Press (New York), pp. 94–104. "Preparation of Higher Plant Chloroplasts."

Experiment **15**

DNA Extraction from Bacterial Cells

RECOMMENDED READING:
Chapter 2, Section E; Chapter 5, Sections A, B, C.

SYNOPSIS
DNA, the genetic substance in biological cells, is a large polymer of nucleotide monomers. This experiment introduces the student to a general method for isolation and partial purification of nucleic acids from microorganisms. The procedure consists of, first, disrupting the cell wall or membrane, second, dissociating bound proteins, and third, separating the DNA from other soluble compounds. The isolated DNA is characterized by ultraviolet spectroscopy.

I. INTRODUCTION

DNA Structure and Function

DNA in all forms of life is a polymer made up of nucleotides containing four major types of heterocyclic nitrogen bases, adenine, thymine, guanine, and cytosine. The nucleotides are held together by 3′, 5′- phosphodiester bonds (Figure E15.1). The quantitative ratio and sequence of bases vary with the source of the DNA. Native DNA exists as two complementary strands held together by hydrogen bonds and arranged in a double helix.

DNA was first isolated from biological material in 1869, but its participation in the transfer of genetic information was not recognized until the mid-1940s. Since that time, DNA has been the subject of thousands of

425

physical, chemical, and biological investigations. A landmark discovery was the elucidation of the three-dimensional structure of DNA by Watson and Crick in 1953. The **double helix** as envisioned by Watson and Crick is now recognized as a significant structural form of native DNA (Figure E15.2). Other discoveries of importance include methods for sequential analysis of the nucleotide bases, regulation of gene expression, cleavage of DNA by specific restriction endonucleases, and recombinant DNA.

DNA in prokaryotic cells (simple cells with no major organelles and a single chromosome) exists as a single molecule in a circular, double-

FIGURE E15.1

The covalent structure of DNA. From Geoffrey Zubay, BIOCHEMISTRY © 1983, Addision-Wesley, Reading, Massachusetts. p. 666, Fig. 18–7. Reprinted with permission.

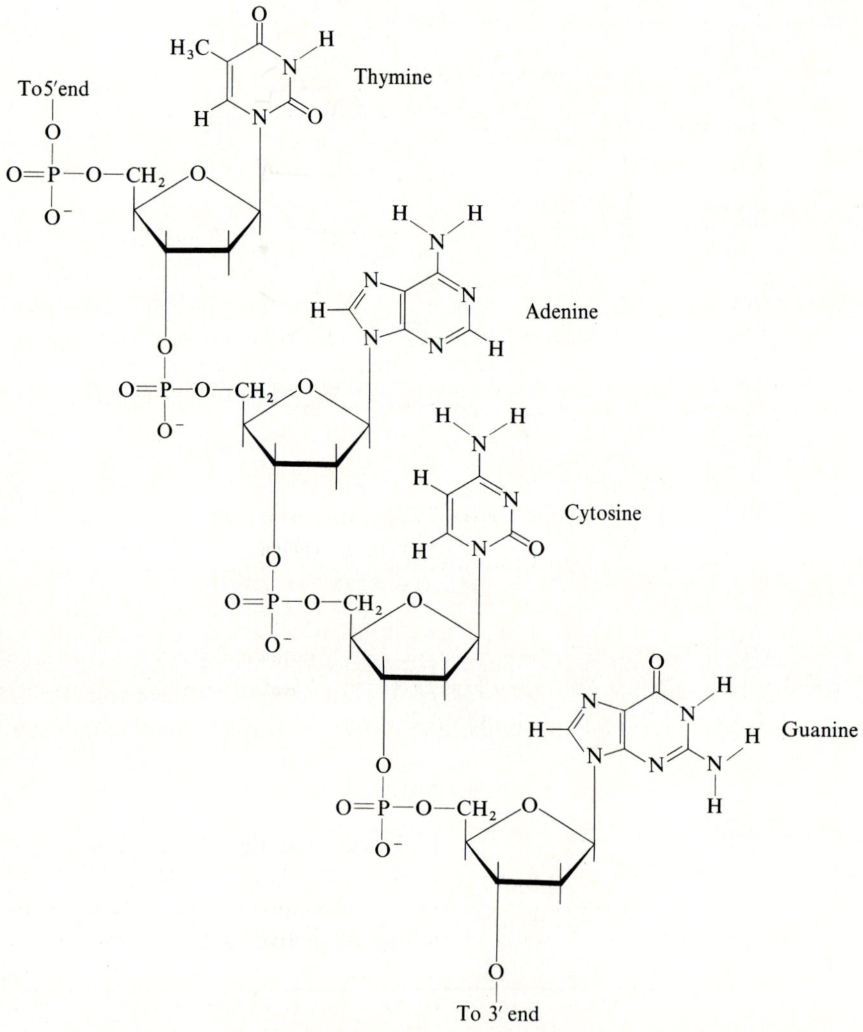

Thymine

Adenine

Cytosine

Guanine

To 5′ end

To 3′ end

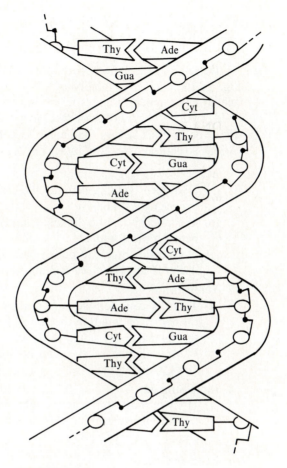

FIGURE E15.2
The helix conformation of DNA showing base pairs. From Geoffrey Zubay, BIOCHEM-
ISTRY © 1983, Addision-Wesley, Reading, Massachusetts. p. 669, Fig. 18–11(b).
Reprinted with permission.

helix form with a molecular weight of at least 2×10^9. Eukaryotic cells
(cells with major organelles) contain several chromosomes and, thus,
several very large DNA molecules.

Isolation of DNA

Because of the large size and the fragile nature of chromosomal
DNA, it is unlikely that anyone has ever isolated it in an intact, undamaged form. Several isolation procedures have been developed that
provide DNA in a biologically active form, but this does not mean it is
completely undamaged. These DNA preparations are stable, of high
molecular weight and relatively free of RNA and protein. Here, a general

method will be described for the isolation of DNA in a stable, biologically active form from microorganisms. The procedure outlined is applicable to many microorganisms and can be modified as necessary.

Designing an isolation procedure for DNA requires extensive knowledge of the chemical stability of DNA as well as its condition in the cellular environment. Figures E15.1 and E15.2 illustrate several chemical bonds that may be susceptible to cleavage during the extraction process. The experimental factors that must be considered and their effects on various structural aspects of intact DNA are outlined below.

1. pH
   (a) Hydrogen bonding between the complementary strands is stable between pH 4 and 10.
   (b) The phosphodiester linkages in the DNA backbone are stable between pH 3 and 12.
   (c) N-glycoside bonds to purine bases (adenine and guanine) are hydrolyzed at pH values of 3 and less.

2. Temperature
   (a) There is considerable variation in the temperature stability of the hydrogen bonds in the double helix, but most DNA will begin to unwind in the range of 80–90°C.
   (b) Phosphodiester linkages and N-glycoside bonds are stable up to 100°C.

3. Ionic Strength
   (a) DNA is most stable and soluble in salt solutions. Salt concentrations of less than 0.1 M weaken the hydrogen bonding between complementary strands.

4. Cellular Conditions
   (a) Before the DNA can be released, the bacterial cell wall must be lysed. The ease with which the cell wall is disrupted varies from organism to organism. In some cases (yeast), extensive grinding or sonic treatment is required, whereas in others (*B. subtilis*), enzyme hydrolysis of the cell wall is possible.
   (b) Several enzymes are present in the cell that may act to degrade DNA, but the most serious damage is caused by the deoxyribonucleases. These enzymes catalyze the hydrolysis of phosphodiester linkages.
   (c) Native DNA is present in the cell as DNA-protein complexes. The proteins (basic proteins called **histones**) must be dissociated during the extraction process.

5. Mechanical Stress on the DNA
   (a) Gentle manipulations may not always be possible during the isolation process. Grinding, shaking, stirring, and other disruptive procedures may cause cleavage (shearing or scission) of the DNA

chains. This usually does not cause damage to the secondary structure of the DNA, but it does reduce the length of the molecules.

Now that these factors are understood, a general procedure of DNA extraction may be outlined:

Step 1. Disruption of the cell membrane and release of the DNA into a medium in which it is soluble and protected from degradation

The isolation procedure described here calls for the use of an enzyme, lysozyme, to disrupt the cell membrane. Lysozyme catalyzes the hydrolysis of glycosidic bonds in cell wall carbohydrates, thus causing destruction of the outer membrane and release of DNA and other cellular components. The medium for solution of DNA is a buffered, saline solution containing EDTA. DNA, because it is ionic, is more soluble and stable in salt solution than in distilled water. The EDTA serves at least two purposes. First, it binds divalent metal ions (Cd^{2+}, Mg^{2+}, Mn^{2+}) that could form salts with the anionic phosphate groups of the DNA. Second, it inhibits deoxyribonucleases that have a requirement for Mg^{2+} or Mn^{2+}. Citrate has occasionally been used as a chelating agent for DNA extraction; however, it is not an effective agent for binding Mn^{2+}. The mildly alkaline medium (pH 8) acts to reduce electrostatic interaction between DNA and the basic histones and the polycationic amines, spermine and spermidine (see Experiment 21). The relatively high pH also tends to diminish nuclease activity and denature other proteins.

Step 2. Dissociation of the protein-DNA complexes

Detergents are used at this stage to disrupt the ionic interactions between positively charged histones and the negatively charged backbone of DNA. Sodium dodecyl sulfate (SDS), an anionic detergent, binds to proteins and gives them extensive anionic character. A secondary action of SDS is to act as a denaturant of deoxyribonucleases and other proteins. Also favoring dissociation of protein-DNA complexes is the alkaline pH, which reduces the positive character of the histones. To ensure complete dissociation of the DNA-protein complex and to remove bound cationic amines, a high concentration of a salt (NaCl or sodium perchlorate) is added. The salt acts by diminishing the ionic interactions between DNA and cations.

Step 3. Separation of the DNA from other soluble cellular components

Before DNA is precipitated, the solution must be deproteinized. This is brought about by treatment with chloroform/isoamyl alcohol and followed by centrifugation. Upon centrifugation, three layers are produced:

an upper aqueous phase, a lower organic layer, and a compact band of denatured protein at the interface between the aqueous and organic phases. Chloroform causes surface denaturation of proteins. Isoamyl alcohol reduces foaming and stabilizes the interface between the aqueous phase and the organic phases where the protein collects.

The upper aqueous phase containing nucleic acids is then separated and the DNA precipitated by addition of ethanol. Because of the ionic nature of DNA, it becomes insoluble if the aqueous medium is made less polar by addition of an organic solvent. The DNA forms a threadlike precipitate that can be collected by "spooling" onto a glass rod. The isolated DNA may still be contaminated with protein and RNA. Protein can be removed by dissolving the spooled DNA in saline medium and repeating the chloroform/isoamyl alcohol treatment until no more denatured protein collects at the interface.

RNA does not normally precipitate like DNA, but it could still be a minor contaminant. RNA may be degraded during the procedure by treatment with ribonuclease after the first or second deproteinization steps. Alternatively, DNA may be precipitated with isopropanol, which leaves RNA in solution. Removal of RNA sometimes makes it possible to denature more protein using chloroform/isoamyl alcohol. If DNA in a highly purified state is required, several deproteinization and alcohol precipitation steps may be carried out. It is estimated that up to 50% of the cellular DNA is isolated by this procedure. The average yield is 1 to 2 mg per gram of wet packed cells. The isolated DNA has a molecular weight on the order of 10×10^6.

Characterization of DNA

The DNA isolated in this experiment is of sufficient purity for characterization studies. Only a measurement of UV absorbance will be made in this experiment, but Experiment 16 will be dedicated to a more complete analysis of DNA structural characteristics.

DNA has significant absorption in the UV range because of the presence of the aromatic bases, adenine, guanine, cytosine, and thymine. This provides a useful probe into DNA structure because structural changes such as helix unwinding affect the extent of absorption. In addition, absorption measurements are used as an indication of DNA purity. The major absorption band for purified DNA peaks at about 260 nm. Protein material, the primary contaminant in DNA, has a peak absorption at 280 nm. The ratio A_{260}/A_{280} is often used as a relative measure of the nucleic acid/protein content of a DNA sample. The typical A_{260}/A_{280} for isolated DNA is 1.9. A smaller ratio indicates increased contamination by protein. Compare this measurement of DNA purity with the spectrophotometric method for protein concentration in Chapter 2.

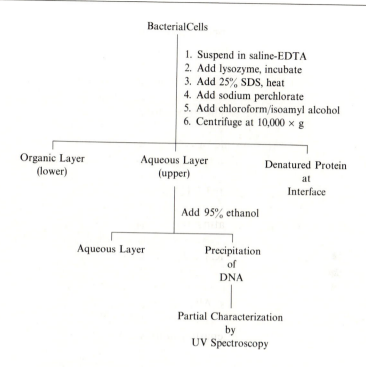

FIGURE E15.3
Flowchart for isolation of DNA from bacterial cells.

Overview of the Experiment

This experiment describes an isolation procedure for DNA that can be applied to most microorganisms. A flow chart of the procedure is shown in Figure E15.3. The general procedure was first reported by J. Marmur (1961, 1963). Recommended organisms are *Bacillus subtilus*, *Escherichia coli*, and *Clostridium welchii*. Other microorganisms may be used, but modifications in the procedure may have to be made as described by Marmur.

After extraction of the DNA, the product will be analyzed by spectral measurement in the ultraviolet range. Further characterization is described in Experiment 16.

The isolation and analysis of DNA may be completed during a 3-hour laboratory period. If more time is available, additional extractions by chloroform/isoamyl alcohol may be completed to remove contaminating protein and achieve a highly purified form of DNA.

II. EXPERIMENTAL

Materials and Supplies

Bacterial cells, wet packed, 2 to 3 g. Use *B. subtilus*, *E. coli*, or *C. welchii*. If lyophilized cells are available, use 0.5 g.

Saline-EDTA, 0.15 M NaCl plus 0.1 M ethylenediaminetetraacetate, pH 8

Sodium dodecyl sulfate (SDS), 25% in water

Lysozyme solution, 10 mg/ml in water

Sodium perchlorate, 5 M

Chloroform/isoamyl alcohol, 24:1

95% ethanol

Saline-citrate, 0.15 M NaCl plus 0.015 M citrate, pH 7.0

Quartz cuvets

Centrifuges, one capable of speeds up to 10,000 rpm and a clinical centrifuge

Spectrophotometer for UV measurements

Magnetic stirrer

Water bath at 60°C

Procedure (Marmur, 1961, 1963)

A. Isolation of DNA

CAUTION:
Experimental work with bacterial cells presents a potential biohazard. Some strains of the bacteria recommended for use may cause gastro-enteritis (abdominal pain and diarrhea). Do not pipet any solutions by mouth. Wash your hands well with hot water and soap before eating, drinking, or smoking.

Ethanol is volatile and flammable. There should be no flames in the laboratory during this experiment.

Suspend 2 to 3 g of wet packed bacterial cells or 0.5 g of lyophilized cells in 25 ml of saline-EDTA solution in a 125 ml Erlenmeyer flask. Add 1.0 ml of the lysozyme solution (10 mg of lysozyme) and incubate the mixture at 37° for 30 to 45 min. To bring about complete cell lysis, add 2.0 ml of 25% SDS solution and heat the mixture in a waterbath (60°C) for 10 min. To avoid excessive foaming, stir the mixture very gently. As the nucleic acid is released from the lysed cells, the solution will show an increase in viscosity and a decrease in cloudiness.

After the heating period, allow the mixture to cool to room temperature or cool the flask in cold running water. Add 9.0 ml of sodium perchlorate solution (5 M), which brings the final salt concentration to about 1 M. Mix well, but gently, and then add a volume of chloroform/isoamyl alcohol (24:1) equal to the total volume of the extraction mixture (40 to

45 ml). Shake the solution in a separatory funnel, a tightly stoppered flask, or a screwcap centrifuge tube for 20 to 30 min. Then, centrifuge the emulsion for 5 min at $10,000 \times g$. Three layers should be visible: a bottom organic phase, a middle band of denatured protein at the interface, and an upper aqueous layer containing the nucleic acids. Carefully transfer the aqueous layer to a 125 ml beaker or large test tube. Precipitate the nucleic acids by carefully layering about 2 volumes of 95% ethanol over the aqueous phase. The ethanol should be poured down the side of the beaker or tube using a glass stirring rod. Mix the two layers very gently with a circular motion of the stirring rod and "spool" all the fibers of nucleic acid onto the rod. Press the spooled DNA against the inside of the glass vessel to remove solvent. (If the nucleic acid does not form fibrous threads upon addition of ethanol, centrifuge the mixture for 10 min at 2000 to 3000 rpm in a clinical centrifuge. Transfer the sediment to about 10 ml of saline-EDTA solution and reprecipitate with 95% ethanol.)

If desired, the isolated nucleic acid sample may be treated again with chloroform/isoamyl alcohol for further deproteinization. However, if time does not allow for this, dissolve the spooled nucleic acid in 10 to 15 ml of saline-citrate. Save the DNA solution for Part B and Experiment 16.

B. Ultraviolet Measurement of Isolated DNA

Dilute a 0.5 ml aliquot of the DNA solution with 4.5 ml of saline citrate. Transfer some of the solution to a quartz cuvet and determine the A_{260}, using a cuvet containing saline-citrate as reference. If the A_{260} is greater than 1, quantitatively dilute the sample until the absorbance reading is between 0.5 and 1.0. Determine and record the A_{280} on the same DNA sample.

III. ANALYSIS OF RESULTS

A. Isolation of DNA

Write all observations of your nucleic acid preparation in your notebook. Explain the purpose of each step and the role played by each reagent. Speculate on any damage that may have altered the molecular structure of the DNA.

B. Ultraviolet Measurement of Isolated DNA

Calculate the A_{260}/A_{280} ratio. Extensively purified DNA has a ratio of about 1.9. Comment on the protein content of your isolated nucleic acid. Calculate the DNA concentration (μg/ml) in the saline-citrate solution and determine the total yield of isolated DNA. Assume the $E_{260}^{1\%}$ of native DNA is about 200.

IV. QUESTIONS

1. Why does the solution become more viscous during lysozyme-SDS treatment?

2. How could RNA be separated from the nucleic acid preparation?

3. How does SDS help disrupt the cell membrane?

4. If extremely high purity DNA is desired, what further steps could be followed in addition to those described here?

5. A solution of purified DNA gave an A_{260} of 0.55 when measured in a quartz cuvet of 1 cm pathlength. What is the concentration of DNA in μg/ml?

V. REFERENCES

General

R.C. Bohinski, *Modern Concepts in Biochemistry*, Fourth Edition (1983), Allyn and Bacon (Boston), pp. 163–197. An introduction to the structure and function of DNA.

J.D. Rawn, *Biochemistry* (1983), Harper & Row (New York), pp. 320–394. An introductory chapter on the structure and function of DNA.

E. Smith, R. Hill, I. Lehman, R. Lefkowitz, P. Handler, and H. White, *Principles of Biochemistry: General Aspects*, Seventh Edition (1983), McGraw-Hill (New York), pp. 126–155. Structure and properties of the nucleic acids.

L. Stryer, *Biochemistry*, Second Edition (1981), Freeman (San Francisco), pp. 559–596. Structure and properties of DNA.

W.B. Wood, J.H. Wilson, R.M. Benbow, and L.E. Hood, *Biochemistry*, *A Problems Approach*, Second Edition (1981), Benjamin/Cummings (Menlo Park, CA), pp. 315–343. A short discussion on DNA structure with excellent problems.

G. Zubay, *Biochemistry* (1983), Addison-Wesley (Reading, MA), pp. 661–695. A well-written, extensive discussion of nucleic acid structure and function.

Specific

J. Marmur in *Methods in Enzymology*, S.P. Colowick and N.O. Kaplan, Editors, Vol. VI, (1963), Academic Press (New York), pp. 726–738. "A Procedure for the Isolation of Deoxyribonucleic Acid from Microorganisms."

J. Marmur, *J. Mol. Biol.*, *3*, 208 (1961). "A Procedure for the Isolation of Deoxyribonucleic Acid from Microorganisms."

Experiment **16**

Characterization of DNA by Ultraviolet Absorption Spectrophotometry

RECOMMENDED READING:
Chapter 5, Sections A, B, C; Introduction to Experiment 15.

SYNOPSIS
Native DNA exists in a double helix held together by base-base interactions between the two polynucleotide strands. When DNA is exposed to denaturating conditions, the interactions between the two strands weaken and the polynucleotide backbones separate. The process of DNA dissociation (denaturation) is accompanied by a significant increase in absorbance at 260 nm. The change in absorbance is used to determine the melting temperature (T_m) of DNA, to estimate % GC content, and as an indication of purity.

I. INTRODUCTION

Denaturation of DNA

Solutions of native DNA display strong absorption in the ultraviolet region between 240 and 310 nm. The wavelength of maximum absorption is about 260 nm for most DNA. In Experiment 15, it was pointed out that this absorption is due primarily to the presence of the purine and pyrimidine bases in the DNA. If DNA solutions are treated with denaturing agents (heat, alkali, organic solvents) their ultraviolet absorbing properties are strikingly changed. Figure E16.1 shows the effect of heat on the UV absorption of DNA. The curve is obtained by plotting $A_{260(T)}/A_{260(25°)}$ vs. temperature (T). Heating through the temperature range of 25° to about 80°C results only in minor increases in absorption. However, as the temperature is increased, there is a sudden increase in UV absorption which is then followed by a constant A_{260}. The total increase in absorption is usually on the order of 40% and occurs over a small temperature range. Figure E16.1 is called a **thermal denaturation curve**, a **temperature profile**, or a **melting curve**. The temperature corresponding to the midpoint of each absorption increase is defined as T_m,

FIGURE E16.1
A typical thermal denaturation curve for DNA. T_m is measured as the temperature at the midpoint of the absorbance increase.

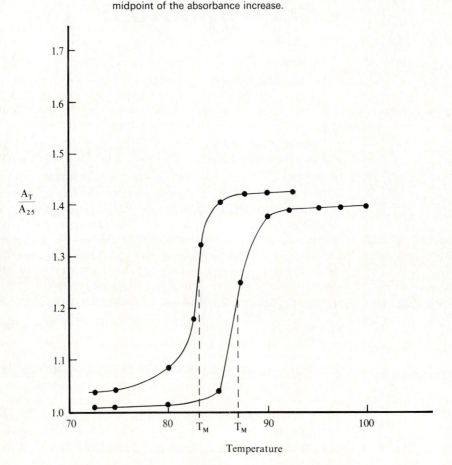

the **transition temperature** or **melting temperature**. (This should not be confused with melting point, transformation of a substance from solid to liquid, as was routinely studied in organic laboratory.) Each species of DNA has a characteristic T_m value that can be used for identification and characterization purposes.

The origin of the absorption increase, called a **hyperchromic effect**, is well understood. The absorption changes are those that result from the transition of an ordered double helix DNA structure to a denatured state or random, unpaired DNA strands. Native DNA in solution exists in the double helix held together primarily by hydrogen bonding between complementary base pairs on each strand (see Figure E15.2, Experiment 15). Hydrophobic and π-π interactions between stacked base pairs also strengthen the double helix. Agents that disrupt these forces (hydrogen bonding, hydrophobic and π-π interactions) cause a dissociation or unwinding of the double helix. In a random coil arrangement, base-base interactions are at a minimum; this alters the resonance behavior of the aromatic rings, thus causing an increase in absorption. The process of DNA dissociation can, therefore, be characterized by monitoring the UV absorbing properties of DNA under various conditions. Measurements of T_m and studying conditions that lead to denaturation provide valuable information about the nucleotide content of the DNA and also give an estimate of the purity or quality of the DNA.

Many agents have been evaluated as potential denaturants of DNA. Heat has received the most attention, but chemical agents such as organic solvents, urea, salts, and alkali are known to transform the double helix of DNA to a random coil. In addition, the action of deoxyribonuclease, an enzyme that catalyzes the hydrolysis of phosphodiester bonds, can be assayed by monitoring the change in A_{260}.

Using the Hyperchromic Effect to Characterize DNA

The hyperchromic effect is a relatively simple process to measure, but it can be very useful in characterizing DNA. Preparation of a temperature profile is one of the most convenient methods for measuring the increase in UV absorption. The results may be used for several purposes.

1. Determination of T_m. As was noted earlier in this experiment, T_m is a valuable physical constant that can be used in the identification and characterization of DNA. It can be measured as the temperature corresponding to the midpoint of the absorption increase. The T_m for most DNA is in the range of $80°$ to $90°C$.

2. Calculation of Base Composition. The stability of the DNA double helix depends upon the base composition of the DNA. The greater the guanine plus cytosine content (% GC base pairs), the more stable the double helix and the higher the T_m. GC pairs are held together by three hydrogen bonds; in contrast, AT pairs have just two hydrogen bonds. A linear relationship has been described that correlates the T_m value for a DNA sample with the % GC content (Equation 1, Marmur and Doty, 1962).

$$\%\text{GC} = 2.44 \, (T_m - 69.3) \qquad\qquad\qquad \text{(Equation E16.1)}$$

This has been derived by measuring the T_m for numerous DNA samples for which the base composition was known. Equation 1 is valid for GC percentages of 30 to 70%.

3. Purity of DNA. The greater the purity of a DNA sample, the greater the extent of hyperchromicity. Highly purified DNA shows a 40% increase in absorbance upon complete thermal denaturation. Although a quantitative estimate of purity is not possible, a hyperchromic effect of significantly less than 40% indicates relatively impure DNA. This could also mean that the DNA is already partially denatured or contaminated with UV absorbing materials. The shape of the temperature profile can also be used as an indication of purity. Impure or partially denatured DNA often yields a broad temperature profile, with the hyperchromic effect occurring over a wide temperature range.

Overview of the Experiment

The temperature profile for the DNA isolated in Experiment 15 will be constructed. Two experimental options are presented for the thermal denaturation study. Option 1 requires a spectrophotometer equipped with a constant temperature bath with a temperature range of 25° to 100°C. A high-boiling solvent such as ethylene glycol must be used as the circulating fluid. The quartz cuvet containing DNA in a standard saline solution is placed in the spectrophotometer cell holder and the temperature is increased. Measurements of A_{260} are made at various temperature intervals after equilibration. During the heating process, the volume of water increases, causing a concentration dilution of DNA. Table E16.1 gives a list of volume changes occurring with water over a temperature range of 25°C to 100°C. Absorption increases up to 40% are expected for these measurements.

Option 2 consists of heating several tubes of the same DNA solution to different temperatures in constant temperature baths. Each tube is rapidly cooled in an ice bath and the A_{260} is measured for each tube at 25°C. The latter method has a major disadvantage in that quick-cooling of denatured DNA leads to partial reassociation of the two strands. Hence, the absorbance is usually only 10 to 12% greater than that of native DNA.

The effect of the organic solvent dimethylformamide (DMF) on DNA structure will also be determined. DNA in standard saline solution will

TABLE E16.1

Volume of water relative to 25°C. From M. Mandel, *Methods in Enzymology*, L. Grossman and K. Moldave, Eds., Vol. XIIB (1968), p. 200. Academic Press (New York). Copyright 1968

T	V_T/V_{25}	T	V_T/V_{25}
25	1.0000	66	1.0174
26	1.0003	67	1.0180
27	1.0005	68	1.0185
28	1.0008	69	1.0191
29	1.0011	70	1.0197
30	1.0014		
31	1.0017	71	1.0203
32	1.0020	72	1.0209
33	1.0024	73	1.0215
34	1.0027	74	1.0221
35	1.0030	75	1.0228
36	1.0034	76	1.0234
37	1.0037	77	1.0240
38	1.0041	78	1.0247
39	1.0045	79	1.0253
40	1.0049	80	1.0260
41	1.0053	81	1.0266
42	1.0057	82	1.0273
43	1.0061	83	1.0280
44	1.0065	84	1.0287
45	1.0069	85	1.0293
46	1.0073	86	1.0300
47	1.0078	87	1.0308
48	1.0082	88	1.0314
49	1.0087	89	1.0321
50	1.0091	90	1.0329
51	1.0096	91	1.0336
52	1.0100	92	1.0343
53	1.0105	93	1.0351
54	1.0110	94	1.0358
55	1.0115	95	1.0365
56	1.0120	96	1.0373
57	1.0125	97	1.0380
58	1.0131	98	1.0388
59	1.0135	99	1.0396
60	1.0141	100	1.0404
61	1.0146	101	1.0411
62	1.0152	102	1.0419
63	1.0157	103	1.0426
64	1.0162	104	1.0433
65	1.0168	105	1.0441

be treated with DMF at room temperature and the hyperchromic effect measured.

Option 1 for thermal denaturation of DNA will require a 3- or 4-hour laboratory period. Thermal denaturation by option 2 and DMF denaturation can also be completed in a 3- or 4-hour laboratory period.

II. EXPERIMENTAL

Materials and Supplies

DNA, isolated in Experiment 15 or commercially obtained. The product from Experiment 15 is already dissolved in standard saline buffer. For Part A or B, a solution of commercial DNA is prepared in standard saline buffer at a concentration of 20 μg/ml, $A_{260} = 0.4$. For Part C, the solution should have an A_{260} of 0.8.

Standard saline citrate, 0.15 M NaCl, 0.015 M sodium citrate, pH 7.0

Spectrophotometer for UV measurements

Constant temperature circulating bath with ethylene glycol as fluid (Part A)

Quartz cuvet with glass covers

Several constant temperature baths set at 35°, 50°, 70°, 75°, 80°, 85° 90°, 95° and 100°C (Part B)

Saline citrate in dimethylformamide
(a) 0.15 M NaCl, 0.015 M sodium citrate, pH 7.0 in 20% DMF
(b) 0.15 M NaCl, 0.015 M sodium citrate, pH 7.0 in 40% DMF
(c) 0.15 M NaCl, 0.015 M sodium citrate, pH 7.0 in 60% DMF

Several 10 × 100 nm test tubes

Procedure

A. Measurement of T_m with Circulating Temperature Bath

Turn on the spectrophotometer and UV lamp for a 20-minute warm-up. Set the circulating bath for 25°C for the initial absorbance measurement. If isolated DNA from Experiment 15 is used, measure the A_{260} of the solution at 25°C. Dilute the solution with standard saline citrate to a final A_{260} of 0.4. Dilute just enough DNA solution for a 1 ml or 3 ml cuvet, whichever is available. Place a cuvet containing standard saline citrate buffer in the sample holder of the spectrophotometer. Ajust the absorbance at 260 nm to zero. Replace the buffer in the cuvet with the diluted DNA solution. Place the cuvet into the spectrophotometer and wait about three minutes for temperature equilibration. Record the A_{260} in your lab book. This measurement should be taken with special care, since all other readings will be related to this one at 25°C. Now raise the temperature of the bath to about 50°C. Remove the cuvet, invert and tap gently to remove air bubbles on the inside walls. Replace in the cell holder and continue heating to about 80°C. Allow 5 minutes for equilibration and take an absorbance reading. Now, raise the temperature in increments of 2°C, allow for temperature equilibration (about 5 minutes), and record the A_{260}. Repeat this process until a constant A_{260} is obtained. You should have noted a temperature range of 5 to 10°C where there was a large increase in A_{260}.

B. Measurement of T_m with Constant Temperature Baths.

A minimum of six constant temperature baths should be used. If only six are available, they should be set at the temperatures underlined in the Materials and Supplies section. Turn on the spectrophotometer and UV lamp for a 20-minute warm-up. Obtain test tubes (two more than the number of constant temperature baths). If you are using your isolated DNA, dilute it to an A_{260} of 0.4. If commercial DNA is used, the solution is already prepared at the proper concentration. Transfer 3.0 ml of the DNA solution into each of the test tubes. Place a marble over the top of each tube and allow the tubes to incubate for 15 minutes. Maintain one tube at room temperature, two in the 100°C bath, and one tube in each of the other constant temperature baths. After the incubation period, quick-cool all the tubes except one in the 100°C bath and the one at room temperature, by placing them in an ice water bath for ten minutes. One of the tubes in the 100°C bath is allowed to slowly cool to room temperature. Measure and record the A_{260} of each DNA solution. Allow the slow-cooled tube to remain at room temperature for about one hour before the A_{260} measurement.

C. Dimethylformamide as a Denaturant of DNA

CAUTION:
Dimethylformamide (DMF) is a toxic agent. Do not allow any solutions to come into contact with your skin. In the event of contact, wash immediately with soap and water. Do not mouth pipet any solutions.

The DNA solution for this part should have an initial A_{260} of approximately 0.8 to 0.9. Obtain four test tubes and pipet 2.0 ml of the DNA solution into each as outlined in Table E16.2. Add saline-DMF as described in Table E16.2. Allow the tubes to incubate at room temperature for 15 minutes. Measure and record the absorbancy of each mixture at **270** nm.

TABLE E16.2
Preparation for Part C, DMF as a Denaturant of DNA

| Reagent | Tube Number | | | |
	1	2	3	4
DNA ($A_{260} = 0.8 - 0.9$)	2.0	2.0	2.0	2.0
Saline-DMF, 0%	2.0	—	—	—
Saline-DMF, 20%	—	2.0	—	—
Saline-DMF, 40%	—	—	2.0	—
Saline-DMF, 60%	—	—	—	2.0

III. ANALYSIS OF RESULTS

A. Measurement of T_m with Circulating Temperature Bath

Correct the A_{260} measurements taken at each temperature by multiplying by the appropriate value of V_T/V_{25} from Table E16.1. This corrects each reading for dilution of the DNA caused by heat expansion of the water solvent. Calculate $A_{260(T)}/A_{260(25°)}$ for each temperature and plot the absorbance ratio against the temperature of the measurement (see Figure E15.1). Connect the points with a continuous line and determine the midpoint of the absorbance increase. The corresponding temperature is defined as the T_m.

What was the percentage increase in the overall hyperchromic effect? What does this indicate about the purity of the DNA sample you used?

Calculate the % GC content of your DNA sample, using Equation 1.

B. Measurement of T_m with Constant Temperature Baths

There is no need to correct the absorbance readings for volume expansion because all readings were taken at 25°C. Calculate the ratio $A_{260(T)}/A_{260(25°)}$ at each temperature. Plot the absorbance ratio versus the temperature. Do not include the tube that was allowed to cool slowly. Connect the points for a smooth temperature profile. Measure T_m by determining the temperature at the midpoint of the absorbance increase.

Compare the $A_{260(T)}/A_{260(25°)}$ for the two tubes in the 100°C bath. Why are the ratios not the same? Explain the experimental observation in terms of the molecular structure DNA.

What was the percentage increase in the overall hyperchromic effect. Give a qualitative answer regarding the purity of the DNA sample you used.

Calculate the % GC content of your DNA sample, using Equation 1.

C. Dimethylformamide as a Denaturant of DNA

Prepare a plot of A_{270} (y-axis) versus concentration of DMF in the buffer. Explain the shape of the curve. What is the effect of DMF on DNA structure?

IV. QUESTIONS

1. Predict and explain the action of each of the following conditions on the T_m of native DNA.

    (a) Measure T_m in pH 12 buffer.
    (b) Measure T_m in distilled water.
    (c) Measure T_m in 50% methanol-water.
    (d) Measure T_m in standard saline citrate solution containing sodium dodecyl sulfate.

2. A sample of highly purified DNA gave an A_{260} at 25°C of 0.45. The sample was divided into two samples, and each was treated as follows:

Sample I	Sample II
1. Heat to 100°C	1. Heat to 100°C
$A_{260(100°C)} = 0.65$	$A_{260(100°C)} = 0.65$
2. Quick-cool in ice water	2. Very slow cool to 25°C
$A_{260(25°C)} = 0.55$	$A_{260(25°C)} = 0.46$

Explain why different final A_{260} readings at 25°C are obtained.

3. A sample of DNA was chemically degraded and found to have an AT content of 60%. Estimate the T_m for the DNA sample.

4. A sample of purified DNA was incubated with deoxyribonuclease at 37°C. An aliquot was removed from the reaction mixture every minute for five minutes and the A_{260} recorded. The following data were obtained.

Time (min)	A_{260}
0	0.60
1	0.64
2	0.67
3	0.70
4	0.72
5	0.73

Describe the action of DNAase on DNA and explain the increase in A_{260}.

5. Why were absorbance measurements taken at 270 nm rather than 260 nm for Part C?

V. REFERENCES

General

R.C. Bohinski, *Modern Concepts in Biochemistry*, Fourth Edition (1983), Allyn and Bacon (Boston), pp. 186–187. A discussion on denaturation of DNA.

J.D. Rawn, *Biochemistry* (1983), Harper & Row (New York), pp. 333–355. An introduction to DNA structure and physical properties.

E. Smith, R. Hill, I. Lehman, R. Lefkowitz, P. Handler, and A. White, *Principles of Biochemistry: General Aspects*, Seventh Edition (1983), McGraw-Hill (New York), pp. 130–145. DNA structure and physical properties.

L. Stryer, *Biochemistry*, Second Edition (1981), Freeman (San Francisco), pp. 559–573. Physical and structural properties of DNA.

G. Zubay, *Biochemistry* (1983), Addison-Wesley (Reading, MA), pp. 665–685. An introduction to the structural properties of DNA.

Specific

M. Mandel and J. Marmur in *Methods in Enzymology*, L. Grossman and K. Moldave, Editors, Vol. XIIB, (1968), Academic Press (New York), pp. 195–206. "Use of Ultraviolet Absorbance-Temperature Profile for Determining the Guanine plus Cytosine Content of DNA."

Experiment **17**

Growth of Bacteria and Amplification of ColE1 Plasmids

RECOMMENDED READING:
Chapter 9.

SYNOPSIS
Extrachromosomal DNA molecules called plasmids are harbored in some strains of *E. coli*. The normal copy number of the plasmids is small, between 20 and 30; however, if these strains of *E. coli* are grown in the presence of chloramphenicol, up to 300 copies may be replicated per cell. ColE1 plasmids have been demonstrated to be useful vehicles in molecular cloning. This experiment describes a method for the growth of *E. coli* and amplification of the ColE1 plasmids.

I. INTRODUCTION

Bacterial plasmids are extrachromosomal DNA molecules that replicate autonomously. These molecules are double-stranded and closed-circular, and range in molecular weight from 2 to 20×10^6. The plasmids confer specific characteristics (phenotypes) to the host cell. Many of these characteristics represent a protective function such as resistance to antibiotics, production of toxins, or production or enzymes that degrade foreign organic molecules.

Plasmids as Cloning Vehicles

Recent advances in bacterial genetics have led to the use of plasmids as cloning vehicles. Natural plasmids can be isolated from bacterial cells, modified by insertion of a segment of foreign DNA, and transferred to a bacterial host. Inside the host cell, copies of the hybrid plasmid molecules are replicated. As discussed in Chapter 9, plasmid replication may be under stringent control (only a few copies are made) or under relaxed control (many copies are made). The plasmids with most value in molecular cloning experiments are those under relaxed control. Many copies of the hybrid DNA are, therefore, available for analysis, characterization, and transformation experiments. In addition, it has been demonstrated that the replication of relaxed plasmids continues after chromosomal DNA synthesis has been inhibited by the addition of chloramphenicol (Clewell, 1972). This discovery provided a major breakthrough in recombinant DNA research because it greatly improved techniques for the preparative-scale isolation and purification of plasmids. Under normal conditions, plasmids may replicate only 20 or 30 times. Growth of bacterial cells in the presence of chloramphenicol may increase the plasmid copy number up to a hundredfold. At these levels, the plasmid DNA comprises 30 to 45% of the total cellular DNA.

The first class of natural plasmids to be recognized as a potential cloning vehicle was ColE1 (Hershfield et al., 1974). These plasmids, which are present in only some *E. coli* strains, confer an ability to produce the antibiotic protein colicin E1, and immunity to the bacteriocidal action of colicin. Several properties of the ColE1 plasmids make them attractive as cloning vehicles:

1. They are relatively small, MW $= 4.2 \times 10^6$.

2. They are under relaxed control. Normally 20 to 30 copies are present in each cell. This number can be increased to about 3000 under appropriate growth conditions. These plasmids then make up about 45% of the total cellular DNA.

3. ColE1 plasmids have at least two identifiable markers, ability to produce colicin E1 and resistance to colicin E1.

4. They have a single restriction site for EcoRI and other restriction enzymes.

In vitro insertion of a foreign DNA fragment into the EcoRI cleavage site and incorporation into a host cell (transformation) leads to an immunity of the host cell toward colicin E1, but the cell is unable to produce colicin. ColE1 plasmids have been modified by addition or deletion of nucleotide bases. One especially useful ColE1 derivative is the widely used cloning vehicle pBR322. This plasmid has several desirable properties, including small size (MW $= 2.8 \times 10^6$), two markers for screening purposes (ampicillin and tetracycline resistance), several single restriction sites, and a restriction site (BamHI) that destroys tetracycline resistance.

Growth of Bacteria

Most plasmids currently used as cloning vehicles are of *E. coli* origin. Strains of *E. coli* are relatively easy to grow and maintain. The rate of growth is exponential and can be monitored by measuring the absorbance of a culture sample at 600 nm. One absorbance unit corresponds to a cell density of approximately 8×10^8 cells/ml.

Some *E. coli* strains that harbor the ColE1 plasmids are RR1, HB101, GM48, 294, SK1592, JC411Thy$^-$/ColE1, and CR34/ColE1. The typical procedure for growth and amplification of plasmids is, first, to establish the cells in normal media for several hours. An aliquot of this culture is then used to inoculate media containing the appropriate antibiotic. After overnight growth, a new portion of media containing the antibiotic is inoculated with an aliquot of the overnight culture. After the culture has been firmly established, a solution of chloramphenicol is added to inhibit chromosomal DNA synthesis. The ColE1 plasmids continue to replicate. The culture is then incubated for 12 to 18 hours, and harvested by centrifugation.

Overview of the Experiment

This experiment introduces students to some of the microbiological procedures required in molecular cloning. The procedures and equipment are standard for microbiology. All media and supplies, except antibiotics, must be autoclaved before use. Sterile transfers are required at all stages.

The extensive time requirements and the unusual schedule commitments may make this experiment cumbersome for students to complete. The total amount of actual working time is small, but several hours of incubation time are required for a suitable culture. Experiments 18, 19, and 20 (Isolation and Analysis of Plasmid DNA) work well with chloramphenicol amplified or unamplified cells; however it is recommended that amplified cultures be used.

The experiment may be carried out in one of the following ways:

1. Complete as a class project, preparing enough cell culture for all members to use in future experiments.
2. Complete in groups of 4 to 6 students. At least one student in each group should be experienced in standard microbiological procedures.
3. The instructor or teaching assistant can prepare the cultures for plasmid DNA isolation in Experiments 18 or 20.

II. EXPERIMENTAL

Materials and Supplies

Escherichia coli slant. The strain must contain the ColE1 plasmids. Recommended strains are RR1, HB101, GM48, 294, SK1591, JC411Thy⁻/ColE1, or CR34/ColE1

Luria-Bertani (LB) Medium. Each liter contains 10 g Bacto-tryptone, 5 g Bacto-yeast extract, and 10 g NaCl. Adjust to pH 7.5 with NaOH.

Ampicillin. A solution of the sodium salt is prepared in water (25 mg/ml) and sterilized by passage through a microfilter. Store the solution in a freezer.

Chloramphenicol, 25 mg/ml in 100% ethanol

Autoclave

Flask shaker in environmental room at 37°C

Spectrophotometer and cuvets for absorbance measurements at 600 nm

Standard microbiological equipment: sterile flasks, pipets, etc.

Centrifuge, capable of $4000 \times g$

Standard Tris buffer, 0.01 M Tris-HCl, 0.001 M EDTA, and 0.1 M NaCl, pH 7.8. Keep cold.

Procedure (Maniatis, Fritsch, and Sambrook, 1982)

CAUTION:
Experimentation with bacterial cells presents a potential biohazard. Do not pipet any solutions by mouth. Wash your hands well with hot water and soap before eating, drinking, or smoking.

1. Using sterile techniques, transfer the appropriate bacterial cells from a slant to a flask containing 10 ml of sterilized LB medium and 40 μl of the ampicillin solution. Incubate the culture with vigorous shaking at 37°C for 12 to 15 hours or overnight.

2. Prepare a 125 ml Erlenmeyer flask for transfer by adding 25 ml of sterile LB medium and 0.10 ml of the ampicillin solution. Transfer 0.1 ml of the overnight culture (from step 1) to the flask. Incubate at 37°C with vigorous shaking. At various time periods, remove 1 ml aliquots of the culture, transfer to a cuvet, and determine the cell density by measuring the absorbance at 600 nm.

3. When A_{600} reaches 0.6, transfer the culture from step 2 to a flask with 500 ml of sterile LB medium containing 2 ml of the ampicillin solution. Incubate at 37°C for $2\frac{1}{2}$ hours with vigorous shaking.

4. When A_{600} reaches about 0.4, add 4.0 ml of the chloramphenicol/ethanol solution to the growing culture.

5. Finally, incubate, with shaking, at 37°C for 12 to 16 hours.

6. Isolate the bacterial cells by centrifugation (4000 × g, 10 min, 4°C).

7. Pour off and discard the supernatant. Wash the cells by adding 100 ml of cold Tris/NaCl/EDTA solution. Mix gently, but well, and centrifuge again as in step 6.

8. Pour off the supernatant and, if necessary, store the cells in a freezer until Experiment 18 or 20 is started.

III. ANALYSIS OF RESULTS

Prepare a flowchart showing each step carried out in the growth of the bacteria. Construct a graph of A_{600} vs. time, and plot on it the data obtained from step 2.

IV. QUESTIONS

1. Explain the action of ampicillin as an inducer of plasmid replication.

2. How does chloramphenicol inhibit protein synthesis?

V. REFERENCES

General

Reference books on microbiology/microbiological techniques may be consulted for further information.

Specific

D. Clewell, *J. Bacteriol.*, *110*, 667–676 (1972). "Nature of ColE1 Plasmid Replication in *Escherichia coli* in the Presence of Chloramphenicol."

V. Hershfield, H. Boyer, C. Yanofsky, M. Lovett, and D. Helinski, *Proc. Natl. Acad. Sci.*, *71*, 3455 (1974). "Plasmid ColE1 as a Molecular Vehicle for Cloning and Amplification of DNA."

T. Maniatis, E. Fritsch, and J. Sambrook, *Molecular Cloning, A Laboratory Manual* (1982), Cold Spring Harbor Laboratory (Cold Spring Harbor, NY), pp. 88–89. "Growth of Bacteria and Amplification of the Plasmid."

Preparative-Scale Isolation and Purification of ColE1 Plasmids

RECOMMENDED READING:
Chapter 7, Sections A, B, C; Chapter 9; Introduction to Experiments 15 and 17.

SYNOPSIS

Plasmid DNA, which is a useful cloning vehicle, is present in some special strains of bacterial cells. It can be separated from chromosomal DNA by exposing a cell lysate to conditions that irreversibly denature the chromosomal DNA, but leave the plasmid DNA intact. In this experiment, ColE1 plasmids will be isolated from *E. coli* cells and purified by density gradient ultracentrifugation.

I. INTRODUCTION

Isolation of Plasmid DNA

The presence of ColE1 and other plasmids in a bacterial cell may be confirmed by genetic screening of antibiotic resistance. However, it is sometimes necessary to isolate plasmid DNA for further characterization and manipulation. Isolation and purification of plasmids is usually carried out for one of the following reasons:

1. construction of new recombinant plasmids,
2. analysis of molecular weight by agarose gel electrophoresis,
3. electrophoretic analyses of restriction enzyme digests,
4. construction of a restriction enzyme map,
5. sequence analysis of nucleotides by the Sanger or Maxam-Gilbert methods, or
6. hybridization experiments.

Several methods for the isolation of plasmid DNA have been developed; some lead to a more highly purified product than others. Surprisingly, purity of the plasmid is not always necessary for the purposes listed above. For instance, molecular cloning of a hybrid plasmid does not require a highly purified vehicle. If nucleotide sequence, restriction mapping, or hybridization studies are contemplated, however, a more purified plasmid is desirable.

All isolation methods have the same objective, separation of plasmid DNA from chromosomal DNA. Plasmid DNA has two major structural differences from chromosomal DNA:

1. Plasmid DNA is almost always extracted in a covalently closed circular configuration, whereas isolated chromosomal DNA usually consists of sheared linear fragments.
2. ColE1 plasmids and other potential vehicles are much smaller than chromosomal DNA.

The structural differences cause physicochemical differences that can be exploited to separate the two types of molecules. Methods for the isolation of plasmid DNA fall into three major categories:

1. Those methods that rely on specific interaction between plasmid DNA and a solid support. Examples are adsorption to nitrocellulose microfilters and hydroxyapatite columns.
2. Those methods that cause selective precipitation of chromosomal DNA by various agents. This method exploits the relative resistance of covalently closed circular DNA to extremes of pH, temperature, or other denaturing agents.
3. Those methods based on differences in sedimentation behavior when the two types of DNA are centrifuged in the presence of an intercalating dye. This is the method of choice if highly purified plasmid DNA is required.

Two isolation procedures will be described in this experiment. One procedure (from category 2) yields plasmid DNA that is sufficiently pure for digestion by restriction enzymes, analysis by agarose electrophoresis, and construction of new hybrid plasmid. A method of more substantial purification by ultracentrifugation will also be outlined.

Separation of Plasmid DNA by Boiling (Holmes and Quigley, 1981)

The total cellular DNA must first be released by lysis of the bacterial cells. This is brought about by incubation with lysozyme in the presence of reagents that inhibit nucleases (see Experiment 15). Chromosomal DNA is then separated from the plasmid DNA by boiling the lysis mixture for a brief period, followed by centrifugation. In contrast to closed circular plasmid DNA, linear chromosomal DNA becomes irreversibly denatured by heating and forms an insoluble gel, which sediments during centrifugation. Even though plasmid DNA may become partially denatured during boiling, the closed circular double helix reforms upon cooling. The boiling serves a second purpose, that of denaturing deoxyribonucleases and other proteins.

Purification of Plasmid DNA by Ultracentrifugation (Radloff, Bauer, and Vinograd, 1967)

The plasmids obtained as described above are sufficiently pure for some studies; however, they are contaminated by RNA and a small amount of linear DNA. The RNA may be removed by treatment of the lysate with ribonuclease, but ultracentrifugation is more often the method of choice.

Ethidium bromide and other intercalating dyes bind to double-stranded nucleic acids, causing partial unwinding of the helix and other topologic changes. Closed circular DNA binds less dye than linear DNA, which leads to a difference in the buoyant density of plasmid DNA:dye versus linear DNA:dye. The density of the DNA:dye complex is related inversely to the amount of bound dye; therefore, the plasmid DNA:dye complex sediments to a region of higher density on a cesium chloride density gradient. Figure E18.1 shows the results of ultracentrifugation of a cleared lysate. Two red bands are observed; the upper corresponds to linear DNA, the lower to plasmid DNA. Ribonucleic acids sediment as a pellet in the tube, and any contaminating protein floats to the top of the gradient. The two DNA bands may be isolated by fractional collection.

The ethidium bromide dye must be removed from the plasmid solution before restriction enzyme digestion. This may be accomplished by an ion-exchange column or extraction with an organic solvent.

If an ultracentrifuge is not available and highly purified plasmid DNA is desired, contaminating RNA may be removed by treatment of the lysate with ribonuclease (Procedure, Part C; Birnboim, 1983).

FIGURE E18.1
Ultracentrifugation of plasmid DNA on a cesium chloride gradient.

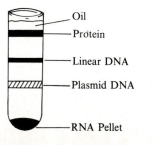

Oil

Protein

Linear DNA

Plasmid DNA

RNA Pellet

Overview of the Experiment

The isolation of plasmid DNA by the boiling method (Part A) yields a product of sufficient purity for analysis in Experiment 19. The procedure works well with chloramphenicol-amplified or untreated cell cultures. If cell cultures are available, the isolation method can easily be completed in about 2 hours.

If an ultracentrifuge is available, it is recommended that students further purify the plasmid preparation on a cesium chloride gradient (Part B). Only about 1 hour is necessary to prepare the sample for centrifugation; however, 36 to 40 hours of ultracentrifugation are required.

Part C, digestion of RNA by ribonuclease, can be done directly on the cleared lysate from Part A. Part C can be completed in approximately 1 hour.

II. EXPERIMENTAL

Materials and Supplies

A. Isolation of Plasmid DNA
   E. coli cells, 100 ml of culture grown as described in Experiment 17
   Standard Tris buffer, 0.01 M Tris-HCl, containing 0.1 M NaCl and 0.1 mM EDTA, pH 8.0. Keep cold.
   Lysozyme solution, 10 mg/ml in 0.01 M Tris-HCl, pH 8.0. Must be freshly prepared.
   Bunsen burner
   Boiling water bath in 1-liter beaker
   Centrifuge, capable of $12,000 \times g$
B. Ultracentrifugation of the Plasmid Lysate on a CsCl Gradient
   Cleared lysate from Part A
   Solid cesium chloride
   Tris-EDTA buffer, 0.01 M Tris-HCl containing 1 mM EDTA, pH 8.0
   Ethidium bromide, 10 mg/ml in H_2O
   Light mineral oil
   Ultracentrifuge, rotor and appropriate tubes. The recommended rotors are the SW 40, 50, or 65 for the Beckman Spinco Model L instrument or the TV865 for the Sorval instrument.
   Hypodermic needle, 18 or 21 gauge
   Isoamyl alcohol, saturated with water and solid CsCl
   Dialysis tubing, 1 cm × 20 cm
C. Digestion of RNA by Ribonuclease
   Cleared lysate from Part A
   Ribonuclease A, 1 mg/ml in 5 mM Tris-HCl, pH 8.0. This must be heated in an 80°C bath for 10 minutes to denature deoxyribonucleases.

Ribonuclease T1, 500 units/ml in 0.05 M Tris-HCl, pH 8.0. This must be heated as described for ribonuclease A.

Sodium dodecyl sulfate, 10% in water

Acetate-MOPS buffer, 0.1 M Na acetate, 0.05 M MOPS, pH 8.0

Isopropanol

Ethanol

Tris-EDTA, 0.01 M Tris-HCl, 0.01 M EDTA, pH 7.5

Procedure

A. Isolation of Plasmid DNA

The procedure outlined here can be used for bacterial culture ranging from 1 ml to 1 liter. Instructions will be given for 100 ml of cell culture. If only very small cultures are available, follow the procedure in Experiment 20.

1. Obtain 100 ml of cell culture and centrifuge at $4000 \times g$ for 10 minutes at 4°C. Discard the supernatant.

2. Wash the cells by adding 100 ml of standard Tris buffer containing NaCl and EDTA. Centrifuge as in step 1 and discard the supernatant.

3. Suspend the pellet in 15 ml of standard Tris buffer and transfer to a 50 ml Erlenmeyer flask.

4. Add 2.0 ml of lysozyme solution.

5. With a clamp, hold the flask over a Bunsen burner flame until the solution begins to boil.

6. Incubate the flask in a boiling water bath for 40 seconds.

7. Cool the flask in ice water.

8. Transfer the lysate to a centrifuge tube and centrifuge at $12,000 \times g$ for 20 minutes at 4°C.

9. Remove the cleared lysate (supernatant) and transfer it to a plastic test tube for storage. The present lysate containing plasmids may be analyzed directly (Experiment 19) without further purification. Further purification of the lysate by ultracentrifugation is described in Part B.

B. Ultracentrifugation of the Plasmid on a CsCl Gradient

CAUTION:

Ethidium bromide is a potential carcinogen. Always wear gloves when handling solutions or reaction mixtures containing the substance.

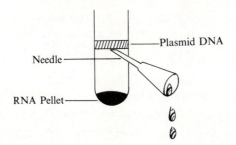

FIGURE E18.2
Collection of the plasmid fraction after centrifugation. The plastic tube is punctured with a needle and the plasmid is collected in a tube.

1. Measure the volume of the cleared lysate from Part A.

2. Add 1 g of solid CsCl for each milliliter of lysate.

3. Add 0.6 ml of ethidium bromide for each ml of CsCl-lysate solution.

4. Transfer the mixture to an ultracentrifuge tube, fill the remaining space in the tube with light mineral oil, and centrifuge at 40,000 rpm for 40 hours. Since several ultracentrifuge models are available, specific instructions will not be outlined here. Have your instructor assist in the operation of the instrument.

5. After the centrifugation, recover the purified plasmid DNA by piercing the side of the tube with an 18- or 21-gauge needle. Insert the needle at a location just below the lower red band and collect the plasmid fraction as shown in Figure E18.2.

6. Remove the ethidium bromide by extracting with isoamyl alcohol. Mix the plasmid fraction with an equal volume of isoamyl alcohol that has been equilibrated with water and solid cesium chloride. The two phases may be readily separated by centrifugation at $2000 \times g$ for 5 minutes.

7. Transfer the aqueous phase containing the plasmid (lower phase) to a new test tube and repeat the extraction process at least three times with fresh isoamyl alcohol. There should no longer be a pink color associated with the aqueous layer containing the plasmid.

8. Finally, transfer the aqueous phase into dialysis tubing and dialyze against standard Tris-HCl/EDTA buffer, pH 8.0. Dialyze at least 12 to 15 hours, with buffer changes every 4 to 5 hours. This step removes CsCl from the plasmid preparation.

9. Transfer the dialyzed extract to a plastic tube and store for Experiment 19.

C. Digestion of RNA by Ribonuclease (Birnboim, 1983)

1. Treat the cleared lysate from Part A as follows. Add 0.2 ml of ribonuclease A and 0.2 ml of ribonuclease T1.

2. Incubate the reaction mixture at 37°C for 15 minutes.

3. Add 0.04 ml of 10% SDS and 2 ml of acetate-MOPS buffer, pH 8.0.

4. Precipitate the plasmid DNA by dropwise addition of 4 ml of isopropanol. Mix well while adding the isopropanol.

5. Allow the solution to sit at room temperature for 15 minutes, and then centrifuge at 6000 rpm for 15 minutes.

6. Dissolve the pellet (plasmid DNA) in 2 ml acetate-MOPS buffer and precipitate the plasmid DNA with two volumes of ethanol.

7. Centrifuge as in step 5 and dissolve the pellet in 2 ml of Tris-HCl/EDTA buffer, pH 7.5.

III. ANALYSIS OF RESULTS

A. Isolation of Plasmid DNA

Prepare a flowchart outlining each isolation step. Write a brief explanation of the purpose of each reagent and procedural step.

B. Ultracentrifugation of the Plasmid on a CsCl Gradient

Draw a picture of the centrifuge tube after centrifugation. Identify the components present in each region of the tube.

IV. QUESTIONS

1. Compare the isolation procedure used in this experiment to that used for total cellular DNA in Experiment 15.

2. Why are the ribonuclease solutions heated before use in this experiment?

V. REFERENCES

General

T. Maniatis, E. Fritsch, and J. Sambrook, *Molecular Cloning, A Laboratory Manual*, (1982), Cold Spring Harbor Laboratory (Cold Spring Harbor, NY), pp. 86–94. An excellent reference book for use in the laboratory.

Specific

H. Birnboim in *Methods in Enzymology*, R. Wu, L. Grossman, and K. Moldave, Editors, Vol. 100 (1983), Academic Press (New York), pp. 243–255. "A Rapid Alkaline Extraction Method for the Isolation of Plasmid DNA."

D. Holmes and M. Quigley, *Anal. Biochem., 114*, 193–197 (1981). "A Rapid Boiling Method for the Preparation of Bacterial Plasmids."

R. Radloff, W. Bauer, and J. Vinograd, *Proc. Natl. Acad. Sci. USA, 57*, 1514–1521 (1967). "A Dye-Buoyant-Density Method for the Detection and Isolation of Closed Circular Duplex DNA."

Experiment **19**

The Action of Restriction Enzymes on a Bacterial Plasmid or Viral DNA

RECOMMENDED READING:
Chapter 4, Section B; Chapter 9; Introduction to Experiments 17 and 18.

SYNOPSIS
Restriction endonucleases catalyze the cleavage of specific phosphodiester bonds in double-stranded DNA. These enzymes are often used to linearize a plasmid for hybrid DNA construction. They have also found use in the analysis of DNA. Fragments obtained by restriction enzyme action may be separated and sized by agarose gel electrophoresis. In this experiment, students will incubate various restriction enzymes with plasmid or viral DNA and analyze the products by electrophoresis.

I. INTRODUCTION

Experiments 17 and 18 introduced students to some principles and techniques involved in recombinant DNA research. Specifically, the experiments outlined the replication, isolation, and purification of bacterial plasmid vehicles for molecular cloning experiments. This experiment describes two more tools that are essential in hybrid plasmid construction and analysis—restriction enzyme action and agarose gel electrophoresis. The procedures introduced here have also found widespread use in the analysis and characterization of all DNA molecules.

The Action of Restriction Endonucleases

Bacterial cells produce many enzymes that act to degrade various forms of DNA. Of special interest are the **restriction endonucleases**, enzymes that recognize specific base sequences in double-stranded DNA and catalyze cleavage of the two strands in or near that specific region. The biological function of these enzymes is to degrade or restrict foreign DNA molecules. The action of an restriction enzyme, EcoRI, is shown in Reaction 1.

$$5' \ldots G \overset{\downarrow}{-} G-A-T-C-C \ldots 3' \xrightarrow[\text{H}_2\text{O}]{\text{EcoRI}} \quad 5' \ldots G$$
$$3' \ldots C-C-T-A-G \underset{\uparrow}{-} G \ldots 5' \qquad\qquad 3' \ldots C-C-T-A-G$$

$$+ \qquad \begin{array}{l} G-A-T-C-C \ldots 3' \\ G \ldots 5' \end{array} \qquad \text{(Reaction E19.1)}$$

The site of action of EcoRI is a specific hexanucleotide sequence. Two phosphodiester bonds are hydrolyzed (see arrows), resulting in fragmentation of both strands. Note the twofold rotational symmetry feature at the recognition site and the formation of cohesive ends (see Chapter 9). The weak base pairing between the cohesive ends is not sufficient to hold the two fragments together.

More than one hundred restriction enzymes have been isolated and characterized. Nomenclature for the enzymes consists of a three-letter abbreviation representing the source (Eco = *E. coli*), the strain (R), and a Roman numeral designating the order of discovery. EcoRI is the first to be isolated from *E. coli* (strain R) and characterized. Table E19.1 lists several other restriction enzymes and their recognition sequence.

Restriction enzymes are used extensively in nucleic acid chemistry. They may be used to cleave large DNA molecules into smaller fragments that are more amenable to analysis. For example, λ phage DNA, a linear, double-stranded molecule of 50,000 base pairs (MW $= 31 \times 10^6$), is cleaved into six fragments by EcoRI, or into more than 50 fragments by HinfI (*Haemophilus influenzae*, serotype f). The base sequence recognized by a restriction enzyme is likely to occur only a very few times in any particular DNA molecule; therefore, the smaller the DNA molecule, the fewer the number of specific sites. The λ phage DNA is cleaved into 0 to 50 or more fragments depending on the restriction enzyme used, whereas larger bacterial or animal DNA will, most likely, have many recognition sites and be cleaved into hundreds of fragments. Smaller DNA molecules, therefore, have a much greater chance of producing a unique set of fragments with a particular restriction enzyme. It is unlikely that this set of

TABLE E19.1
Specificity and Optimal Conditions for Several Restriction Endonucleases.

Name	Recognition Sequence 5′.......3′	T	pH	[Tris] mM	[NaCl] mM	[MgCl$_2$] mM	[ME] mM
AluI	A—G↓C—T	37°	7.5	10	50	10	10
BalI	T—G—G↓C—C—A	37°	7.9	6	—	6	6
BamHI	G↓G—A—T—C—C	37°	8.0	20	100	0.7	1
BclI	T↓G—A—T—C—A	60°	7.5	10	50	10	1
EcoRI	G↓A—A—T—T—C	37°	7.5	10	100	10	1
HaeII	Pu—G—C—G—C↓Py*	37°	7.5	10	50	10	10
HindIII	A↓A—G—C—T—T	37–55°	7.5	10	60	10	1
HpaI	G—T—T↓A—A—C	37°	7.5	10	50	10	1
SalI	G↓T—C—G—A—C	37°	8.0	10	150	10	1
TaqI	T↓C—G—A	65°	8.4	10	100	10	10

* Pu = a purine base

Py = a pyrimidine base

fragments will be the same for any two different DNA molecules, so the fragmentation pattern is unique and can be considered a "fingerprint" of the DNA substrate. The fragments are readily separated and sized by agarose gel electrophoresis as described later in this experiment.

Restriction endonucleases are also valuable tools in the construction of hybrid DNA molecules. Several restriction enzymes act at a single site on bacterial plasmid vehicles. This linearizes the circular plasmid and allows for the insertion of a foreign DNA fragment. For example, the popular plasmid vehicle pBR322 has a single restriction site for BamHI (*Bacillus amyloliquefaciens*, H) that is within the tetracycline resistance gene. Not only does the enzyme open the plasmid for insertion of a DNA fragment, but it also destroys a phenotype; this fact aids in the selection of transformed bacteria (see Chapter 9).

Practical Aspects of Restriction Enzyme Use

Restriction enzymes are heat-labile and expensive biochemical reagents. Their use requires considerable planning and care. Each restriction nuclease has been examined for optimal reaction conditions in regard to specific pH range, buffer composition, and incubation temperature. This information for each enzyme is readily available from the commercial supplier of the enzyme or from the literature (Fuchs and Blakesley, 1981, or Table E19.1). The temperature range and pH optima for most restriction nucleases are similar (37°C, 7.5–8.0); however, optimal buffer composition is variable. Typical buffer components are Tris, NaCl, $MgCl_2$, and a sulfhydryl reagent (β-mercaptoethanol or dithiothreitol). Proper reaction conditions are crucial for optimal reaction rate; but, more important, changing reaction conditions has been shown to alter the specificity of some restriction enzymes.

Although restriction enzymes are very unstable reagents, they can be stored at $-20°C$ in buffer containing 50% glycerol. They are usually prepared in an appropriate buffer and shipped in packages containing dry ice. Because of cost factors, digestion by restriction enzymes is carried out on a microscale level. A typical reaction mixture will contain about 1 μg or less of DNA and 1 unit of enzyme in the appropriate incubation buffer. One unit is the amount of enzyme that will degrade 1 μg of λ phage DNA in 1 hour at the optimal temperature and pH. The total reaction volume is usually between 20 and 50 μl. Incubation is most often carried out at the recommended temperature for about 1 hour. The reaction is stopped by adding EDTA solution, which complexes divalent metal ions essential for nuclease activity.

Reaction mixtures from restriction enzyme digestion may be analyzed directly by agarose gel electrophoresis. This technique combines high resolving power and sensitive detection to allow for the analysis of minute amounts of DNA fragments.

Characterization of DNA by Agarose Gel Electrophoresis

Agarose gels are the standard media used to detect, separate, and characterize small DNA molecules. If a low concentration of agarose is used (0.3%), DNA molecules up to 150×10^6 in molecular weight can be separated by electrophoresis. It is more common, however, to use higher concentrations of agarose (0.5 to 2%), which allow for analysis of DNA fragments in the range of 1 to 90×10^6.

As previously discussed in Chapter 4 the mobility of nucleic acids in agarose gels is influenced by the **agarose concentration**, the **molecular size of the DNA**, and the **molecular shape of the DNA**. In general, the lower the agarose concentration in the gels, the larger the DNA that can be analyzed. However, this reaches a practical limit because agarose concentrations less than 0.2 to 0.3% form gels that are too fragile for ordinary use. For most routine analyses of restriction fragments, agarose concentrations in the range of 0.7 to 1% are most appropriate. Nucleic acids migrate in an agarose medium at a rate that is inversely proportional to their molecular weights. In fact, a linear relationship exists between mobility and the log of the molecular weight of a DNA fragment. A standard curve is readily prepared by analyzing a restriction enzyme digest containing DNA fragments of known molecular weight (Figure E19.1). A typical standard mixture is the EcoRI digestion of λ phage DNA. This mixture contains fragments ranging from about 2×10^6 to 14×10^6 in molecular weight.

The influence of molecular shape on electrophoretic mobility is not a critical factor to consider in the experiment. The action of restriction enzymes on both covalent closed circular DNA and linear DNA results in the formation of linear fragments.

Agarose gel electrophoresis is an ideal technique for analysis of DNA fragments. In addition to the positive characteristics discussed, the technique is simple, rapid, and relatively inexpensive. Fragments that differ in

FIGURE E19.1

Standard curve of λ DNA fragments from EcoRI cleavage. The electrotrophoresis conditions were: 0.7% agarose gel, Tris-borate buffer (pH 8.2), and 70 V for 6 hours.

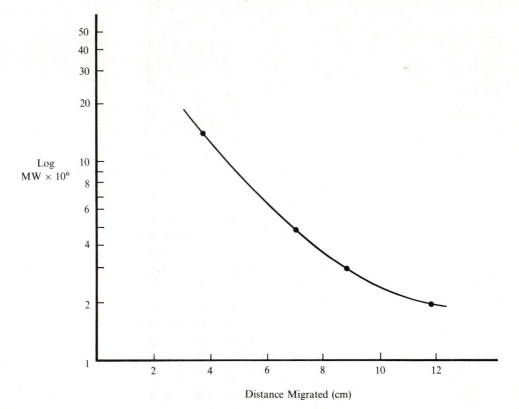

molecular weight by as little as 1% can be resolved on agarose gels, and as little as 1 ng of DNA can be detected on a gel. Nucleic acids are visualized after electrophoresis by soaking the gel in ethidium bromide, or by performing the electrophoresis with the dye incorporated in the gel and buffer.

Overview of the Experiment

The objective of the experiment is to evaluate the action of restriction enzymes on bacterial plasmids, λ phage DNA, or viral DNA. The DNA will be incubated under the appropriate conditions with various restriction enzymes. The reaction mixtures will be subjected to agarose gel electrophoresis in order to determine the number and molecular size of the restriction fragments.

The following options are available for this experiment:

1. Students may use isolated and purified ColE1 plasmids from Experiments 17 and 18. Alternatively, several purified plasmids, including ColE1, pBR322, and pUB110, are available from commercial sources. The plasmids may be cleaved with various restriction enzymes and the products analyzed by agarose gel electrophoresis.

2. Students may characterize λ phage of Adenovirus 2 DNA. Both are standard substrates for assays of restriction endonucleases.

It may be more convenient to supply students with DNA rather than have them isolate and purify plasmids as described in Experiments 17 and 18.

Approximately 2 hours are required to prepare and incubate the restriction enzyme reaction mixtures. The agarose slab gel may be prepared during the incubation period. Up to 1 to 2 hours are required to prepare the gel slab. The electrophoresis may be completed as a group project. Most electrophoresis chambers will accommodate slab gels with 15 to 20 sample wells.

Relatively inexpensive slab gel electrophoresis chambers are available commercially or they may be constructed from readily available materials. References for construction are given in Chapter 4.

II. EXPERIMENTAL

Materials and Supplies

Restriction Enzyme Digestion
    DNA: plasmid, λ phage or Adenovirus 2, 0.5 mg/ml in Tris-HCl/ NaCl/EDTA, pH 7.0.

Restriction enzymes, 1 unit/μl. Keep stored in freezer until ready to use. Recommended enzymes are BamHI and EcoRI for λ phage DNA, EcoRI and TaqI for plasmids (pBR322 or ColE1), and HpaI for Ad 2 DNA.

Sterilized H_2O

Incubation buffers

BamHI: 0.2 M Tris-HCl, pH 8.0, 0.07 M $MgCl_2$, 1 M NaCl, 0.01 M mercaptoethanol

EcoRI: 0.1 M Tris-HCl, pH 7.5, 0.1 M $MgCl_2$, 1 M NaCl, 0.01 M mercaptoethanol

HpaI: 0.1 M Tris-HCl, pH 7.5, 0.1 M $MgCl_2$, 0.5 M NaCl, 0.01 M mercaptoethanol

Quench buffer, 0.1 M EDTA, pH 7

Gel-loading buffer, Tris-acetate containing 50% glycerol and 0.25% bromphenol blue

Agarose Gel Electrophoresis

Agarose, electrophoresis grade

Agarose slurry buffer, 0.04 M Tris-acetate, 0.002 M EDTA, pH 7.8

Ethidium bromide solution, 10 mg/ml in H_2O

Electrophoresis buffer, 0.04 M Tris-acetate, 0.002 M EDTA, pH 7.8, containing 0.5 μg/ml ethidium bromide

Incubation baths, one at 37°C and one at 65°C

Agarose gel electrophoresis equipment

UV light

Standard restriction enzyme digest, EcoRI of λ phage DNA

Procedure

Restriction Enzyme Digestion (Sharp, Sugden, and Sambrook, 1973)

BamHI

1. Obtain a small test tube for each reaction mixture and add the following components:

    15 μl of sterile H_2O

    2 μl of the DNA solution

    2 μl of BamHI incubation buffer; mix well by tapping.

    1 unit of BamHI restriction enzyme. The actual volume depends upon the concentration of enzyme. In most cases this will be 1 μl of solution. Mix the tube by gentle tapping.

2. Incubate the reaction mixture at 37°C for 1 hour.

3. Stop the reaction by adding 2 μl of EDTA quench buffer.

4. Save the reaction mixture for agarose gel electrophoresis.

EcoRI

1. Obtain a small test tube for each reaction mixture and add the following components:

    15 μl of sterile H_2O

    2 μl of DNA solution

    2 μl of EcoRI incubation buffer; mix well by tapping.

    Add 1 unit of EcoRI restriction enzyme. The volume depends upon the concentration of enzyme. For most enzymes this will be 1 μl of solution. Mix the tube with gentle tapping.

2. Incubate the reaction mixture at 37°C for 1 hour.

3. Add 2 μl of EDTA quench buffer.

4. Save the reaction mixture for agarose gel electrophoresis.

HpaI

1. Obtain a small test tube for each reaction mixture and add the following components:

    15 μl of sterile H_2O

    2 μl of the DNA solution

    2 μl of HpaI incubation buffer; mix well by tapping.

    1 unit of HpaI restriction enzyme. In most cases this will be 1 μl. Mix well by gentle tapping.

2. Incubate the reaction mixture at 37°C for 1 hour.

3. Add 2 μl of EDTA quench buffer.

4. Save the reaction mixture for agarose gel electrophoresis.

Agarose Gel Electrophoresis

Many varieties of electrophoresis chambers are available. Specific details for the use of the electrophoresis unit should be obtained from the accompanying instructions or from your instructor.

CAUTION:

Do not touch the electrophoresis chamber or wires while the electrophoretic operation is in progress. Voltages may reach as high as 100 volts and shocks may be fatal.

Ethidium bromide is a mutagen. Always wear gloves when handling solutions or agarose gels containing the compound.

Avoid looking into UV light during the detection of DNA fragments on the agarose gel.

---

1. Prepare a 1% (w/v) slurry of agarose in the agarose slurry buffer. The actual amount of solution to use depends on the size of the slab gel to be prepared. Follow the instructions given by your instructor.

2. Heat the slurry in a boiling water bath until the agarose dissolves.

3. After the agarose solution has cooled to 50°C, add ethidium bromide solution to give a final dye concentration of 0.5 μg/ml.

4. Quickly pour the agarose solution over the glass plate to a depth of 2 to 4 mm. Insert the comb to make the sample wells.

5. Allow the gel to set for approximately 30 minutes and carefully remove the comb.

6. Clamp the agarose slab into the electrophoresis chamber and allow it to set for another 30 minutes.

7. Load one sample into each of the sample wells in the following manner. Mix each 20 μl of reaction mixture from the restriction enzyme with 10 μl of gel-loading buffer and apply one sample to each well. Put EcoRI standard digest into one sample well.

8. Add electrophoresis buffer to the reservoirs and connect the power supply.

9. Carry out the electrophoresis at approximately 3 V/cm or about 50 to 70 volts total.

10. Turn off the power supply when the tracking dye has migrated to the edge of the gel.

11. Remove the gel from the chamber and glass plate, and examine it under a UV light. DNA fragments will appear as red-orange fluorescent bands.

12. Draw a picture of the slab gel, showing the position of each DNA band.

III. ANALYSIS OF RESULTS

Study the picture of the slab gel after electrophoresis. Measure the distance migrated by each DNA band. Use the EcoRI standard digest of λ phage DNA to prepare a curve for molecular weight determination. Plot log molecular weight vs. the relative migration. The fragments from EcoRI cleavage of λ DNA are 13.7×10^6, 4.68×10^6, 3.7×10^6, 3.56×10^6, 3.03×10^6, and 2.09×10^6 in molecular weight. Use the standard curve to estimate the molecular weight of restriction enzyme fragments in each digest. Compare with Figure E19.1.

IV. QUESTIONS

1. How does the EDTA quench buffer stop the restriction enzyme reaction?

V. REFERENCES

General

See the list of references at the end of Chapter 9.

R. Fuchs and R. Blakesley in *Methods in Enzymology*, R. Wu, L. Grossman, and K. Moldave, Editors, Vol. 100 (1983), Academic Press (New York), pp. 3–38. "Guide to the Use of Type II Restriction Endonucleases."

T. Maniatis, E. Fritsch, and J. Sambrook, *Molecular Cloning, A Laboratory Manual* (1982), Cold Spring Harbor Laboratory (Cold Spring Harbor, NY), pp. 98–106, 150–153. Details are given for the use of restriction enzymes and agarose gel electrophoresis.

R. Roberts in *Methods in Enzymology*, R. Wu, Editor, Vol. 68 (1979), Academic Press (New York), pp. 27–41. "Directory of Restriction Endonucleases."

H. Smith, *Science, 205*, 455–462 (1979). "Nucleotide Sequence Specificity of Restriction Endonucleases."

Specific

R. Fuchs and R. Blakesley (see General References).

S. Sharp, B. Sugden, and J. Sambrook, *Biochemistry, 12*, 3055–3063 (1973). "Detection of Two Restriction Endonuclease Activities in *Haemophilus parainfluenzae* Using Analytical Agarose-Ethidium Bromide Electrophoresis."

Experiment **20**

Rapid, Microscale Isolation and Electrophoretic Analysis of Plasmid DNA

RECOMMENDED READING:
Chapter 4; Chapter 9; Introduction to Experiments 17, 18, and 19

SYNOPSIS
This experiment describes a simple and rapid method for evaluation of plasmid DNA. Students are introduced to several procedures that are routine in recombinant DNA research. They will isolate plasmid DNA, digest it with a restriction endonuclease, and analyze the fragments by agarose gel electrophoresis.

I. INTRODUCTION

Microscale Isolation of Plasmids

Often it is necessary to detect and analyze plasmid DNA in a large number of small bacterial samples. Rapid screening of single bacterial colonies or small cultures may have several objectives:

1. detect the presence of plasmids,
2. estimate the size of plasmids,
3. analyze plasmids for restriction enzyme cleavage,
4. perform restriction mapping of plasmids,
5. analyze nucleotide sequence of plasmids,
6. determine optimal conditions for the preparation of hybrid DNA,
7. search for recombinant DNA inserts.

The procedures of plasmid isolation, purification, and analysis outlined in Experiments 17, 18 and 19 are suitable for small and large samples (50 ml to 5 liter cultures), but the methods are time-consuming and expensive, require costly instrumentation, and are not applicable for surveying many samples. However, they must be completed if a relatively large quantity of highly purified plasmid DNA is required. Typical yields from the large-scale procedures are 2 to 3 mg of plasmid DNA per liter of culture.

Several procedures have been developed for the rapid, small-scale isolation of plasmid DNA. Many of these provide plasmids that are of sufficient quantity and quality for characterization by agarose gel electrophoresis and restriction enzyme digestion. Besides providing rapid screening of plasmids from several bacterial samples, small-scale procedures require only minimal handling of potentially hazardous plasmids.

The procedure outlined in this experiment is a micro version of Experiment 18. Bacterial cells grown in small cultures (3 to 5 ml) or in single colonies on agar plates are boiled for a brief period in the presence of reagents that disrupt the cell wall and inhibit nuclease activity. Chromosomal DNA is denatured under these conditions, and it precipitates as a gel that can be centrifuged from the supernatant containing plasmid DNA and bacterial RNA. The RNA, which may interfere with electrophoretic analysis of the plasmids, is removed by isopropanol precipitation of the plasmid DNA and treatment with ribonuclease (RNase). The plasmid DNA is then ready for direct analysis by agarose gel electrophoresis or restriction enzyme digestion followed by analysis of the fragments by electrophoresis.

Analysis of Plasmids and Plasmid Fragments

Several techniques for the characterization of nucleic acids have been introduced in this manual, but the standard method for the separation and analysis of plasmid and other smaller DNA molecules is agarose gel electrophoresis. The method has several advantages, including ease of operation, sensitive staining procedures, high resolution, and a wide range of molecular sizes (0.6 to 100×10^6) that can be analyzed. These characteristics were emphasized in Chapter 4 and in Experiment 19.

Overview of the Experiment

In this experiment, agarose gel electrophoresis will be used to detect the presence of plasmids, evaluate the action of restriction endonucleases on the intact plasmids, and estimate the molecular weight of restriction fragments. The procedures for the isolation and analysis of plasmids are typical of those in recombinant DNA research and may be used in all stages of molecular cloning projects.

This experiment provides an alternative for Experiments 18 to 20. Students completing this modification will be exposed to essentially all of the same principles and techniques except ultracentrifugation and density gradient analysis. However, the microscale version is more convenient to set up and complete, especially if facilities are limited. A single culture of cells can be grown for the whole class as described in Experiment 17. The isolation procedure works well with unamplified or chloramphenicol-amplified cultures. Five milliliters of an unamplified, overnight culture will yield approximately 6 to 10 μg of plasmid DNA.

The isolation procedure requires $1\frac{1}{2}$ to 2 hours of student lab time if a cell culture is prepared in advance. If restriction enzyme digestion is to be carried out before electrophoresis, an additional 1 to $1\frac{1}{2}$ hours is needed. Slow electrophoresis is usually recommended for microscale analysis. Typical conditions are 1 to 3 volts/cm for 9 to 12 hours. It is recommended that the electrophoresis be carried out as a class project. A 20×20 cm agarose slab can hold between 15 and 20 samples. All students can be instructed to isolate plasmid DNA and carry out two digestions with the two restriction enzymes recommended in the procedure.

II. EXPERIMENTAL

Materials and Supplies

Part A

Bacterial culture (5.0 ml) or a colony on an agar plate (4 mm in diameter). The strains of bacteria recommended for this experiment have been listed in Experiment 17. Amplified or unamplified cell cultures are appropriate.

STET Buffer, containing 8% sucrose, 0.5% Triton X-100, 0.05 M EDTA, and 0.05 M Tris-HCl, pH 8.0

Lysozyme solution, 10 mg/ml in 0.01 M Tris-HCl, pH 8.0, freshly prepared

Boiling water bath in a 150 to 200 ml beaker

Centrifuge, capable of $12,000 \times g$

Sodium acetate, 2.5 M

Isopropanol

Dry-ice/acetone bath, approximately $-18°C$

Tris-EDTA-RNase solution, 0.01 M Tris-HCl, 0.001 M EDTA, pH 8.0, containing RNase (50 μg/ml). Heat this solution in an 80°C bath for 10 minutes to denature deoxyribonucleases.

Part B

Restriction enzymes, EcoRI (1 unit/μl) and/or TaqI (1 unit/μl); follow instructions accompanying the product. See Experiment 19.

Incubation buffers for restriction enzymes

EcoRI: 0.1 M Tris-HCl, 0.05 M $MgCl_2$, 0.5 M NaCl, and 0.05 M dithiothreitol, pH 7.5

TaqI: 0.1 M Tris-HCl, 0.1 M $MgCl_2$, 1 M NaCl, and 0.1 M dithiothreitol, pH 7.4

Quenching solution for restriction enzymes, 0.1 M EDTA, pH 7.5

Part C

Agarose powder, electrophoresis grade

Agarose slurry buffer, 0.04 M Tris-acetate, 0.002 M EDTA, pH 7.8

Ethidium bromide solution, 10 mg/ml in H_2O

Electrophoresis buffer, Tris-acetate buffer listed above containing ethidium bromide, 0.5 μg/ml

Gel-loading buffer, Tris-acetate containing 50% glycerol and 0.25% bromphenol blue tracking dye

Electrophoresis chamber. A set-up for horizontal slab gel is recommended (see Experiment 19); however, vertical slab gel or even glass tubes as described in Experiment 5 may be used.

Power supply

UV lamp

Standard restriction enzyme digest, λ phage DNA digested by EcoRI

Procedure

CAUTION:

Experimentation with bacterial cells presents a potential biohazard. Do not mouth-pipet any solutions. Wash hands with hot water and soap before eating, drinking, or smoking.

Ethidium bromide is a mutagen. Always wear gloves when handling solutions or gels.

A. Isolation of Plasmids (Holmes and Quigley, 1981; Maniatis, Fritsch, and Sambrook, 1982)

1. Transfer 5 ml of an overnight culture to a conical centrifuge tube. Alternatively, a single colony (about 4 mm in diameter) of bacteria can be removed from an agar plate with a flat toothpick and suspended in 5.0 ml of STET buffer in a conical tube. Centrifuge the tube at 4000 × g for 5 minutes.

2. Remove as much of the supernatant as possible with a disposable pipet attached through rubber tubing to an aspirator.

3. Resuspend the cells in 0.35 ml of STET buffer.

4. Add 25 μl of freshly prepared lysozyme solution and mix well.

5. Place the suspension in a boiling-water bath for 40 seconds.

6. Centrifuge immediately at 12,000 × g for 10 minutes at room temperature or 4°C to sediment gelatinous chromosomal DNA.

7. Remove the supernatant and transfer to a small conical tube or an Eppendorf tube. Alternatively, the pellet may be removed and discarded with a flat toothpick.

8. To the supernatant, add an equal volume of isopropanol (approximately 0.4 to 0.5 ml) and 40 μl of 2.5 M sodium acetate. Mix well.

9. Cool the mixture in a dry-ice/acetone bath at −18°C for 15 minutes.

10. Centrifuge the mixture at 12,000 × g for 5 minutes to sediment the plasmid DNA.

11. Carefully and completely remove and discard the supernatant. Allow the precipitate to air dry for a few minutes.

12. Resuspend the pellet in 0.05 ml of Tris-EDTA-RNase solution and incubate for 10 minutes at 37°C.

13. The plasmid preparation may be used directly in Parts B and C or stored in a freezer.

B. Digestion of Plasmids with Restriction Enzymes

The reaction conditions are specific for each restriction enzyme. The commercial supplier of the enzyme will provide an information sheet giving details of the reaction conditions. For a detailed listing of optimal conditions for many restriction enzymes, see Fuchs and Blakesley (1983) or Table E19.1, Experiment 19. Optimal conditions for two enzymes, EcoRI and TaqI, will be described here. For instructions and precautions in the use of restriction enzymes, see Experiment 19.

EcoRI Digestion of Plasmids

1. Transfer 5 μl of the plasmid preparation to a small test tube.

2. Add 2 μl of H$_2$O and 2 μl of the EcoRI digestion buffer and mix well.

3. Add 1 unit (1 μl) of EcoRI to the tube. Mix gently by tapping.

4. Incubate at 37°C for 1 hour.

5. Quench the reaction by adding 2 μl of 0.1 M EDTA, pH 7.5 solution.

6. Go directly to Part C, Agarose Gel Electrophoresis.

TaqI Digestion of Plasmids

1. Transfer 5 μl of the plasmid preparation to a small test tube.

2. Add 2 μl of H$_2$O and 2 μl of TaqI digestion buffer. Mix well.

3. Add 1 unit (1 μl) of TaqI to the tube containing plasmid and buffer. Mix gently.

4. Incubate at 37°C for 1 hour.

5. Quench the reaction by adding 2 μl of 0.10 M EDTA, pH 7.5.

C. Agarose Gel Electrophoresis

The details for preparing the agarose gel slab and setting up the electrophoresis chamber were discussed in Experiment 19. The same procedure should be followed here.

CAUTION:

Do not touch the electrophoresis chamber or wires while the electrophoretic operation is in progress. Voltages may reach as high as 100 volts, and shocks may be fatal.

---

1. An agarose concentration of 1% is recommended for this experiment. Prepare the gel with as many sample slots as possible.

2. Mix 10 μl of the plasmid preparation with 6 μl of loading buffer containing bromphenol tracking dye, and transfer it to a gel slot.

3. Mix each restriction enzyme digest with 6 μl of loading buffer and transfer each sample to a separate gel slot.

4. Load at least one slot with a solution containing 10 μl of standard EcoRI digest of λ phage DNA and 6 μl of loading buffer. This digest will be used as markers for molecular weight determination. Fragments of the following molecular weight are present in the digest: 13.7, 4.7, 3.7, 3.6, 3.0 and 2.1 × 10^6.

5. Connect the electrophoresis apparatus to the power supply and apply current to the slab gel. Recommended voltage is 3 V/cm. Continue to pass current through the gel until the tracker dye is near the bottom.

6. *Turn off the power supply* and disconnect the wires to the chamber. Remove the slab gel.

7. Examine the agarose gel under a UV lamp. DNA fragments will appear as red-orange fluorescent bands. Draw a picture of the slab gel, showing the position of each DNA band.

If equipment for agarose slab gel is not available, glass tubes as described in Experiment 5 may be used. Prepare the agarose slurry as described in Experiment 19 and transfer to glass tubes. Apply plasmid samples in loading buffer, place the tubes in the electrophoresis chamber, and apply current at levels of 5 mA/gel until the tracking dye is near the bottom of the tube. Remove the gels from the glass tubes and examine under a UV light.

III. ANALYSIS OF RESULTS

Part A. Construct a flow chart outlining the isolation procedure followed in this experiment. Explain the function of each reagent and step.

Part C. Study your picture of the completed slab gel. Measure the distance migrated by each plasmid and fragment. Use the EcoRI digest of λ phage DNA to prepare a standard curve for molecular weight (see Figure E19.1, Experiment 19). Plot relative migration of each standard fragment vs. the log of the known molecular weight. Use the standard curve to estimate the molecular weight of the intact ColE1 plasmid and restriction enzyme fragments. Would any of the restriction enzymes be suitable for preparing a recombinant plasmid?

IV. QUESTIONS

1. What are some of the properties required of a restriction enzyme used to cleave plasmid DNA for insertion of foreign DNA?

2. What is the purpose of the isopropanol precipitation step in the preparation of plasmids?

3. Describe how you would determine the T_m of plasmid DNA.

4. Why is polyacrylamide electrophoresis not suitable for analysis of DNA and restriction enzyme fragments?

5. How does the EDTA solution quench the restriction enzyme reaction?

V. REFERENCES

General

D. Freifelder, *Physical Biochemistry*, Second Edition (1982), Freeman (San Francisco), pp. 292–308. Agarose-gel electrophoresis applied to DNA molecules.

Specific

R. Fuchs and R. Blakesley in *Methods in Enzymology*, R. Wu, L. Grossman, and K. Moldave, Editors, Vol. 100 (1983), Academic Press (New York), pp. 3–38. "Guide to the Use of Type II Restriction Endonucleases."

D. Holmes and M. Quigley, *Anal. Biochem.*, *114*, 193–197 (1981). "A Rapid Boiling Method for the Preparation of Bacterial Plasmids."

T. Maniatis, E. Fritsch, and J. Sambrook, *Molecular Cloning, A Laboratory Manual*, (1982), Cold Spring Harbor Laboratory (Cold Spring Harbor, NY), pp. 366–67. "Rapid, Small-Scale Isolation of Plasmid DNA."

Experiment **21**

Ethidium Fluorescence Assay of Nucleic Acids: Binding of Polyamines to DNA

RECOMMENDED READING:
Chapter 5, Section D, E, F; Chapter 7, Sections A, B, C; Chapter 9; Introduction to Experiments 15 and 16.

SYNOPSIS
The central role of the nucleic acids in biochemistry has prompted the development of innumerable techniques for their physical, chemical, and biological characterization. The binding of fluorescent dyes to double-helix DNA forms the basis for many of the analytical methods. In this experiment, ethidium bromide interaction with duplex DNA, which results in enhanced fluorescence of the dye, is applied to (1) the measurement of the concentration of DNA in solution and (2) ligand (polyamine) binding to DNA.

I. INTRODUCTION

The nucleic acids are among the most complex molecules that you will encounter in your biochemical studies. When the dynamic role that is played by DNA in the life of a cell is realized, the complexity is understandable. It is difficult to comprehend all the structural characteristics that are inherent in the DNA molecules, but most biochemists are familiar with and accept the double-helix model of Watson and Crick. In fact, the discovery of the double-helical structure of DNA is one of the most significant breakthroughs in our understanding of the chemistry of life. Experiments 15 and 16 introduced you to the basic structural characteristics of the DNA molecule and to the forces that help establish the complementary interactions between the two polynucleotide strands. In this experiment, we will be concerned with a more dynamic property of DNA, the binding of natural polyamines.

Circular Supercoiled DNA

Most diagrams of DNA illustrate the linear nature of the double helix and imply that DNA molecules are extended rods with two ends. The chromosomal and plasmid DNA of many microorganisms, however, is a closed circle, a result of covalent joining of the two ends of a linear double helix. Circular DNA has been discovered in Simian virus 40 (SV 40), bacteriophages (for example, $\phi \chi 174$), bacteria (*E. coli*, for example), and several species of animals.

Closed, circular, duplex DNA has a unique structural feature. It is found twisted into a new configuration, **supercoiled DNA**. Other terms used to define this form are **superhelical DNA** and **supertwisted DNA**. Figure E21.1 illustrates both relaxed (A) and supercoiled (B) DNA. Form (B) is not simply a coiling of the completed, circular DNA, but is the result of extra twisting in the linear duplex form just before covalent joining of the two strands. The supercoiled form is less stable than the relaxed form; however, there is sufficient evidence to conclude that supercoiled DNA may exist *in vivo*. The most convincing evidence is the isolation and characterization of proteins that catalyze the interconversion of relaxed and supercoiled DNA. These catalytic proteins have received various names in the literature, including DNA relaxing enzyme (Keller, 1975), ω protein (Wang, 1971) and the nicking-closing enzyme (Champoux and

FIGURE E21.1
Structures of closed, circular duplex DNA. (A) Relaxed. (B) Supercoiled.

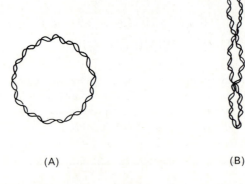

(A)   (B)

Dulbecco, 1972); they are now classified as **topoisomerases** because they catalyze changes in the topology of DNA.

The origin of supercoiling is not completely understood, but may be caused by the presence of more than the standard number of bases per unit length of a DNA helix. Supercoiling may have biological significance, in that the DNA molecule becomes more compact and is more easily stored in the cell. Supercoiled DNA possesses less intrinsic viscosity than relaxed or linear DNA; hence, it sediments at a faster rate in the ultracentrifuge, and can be readily separated and characterized by this technique.

Methods for the Characterization of DNA

A complete understanding of the biochemical functions of DNA requires a clear picture of its structural and physical characteristics. Nature has built into the genetic molecules many features that make them easy to characterize. (1) The purine-pyrimidine bases absorb light strongly at 260 nm (Experiment 16), and they also display weak fluorescence spectra. Minor physical changes in DNA structure modify the extent of absorption and fluorescence. Hence, the bases are natural spectrophotometric probes for the study of DNA structure. (2) The inherent polarization properties of the double helix make optical rotation and circular dichroism measurements convenient. (3) The presence of unique phosphoester linkages in the polynucleotide backbone opens the door for selective chemical cleavage using specific enzymes. (4) Finally, the compact, close-packed nature of DNA leads to molecules of high density, compared to proteins and other biomolecules. This property has been exploited in density gradient ultracentrifugation studies (see Chapter 7 and Experiment 18).

With so many natural characteristics that lend themselves to physical measurement, how does one choose a technique for evaluating the structural properties of DNA? The most important criteria to use in choosing an assay are sensitivity, accuracy, and convenience. Most of the aforementioned methods fit the criteria, except that some require sophisticated techniques, specialized instrumentation, and long periods of time.

FIGURE E21.2
Structure of the fluorescent inter-calation dye, ethidium bromide.

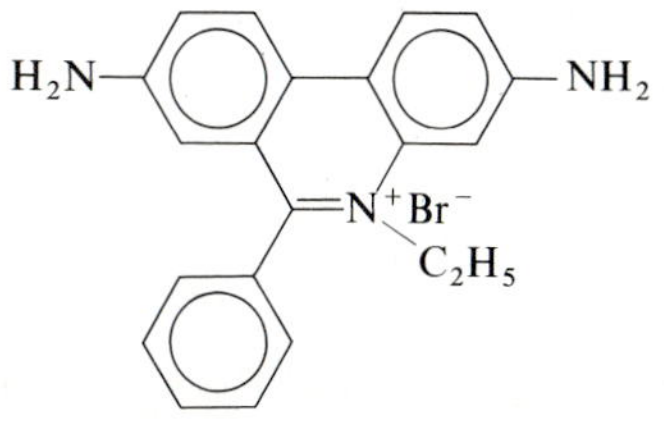

Ethidium Fluorescence Assays of DNA

Fluorescence assays are considered among the most convenient, sensitive, and versatile of all techniques. However, as previously mentioned, the purine-pyrimidine bases yield only weak fluorescence spectra. Le Pecq and Paoletti (1967) showed that the fluorescence of a dye, ethidium bromide, is enhanced about 25-fold when it interacts with DNA. The ethidium bromide, which is a relatively small planar molecule (Figure E21.2), binds to DNA by insertion between stacked base pairs (intercalation). The process of intercalation is especially significant for aromatic dyes, antibiotics, and other drugs. Some dyes, when intercalated into DNA, show an enhanced fluorescence that can be used to detect DNA molecules after gel electrophoresis measurements (see Experiments 19 and 20 and Chapter 4) and for characterization of the physical structure of DNA. Intercalation of dyes into a supercoiled DNA molecule causes unwinding of the superhelix; if a large excess of dye is used, the DNA begins to twist in the opposite direction (negative to positive coiling) (Bauer and Vinograd, 1968).

The enhanced fluorescence of ethidium bromide-DNA complexes has been applied to the measurement of structural, physicochemical, and biochemical properties of DNA and RNA (Morgan et al., 1979). In this experiment two measurements on DNA will be described, (1) determination of the concentration of DNA and RNA in solution and (2) polyamine binding to DNA.

Measurement of the Concentration of DNA and RNA in Solution

Most assays measure both double-stranded and single-stranded DNA. Ethidium interaction with single-stranded DNA does not lead to increased fluorescence, so duplex DNA can be quantified in the presence of dissociated DNA.

Solutions of purified DNA are commonly contaminated with RNA. Since single-stranded RNA can form hairpin loops with base pairing and duplex formation (as in t-RNA), ethidium will also bind with enhanced fluorescence to these duplex regions of RNA. The addition of ribonuclease A results in digestion of RNA and loss of fluorescence due to ethidium binding of RNA. The amount of fluorescence lost is proportional to the concentration of RNA. The ethidium fluorescence remaining after ribonuclease treatment is directly proportional to the concentration of duplex DNA (Equation 1).

$$F_{\text{total}} = F_{\text{DNA}} + F_{\text{RNA}} \qquad\qquad (\text{Equation E21.1})$$

where

F = fluorescence yield due to each type of nucleic acid

When the F_{RNA} term is reduced to zero, the total fluorescence is a direct measure of DNA concentration. The actual concentration of DNA in solution can be calculated by use of a standard solution of DNA or from a standard curve.

Polyamine Binding to DNA

The site of action of many antibiotics is the DNA molecule of an invading organism. In fact, the physiological action of many drugs depends upon interaction with DNA, which often leads to an intercalation complex. It is possible to demonstrate binding and to estimate the association constants for ligand: DNA complexes using the ethidium assay (Morgan et al., 1979). A solution of DNA treated with excess ethidium bromide will give a maximum fluorescence yield. If a ligand is added that can compete with ethidium for the intercalation sites on DNA, the ethidium dissociates from the DNA. The fluorescence yield of ethidium will decrease at a rate that is directly proportional to the concentration of added ligand. That is, the greater the concentration of bound ligand, the greater the extent of ethidium dissociation, and the greater the decrease in ethidium fluorescence. Of course, it must be shown that the ligand does not fluoresce under the conditions of the experiment and that it does not interfere with ethidium fluorescence. An estimate of the binding constant can be obtained from the binding curve of % fluorescence vs. ligand concentration. The ligand concentration that causes a 50% decrease in ethidium concentration is approximately inversely proportional to the binding constant, k_a.

The binding of both synthetic drugs and naturally occurring molecules to DNA is of biochemical significance. There is particularly strong interest in the interactions between polyamines and DNA. Four polyamines, 1, 4-diaminobutane (putrescine), 1,5-diaminopentane (cadaverine), spermidine, and spermine (Figure E21.3), are metabolic products in many cells and are found in relatively high concentrations. Putrescine and cadaverine are derived from the decarboxylation of ornithine and lysine, respectively. The other two polyamines, spermidine and spermine, are synthesized in rather complex processes requiring S-adenosylmethionine. The biochemical functions of these unusual compounds are not completely clear. Polyamines are cationic, so they bind strongly to nucleic acids. They interact by forming salt bridges between their positively charged amino groups and negatively charged phosphate anions of the nucleotides. In fact, this interaction may stabilize supercoiled DNA. The polyamines may play regulatory roles in nucleic acid and protein synthesis. The ethidium assay is particularly convenient for demonstrating polyamine binding to DNA because the analysis is rapid; also, since the ligands are not

FIGURE E21.3

Structures for several significant polyamines.

$$H_2N-CH_2(CH_2)_2CH_2-NH_2$$

1,4-Diaminobutane

$$H_2N-CH_2(CH_2)_3CH_2-NH_2$$

1,5-Diaminopentane

$$H_2N-CH_2(CH_2)_2-CH_2-\overset{\overset{\displaystyle H}{|}}{N}-CH_2-CH_2-CH_2-NH_2$$

Spermidine

$$H_2N-CH_2-CH_2-CH_2-\overset{\overset{\displaystyle H}{|}}{N}-CH_2(CH_2)_2CH_2-\overset{\overset{\displaystyle H}{|}}{N}-CH_2-CH_2-CH_2-NH_2$$

Spermine

aromatic, they display no measurable interference of ethidium fluorescence. However, the measurements lead only to approximate values for the binding constants. For more precise measurements of binding, Scatchard analysis as described in Experiment 6 must be completed.

Overview of the Experiment

This experiment illustrates two of the many applications of the ethidium fluorescence assay to studies on DNA. The concentrations of DNA solutions from Experiment 15 may be determined, or unknown DNA solutions may be prepared. The average time required for Part A, Measurement of DNA Concentration, is 30 to 45 minutes; that for Part B, Binding of Polyamines, is $1\frac{1}{2}$ to 2 hours.

There are several additional experiments that can be completed if time allows. Some of the more interesting are (1) kinetics of DNA reassociation (Britten and Kohne, 1968), (2) measurement of the percentage of supercoiled DNA in a sample (Morgan et al., 1979A; Kowalski, 1979), (3) the effect of covalent crosslinking on reassociation kinetics (Morgan et al., 1979A), (4) assay of DNA in whole cells (Morgan et al., 1979B), and (5) various enzyme assays including DNA polymerase, nucleases, and topoisomerases (Morgan et al., 1979B).

II. EXPERIMENTAL

Materials and Supplies

DNA Standard Solution, 50 μg/ml in 0.01 M Tris-HCl and 0.05 M NaCl, pH 8.0

Unknown DNA solution, prepared in Tris-HCl buffer

Ethidium bromide solution, pH 8.0, 5 mM Tris-HCl, 0.5 μg/ml ethidium bromide, and 0.5 mM EDTA

Ribonuclease A, 20 mg/ml in 0.10 M Tris-HCl, pH 8

Constant temperature bath at 37°C

Spermine solution, 1×10^{-4} M in Tris-HCl buffer, pH 8.0

Spectrofluorimeter and fluorescence cuvets

Excitation wavelength	525 nm
Emission wavelength	590 nm
Slitwidth	20 nm
Scale	$\times 100$

Procedure

A relative fluorescence intensity scale will be defined in Part A of the experiment. The cuvet holder should be provided with a water jacket to maintain the temperature within $\pm 0.5°$. It is recommended that you use the same cuvet for all measurements unless a pair is available that is very well matched. Fluorescence measurements are temperature-dependent. It is necessary to maintain all reagents, except the ribonuclease solution, at 37°C, the same as the setting for the cuvet holder in the fluorimeter.

CAUTION:

Ethidium bromide is a mutagen and a potential carcinogen. Always wear gloves and a face mask while weighing or handling the pure chemical. Gloves should always be worn while using solutions of the dye. Do not allow the solutions of ethidium bromide to come into contact with your skin. **Do not mouth-pipet any ethidium bromide solutions**. Save all assay mixtures and excess ethidium bromide solutions after measurements are taken, and pour them into a container marked "*Waste Ethidium.*"

A. Concentration of DNA Solutions

Turn on the fluorimeter lamp and allow it to warm up for 15 minutes. Set the fluorescence scale for the standard DNA in the following manner. Pipet 3 ml of pH 8 ethidium bromide buffer solution into a fluorescence cuvet. With a micropipet, transfer 15 μl of the standard DNA into the cuvet and mix well. Place the cuvet in the spectrofluorimeter and adjust the fluorescence intensity to "100", F_{std}. This is, of course, an arbitrary setting. The actual setting that is used is not especially important, but you must know what the setting is. The number 100 is convenient for comparison purposes and is customary in fluorescence measurements. Pour the contents of the cuvet into the waste ethidium container provided by your instructor. Clean the cuvet carefully with warm water, rinse several times with distilled water, and dry.

Again, transfer 3 ml of the pH 8 ethidium bromide buffer solution into the cuvet. Add 15 μl of the unknown DNA solution. Mix well and place in the spectrofluorimeter. Record the fluorescence intensity, F_{total}. The measured fluorescence is due to both DNA and RNA that may be present in the isolated DNA. The amount of fluorescence due to ethidium:RNA can be eliminated by digesting the RNA with ribonuclease. To the cuvet containing pH 8 ethidium bromide and unknown DNA, add 2 μl of ribonuclease solution. Mix well and incubate the cuvet at 37°C for 20 minutes in a water bath. After incubation, measure the fluorescence intensity, F_{total}. Compare this with the original measurement taken on the unknown DNA (F_{total}). Is there RNA present in your DNA sample? Calculate the concentration of DNA in your unknown solution as described in the Analysis of Results. The possibility exists that the solution of ribonuclease may contain a fluorescent impurity. What control experiment should be done to correct this situation?

B. Binding of Spermine to DNA

Standardize the spectrofluorimeter in the following way. Pipet 2 ml of the pH 8 ethidium bromide buffer into a cuvet. Add 1.0 ml of Tris-HCl buffer and 20 μl of standard DNA solution. Mix and place in the fluorimeter. Adjust the fluorescence intensity to "100." Clean the cuvet as described in Part A and repeat the assay using various concentrations of spermine. Prepare a table displaying the amount of each component to be added. Four reagents must be in the table: pH 8 ethidium bromide buffer, DNA solution, spermine, and pH 8.0 Tris buffer. Maintain the volume of DNA at 20 μl and ethidium bromide solution at 2.0 ml for all assays. Use 0.05, 0.1, 0.2, 0.4, 0.6, 0.8, and 1.0 ml of spermine in the assays. Remember that the total volume of all constituents in the cuvet must remain constant at 3.02 ml for all the assays. Therefore, the amount of Tris-HCl buffer must change with the amount of spermine added. Prepare each assay separately by adding the proper amount of each component to the cuvet. Mix well and record the fluorescence intensity. Clean the fluorescence cuvet carefully after each assay.

III. ANALYSIS OF RESULTS

A. Concentration of DNA Solution

The concentration of the unknown DNA solution can be calculated by a simple comparison of the fluorescence intensity obtained from the standard DNA reaction mixture (F_{std}) and the fluorescence of the unknown DNA mixture (Equation 2).

$$[DNA]_x = \frac{[DNA]_{std} F_{DNA-x}}{F_{DNA-std}}$$

(Equation E21.2)

where

$[DNA]_x$ = the concentration of the unknown solution of DNA in $\mu g/ml$

$[DNA]_{std}$ = the concentration of the standard solution of DNA in $\mu g/ml$

$F_{DNA-std}$ = the fluorescence yield of the standard DNA solution

F_{DNA-x} = the fluorescence yield of unknown DNA solution after incubation with ribonuclease

To correct for possible fluorescent contamination in the RNase solution, place 3 ml of pH 8 ethidium bromide buffer in a cuvet. Add 15 μl of Tris-HCl buffer, pH 8 and 2 μl of ribonuclease. Mix well and incubate at 37°C for 20 minutes. Record the fluorescence intensity, if any. If the fluorescence is significant (> 1 intensity unit), subtract from F_{DNA-x} before calculation of unknown DNA. Calculate the concentration of DNA in your sample in units of $\mu g/ml$. Was RNA present in your unknown DNA solution?

B. Binding of Spermine to DNA

In Experiment 6, ligand binding to protein was evaluated using Scatchard plots. A reasonably accurate value of the formation constant for the complex, k_a, was determined, but extensive experimentation and data analysis were required. For rapid screening of drug and ligand binding to DNA, the ethidium assay is especially good. The main disadvantage of this technique is that the value for k_a is just an approximation. However, the convenient and rapid assay will demonstrate those ligands that deserve more detailed binding studies by Scatchard analysis.

Prepare a table of fluorescence reading vs. drug concentration in the cuvet in μM. If the fluorescence of untreated DNA was set to 100 on the fluorimeter, then each fluorescence reading taken can be considered a percentage of the maximum possible. Plot the fluorescence reading (y axis) against the spermine concentration (x axis) that caused that particular fluorescence reading. Connect the points with a straight line. The binding constant, k_a, is inversely proportional to the spermine concentration that produces 50% of the fluorescence of untreated DNA.

IV. QUESTIONS

1. Outline a procedure for quantitative analysis of DNA in a crude cell extract. Remember that nucleases present in the extract will continue to digest the DNA present and lead to a low result for DNA concentration. What should be done to decrease this error? DNA and RNA both give enhanced fluorescence with ethidium.

2. Design an ethidium assay for the nicking-closing enzyme, a topoisomerase. (For answer, see Kowalski, 1979.)

3. Compare the structure of ethidium bromide (Figure E21.2) with those of the polyamines (Figure E21.3). Would you expect the binding of spermine or spermidine to DNA to be competitive with the binding of ethidium? Explain.

4. Explain how this experiment could be modified to measure the concentration of an RNA solution.

V. REFERENCES

General

W.R. Bauer, F.H.C. Crick, and J.H. White, *Sci. Amer.*, July 1980, pp. 118–133. "Supercoiled DNA."

R.C. Bohinski, *Modern Concepts in Biochemistry*, Fourth Edition (1983), Allyn and Bacon (Boston), pp. 163–192. Introduction to DNA structure.

J.D. Rawn, *Biochemistry* (1983), Harper & Row (New York), pp. 341–346. Introduction to DNA structure and supercoiling.

T.T. Sakai and S. Cohen in *Progress in Nucleic Acid Research and Molecular Biology*, W.E. Cohn, Ed., Vol. 17 (1976), Academic Press (New York), pp. 15–41. "Effects of Polyamines on the Structure and Reactivity of *t*-RNA."

E. Smith, R. Hill, I. Lehman, R. Lefkowitz, P. Handler, and A. White, *Principles of Biochemistry: General Aspects*, Seventh Edition (1983), McGraw-Hill (New York), pp. 126–151. Introduction to structure and function of DNA.

L. Stryer, *Biochemistry*, 2nd Edition (1981), Freeman (San Francisco), pp. 573–574. "Circular DNA."

G. Zubay, *Biochemistry* (1983), Addison-Wesley (Reading, MA), pp. 676–685. Introduction to DNA supercoiling and denaturation.

Specific

W. Bauer and J. Vinograd, *J. Mol. Biol.*, *33*, 141 (1968). "The Interaction of Closed Circular DNA with Intercalative Dyes."

R.J. Britten and D.E. Kohne, *Science*, *161*, 529 (1968). "Repeated Sequences in DNA."

J.J. Champoux and R. Dulbecco, *Proc. Natl. Acad. Sci.*, *69*, 143 (1972). "An Activity from Mammalian Cells that Untwists Superhelical DNA."

W. Keller, *Proc. Natl. Acad. Sci.*, *72*, 2550 (1975). "Characterization of Purified DNA-relaxing Enzyme from Tissue Culture Cells."

D. Kowalski, *Anal. Biochem.*, *93*, 346 (1979). "A Procedure for the Quantitation of Relaxed Closed Circular DNA in the Presence of Superhelical DNA: An Improved Fluorometric Assay for Nicking-Closing Enzyme."

J.B. Le Pecq and C. Paoletti, *J. Mol. Biol.*, *27*, 87 (1967). "A Fluorescent Complex between Ethidium Bromide and Nucleic Acids."

A.R. Morgan, J.S. Lee, D.E. Pulleyblank, N.L. Murray, and D.H. Evans, *Nucl. Acids Res.*, *7*, 547 (1979A). "Review: Ethidium Fluorescence Assays. Part 1. Physicochemical Studies."

A.R. Morgan, D.H. Evans, J. S. Lee, and D.E. Pulleyblank, *Nucl. Acids Res.*, *7*, 571 (1979B). "Review: Ethidium Fluorescence Assay: Part II. Enzymatic Studies and DNA-protein Interactions."

J.C. Wang, *J. Mol. Biol.*, *55*, 523 (1971). "Interaction between DNA and an *Escherichia coli* Protein ω."

Experiment **22**

The Preparation and Characterization of a Beef Heart Mitochondrial Fraction

RECOMMENDED READING:
Chapter 2, Section E; Chapter 7, Sections A, B, and C

SYNOPSIS

Many of the biochemical processes that generate chemical energy for the cell take place in the mitochondria. These organelles contain the biochemical equipment necessary for fatty acid oxidation, di- and tricarboxylic acid oxidation, amino acid oxidation, electron transport, and oxidative phosphorylation. In this experiment, a mitochondrial fraction will be isolated from beef heart muscle. The mitochondria will be analyzed for protein content and malate dehydrogenase activity.

I. INTRODUCTION

Mitochondrial Structure and Function

Mitochondria are intracellular centers for aerobic metabolism. They are cell organelles that are identified by well-defined structural and biochemical properties. In morphological terms, mitochondria are relatively large particles that are characterized by the presence of two membranes,

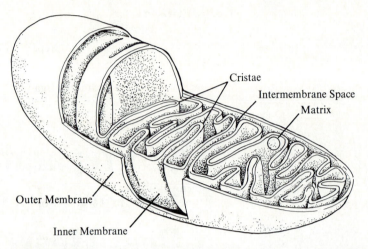

FIGURE E22.1
Structure of the mitochondrion. From BIOLOGY OF THE CELL, Second Edition, p. 135,
by Stephen L. Wolfe. © 1981 by Wadsworth, Inc. Reprinted by permission of Wadsworth
Publishing Company, Belmont, California 94002.

a smooth outer membrane that is permeable to most important metabolites and an inner membrane that has unique transport properties. The
inner membrane is highly folded, which serves to increase its surface area.
The structure of a typical mitochondrion is shown in Figure E22.1.

Biochemically, mitochondria are characterized by the presence of the
proteins and enzymes of respiratory metabolism, many of which are present in the folded inner membrane. Hence, this is the location for the
processes of electron transport, oxidative phosphorylation, and others.
The enzymes involved in the tricarboxylic acid cycle, fatty acid oxidation,
and amino acid metabolism are present in the matrix region.

Isolation of Mitochondria

Because of the biological significance of these intracellular particles,
biochemists have devised several methods for the isolation of mitochondria. Most isolation methods take advantage of the relatively large size
of mitochondria. In animal cells only the nucleus is larger. The standard
method for the preparation of a mitochondrial fraction is differential centrifugation of homogenized tissue (Figure 7.9, Chapter 7). It should be
pointed out that only rarely does one isolate and characterize a preparation of pure mitochondria. Rather, the isolated subcellular fraction is a
mitochondrial fraction, which indicates that it contains mitochondria as
the major components. Other cellular components that may be present
are lysosomes, cell fragments, nuclear fragments, and microbodies (peroxi-

somes). The purity of the fraction depends on the source of the extract and the method chosen for isolation.

The general procedure for preparation of a mitochondrial fraction consists of the following steps:

1. Select and procure a suitable organ or other biological material. Mitochondrial fractions have been prepared from many animal organs including liver, heart, muscle, and brain, many plant tissues including spinach and avocado, and yeast.

2. The tissue is minced or ground and suspended in sucrose buffer. The choice of buffer conditions is critical in order to avoid loss of protein and other components that may leak from the mitochondria. Cytochrome c is one of the easiest of mitochondrial proteins to dislodge, and some is almost always lost. The buffer should be isotonic or hypotonic with a low ionic strength. Sucrose has been found to be an ideal buffer component because mitochondria are especially stable and remain intact under these conditions.

3. Homogenize the tissue. This is carried out with a glass-Teflon homogenizer or a blender. This procedure gently breaks open the cells and allows the release of the subcellular organelles into the buffer. The mechanical disruption of cells should be gentle so the fragile mitochondria are not damaged.

4. Centrifuge the homogenate at low speeds. The heavier particles such as whole and fragmented cells, nuclei and nuclear fragments, and membrane fragments are sedimented during this step.

5. Centrifuge the supernatant from Step 4 at a higher speed. This step sediments lighter particles including mitochondria. Two distinct layers usually are observed in the pellet: (a) a loosely packed, fluffy upper layer, which consists of damaged mitochondria and is sometimes called **light mitochondria**, and (b) a dark brown layer that consists of **heavy mitochondria**.

6. The heavy mitochondrial fraction is suspended in buffer and homogenized for a brief period in a glass-Teflon homogenizer to complete cell lysis.

7. The mitochondrial fraction is then obtained by centrifugation at relatively high speeds (15,000 to 20,000 rpm; $25,000 \times g$).

8. The pellet from Step 7 now consists primarily of heavy mitochondria suspended in sucrose buffer. Typical yields are approximately 1 mg of protein per gram of starting minced tissue.

Several variations of this procedure may be found in the literature. Often, homogenization is facilitated by adding a proteolytic enzyme to the original suspension in Step 2, but the proteolysis must be carefully controlled to avoid degradation of mitochondrial enzymes. The various methods lead to mitochondrial fractions of different quality and biochemical characteristics. The choice of methods depends to a great extent on what is to be done with the mitochondria.

This experiment describes the preparation of a mitochondrial fraction from beef heart muscle. Heart muscle is an excellent choice of tissue because isolated mitochondria are stable and most enzyme activities remain high for a long period of time. Since heart muscle is more fibrous than other tissues, some problems are encountered in homogenization of the tissue. The preparation described in this experiment is suitable for the study of characteristic enzymatic activity, electron transport, and oxidative phosphorylation.

Characterization of the Mitochondrial Fraction

Mitochondria may be characterized by testing for the presence of known enzyme activities. A variety of enzymes is present in mitochondria, including dehydrogenases, transaminases, cytochromes, and phosphate transferases. The enzyme content is usually described in terms of an **enzyme profile** or **enzyme activity pattern**. This consists of measuring the activity of several enzymes. One enzyme that is present in relatively high concentration is malate dehydrogenase, the enzyme that catalyzes the interconversion of malate and oxaloacetate (Reaction 1).

$$\text{malate} + \text{NAD}^+ \rightleftharpoons \text{oxaloacetate} + \text{NADH} + \text{H}^+ \qquad \text{(Reaction E22.1)}$$

The activity of this enzyme in mitochondrial fractions may be estimated by a spectrophotometric assay. Oxaloacetate and NADH are incubated, and the disappearance of NADH is monitored at 340 nm. NAD^+ does not have strong absorption at this wavelength. Note that the reverse reaction is studied because the reaction as shown above is very unfavorable in thermodynamic terms ($\Delta G = +7.1$ kcal/mole).

Overview of the Experiment

In this experiment, students will isolate a mitochondrial fraction from beef heart muscle. The fraction will be characterized by making two measurements:

1. An estimate of protein content by the biuret or Bradford assay. The mitochondrial protein must be solubilized by treatment with a detergent, deoxycholate, before analysis.
2. Measurement of the activity of a mitochondrial "marker enzyme," malate dehydrogenase, using a spectrophometric assay.

The present experiment has the following time requirements:

A. Isolation of the Mitochondrial Fraction: 2 hours. It is recommended that the isolation be done as a class project.

B. Determination of Protein: 30 minutes.

C. Measurement of Malate Dehydrogenase Activity: 30 minutes.

II. EXPERIMENTAL

Materials and Supplies

A. Preparation of a Mitochondrial Fraction

Beef heart, obtain from a local slaughterhouse. It should be obtained 1 to 2 hours after the animal is slaughtered and transported in ice to the laboratory.

Sharp knife

Meat grinder, prechilled; plate holes: 4 to 5 mm

Waring blender

Sucrose-Tris homogenizing solution, 0.25 M sucrose and 0.01 M Tris-Cl, pH 7.8. Keep ice-cold.

2 M Tris base solution, unneutralized, pH 10.8

Cheesecloth

Sucrose-Tris isolation solution, 0.25 M sucrose, 0.01 M Tris-Cl, pH 7.8, 0.001 M succinic acid, and 0.2 mM EDTA. Keep ice-cold.

Glass-Teflon homogenizer, manual or motorized

Centrifuge, capable of $1200 \times g$ and $26,000 \times g$

B. Determination of Protein

Mitochondrial fraction from Part A

Sucrose-Tris isolation solution

10% sodium deoxycholate

Five cuvets

Bovine serum albumin solution, 10 mg/ml

Biuret reagent

Spectrophotometer

C. Measurement of Malate Dehydrogenase Activity

Mitochondrial fraction from Part A

Phosphate buffer, 0.2 M, pH 7.4

Oxaloacetic acid, 0.006 M in phosphate buffer, freshly prepared

NADH, 0.00375 M in phosphate buffer, freshly prepared

Spectrophotometer, suitable for measurements at 340 nm

Procedure

A. Preparation of the Mitochondrial Fraction (Smith, 1967)

All of the following procedures must be carried out at 2–4°C.

1. With a sharp knife, trim all fat and connective tissue from the ice-cold beef heart.

2. Cut the tissue into cubes 4 to 5 cm wide.

3. Quickly weigh 200 to 300 grams of the cubes and pass them through a prechilled meat grinder.

4. Suspend the minced tissue in 400 ml of ice-cold sucrose-Tris homogenizing solution.

5. Adjust the pH to 7.5 ± 0.1 by adding 2 M unneutralized Tris.

6. Pour the neutralized heart mince through two layers of cheesecloth and squeeze out the sucrose solution.

7. Suspend 200 g of the solid mince in 400 ml of ice-cold sucrose-Tris isolation solution.

8. (a) Homogenize 25 to 50 ml portions of the suspension in a glass homogenizing vessel with a motorized pestle. Homogenize each portion for about 10 seconds and then twice more for 5 seconds each. Combine all the homogenate solutions and adjust to pH 7.8 with 2 M Tris base. (b) Alternatively, the minced tissue may be homogenized in a blender. For this procedure, suspend 200 grams of mince in 400 ml of sucrose-Tris isolation solution and add 3 ml of 2 M Tris base. Turn on the blender at high speed for 15 seconds. Add 3 ml of 2 M Tris base and blend for another 5 seconds. Adjust the pH of the solution to 7.8 with 2 M Tris base.

9. Centrifuge the homogenate for 20 minutes at $1200 \times g$. The pellet consists of unfragmented cells and nuclei.

10. Carefully decant the supernatant without disturbing the loosely packed pellet.

11. Filter the supernatant through two layers of cheesecloth. This process removes lipid granules.

12. Adjust the pH of the filtrate to 7.8 with 2 M Tris base.

13. Centrifuge the pH-adjusted suspension for 15 minutes at $26,000 \times g$. At least two distinct layers will be observed in the pellet. The top layer consists of loosely packed light mitochondria (damaged mitochondria). The bottom layer of the pellet is brown and consists of heavy mitochondria. Occasionally a very small black pellet is present beneath the heavy mitochondria.

14. Remove and discard the light mitochondria by pouring off about half of the supernatant and gently shaking the centrifuge tube to release the top layer. Pour off and discard the suspension of light mitochondria.

15. If no tiny black pellet is present, add 10 ml of sucrose-Tris isolation solution and stir the heavy mitochondria with a glass stirring rod. If a black pellet is present, remove the brown heavy mitochondria from each tube with a glass rod and suspend in 10 ml of ice-cold sucrose-Tris isolation solution.

16. Homogenize the mitochondrial suspension in a motorized glass-Teflon homogenizer. Make two passes of the rotating pestle through the suspension, 5 seconds each.

17. Adjust the pH of the homogenate to 7.8 with 2 M Tris base and add sucrose-Tris isolation solution to a total volume of 180 ml.

18. Centrifuge the suspension at $26,000 \times g$ for 15 minutes. The pellet this time consists primarily of a dark brown layer of heavy mitochondria. If an upper layer of light mitochondria is present, remove as in Step 14.

19. Suspend the heavy mitochondria in about 60 ml of sucrose-Tris isolation solution and adjust the pH to 7.8 with 2 M Tris.

20. Store the mitochondrial fraction on cracked ice and begin Part B.

B. Determination of Protein

Before the mitochondrial fraction can be biochemically characterized, the protein content must be measured. The biuret method will be described, although the Lowry or Bradford methods may be used.

1. Obtain five small test tubes and set up the protein assay according to Table E22.1. Tubes 1 and 2 contain two different concentrations of mitochondrial protein. Tubes 3 and 4 are duplicates of a standard protein, bovine serum albumin. Tube 5 is used as a blank for the spectrophotometer. The purpose of sodium deoxycholate is to disrupt the mitochondria and release the protein material into solution. For a discussion of the biuret analysis of proteins, see Chapter 2.

2. Add the appropriate amount of each of the first four reagents (sucrose solution, mitochondria, sodium deoxycholate, and BSA) to each of the five test tubes. If the solutions are not clear, add 10% sodium deoxycholate with a graduated pipet. Be sure you note the exact amount of deoxycholate added. Do not add more than a total of 0.4 ml.

TABLE E22.1

Preparation of Tubes for the Biuret Protein Assay of the Mitochondrial Fraction

Reagents	Tube				
	1	2	3	4	5
Sucrose-Tris isolation solution	0.5	0.5	0.5	0.5	0.5
Mitochondrial fraction	0.5	0.25	—	—	—
10% Sodium deoxycholate	0.2	0.2	0.2	0.2	0.2
Bovine serum albumin	—	—	0.1	0.1	—
H_2O	0.3	0.55	0.70	0.70	0.80
Biuret reagent	1.5	1.5	1.5	1.5	1.5

3. Add sufficient water to each tube so that the total volume of all liquids is 1.5 ml. Table E22.1 is set up assuming that 0.2 ml of deoxycholate solution is sufficient. If, for example, tube 1 requires a total of 0.4 ml of deoxycholate, then only 0.1 ml of water should be added to the tube.

4. Add 1.5 ml of biuret reagent to each tube and mix well by inverting several times while holding a piece of hydrocarbon foil over the opening.

5. Incubate the tubes for 15 min at 37°C.

6. Measure and record the absorbance at 540 nm of tubes 1 through 4, using tube 5 to adjust the A_{540} to 0.0 absorbance.

7. If the absorbance readings of tubes 1 and 2 are greater than twice the A_{540} of tubes 3 or 4, repeat the assay using less mitochondrial fraction in tubes 1 and 2. Be sure the new assays contain a total of 3.0 ml of liquid in each tube.

8. If necessary, dilute the mitochondrial fraction to 1 mg protein/ml with ice-cold sucrose-Tris isolation solution and store on ice for Part C.

C. Measurement of Malate Dehydrogenase Activity (Decker, 1977).

1. Turn on the spectrophotometer and UV lamp and allow to warm up for 15 minutes.

2. Obtain two quartz cuvets and add the following reagents to each.

Cuvet 1	Reagent	Cuvet 2
1.3 ml	0.2 M phosphate, pH 7.4	1.3 ml
—	NADH	0.2 ml
0.1 ml	oxaloacetic acid solution	0.1 ml
1.5 ml	H_2O	1.3 ml

3. Add 0.1 ml (0.1 mg protein) of the diluted mitochondrial fraction (1.0 mg/ml) from Part B to cuvet 1. Cover the cuvet with hydrocarbon foil and gently invert 2 or 3 times.

4. Insert the cuvet into the sample beam of the spectrophotometer and adjust the absorbance at 340 nm to 0.00.

5. Add 0.1 ml (0.1 mg protein) to cuvet 2, mix as before, and immediately place it in the spectrophotometer.

6. If a recorder is available, monitor the change in absorbance at 340 nm for 5 minutes. If no recorder is available, read and record in your notebook the A_{340} at 30-second intervals for 5 minutes.

7. Prepare a plot of A_{340} (y axis) vs. time (x axis).

III. ANALYSIS OF RESULTS

A. Preparation of the Mitochondrial Fraction

Prepare a flow chart of the procedures followed in the isolation of mitochondria. Briefly explain the purpose of each step.

B. Determination of Protein

Calculate the protein concentration in the mitochondrial fraction using Equation 1.

$$\frac{C_{std}}{A_{std}} = \frac{C_{unk}}{A_{unk}} \qquad \text{(Equation 22.1)}$$

where

A_{std} = absorbance at 540 nm of the standard bovine serum albumin (tubes 3 and 4)

A_{unk} = absorbance at 540 nm of the protein in the mitochondrial fraction

C_{std} = concentration of standard bovine serum albumin in mg/ml

C_{unk} = concentration of protein in the mitochondrial fraction in mg/ml

The biuret assay is linear up to protein concentrations of about 2 mg/ml.

C. Measurement of Malate Dehydrogenase Activity

Use the plot of A_{340} vs. time to calculate ΔA/min over the linear portion of the curve. Convert the rate in absorbance terms to activity units. One enzyme unit is the amount of malate dehydrogenase that catalyzes the reduction of one micromole of oxaloacetate to L-malate in one minute under the described assay conditions. The reduction of one micromole of oxaloacetate leads to the oxidation of one micromole of NADH; therefore, Equation 2 may be used to calculate the activity of malate dehydrogenase per mg of mitochondrial protein.

$$\text{units/mg protein} = \frac{\Delta A_{340}/\text{min}}{6.2 \times \text{mg protein/ml reaction mixture}}$$

$$\text{(Equation E22.2)}$$

The millimolar extinction coefficient of NADH is 6.2 $mM^{-1}\, cm^{-1}$.

IV. QUESTIONS

1. Why is sucrose used in the isolation buffer?

2. Describe the function and mode of action of sodium deoxycholate in the protein assay in Part B.

3. Explain the derivation of Equation 2 for malate dehydrogenase activity calculation.

4. Briefly explain how the mitochondrial fraction prepared here could be used to study electron transport and oxidative phosphorylation.

5. How would you prepare more highly purified mitochondria than described in this experiment? Hint: Study Chapter 7.

V. REFERENCES

General

R.C. Bohinski, *Modern Concepts in Biochemistry*, Fourth Edition (1983), Allyn and Bacon (Boston), pp. 16–18; 311–319; 336–356. An introduction to mitochondrial structure and function.

J.D. Rawn, *Biochemistry* (1983), Harper & Row (New York), pp. 613–638. Structure and function of mitochondria.

E. Smith, R. Hill, I. Lehman, R. Lefkowitz, P. Handler, and A. White, *Principles of Biochemistry: General Aspects*, Seventh Edition (1983), McGraw-Hill (New York), pp. 332–337. Structure and function of mitochondria.

L. Stryer, *Biochemistry*, Second Edition (1981), Freeman (San Francisco), pp. 307–331. A complete chapter on oxidative phosphorylation.

G. Zubay, *Biochemistry* (1983), Addison-Wesley (Reading, MA), pp. 395–402. Structure and function of mitochondria.

Specific

A. Smith in *Methods in Enzymology*, R.W. Estabrook and M.E. Pullman, Editors, Vol. X, (1967), Academic Press (New York), pp. 81–86. "Preparation, Properties and Conditions for Assay of Mitochondria: Slaughterhouse Material, Small-Scale."

L. Decker, Editor, *Worthington Enzyme Manual*, (1977), Worthington Biochemical Corporation (Freehold, NJ), pp. 23–26. "Malate Dehydrogenase."

Experiment **23**

Enzymes as Diagnostic Reagents in Medicine

RECOMMENDED READING:
Chapter 5, Sections A, B, C; Introduction to Experiments 7, 8.

SYNOPSIS
The specificity and efficiency of enzymes can be used to great advantage in the quantitative analysis of biological fluids for medical diagnosis. Two analyses that are common in the clinical laboratory are the measurements of serum cholesterol and uric acid. Cholesterol is measured by coupling the enzyme-catalyzed oxidation of cholesterol (generation of H_2O_2) to the peroxidase-catalyzed formation of a chromogen. Uricase-catalyzed oxidation of uric acid is the basis of urate determination.

I. INTRODUCTION AND THEORY

One of the most beneficial and interesting applications of biochemical methods is in the diagnosis of disease states. The majority of procedures in the clinical laboratory are used to measure the concentrations of various constitutents in biological fluids and tissues. An abnormally high or low concentration of a biochemical (enzyme, metabolite, etc.) in a patient's blood or urine specimen is often a signal to the patient's physician that a pathologic condition may exist. Biochemical measurements aid clinicians in at least two ways: (1) they assist in the diagnosis (recognition and identification) of a diseased condition, or (2) they may be used for the confirmation of a suspected diseased condition.

Enzymes as Diagnostic Reagents

The availability of purified enzyme preparations and their unique substrate specificity make enzymes particularly attractive as diagnostic reagents. Since most enzymes act upon a single type of reactant and since the extent of an enzyme-catalyzed reaction depends upon substrate concentration, an estimate of the concentration of a single molecular species (the substrate) is possible. It is, of course, essential that the desired reaction be carried out under optimal conditions, that is, proper pH, temperature, and ionic strength, and with assured absence of interfering substances. Of equal importance are the concentrations of enzyme, cofactor, substrate to be measured, and other required reagents. **All reagents, except the substance to be measured, must be present in excess, so that the rate and extent of the enzyme-catalyzed reaction depend upon only the concentration of the substance to be determined.**

The initial velocity of an enzyme-catalyzed reaction is expressed by the Michaelis-Menten equation (Equation 1).

$$v_0 = \frac{V_{max}[S]}{[S] + K_M}$$ (Equation E23.1)

The maximum rate of a reaction (V_{max}) is attained when all the enzyme active sites are saturated with substrate molecules. In order for the rate to approach V_{max}, the substrate concentration must be high; in fact, it must be much greater than K_M. When $[S] \gg K_M$, the Michaelis-Menten equation becomes:

$$[S] \gg K_M; \qquad v_0 = \frac{V_{max}[S]}{[S]} = V_{max}$$ (Equation E23.2)

In words, under substrate-saturating conditions, the initial rate of the enzyme-catalyzed reaction (v_0) is independent of substrate concentration, $[S]$ (Equation 2). However, when the substrate is present in less than saturating amounts ($[S] \ll K_M$), the Michaelis-Menten equation becomes:

$$[S] \ll K_M; \qquad v_0 = \frac{V_{max}[S]}{K_M}$$ (Equation E23.3)

The initial rate of the enzyme-catalyzed reaction is directly proportional to $[S]$ (Equation 3). Most clinical assays using enzymes are performed under the conditions of Equation 3. From further study of this equation, you will note that v_0 also depends upon enzyme concentration, since there is an enzyme concentration term hidden in V_{max}. (If you have forgotten this, review the derivation of the Michaelis-Menten equation in your biochemistry textbook). This can be used to advantage, because if a reaction used for a clinical analysis is very slow (it probably will be, since $[S]$ is low), excess enzyme can be used so that the reaction will proceed to completion in a reasonable period of time.

In general, the principle behind the clinical measurement of cholesterol and uric acid is the following. A biological fluid (serum, urine, etc.) containing the substance to be measured is incubated with an enzyme system that will interact only with that substance. The extent of the enzyme-catalyzed reaction is determined by monitoring some change associated with conversion of substrate to product. Ideally, there is a change in UV or VIS absorbance that can be measured at a convenient wavelength with a spectrophotometer. The spectrophotometric method might involve measurement of a product that appears in solution or the disappearance of substrate. The substance to be determined is incubated for a time interval sufficient to transform it completely into product or to attain equilibrium between substrate and product. Here you should recognize the importance of the enzyme concentration, a point raised earlier. Sufficient enzyme must be present so that an end point (complete conversion of substrate to product or attainment of equilibrium) is reached in the desired amount of time. Once the extent of reaction has been quantitatively measured, knowledge of the stoichiometry of the conversion of substrate to product allows one to calculate directly the initial concentration of the reactant.

Measurement of Cholesterol

The measurement of serum cholesterol is one of the most common tests performed in the clinical laboratory. Hypercholesterolemia (high blood cholesterol levels) can be the result of a variety of medical conditions (Ellefson and Caraway, 1976). Among those conditions implicated are diabetes mellitus, atherosclerosis, and diseases of the endocrine system, liver, or kidney. High blood cholesterol levels do not point to a specific disease; determination of cholesterol is used in conjunction with other clinical measurements mainly for confirmation of a particular diseased condition, rather than for diagnosis of a specific ailment.

Of current interest is the positive correlation between cholesterol levels and heart disease. Cholesterol in the blood is found associated with various lipoproteins. There are two major types of cholesterol-carrying lipoproteins, high-density lipoproteins (HDL) and low-density lipoproteins (LDL). High serum levels of cholesterol-bearing LDLs show positive correlation with the development of atherosclerosis. In contrast, high levels of HDL cholesterol are inversely related to a predisposition of coronary artery disease. In general, the higher the ratio of HDL-cholesterol to LDL-cholesterol, the lower the incidence of heart disease. In order to characterize the LDLs and HDLs and determine the amount of cholesterol associated with each, it is essential to separate them physically. They differ in density and size and may be separated by ultracentrifugation, but they can also be separated by electrophoresis (see Experiment 25) and selective precipitation of LDLs with divalent cations and polyanions.

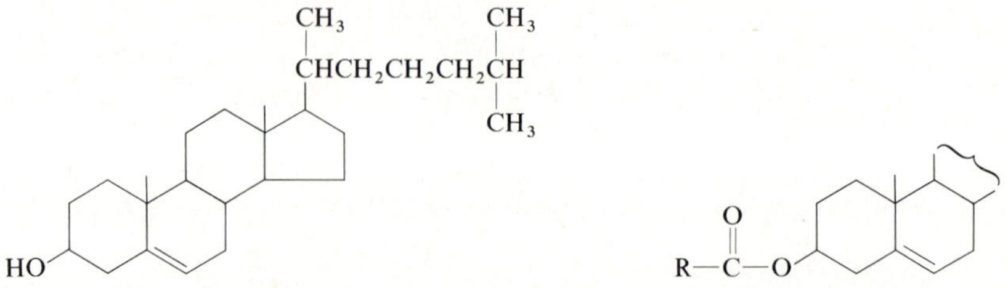

$$CH_3$$
$$CH_3$$
$$CHCH_2CH_2CH_2CH$$
$$CH_3$$

HO—

FIGURE E23.1
Structures of cholesterol and a cholesterol ester.

Cholesterol

$$R—C—O$$
$$\overset{O}{\parallel}$$

Cholesterol Ester

Clinical measurements of **total cholesterol** in serum or plasma detect cholesterol esters in addition to cholesterol. Between 60 and 70% of the cholesterol transported in blood is in an esterified form, where the β-3-OH group on the steroid skeleton is covalently linked to a naturally occurring fatty acid, shown as an R group in Figure E23.1. A sensitive and reproducible analysis of cholesterol and cholesterol esters is based on the three reactions shown below.

$$\text{cholesterol esters} + H_2O \xrightarrow[\substack{\text{or} \\ \text{cholesterol esterase} \\ \text{(EC 3.1.1.13)}}]{H^+ \text{ or } OH^-} \text{cholesterol} + \text{fatty acids}$$

(Reaction E23.1)

$$\text{cholesterol} + O_2 \xrightarrow[\substack{\text{cholesterol} \\ \text{oxidase}}]{} \text{cholest-4-ene-3-one} + H_2O_2$$

(Reaction E23.2)

$$H_2O_2 + \text{phenol} + \text{4-aminoantipyrine} \xrightarrow[\text{peroxidase}]{} \text{quinoneimine chromogen}$$
$$(\lambda_{max} = 510 \text{ nm})$$

(Reaction E23.3)

Reaction 1, catalyzed by cholesterol esterase, shows the hydrolysis of cholesterol esters in the sample. In Reaction 2, cholesterol is then oxidized to cholest-4-ene-3-one. Unfortunately, Reaction 2 does not lead to a major absorbance change at an accessible wavelength, so the rate of the reaction cannot be directly measured. When it is not convenient or possible to directly monitor the progress of a reaction, it may be possible to "couple" it to another reaction that offers a measurable absorbance change. (This was also done in the lactose synthetase assay, Experiment 8.) The oxidation of cholesterol in Reaction 2 is accompanied by production of hydrogen peroxide. In a coupled reaction, H_2O_2 rapidly reacts with phenol and 4-aminoantipyrine in the presence of horseradish peroxidase to produce a quinoneimine chromogen, which has a maximum absorbance at 510 nm (Reaction 3, Figure E23.2). In order for Reactions 2 and 3 to be a properly coupled system, the coupled reaction must be fast enough to decompose H_2O_2 as rapidly as it is produced in Reaction 2. This condition exists for the coupled reactions, so the absorbance change due to the

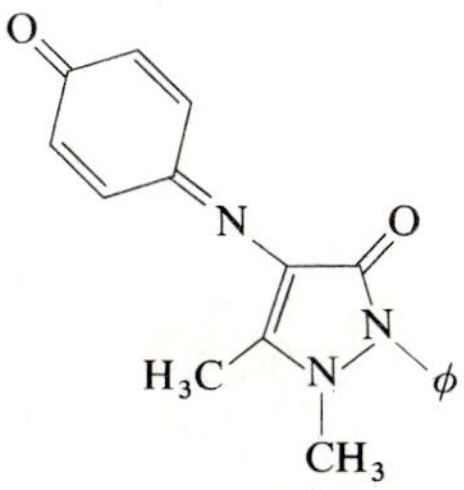

Quinoneimine
Chromogen

$\lambda_{max} = 510$ nm

FIGURE E23.2
Structure of quinoneimine chromogen formed in the assay of cholesterol (Reactions 1, 2, and 3).

production of quinoneimine is directly proportional to the concentration of total cholesterol in the sample analyzed. A more detailed kinetic analysis of coupled assays is beyond the scope of this chapter. Interested students should refer to Cornish-Bowden (1979).

In this experiment, two cholesterol measurements will be made: (1) total serum cholesterol and (2) HDL serum cholesterol, the amount of cholesterol associated with the HDL fraction. The following relationship leads to an estimate of LDL (Equation 4).

$$C_{LDL} = C_{serum} - C_{HDL} \qquad \text{(Equation E23.4)}$$

where

C_{LDL} = concentration of cholesterol in the low-density lipoproteins and very low-density lipoproteins

C_{serum} = total concentration of serum cholesterol

C_{HDL} = the concentration of cholesterol in the HDL fraction

C_{serum} will be determined by performing the described cholesterol assay directly on serum (Sigma Chemical Co., 1981). C_{HDL} will be determined on a separated, soluble HDL fraction of serum. Very low-density lipoproteins and low-density lipoproteins are selectivity removed from serum by precipitation with magnesium-phosphotungstate reagent.

The raw data collected in the experiment are in the form of absorbance measurements made at 510 nm. These numbers are then converted to serum cholesterol concentration, which is reported in mg/100 ml serum.

$$\text{cholesterol concentration (mg/100 ml)} = \frac{A_{510(x)}}{A_{510(s)}} \times C_s$$

$$\text{(Equation E23.5)}$$

where

$A_{510(x)}$ = Absorbance at 510 nm obtained with unknown serum sample

$A_{510(s)}$ = Absorbance at 510 nm obtained with standard cholesterol solution

C_s = Concentration of cholesterol in standard (mg/100 ml)

Normal levels of serum cholesterol vary widely. For males, total cholesterol is in the range from 130 to 320 mg/100 ml and HDL cholesterol ranges from 30 to 70 mg/100 ml. For females, the corresponding ranges are 130 to 295 and 35 to 80 mg/100 ml, respectively. Cholesterol levels depend on such factors as sex, age, diet, and emotional stress. Typical ranges at various age intervals are shown in Table E23.1 (Ellefson and Caraway, 1976).

TABLE E23.1

Average Total Serum Cholesterol Levels as a Function of Age

Age, years	Total Cholesterol, mg/100 ml serum
0–19	120–230
20–29	120–240
30–39	140–270
40–49	150–310
50–59	160–330

Measurement of Uric Acid

Normal levels of human serum uric acid are 3.5 to 7.2 mg/100 ml for men and 2.6 to 6.0 mg/100 ml for women. However, these levels are influenced by diet, age, and disease states. Uric acid is derived from the metabolism of purines and is normally eliminated by urinary excretion. A diet of purine-rich foods (beef liver, kidney, anchovies) increases serum uric acid levels. Increased levels of serum uric acid are associated with many disease states, including acute and chronic nephritis (renal failure), urinary obstruction, uncompensated hypertension, gout, leukemia, malignant tumors with extensive necrosis, acute infections, and Lesch-Nyhan syndrome (Faulkner and King, 1976). Although the classic medical condition associated with elevated serum levels of uric acid is gout, patients with renal failure show much higher levels. Serum uric acid levels in gouty patients are usually between 7 and 10 mg/100 ml, whereas in chronic renal failure, levels may reach 25 to 30 mg/100 ml. The biochemical characteristic of gout is elevated levels of urate in serum, which results in precipitation of sodium urate crystals in joints, causing them to become inflamed.

Uric acid is the end product of purine metabolism in humans, some apes, and the Dalmation. Most other mammals have the enzyme uricase (urate:O_2 oxidoreductase, EC 1.7.3.3), which catalyzes the oxidation of urate to allantoin (Reaction 4).

uric acid   allantoin   (Reaction E23.4)

This reaction forms the basis of a quantitative determination of uric acid in biological fluids. Uric acid absorbs strongly in the ultraviolet region ($\lambda_{max} = 292$ nm, $\varepsilon = 12,300$), whereas the product allantoin has minimal absorption at this wavelength. The conversion of uric acid to allantoin is therefore associated with a significant change in absorbance, so no coupling reaction is necessary for assay of uric acid. Uricase and the sample containing uric acid are incubated in constant pH buffer, and the absorbance change at 292 nm is monitored. The overall decrease in absorbance ($A_{initial} - A_{final}$) is directly proportional to the original concentration of uric acid in the specimen. This assay is particularly advantageous because uricase has high specificity for uric acid; however, the enzyme also catalyzes the oxidation of 6-thiouric acid, a metabolic product of an anticancer drug, 6-mercaptopurine.

The data collected in the analysis of serum uric acid will consist of absorbance changes at 292 nm or ΔA_{292} ($A_{initial} - A_{final}$). In order to relate absorbance changes to concentration of uric acid, Beer's law is used:

$$A = \varepsilon b C$$

where

A = absorbance

ε = molar extinction coefficient of uric acid, 12,300 M^{-1} cm^{-1}

b = cell path length, usually 1 cm

C = concentration of uric acid in mmole/ml

Rearranging, we have the relationship,

$$C = \frac{A}{b\varepsilon}$$

This leads to the concentration of uric acid in mmole/ml. This must be converted to mg/100 ml serum (Sigma Chemical Co., 1976).

$$C' = \frac{\Delta A_{292} \times 168.1 \times 100 \times 3.05}{12,300 \times 0.0833} \qquad \text{(Equation E23.6)}$$

where

C' = concentration of uric acid in mg/100 ml

ΔA_{292} = absorbance change for reaction, $A_{initial} - A_{final}$

168.1 = molecular weight of uric acid

100 = factor to convert mg/ml of uric acid to mg/100 ml serum

3.05 = correction for reaction volume

0.0833 = volume of serum used in analysis

12,300 = molar extinction coefficient of uric acid

Equation 6 illustrates the relationship between the decrease in absorbance at 292 nm associated with Reaction 4 and the concentration of uric acid in mg/100 ml.

Overview of the Experiment

Clinical methods based on enzyme reactions are now used routinely for the determination of glucose, cholesterol, urea, uric acid, and many other metabolites in blood, urine, and other biological fluids as well as

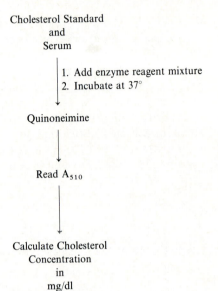

Cholesterol Standard
and
Serum

1. Add enzyme reagent mixture
2. Incubate at 37°

Quinoneimine

Read A_{510}

Calculate Cholesterol
Concentration
in
mg/dl

FIGURE E23.3
The flowchart for measurement
of total serum cholesterol.

FIGURE E23.4
The flowchart for measurement
of total serum HDL cholesterol.

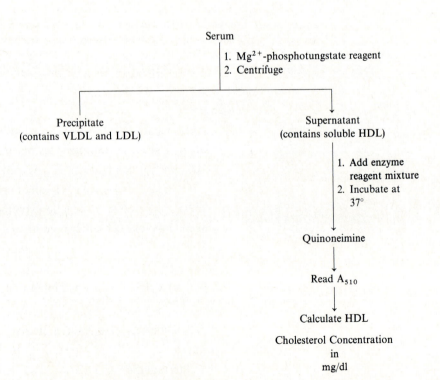

Serum

1. Mg^{2+}-phosphotungstate reagent
2. Centrifuge

Precipitate
(contains VLDL and LDL)

Supernatant
(contains soluble HDL)

1. Add enzyme
   reagent mixture
2. Incubate at
   37°

Quinoneimine

Read A_{510}

Calculate HDL

Cholesterol Concentration
in
mg/dl

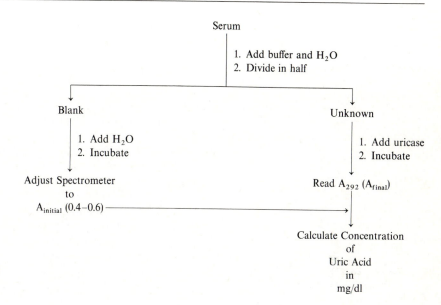

FIGURE E23.5
The flowchart for measurement of serum uric acid.

tissue specimens. Diagnostic kits containing all the required reaction components are commercially available at reasonable cost. These kits make it possible for measurements to be made in a simple, accurate, rapid, and reproducible manner. In modern hospital laboratories, much of this testing is completely automated, which greatly improves the speed and reproducibility, but not necessarily the accuracy.

In this experiment, two measurements that are common in the clinical laboratory will be performed. Both procedures rely on spectrophotometric techniques, but they differ in several regards. The measurement of cholesterol introduces coupled enzyme assays. Uric acid determination illustrates the direct measurement of a molecular species and use of Beer's law for the calculation of concentration. Study Figures E23.3, E23.4, and E23.5 for a summary of the procedures. All parts of this experiment (total cholesterol, HDL cholesterol, and uric acid) can be completed in 3 hours.

II. EXPERIMENTAL

Materials and Supplies

A. Measurement of Cholesterol

Blood serum. Samples can often be obtained from a local hospital laboratory. To collect these samples, blood was probably drawn into a tube not containing anticoagulant or sodium fluoride, and was centrifuged to separate plasma and serum. The samples should be kept at 4°C up to the time of use. Cholesterol is stable in serum for at least 1 week at room temperature, but at 4°C stability continues for 6 months.

Diagnostic kit for cholesterol. This may be obtained from Calbiochem-Behring or Sigma Chemical Co. The stock reagent contains pancreatic cholesterol esterase, microbial cholesterol oxidase, horseradish peroxidase, 4-amino-antipyrine, and phenol. Your instructor will reconstitute the stock reagent by addition of water.

Cholesterol aqueous standard. This contains cholesterol (200 mg/100 ml) in saline solution.

Precipitating reagent. Contains phosphotungstate and magnesium ions.

Spectrophotometer and cuvets. Any spectrometer that measures in the range of 510 ± 10 nm is suitable.

Constant temperature bath at 37°C

Centrifuge, capable of speeds up to 2000 rpm

Centrifuge tubes, 10 ml, conical

Saline solution, 0.15 M NaCl in water

Hydrocarbon foil

B. Measurement of Uric Acid

Blood serum. Same as for cholesterol. Uric acid is stable for 3 to 5 days in refrigerated serum.

Glycine buffer, 0.7 M, pH 9.4

Uricase solution. Activity should be between 0.2 and 0.4 units/ml. A uric acid diagnostic kit is available from Sigma Chemical Co.

Spectrophotometer for measurements in the ultraviolet

Quartz cuvets, matched set

Procedure

CAUTION:

Experimentation with human blood presents a potential biohazard. Do not pipet any solutions by mouth. Wash hands with hot water and soap before eating, drinking, or smoking. If students desire to test their own blood samples, the campus health service or other authorized agency should.be present to supervise the collection of blood.

A. Measurement of Cholesterol

1. Measurement of Total Cholesterol. The procedure will be described for the analysis of two different serum samples. Turn on the spectrometer and allow warmup for 15 to 20 minutes. Set the wavelength to 510 nm. Obtain two 3 ml cuvets or Bausch and Lomb colorimeter tubes for the blank and standard. In addition, one cuvet will be needed for each serum sample to be tested. Label the cuvets Blank, Standard, Serum$_1$, and Serum$_2$. Make the following additions:

Blank: 0.02 ml of water

Standard: 0.02 ml of cholesterol standard

Serum$_1$: 0.02 ml of Serum$_1$

Serum$_2$: 0.02 ml of Serum$_2$

Pipet 1.0 ml of cholesterol enzyme reagent into each tube. Mix well, but do not shake. Shaking will cause foaming and protein denaturation. Mixing is best done by tightly covering the cuvet with hydrocarbon foil and gently inverting 3 to 5 times. Incubate the cuvets at 37°C in a constant temperature water bath for 20 minutes. Remove all the cuvets from the bath and wipe dry with a tissue. Add 2.0 ml of saline water to each cuvet and mix well. Place Blank in the spectrometer and adjust A_{510} to 0.0. Read and record in your notebook A_{510} for the Standard, Serum$_1$, and Serum$_2$. Take the absorbance readings within 30 minutes after removal from the bath to avoid color fading.

This procedure does not take into account any absorbance due to the serum. If the serum is turbid, a correction should be made by measuring the absorbance at 510 nm of a 0.02 ml sample of blood serum in 3.0 ml of saline water. Read the A_{510} of this solution using saline water as blank. Record this reading in your notebook as A_c for correction. The calculation for cholesterol concentration will be described in the Analysis of Results.

2. Measurement of HDL Cholesterol. Obtain a conical centrifuge tube for each serum sample you wish to analyze. To each tube add 0.4 ml of serum and 0.05 ml of phosphotungstate precipitating reagent. Mix well. Centrifuge the tubes for 10 minutes at 2000 rpm. Obtain four cuvets, one each for Blank, Standard, Serum$_1$, and Serum$_2$. Label the cuvets and add the following reagents:

Blank: 0.15 ml of water

Standard: 0.02 ml of cholesterol standard plus 0.13 ml of saline water

Serum$_1$: 0.15 ml of supernatant$_1$ from centrifugation

Serum$_2$: 0.15 ml of supernatant$_2$ from centrifugation

Add 1.0 ml of cholesterol enzymatic reagent to each cuvet, cover with hydrocarbon foil, and mix well. Incubate for 20 minutes at 37°C. Add 2.0 ml of saline water to each cuvet. Adjust spectrometer to zero with Blank and read A_{510} for each sample. Record in your notebook for further calculations.

B. Measurement of Uric Acid

Turn on the spectrophotometer and UV lamp. Pipet into a clean test tube 0.2 ml of serum, 1.0 ml of glycine buffer, pH 9.4, and 6.0 ml of water.

Mix well. Obtain two test tubes; label one tube Blank and the other Unknown. Add reagents to the two labeled tubes as follows:

Blank: 3.0 ml of serum in glycine buffer prepared above, plus 0.05 ml of water

Unknown: 3.0 ml of serum in glycine buffer prepared above, plus 0.05 ml of uricase solution

Gently mix each tube well and let stand at room temperature for 15 minutes. Transfer the two solutions to two separate matched quartz cuvets. Set the spectrophotometer to 292 nm and, with the Blank cuvet in the light path, set the meter to an initial absorbance setting of about 0.40 by adjusting slit width or energy level. The actual $A_{initial}$ is not critical; it may be 0.30, 0.40, or 0.50. However, it should be high enough so the A_{final} is not less than 0. Be sure to record the $A_{initial}$ in your notebook. Remove the Blank cuvet from the light path and read A_{292} for the Unknown. After 5 more minutes of incubation, adjust the spectrophotometer with Blank to $A_{initial}$ as before and read the absorbance of the Unknown cuvet. If this absorbance has decreased, wait another 5 minutes and then read A_{292} of the unknown. Be sure to adjust to $A_{initial}$ with Blank. Continue taking A_{292} readings for the unknown until the uricase reaction is complete (until A_{292} readings are constant). Read and record the final absorbance of the Unknown cuvet.

The uricase solution may contribute some absorbance to the reaction mixture that is not associated with the water blank. To correct for this, pipet 0.05 ml of uricase solution into 3.0 ml of glycine buffer, pH 9.4, in a quartz cuvet. Mix well and read the A_{292} for this solution, using glycine buffer in a quartz cuvet as reference. The A_{292} ($A_{uricase}$) will be approximately 0.005.

In summary, the above procedure should result in three absorbance readings,

1. $A_{initial}$

2. A_{final} depends on uric acid concentration

3. $A_{uricase}$ = approximately 0.005

The calculations for uric acid determination are described in the Analysis of Results.

III. ANALYSIS OF RESULTS

A. Measurement of Cholesterol

1. Calculation of Total Serum Cholesterol. The cholesterol assay as used in this experiment is linear up to 500 mg/100 ml serum. A calibration curve is not essential and a single standard can be used. Use Equation 5 to calculate the cholesterol concentration. If the serum sample was turbid, you should have determined A_{510} of serum in saline water, A_c. If so, subtract this from $A_{510(x)}$. Equation 5 then becomes

$$\text{mg/100 ml} = \frac{A_{510(x)} - A_c}{A_{510(s)}} \times C_s$$

The concentration of standard cholesterol is 200 mg/100 ml. Calculate total cholesterol in the serum samples you tested. Compare your results with normal values of serum cholesterol.

2. Calculation of HDL Cholesterol Concentration. The serum sample is treated differently than in procedure 1 so the calculation is modified.

$$\text{HDL cholesterol (mg/100 ml)} = \frac{A_{\text{serum}}}{A_{\text{standard}}} \times C_s \times 1.125$$

where

A_{serum} = absorbance at 510 nm from serum containing HDL cholesterol

A_{standard} = absorbance at 510 nm from known amount of cholesterol standard II

C_s = concentration of cholesterol standard. The standard is 200 mg/100 ml; however, this was diluted before analysis:

$$\frac{0.02 \text{ ml standard}}{0.15 \text{ ml total volume}} \times 200 \text{ mg/100 ml} = 26.6 \text{ mg/100 ml}$$

1.125 = factor to correct for serum dilution; 0.4 ml of serum was treated with 0.05 ml of precipitating reagent, so the factor is

$$\frac{0.4 + 0.05}{0.4} = \frac{0.45}{0.4} = 1.125$$

Calculate the concentration of serum HDL cholesterol in the samples you tested. Also, calculate the concentration of cholesterol associated with LDL for each serum sample.

B. Measurement of Uric Acid

The relationship derived from Beer's law was described in the Introduction. This may be simplified to:

$$C' = \Delta A_{292} \times 50.04$$

where

C' = concentration of uric acid in mg/100 ml serum

$\Delta A_{292} = A_{\text{initial}} - A_{\text{final}}$

If you corrected for uricase absorption, then this term becomes $(A_{initial} - A_{final}) + A_{uricase}$. Calculate the concentration of uric acid in each serum sample. Compare these values to normal values.

IV. QUESTIONS

1. What is the structure of uric acid under physiological conditions?

2. Derive the equation used for the calculation of uric acid concentration. Hint: Begin with the Beer's law relationship.

3. Assume you are to determine the blood serum level of cholesterol by the single standard method. The absorbance data obtained are: $A_{standard} = 0.350$ (250 mg/100 ml); $A_{sample} = 0.390$; and $A_{correction} = 0.02$. Calculate the concentration of cholesterol in mg/100 ml. What would be the % error if $A_{correction}$ were not used?

4. What is the primary assumption made in the use of a single standard as in Question 3?

5. The cholesterol assay is linear to a level of 500 mg/100 ml. What would you do if a determination for cholesterol showed the level to be about 650 mg/100 ml?

6. Study Reaction 4 (uricase-catalyzed reaction) and suggest an alternate method for the determination of uric acid.

V. REFERENCES

General

M. Gurr and A. James, *Lipid Biochemistry*, 2nd Edition (1975), Halsted Press (London), pp. 165–200. An excellent introductory textbook for undergraduates.

R. Montgomery, R. Dryer, T. Conway, and A. Spector, *Biochemistry, A Case-Oriented Approach*, 3rd Edition (1981), Mosby (St. Louis), pp. 353–391; 513–518. A good mix of biochemistry and medicine.

N. Tietz, Editor, *Fundamentals of Clinical Chemistry*, 2nd Edition (1976), Saunders (Philadelphia), pp. 506–514; 999–1002. A classic reference in the area of clinical chemistry.

Specific

A. Cornish-Bowden, *Fundamentals of Enzyme Kinetics* (1979), Butterworths (London), pp. 42–45. An excellent text for advanced reading in enzyme kinetics.

R.D. Ellefson and W.T. Caraway in *Fundamentals of Clinical Chemistry*, N. Tietz, Editor, 2nd Edition (1976), Saunders (Philadelphia), pp. 506–512. "Cholesterol."

W.R. Faulkner and J.W. King in *Fundamentals of Clinical Chemistry*, N. Tietz, Editor, 2nd Edition (1976), Saunders (Philadelphia), pp. 999–1002. "Uric Acid."

Sigma Chemical Company Technical Bulletin No. 292-UV (1976). "The Ultraviolet Determination of Uric Acid." Sigma Chemical Co., P.O. Box 14508, St. Louis, MO 63178.

Sigma Chemical Company Technical Bulletin No. 350-HDL (1981). "The Enzymatic Determination of HDL and Total Cholesterol." Sigma Chemical Co., P.O. Box 14508, St. Louis, MO 63178.

Experiment **24**

Determination of Vitamin C in Foods and Biological Fluids

RECOMMENDED READING:
Chapter 8

SYNOPSIS
Vitamin C is a nutritional factor that must be present in our diet. Its presence in dietary materials can be detected and quantified by titration with 2,6-dichlorophenolindophenol. In this experiment, serum, plasma, or urine, and several fruits and vegetables are analyzed for vitamin C.

I. INTRODUCTION AND THEORY

Complete lack of vitamin C (ascorbic acid) in the diets of humans and other primates leads to a classic disease, scurvy. This nutritional disease, which was probably the first to be recognized, was widespread in Europe during the fifteenth and sixteenth centuries, but it is rare today.

Ascorbic acid is widely distributed in nature, but it occurs in especially high concentration in citrus fruits and green plants such as green peppers and spinach. Ascorbic acid can be synthesized by all plants and animals with the exception of humans, other primates, and guinea pigs. Therefore, vitamin C must be present in our dietary substances.

The fundamental role of ascorbic acid in metabolic processes is not well understood. There is some evidence that it may be involved in metabolic hydroxylation reactions of tyrosine, proline, and some steroid hormones, and in the cleavage-oxidation of homogentisic acid. Its function in these metabolic processes appears to be related to the ability of vitamin C to act as a reducing agent.

Although there is much controversy about the exact requirement, the adult Recommended Daily Allowance of vitamin C is 70 mg per day. Some scientists and physicians have suggested doses up to 1 to 3 grams per day in order to help resist the common cold. Deficiency of vitamin C results in swollen joints, abnormal development and maintenance of tissue structures, and eventually, scurvy.

Chemically, ascorbic acid is a water-soluble, slightly acidic carbohydrate that exists in an oxidized or reduced form (Reaction 1). Both forms are biologically active.

ascorbic acid		dehydroascorbic acid
(reduced form)		(oxidized form)

(Reaction E24.1)

Because of the clinical significance of vitamin C, it is essential to be able to detect and quantify its presence in various biological materials. Analytical methods have been developed to determine the amount of ascorbic acid in foods and in biological fluids such as blood and urine. Ascorbic acid may be assayed by titration with iodine, reaction with 2,4-dinitrophenylhydrazine, or titration with a redox indicator, 2,6-dichlorophenolindophenol (2,6-DCIP) in acid solution. The latter method will be used in this experiment because it is reasonably accurate, rapid, convenient, and versatile since it can be applied to many different types of samples.

The reaction of 2,6-DCIP with ascorbic acid is shown in Figure E24.1. Ascorbic acid reduces the indicator dye from an oxidized form (red in acid) to a reduced form (colorless in acid). The procedure is simple, beginning with dissolution of the sample to be tested in metaphosphoric acid. An aliquot of the sample is then titrated directly with a solution of 2,6-DCIP. Although the original DCIP solution is blue, it becomes light red in the acid solution. Upon reaction with ascorbic acid in the sample, the dye becomes colorless. Titration is continued until there is a very slight excess of dye added (faint pink color remains in the acid solution).

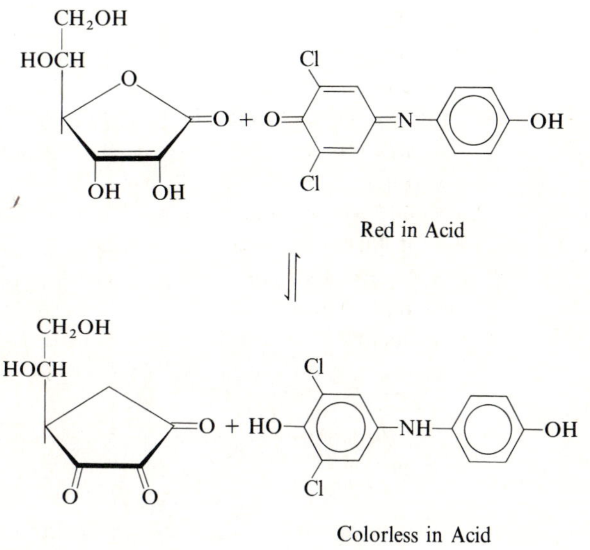

FIGURE E24.1
Reaction of the redox dye, DCIP,
with ascorbic acid.

Samples for analysis often contain traces of other compounds, in addition to ascorbic acid, that reduce DCIP. One method to minimize the interference of other substances is to analyze two identical aliquots of the sample. One aliquot is titrated directly and the total content of all reducing substances present is determined. The second aliquot is treated with ascorbic acid oxidase to destroy ascorbic acid and then titrated with DCIP. The second titration allows determination of reducing substances other than ascorbic acid. A second method that may be used to reduce the effect of other reducing substances is to perform the titration in the pH range of 1 to 3. The interfering agents react very slowly with DCIP under these conditions.

The determination of vitamin C in biological fluids such as blood and urine is more difficult because only small amounts of the vitamin are present and many interfering reducing agents are present. Substances containing sulfhydryl groups, sulfite, and thiosulfate are common in biological fluids and react with DCIP, but much more slowly than ascorbic acid. The interference by sulfhydryl is often minimized by the addition of *p*-chloromercuribenzoic acid.

Overview of the Experiment

In this experiment several biological samples will be prepared, and the amount of vitamin C in each will be determined. Each sample will be titrated with standardized redox indicator, 2,6-DCIP. The vitamin C content will be calculated in terms of milligrams per milliliter (mg/ml) of sample.

This experiment is flexible in that students may choose the samples they wish to analyze. It is suggested that several fruits and vegetables as well as an unknown ascorbic acid sample be available for analysis. If desired, urine and/or serum may be analyzed; however, both samples will require deproteinization. Ascorbic acid in blood remains stable for only about 3 hours, so fresh samples must be analyzed. Blood plasma samples should be collected with EDTA, oxalate, or heparin as an anticoagulant.

The time required for this experiment can be adjusted by controlling the number and type of samples to be analyzed.

II. EXPERIMENTAL

Materials and Supplies

Fruit juices—orange, grapefruit, lemon, or lime

Whole fruits and vegetables—oranges, grapefruit, lemons, limes, green peppers, tomatoes, potatoes, spinach, lettuce, and others

Fresh biological fluids (serum, plasma, or urine)

Metaphosphoric acid/acetic acid solution, 4%

Unknown ascorbic acid in metaphosphoric acid/acetic acid solution, 0.5 to 3.0 mg/ml

Standard ascorbic acid in metaphosphoric acid/acetic acid solution, 1 mg/ml

2,6-Dichlorophenolindophenol solution in H_2O, 25 mg/100 ml

p-Chloromercuribenzoic acid, 2 mg/ml in 0.05 N NaOH

Ascorbic acid oxidase, lyophilized powder

Buret; use a 10-ml microburet, if available

Mortar and pestle

Knife

Filter paper, fast flow

Glass funnel

Procedure

A. Standard Ascorbic Acid Solution. Fill a microburet with 2,6-DCIP solution. The top of the solution should be at or slightly below the zero mark of the buret, and the glass tip must be full of solution. Using a pipet, transfer 1.0 ml of the ascorbic acid standard solution to a 50 ml Erlenmeyer flask containing 5 ml of 4% metaphosphoric acid/acetic acid solution. Read and record the initial reading on the buret. Titrate by rapid, dropwise addition of 2,6-DCIP from the buret while mixing the contents of the flask. Add DCIP solution until a distinct rose-pink color persists for 15 to 20 seconds. Record the final reading on the buret. Repeat this procedure twice more, each time with a fresh 1.0-ml sample of ascorbic acid standard. In a similar fashion, titrate three blanks, each containing 5.0 ml of 4% metaphosphoric acid/acetic acid solution and 1.0 ml of water. Average the results for each series of measurements.

B. Unknown Ascorbic Acid Solution. Obtain a sample containing an unknown amount of ascorbic acid from your instructor. Place 1.0 ml of the unknown in a 50-ml Erlenmeyer flask. Add 5.0 ml of metaphosphoric acid solution to the flask and titrate as before with DCIP. Repeat with two more samples of the unknown.

C. Fruit Juice. Mix the stock solution of juice well and pour about 15 ml into a small beaker. Dilute 10.0 ml of this juice to 50 ml with the metaphosphoric acid solution. Filter this solution through a rapid flow, fluted filter paper. Pipet 10.0 ml of the filtrate into a 50-ml Erlenmeyer and titrate rapidly to a persistent rose-pink color with the standard DCIP solution as previously described. Repeat the titration with two 10.0-ml samples of the diluted juice. For a blank, add 10.0 ml of water to 40 ml of metaphosphoric acid solution and titrate three 10.0-ml samples with DCIP. Record the volume of DCIP necessary to titrate each juice sample and each blank.

To test for the presence of interfering substances in the juice, pipet a 10.0 ml sample of the fresh, undiluted juice into a 50-ml Erlenmeyer. Add a few crystals of ascorbic acid oxidase to destroy the ascorbic acid. Let stand, after gentle mixing, for 10 minutes. Add 40 ml of metaphosphoric acid solution. Titrate three 10.0-ml portions of this diluted juice with DCIP.

D. Whole Fruit. Peel the fruit, weigh to the nearest 0.1 g, and cut a sample of 25 to 50 g from the edible portion. Weigh the sample to the nearest 0.1 g. Slice the sample into many small pieces, transfer to a mortar, and grind with 25 ml of the metaphosphoric acid solution. Pour the liquid from the extract into a 100-ml volumetric flask. Grind the solid residue in the mortar twice using 25 ml of the metaphosphoric acid solution. Add the liquid extract each time to the 100-ml volumetric flask. Add metaphosphoric acid solution to the volumetric flask to a total volume of 100 ml. Filter the solution through a rapid flow, fluted filter paper. Transfer 10.0 ml of the filtrate into a 50-ml Erlenmeyer flask. Titrate rapidly with the DCIP solution. Repeat the titration on two more 10.0-ml samples of the filtrate. Titrate 10.0 ml of metaphosphoric acid solution as a blank.

E. Serum, Plasma, or Urine. A protein-free sample must be prepared before analysis of vitamin C. Add 6.0 ml of metaphosphoric acid solution to 4.0 ml of serum, plasma, or urine. Mix well and centrifuge for 10 minutes at 1000 rpm. The supernatant from serum or plasma samples may be titrated directly with standard DCIP. Transfer 9.0 ml of the sample to a 50-ml Erlenmeyer flask and titrate with DCIP. Repeat the titration on two more samples of protein-free serum or plasma.

The supernatant from urine must be treated with *p*-chloromercuribenzoic acid before titration. Transfer 9.0 ml of the protein-free supernatant into a 50-ml Erlenmeyer and add 1.0 ml of *p*-chloromercuribenzoic acid. Allow the solution to stand for 5 to 10 minutes and then centrifuge for 10 minutes at 1000 rpm. Titrate 9 ml of the supernatant with DCIP.

III. ANALYSIS OF RESULTS

A. Standard Ascorbic Acid Solution. Subtract the volume of titrant required for the blank from the titrant required for the sample. The difference represents the volume of DCIP that is equivalent to 1 mg of vitamin C. Calculate the standard deviation of your answer. What are the limits for 95% confidence? Review Chapter 8 for statistical analysis.

B. Unknown Ascorbic Acid Solution. Calculate the concentration of ascorbic acid in the unknown sample in units of 1 mg/ml. Express your answer with confidence limits at the 95% confidence level.

C. Fruit Juices. Calculate the amount of ascorbic acid in the juices and report in terms of mg/100 ml of pure juice. Again, calculate confidence limits at the 95% confidence level. Remember that you started with 10 ml of pure juice, diluted it with 40 ml of metaphosphoric acid solution, and titrated 10 ml of the diluted juice. Are there reducing substances in addition to ascorbic acid in the juice?

D. Whole Fruit. Calculate the total amount of ascorbic acid in the whole fruit. Use proper statistical analysis.

E. Biological Samples. Calculate the amount of ascorbic acid in units of mg/100 ml of fluid. When calculating the ascorbic acid content for urine, be sure to multiply the final result by 4/3 to correct for the dilution by *p*-chloromercuribenzoic acid solution.

Typical values for vitamin C content are shown below.

Orange juice	20–80 mg/100 ml
	Average = 45 mg/100 ml
Grapefruit juice	35–65 mg/100 ml
	Average = 40 mg/100 ml
Lemon juice	30–70 mg/100 ml
	Average = 45 mg/100 ml
Lime juice	5–40 mg/100 ml
	Average = 15 mg/100 ml
Plasma	0.2–2 mg/100 ml

Description of Nutritional Conditions
    less than 0.1 mg ascorbic acid/100 ml plasma—deficiency
    0.1–0.19 mg ascorbic acid/100 ml plasma—low
    0.2–0.4 mg ascorbic acid/100 ml plasma—acceptable

IV. QUESTIONS

1. Write the reaction for the oxidation of ascorbic acid catalyzed by ascorbic acid oxidase.

2. A 10-ml sample of pure orange juice was diluted with 40 ml of metaphosphoric acid solution. Then 10 ml of the diluted juice was titrated to an endpoint with 9.27 ml of 2,6-DCIP. A blank required 0.25 ml of DCIP. A 1-mg sample of pure ascorbic acid required 6.52 ml (after blank correction) of 2,6-DCIP. What is the concentration of ascorbic acid in the orange juice in mg/100 ml?

3. How does *p*-chloromercuribenzoic acid function to remove sulfhydryl group interference during the DCIP titration?

4. A whole, peeled orange weighed 100 g. Three sections (30 g) were extracted with metaphosphoric acid and the total extract filtered and diluted to 100 ml. A sample of 10 ml of the filtered extract required 4.10 ml of 2,6-DCIP after blank correction. Also, 1 mg of standard ascorbic acid required 7.2 ml of DCIP (blank corrected). How many milligrams of ascorbic acid are present in the whole, peeled orange?

V. REFERENCES

General

J.T. Baker Chemical Co., *Product Information Bulletin*, Commodity No. H114, Phillipsburg, N.J. 08865. Analytical uses of 2,6-DCIP.

J.M. Orten and O.W. Neuhaus, *Human Biochemistry*, Ninth Edition (1975), Mosby (St. Louis), pp. 588–593. Biochemical properties of ascorbic acid.

J.D. Rawn, *Biochemistry* (1983), Harper & Row (New York), p. 397. Biological function of ascorbic acid.

N. Tietz, Editor, *Fundamentals of Clinical Chemistry*, Second Edition (1976), Saunders (Philadelphia), pp. 547–551. A detailed discussion of chemical, biological, and clinical aspects of ascorbic acid.

A. White, P. Handler, E. Smith, R. Hill, and I. Lehman, *Principles of Biochemistry*, Sixth Edition (1978), McGraw-Hill (New York), pp. 1358–1361. Structure and properties of ascorbic acid.

Electrophoresis of Blood Lipoproteins on Cellulose Acetate

RECOMMENDED READING:
Chapter 4

SYNOPSIS
Large nonpolar lipid molecules are transported in the blood stream as components of lipoproteins. The clinical analysis of serum/plasma lipoproteins is valuable for the diagnosis of many metabolic disorders, particularly those associated with the cardiovascular system. In this experiment blood lipoproteins are characterized by electrophoresis on cellulose acetate membranes.

I. INTRODUCTION

Structure and Function of Blood Lipoproteins

The complex lipid molecules that are transported through our veins and arteries are not very soluble in the aqueous blood plasma. These substances, including the triacylglycerols, cholesterol, cholesterol esters, and phospholipids, are carried from one tissue to another as constituents of lipoproteins. Lipoproteins are particles that range in molecular weight from 250,000 to 40,000,000 and in size from 10 to 1000 nm. The lipoproteins are classified according to their densities, which are indicated by their

TABLE E25.1
Properties of Major Serum Lipoproteins

	Chylomicron	VLDL	LDL	HDL
Electrophoretic classification	omega	pre-beta	beta	alpha
Density	< 0.95	0.95–1.006	1.006–1.063	1.063–1.2
Composition (% dry weight)				
protein	1–2	5–10	25	30–35
triacylglycerol	85	52–57	10	5–10
cholesterol/ cholesterol ester	5	20	45	30
phospholipids	8–9	18	20	30

behavior in the ultracentrifuge. In Table E25.1 the physical and chemical characteristics of the common human lipoproteins are reviewed. The least dense lipoproteins (chylomicrons) contain 1 to 2% protein and 98 to 99% lipid material made up primarily of triacylglycerols. In contrast, high-density lipoproteins (HDLs) consist of approximately 33% protein, but only 8% triacylglycerol. The primary lipids associated with the HDLs are the phospholipids (about 30%) and cholesterol/cholesterol esters (also 30%). Note from Table E25.1 that the cholesterol content varies from 5% in the chylomicrons to 45% in the LDLs.

All lipoproteins have the same general structure; they are comprised of a central hydrophobic core of triacylglycerol and cholesterol ester with a surrounding layer composed of protein and amphipathic lipids (lipids that have both hydrophilic functional groups and hydrophobic regions).

The biological function of each of the lipoprotein types is not completely understood, but we do have some knowledge of their action. Chylomicrons and VLDLs transport primarily dietary triacylglycerols from the intestines and liver to the peripheral tissue, where the fatty acids are released and used for energy metabolism or stored for future degradation. The LDLs (which may be formed by removal of triacylglycerols from VLDLs) carry mostly cholesterols. The function of the HDLs is even less understood, but they play some role in the transport of cholesterol from peripheral tissue to the liver.

Clinical Implications of Blood Lipoproteins

There are currently high interest and quite rapid advances in our knowledge of lipoprotein metabolism and function. This is due to evidence that the incidence of atherosclerosis and other heart diseases shows a positive correlation with the level of blood cholesterol and lipoprotein concentration. Individuals with high serum levels of cholesterol-bearing LDLs show an increased incidence of atherosclerosis and other cardiovascular diseases. In contrast, high levels of HDL cholesterol are inversely related to coronary disease. In general, the higher the ratio of HDL-cholesterol to LDL-cholesterol, the lower the incidence of heart disease. The reasons behind these observations are not well understood, but one current explanation is that LDLs carry cholesterol to peripheral tissues where it may be deposited, whereas HDLs carry cholesterol away from peripheral tissues to the liver where it is degraded and excreted.

Clinical Analysis of Blood Lipoproteins

Clinical measurements of serum lipoproteins are of value because they can be used in the recognition and identification of metabolic disorders affecting the cardiovascular system. In Experiment 23, you are introduced to a method for the quantitative analysis of total serum cholesterol and HDL-cholesterol. The results from this analysis may be used as an indicator for heart disease, but more specific information about the type and concentration of serum lipoproteins is needed to recognize specific disorders such as hyperlipoproteinemias, abnormal lipoproteins, or lipoprotein deficiencies.

The ultracentrifuge may be used to separate serum lipoproteins into four fractions based on density (Table E25.1). This technique is of value for the separation and isolation of specific lipoproteins for further characterization, but it does not provide a rapid and satisfactory method for their identification and quantification.

The most valuable technique for the separation and analysis of serum lipoproteins is electrophoresis; in fact, the hyperlipidemias are currently described and identified on the basis of electrophoretic patterns. Each type of lipoprotein disorder leads to a distinct electrophoretic profile due to increased or decreased concentrations of certain lipoproteins. Normal serum is characterized by negligible chylomicrons (omega class, near origin), distinct LDL band (beta class), VLDL (pre-beta class), and HDL (alpha class) as shown in Figure E25.1. The significant characteristics of several types of inherited hyperlipoproteinemia are also shown in Figure E25.1.

Serum lipoproteins have been analyzed on all types of electrophoretic supports including paper, cellulose acetate, polyacrylamide gel, and agarose. Although the relative migration of the lipoproteins is similar in all media, each type of support offers some advantages. Paper and cellulose acetate electrophoresis is simple, convenient, and reproducible; however, it is not very sensitive, sometimes leads to streaking, and gives variable results depending on experimental conditions (time period of electrophoresis, buffer concentration, staining technique, etc.). Agarose gel and polyacrylamide gel are both more sensitive than paper/cellulose acetate, but

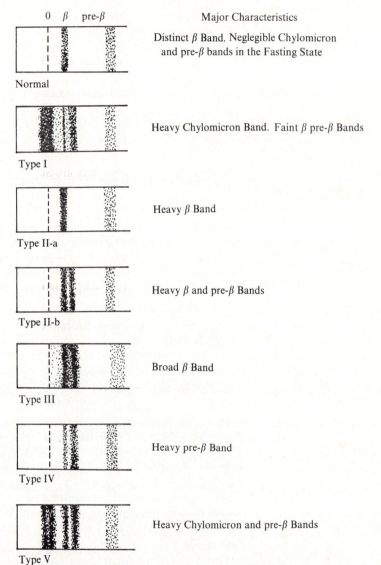

FIGURE E25.1

Electrophoretic patterns of lipoproteins in hyperlipoproteinemia. Migration is from the origin, O, to the right. α = HDL, β = LDL. Chylomicrons (VLDL) remain near the origin. From R. Ellefson and W. Caraway in FUNDAMENTALS OF CLINICAL CHEMISTRY, Second Edition by N. Tietz, W.B. Saunders Co. (Philadelphia), 1976. Reprinted by permission.

the experimental procedure is more time-consuming and VLDLs sometimes migrate in an unpredictable fashion. For reasons of simplicity, convenience, and diversity of techniques, cellulose acetate media will be used in this experiment to separate and characterize serum lipoproteins.

The lipoproteins are colorless and will not be visible on the electrophoresis support. The common stains used in protein electrophoresis (Coomassie blue, amido black, and silver) visualize all proteins in blood serum. The use of these dyes on plasma or serum results in complex patterns of up to 25 protein bands, only some of which are lipoproteins. For specific staining of the lipoproteins, lipophilic dyes are used. Two of the most useful lipoprotein-specific dyes are Sudan black B (a prestaining dye) and Oil Red O (a poststaining dye). Sudan black B is a very sensitive stain, but its use as a prestain requires a 1-hour incubation with the serum sample prior to electrophoresis. Oil Red O is not as sensitive as Sudan black, and it may require as much as 5 hours for the staining-destaining procedure.

Overview of the Experiment

Blood plasma or serum is prepared and analyzed for lipoprotein content by cellulose acetate electrophoresis. Directions for both prestaining with Sudan black and poststaining with Oil Red are given, and the preferred method may be selected. The resulting electrophoregrams may be compared with typical electrophoretic patterns as shown in Figure E25.1.

Completion time for this experiment depends upon the staining procedure chosen. If only a single 3-hour period is available, the instructor or teaching assistants can prepare the Sudan black prestained serum/plasma samples up to 12 to 16 hours before electrophoresis. Spotting the sample and electrophoresis will then require 2 to 3 hours. During this waiting period, the cholesterol measurements described in Experiment 23 may be done on the blood samples. If students are to complete the entire experiment (prestaining of the sample, application, and electrophoresis), approximately 4 hours are required.

If the Oil Red poststaining procedure is done, the electrophoresis can be run during one 3-hour class period and the staining may be done during the next class period. As previously mentioned, cholesterol measurements on the blood samples can be completed during the electrophoresis, and another experiment can also be scheduled during the staining procedure.

II. EXPERIMENTAL

Materials and Supplies

Whole blood, plasma or serum

Sodium oxalate, 0.1 M

Barbital buffer, pH 8.6, 0.058 M, containing 0.001 M EDTA and 1% serum albumin

Oil Red O dye in acetone/H_2O

Cellulose acetate strips

Double wire applicator

Electrophoresis apparatus with power supply

Forceps or tweezers

Tray for wetting and staining

Sudan black B prestain solution

Oven at 110°C

Procedure

Preparation of Blood Sample

The blood specimen should be obtained from a patient who has fasted for 12 hours. Plasma or serum may be used. If plasma is desired, the blood should be collected in EDTA-treated tubes, or it should be treated with sodium oxalate as follows.

Plasma. Immediately mix 9 ml of fresh blood with 1 ml of 0.1 M sodium oxalate. Centrifuge the mixture for 15 to 20 minutes at 5000 × g and pour the plasma (supernatant) into a container. The plasma may be used directly for electrophoresis or stored for up to 1 to 2 days at 0 to 4°C. Frozen specimens often display atypical results.

Serum. Allow freshly drawn blood to clot while standing at room temperature for 15 to 20 minutes. Pour the liquid from the clot into a tube and centrifuge it for 15 to 20 minutes at 5000 × g. Pour off the serum (supernatant) and store no more than 1 to 2 days at 0 to 4°C.

Prestaining with Sudan Black

A plasma or serum sample (0.2 ml) is mixed in a small test tube with 0.1 ml of the Sudan black staining solution. The mixture is allowed to sit in the dark at room temperature for one hour. If desired, the prestained sample can be stored for up to 16 hours at 4°C before application on the cellulose acetate.

Preparation of Cellulose Acetate Strip

Place 100 ml of buffer into a staining tray. Obtain a cellulose acetate strip from the package, using forceps. With a pencil, *very faintly* mark the strip for location of the sample. This mark should be near the center and perpendicular to the length of the strip. Do not touch the strip with your fingers. Float the strip on the surface of the buffer. Do not dunk it in buffer until it is wet. When the strip is completely wet, submerge it in the buffer and allow it to soak for 10 minutes. Meanwhile, fill the electrophoresis chambers with buffer. The levels of buffer in the two electrode chambers must be the same. Remove the cellulose acetate strips from the soaking buffer and remove excess buffer by blotting between two filter papers. Place the strip across the support bridges in the electrophoresis chamber so that the strip is centered and the ends are below the buffer level in each buffer chamber. Most electrophoresis chambers have magnetic pieces to hold the cellulose acetate strips under tension. Immediately proceed to the next step without allowing the strip to dry.

Application of Sample to the Wetted Strip

Obtain a double wire applicator. With a syringe, transfer 10 μl of plasma or serum (unstained or prestained) to the applicator and apply the sample as a thin line on the origin mark of the cellulose strip. Place the sample near the center of the strip away from the edges. Cover the electrode chamber.

CAUTION:

Do not touch the electrophoresis chamber or wires while the electrophoresis operation is in progress. Connect the power supply to the electrophoresis unit and have your instructor check the apparatus before the power supply is turned on.

Electrophoresis Procedure

Turn on the power supply and apply a current of 2 mA for each strip in the chamber. Apply the current for 30 to 45 minutes. Turn off the power supply and remove the strips from the chamber. Quickly dry the strips in an oven at 110°C.

Staining the Electrophoregrams with Oil Red

Soak the unstained strips in the Oil Red solution for 2 to 4 days. During this period, the strips should be lifted with a forceps from the bath and dunked several times in the dye. Soak the stained strips in distilled water or 7.5% acetic acid for 30 to 60 minutes to remove excess stain. Allow the stained strips to dry, and remove additional stain by soaking in hexane for 30 minutes if necessary.

III. ANALYSIS OF RESULTS

Draw a picture of the stained electrophoregram in your lab book. Compare the lipoprotein pattern on your strip with the profiles shown in Figure E25.1. Is there any indication that the experimental lipoprotein profile is abnormal? Identify all the stained bands on the cellulose acetate strip.

IV. QUESTIONS

1. The electrophoresis buffer can be reused several times, but it is important to alternate the polarity of the chamber after each use. Why?

2. The sample application method used in this experiment is the *wet application method*. An alternate method is *dry application* using a dry cellulose acetate strip. Describe advantages/disadvantages of each method.

3. Describe a procedure for the preparative scale separation of lipoproteins using the ultracentrifuge.

4. Compare the advantages/disadvantages of completing this experiment using polyacrylamide gel electrophoresis.

5. Describe a procedure for measurement of the molecular weights of the protein subunits in the lipoproteins.

V. REFERENCES

General

For discussions on the theory and practice of cellulose acetate electrophoresis, refer to Chapter 4 and the general references listed there.

J.D. Rawn, *Biochemistry* (1983), Harper & Row (New York), pp. 460–461. A brief discussion of plasma lipoproteins.

G. Zubay, *Biochemistry* (1983), Addison-Wesley (Reading, MA), pp. 555–560. Structure and function of serum lipoproteins.

Specific

R.D. Ellefson and W.T. Caraway in *Fundamentals of Clinical Chemistry*, N. Tietz, Editor, 2nd Edition (1976), Saunders (Philadelphia), pp. 529–538. "Lipoproteins in Serum."

K.A. Narayan in *Analysis of Lipids and Lipoproteins*, E.G. Perkins, Editor (1975), The American Oil Chemist's Society (Champaign, IL), pp. 225–249. "Electrophoretic Methods for the Separation of Serum Lipoproteins."

P. Nilsson-Ehle, A.S. Garfinkel, and M.C. Schotz, *Ann, Rev. Biochem.*, *49*, 667–693 (1980). "Lipolytic Enzymes and Plasma Lipoprotein Metabolism."

Experiment **26**

The Radioimmunoassay of Prostaglandins F in Plasma

RECOMMENDED READING:
Chapter 6; Chapter 8.

SYNOPSIS

Prostaglandins are fatty acid derivatives that are produced in essentially every human tissue. They have a wide range of physiological activity, but generally they act as hormone-like substances to regulate intracellular processes. Since the compounds are found in minute quantities in biological materials, very sensitive analytical procedures must be used to detect and quantify their presence. This experiment introduces the radioimmunoassay for the analysis of one of prostaglandin, PGF, in blood plasma.

I. INTRODUCTION

The Radioimmunoassay

Most biochemically significant molecules are present in extremely small quantities in biological materials. One goal of this laboratory manual has been to introduce students to analytical procedures that have sufficient sensitivity to detect and quantify chemical compounds in biological extracts. However, even some of the more sensitive procedures previously used in this manual (fluorescence, electrophoresis, etc.) are not applicable to the measurement of many important biomolecules. Compounds such as peptide and steroid hormones, prostaglandins, drugs, and various metabolites may be present at nanogram $(10^{-9}$ g) to picogram $(10^{-12}$ g) levels. Most analytical procedures are useless in this range. One technique available today that probably offers greater sensitivity than any previously discussed is the **radioimmunoassay** (RIA). This method combines the sensitivity of radioisotope counting with the specificity of immunology.

The basic principles of the RIA are outlined in Reaction 1.

$$\text{Ag* + Ag + Ab} \underset{\text{Ag—Ab}}{\overset{\text{Ag*—Ab}}{\rightleftharpoons}} \qquad \text{(Reaction E26.1)}$$

The following definitions will clarify the terms in the equation and explain some of the principles of immunochemistry:

Ag = antigen, a substance that induces the formation of antibodies when injected into an animal

Ag* = radiolabeled antigen

Ab = antibody, a family of serum proteins (immunoglobulins) that is produced in response to the presence of an antigen. Antibodies have the ability to interact in a specific manner with the antigen that stimulated their production.

Ag* — Ab or Ag — Ab = a complex consisting of antigent and antibody that is formed by specific noncovalent interactions.

Reaction 1 is the basis for an experimental procedure that allows the quantitative analysis of an antigenic substance, Ag. The basic steps for the RIA are outlined below and in Figure E26.1.

(1) A limiting amount of antibody, Ab, directed against a specific antigen is first incubated with various saturating levels of antigen. The equilibrium set up between Ab and Ag is under the control of the law of mass action (Step 1 in Figure E26.1).

(2) Radiolabeled antigen, Ag*, is then added to the incubation mixture. Ag* combines with free Ab, which shifts the equilibrium of Reaction 1. Now, two forms of antigen-antibody complex are present, Ag—Ab and Ag*—Ab. The two types of antigen, Ag and Ag*, compete for a limited number of binding sites on Ab. As the ratio of Ag/Ag* increases, the concentration of radiolabeled free antigen, Ag*, increases. This is brought about by an increase in the amount of Ag—Ab formed at the expense of Ag*—Ab. If Ag/Ag* = 1, then one-half of the Ag* is bound; however, if Ag/Ag* = 2, then only one-third of the Ag* is bound (Step 2 in Figure E26.1).

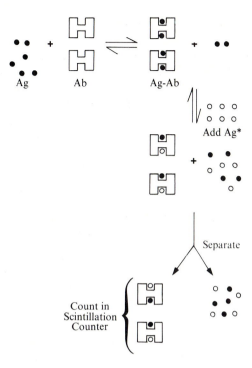

FIGURE E26.1
The basic principles of the RIA.
See text for details.

FIGURE E26.2
A standard curve for the RIA of
an antigen.

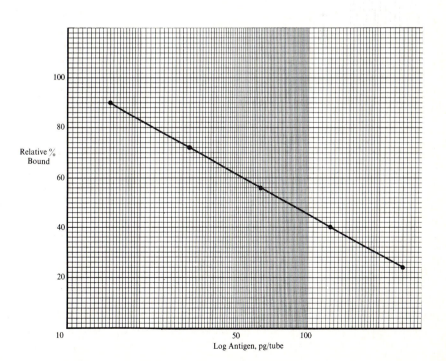

(3) The antibody-antigen complex is then physically separated from free Ag and Ag*. The amount of radioactivity in the Ag*—Ab fraction is experimentally measured in a liquid scintillation counter (Step 3 in Figure E26.1).

(4) The scintillation counting data are used to construct a standard curve. A plot of % bound Ag vs. log of the amount of Ag initially present is used to estimate the amount of Ag in an unknown sample. A typical plot is shown in Figure E26.2.

Experimental Procedures of the RIA

The RIA may be used for the analysis of a great variety of biomolecules. Table E26.1 lists several compounds that have been measured by an RIA method. Some of the necessary procedures are described below.

Preparation of Antigen. Once the substance to be analyzed has been determined (we will call it X), it must be converted into a form that will

TABLE E26.1
Biologically Significant Compounds Measured by RIA

Drugs

mescaline	digitoxin
morphine	barbiturates
LSD	

Steroids

estrogens	progesterone
testosterone	aldosterone

Intracellular Messengers

cyclic AMP	prostaglandins E, A and
cyclic GMP	F and metabolites

Peptide Hormones

luteinizing hormone-release factor	calcitonin
thyrotropin-releasing hormone	gastrin
parathyroid hormone	secretin
insulin	glucagon
bradykinin	angiotensins
ACTH	
vasopressin	

Miscellaneous Compounds
enzyme cofactors
messenger RNA
carcinoembryonic antigen
human immunoglobulin

elicit the production of specific antibodies. If X is a small molecule less than 5000 to 10,000 in molecular weight), it will probably not stimulate the formation of antibodies. X must first be covalently attached to a large immunogenic molecule, usually a protein. One of the most common proteins used is bovine serum albumin (BSA). Various linking methods may be used to produce X—BSA.

Preparation of Reagent Antibodies. The specific reagent antibody against the antigen, X—BSA, is produced by immunizing an animal, usually a rabbit. The animal is injected with the X—BSA according to a schedule, and serum containing the antibodies (antiserum) is collected. The antibodies need not be purified, but the antiserum must contain immunoglobulins with a specific affinity for X and should be checked for the presence of interfering substances. Although the immunogenic substance, X—BSA, has been used to elicit the antibodies, the antibodies will also interact in a specific manner and bind with the compound X.

Preparation of Radiolabeled Antigen. The labeled antigen, X*, must be chemically pure and have a high specific activity. The most commonly used isotopes are ^{3}H and ^{125}I. Many tritiated compounds are commercially available or may be prepared by custom synthesis. ^{125}I may be introduced into some peptide antigens using iodination catalyzed by lactoperoxidase.

Separation of Free and Bound Antigen. After the specific antibody has been incubated with a mixture of X and X*, the free X* must be separated from the antibody-bound X*. The conditions must be mild so that the complex X*—Ab is not disrupted. Potential separation methods may be divided into two types.

1. Adsorption. The antigen molecules, if they are small, may be adsorbed onto various media, including dextran-coated charcoal, florisil, or ion-exchange resin. These substances effectively remove small molecules from solution, but the time of contact between the adsorbent and reaction mixture must be carefully controlled. In this experiment, dextran-coated charcoal is used to adsorb the small unbound antigen. Large molecules, including the antigen-antibody complex, are not able to pass through the porous coating of dextran to the charcoal surface.

2. Precipitation of X—Ab. Various reagents, including organic solvents and ammonium sulfate, are used to cause precipitation of the complex. The precipitation may also be brought about by the **double-antibody method**. Here, a second antibody is prepared by immunization of another species of animal (sheep, horse, or cow) with the original rabbit antibody to X—BSA. The antiserum from the second animal, when added to the original rabbit antibody-antigen mixture, will cause precipitation of the original X-rabbit antibody.

The RIA of Prostaglandins

Prostaglandins are a relatively recently discovered class of biochemicals that have a wide range of physiological activities. These compounds are derived from C_{20} polyunsaturated acids. The structures of two such compounds are shown in Figure E26.3. Although the specific functions for the protaglandins are not known, in general they have hormonelike activities that are exerted in localized regions rather than throughout the entire body. One of the prostaglandins analyzed here, $PGF_{2\alpha}$, has several effects, including contraction of smooth muscle, regulation of cyclic GMP levels, and inhibition of progesterone secretion.

The medical and biochemical significance of the prostaglandins has led to the development of analytical techniques for their detection in biological tissues. Two major problems must be overcome before accurate, reproducible analytical techniques are available. First, although prostaglandins are present in most human biological fluids and tissue, their concentrations are on the order of picograms/ml (1 picogram = 10^{-12} g). Second, prostaglandins are very unstable. Some have a half-life of only a few seconds.

The RIA is becoming widespread in the analysis of prostaglandins. Antibodies to various prostaglandins have been prepared by immunizing animals with prostaglandins covalently linked to proteins such as bovine serum albumin. The antiprostaglandin-BSA serum may be used for the analysis of prostaglandins in various samples, including blood products and tissue. Antisera prepared against a single prostaglandin has very little cross-reactivity with other prostaglandins. For example, rabbit antiprostaglandin F—BSA serum can be used to analyze a sample for prostaglandins $PGF_{2\alpha}$ and $PGF_{1\alpha}$. The presence of other prostaglandins,

FIGURE E26.3
The structures of two prostaglandins.

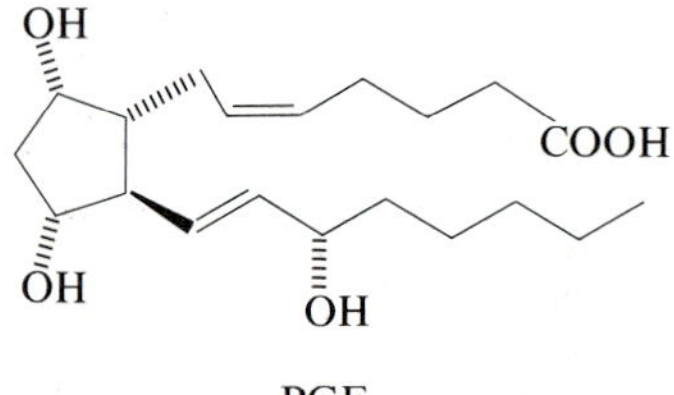

$PGF_{2\alpha}$

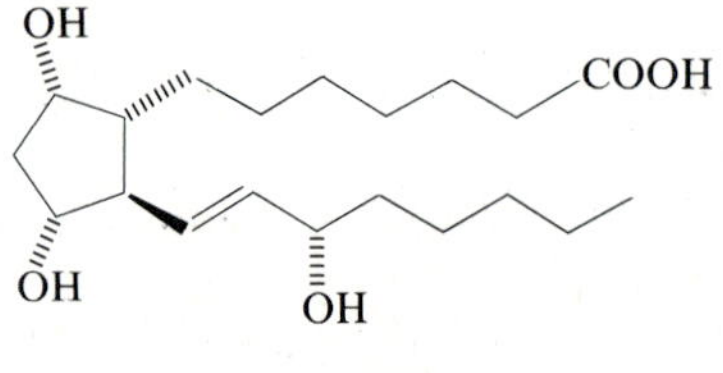

$PGF_{1\alpha}$

including PGA, PGB, or PGE, in the sample does not interfere with the analysis of PGF. The antiserum against PGF does not bind the other prostaglandins at a significant level.

One difficulty that is often encountered with prostaglandin measurement is that nonantibody proteins in the sample to be analyzed strongly bind prostaglandins, competing with their binding to the antibody. To circumvent this problem, the prostaglandins may be extracted from the biological sample. In this experiment, an organic extraction is used to isolate prostaglandins from blood plasma before the analysis.

Overview of the Experiment

In this experiment, the concentration of prostaglandins F ($PGF_{2\alpha}$ and $PGF_{1\alpha}$) in plasma samples will be estimated using rabbit antiprostaglandin F—BSA serum. The antiserum, prepared by injection of prostaglandin $F_{2\alpha}$—BSA into rabbit, is commercially available. The antiserum interacts specifically with PGFs, and only very weakly with other prostaglandins. Tritiated $PGF_{2\alpha}$, also commercially available, will be the form of the radiolabeled PGF.

A standard curve will be prepared by incubating various ratios of $PGF_{2\alpha}/PGF_{2\alpha}$—$^3H$ with the antiserum, separating the bound $PGF_{2\alpha}$—^3H from the unbound, and counting the radioactivity associated with the antibody. The standard curve will be used to determine the concentration of PGF in blood plasma.

Two forms of the plasma sample will be analyzed for PGF. One portion of the plasma will be extracted to isolate the prostaglandins, and another portion will be used directly in the RIA. A comparison of the two procedures will be made.

Approximately 3 to 4 hours are required to complete this experiment.

II. EXPERIMENTAL

Materials and Supplies

Plasma samples. Blood samples should be collected in heparinized or EDTA-treated tubes and used within 1 hour. If the blood sample must be collected early, it is stable at $-20°C$ for about 1 week.

Rabbit antiprostaglandin F—BSA serum. May be obtained commercially in a freeze-dried from. Suppliers will provide instructions for dilution and use. Generally, the diluted solution is stable for 12 hours.

Potassium phosphate buffer, 0.01 M phosphate, pH 7.4, containing 0.15 M NaCl, 0.1% NaN_3 and 0.1% BSA

Dextran-coated charcoal

Standard $PGF_{2\alpha}$. Five standard solutions are required: 2.5 ng/ml, 1.25 ng/ml, 0.63 ng/ml, 0.31 ng/ml, and 0.15 ng/ml.

Tritiated $PGF_{2\alpha}$. Fresh solution of 50,000 to 100,000 dpm/ml of tritiated compound (50–120 Ci/mmole) in phosphate buffer.

Scintillation liquid, contains PPO, POPOP, toluene, and Triton X-100.

Scintillation vials

Glass conical centrifuge tubes

Petroleum ether

Extraction solvent, ethyl acetate, isopropanol, and 0.1 M HCl (3:3:1)

Vortex mixer

Ethyl acetate

Centrifuge, capable of 7500 rpm and refrigerated at 4°C

Several small test tubes

Water bath, 50 to 60°C

Tank of prepurified N_2 gas

Procedure

CAUTION:

The phosphate buffer contains azide, a posion. Do not pipet any solutions by mouth. Experimentation with blood products presents a potential biohazard. Do not mouth pipet blood plasma or any solutions containing plasma. Dispose of excess phosphate buffer down the drain with lots of water.

The tritiated $PGF_{2\alpha}$ is highly radioactive. Observe all precautions, regarding handling of radioactive materials, described in Chapter 6. Immediately notify your instructor of any radioactive spills. Dispose of all radioactively contaminated waste materials in the proper containers.

A. Preparation of Blood Plasma

Centrifuge 3 ml of blood for 15 to 20 minutes at $5000 \times g$ (4°C). Pour the plasma (supernatant) into a glass tube and store on ice.

B. Extraction of Prostaglandins (Jaffe and Behrman, 1974)

Transfer 1.0 ml of plasma into two graduated conical centrifuge tubes. Add 3.0 ml of petroleum ether to each, and shake the tubes well to extract neutral lipids from the plasma. Allow the two layers to separate, remove the petroleum ether layer (upper) with a disposable pipet and discard into an "organic waste" bottle.

To the aqueous plasma phase, add 3.0 ml of ethyl acetate:iso-propanol:0.1 M HCl (3:3:1). Vortex the mixture twice for 15 seconds. Add 2.0 ml of ethylacetate and 3.0 ml of water to the mixture in each tube. Mix well by vortexing. Centrifuge the mixtures at 1000 rpm for 10 minutes to separate the phases.

Transfer the organic layer (upper) to two small test tubes and heat in a water bath at 50 to 60°C under a mild stream of N_2 gas to evaporate the organic solvent. The remaining residues are used in the next part.

C. Analysis of PGF (Miles-Yeda, Ltd)

Obtain 10 test tubes and prepare as described below and in Table E26.2. Tubes 11 and 12 in the table are the tubes containing the residue prepared in Part B.

Pipet 0.1 ml plasma into tubes 9 and 10. Add 0.10 ml of the appropriate unlabeled, standard PGF to tubes 4 through 8:

Tube 4	2.50 ng/ml
Tube 5	1.25 ng/ml
Tube 6	0.63 ng/ml
Tube 7	0.31 ng/ml
Tube 8	0.15 ng/ml

Add the appropriate amounts of phosphate buffer to tubes 1, 2, 3, 11, and 12 (see Table E26.2). Add 0.5 ml of antiserum to all tubes except 1 and 2.

Incubate all the tubes at 4°C for 30 minutes. After the incubation period, add 0.1 ml of ^3H—$PGF_{2\alpha}$ solution to all tubes, 1 to 12. Incubate all the tubes at 4°C for 60 min. Maintain the tubes at 4°C and add 0.2 ml of dextran-coated charcoal solution to all tubes except number 2.

Centrifuge all tubes at 4°C at 3000 rpm for 15 minutes. During the centrifugation, prepare 12 scintillation vials by adding 6 ml of scintillation fluid to each and numbering them 1 to 12. Use a separate pipet for each

TABLE E26.2

Preparation of Tubes for the RIA of Prostaglandins F

Reagents	Tube 1	2	3	4	5	6	7	8	9,10	11,12
Blood plasma	—	—	—	—	—	—	—	—	0.10	—
Plasma extract	—	—	—	—	—	—	—	—	—	residue
Standard $PGF_{2\alpha}$	0.1	——————————————————————————→							—	—
Phosphate buffer	0.60	0.90	0.10	—	—	—	—	—	—	0.10
Antiserum	—	—	0.5	——————————————————————————→						

tube and transfer 0.5 ml of each supernatant to the appropriate scintillation vial. Count each sample for 5 to 10 minutes in a liquid scintillation counter.

III. ANALYSIS OF RESULTS

1. Prepare a table with the headings:

Tube No.	Gross Counts	CPM	Net CPM	% Bound	Relative % Bound

2. Enter the gross counts for each sample into the table and convert to cpm by dividing each gross count by the time of measurement.

3. Calculate the net cpm for all standards and unknowns by subtracting the count of tube 1 (blank) from all tubes except tube 2 (total).

4. Calculate the percent bound for tubes 3 to 8 by dividing the net counts for each by the counts obtained in tube 2 (total). The column marked % bound represents the counts in each tube relative to the total possible counts (tube 2).

5. Calculate the relative percent bound for each standard tube (4 to 8) using the following equation:

$$\text{Relative \% bound} = \frac{\text{net cpm in sample}}{\text{net cpm in zero control}} \times 100$$

zero control = tube 3

6. Construct a standard curve by plotting relative % bound on the y axis vs. the log of the amount of $PGF_{2\alpha}$ in each tube (4 to 8). This amount is the unlabeled standard $PGF_{2\alpha}$ added to each standard tube. For example, tube 4 contained 0.1 ml of a solution of $PGF_{2\alpha}$ containing 2.50 ng/ml. This means that 250 picograms of unlabeled $PGF_{2\alpha}$ were present in the tube. After all the points are plotted, connect them with a smooth curve.

7. Calculate the relative % bound for the unknown samples exactly as for the known standards. Use the standard curve to determine the concentration of PGF in the plasma and plasma extract. Are there differences in your results? Explain.

IV. QUESTIONS

1. Outline the necessary procedure to perform an RIA for the hormone 17-β-estradiol.

2. In this experiment, why was the PGF linked to bovine serum albumin before preparation of the antibodies?

3. The following counting data were obtained from the RIA of the hormone estradiol. Prepare a standard curve and determine the amount of estradiol in samples 1 and 2.

Sample	Net cpm	% Bound	Relative % Bound
Total	42,350		
Unknown 1	11,256		
Unknown 2	14,324		
Standards			
500 ng/tube	2,252		
250 ng/tube	5,612		
125 ng/tube	8,978		
63 ng/tube	12,345		
31 ng/tube	16,162		
0 ng/tube	22,446		

V. REFERENCES

General

R.C. Bohinski, *Modern Concepts in Biochemistry*, 4th Edition (1983), Allyn and Bacon (Boston), pp. 500–501. An introduction to antibodies and RIA.

T.G. Cooper, *The Tools of Biochemistry* (1977), Wiley (New York), pp. 295–304. An introduction to principles and practice of RIA.

D. Freifelder, *Physical Biochemistry*, 2nd Edition (1982), Freeman (San Francisco), pp. 333–350. Applications of RIA to biochemistry.

H. Markowitz and N. Jiang in *Fundamentals of Clinical Chemistry*, N. Tietz, Editor, 2nd Edition (1976), Saunders (Philadelphia), pp. 274–297. "Immunochemical Principles."

D. Orth in *Methods in Enzymology*, B. O'Malley and J. Hardman, Editors, Vol. 37B, Academic Press (New York), pp. 22–38. "General Considerations for Radioimmunoassay if Peptide Hormones."

J.D. Rawn, *Biochemistry* (1983), Harper & Row (New York), pp. 144–149. An introduction to immunoglobulins.

L. Stryer, *Biochemistry*, Second Edition (1981), Freeman (San Francisco), pp. 789–812. Introduction to immunoglobulins.

H. Van Vunakis in *Methods in Enzymology*, H. Van Vunakis and J. Langone, Editors, Vol. 70A (1980), Academic Press (New York), pp. 201–209. "Radioimmunoassays: An Overview."

Specific

B. Jaffe and H. Behrman in *Methods of Hormone Radioimmunoassay*, pp. 19–34 (1974). "Prostaglandins E, A and F."

Miles-Yeda Ltd., Information Bulletin, Miles Laboratories, Inc., Research Products, Elkhart, IN 46514. "Rabbit Anti-Prostaglandin F—BSA Serum."

Appendix **I**

Properties of Common Acids and Bases

Compound	Formula	Molecular Weight	Specific Gravity	% by Weight	Molarity (M)
Acetic acid, glacial	CH_3COOH	60.1	1.05	99.5	17.4
Ammonium hydroxide	NH_4OH	35.0	0.89	28	14.8
Formic acid	$HCOOH$	46.0	1.20	90	23.4
Hydrochloric acid	HCl	36.5	1.18	36	11.6
Nitric acid	HNO_3	63.0	1.42	71	16.0
Perchloric acid	$HClO_4$	100.5	1.67	70	11.6
Phosphoric acid	H_3PO_4	98.0	1.70	85	18.1
Sulfuric acid	H_2SO_4	98.1	1.84	96	18.0

Properties of Common Buffer Compounds

Compound	Abbreviation	Molecular Weight	pK$_1$	(20°C)		
				pK$_2$	pK$_3$	pK$_4$
N-(2-Acetamido)-2-aminoethanesulfonic acid	ACES	182.2	6.9	—	—	—
N-(2-Acetamido)-2-iminodiacetic acid	ADA	212.2	6.60	—	—	—
Acetic acid		60.1	4.76	—	—	—
Arginine		174.2	2.17	9.04	12.48	—
Barbituric acid		128.1	3.79	—	—	—
N,N-bis(2-Hydroxyethyl)-2-aminoethane-sulfonic acid	BES	213.1	7.15	—	—	—
N,N-bis(2-Hydroxyethyl)glycine	Bicine	163.2	8.35	—	—	—
Boric acid		61.8	9.23	12.74	13.80	—
Citric acid		210.1	3.10	4.75	6.40	—
Ethylenediaminetetraacetic acid	EDTA	292.3	2.00	2.67	6.24	10.88
Formic acid		46.03	3.75	—	—	—
Fumaric acid		116.1	3.02	4.39	—	—
Glycine		75.1	2.45	9.60	—	—
Glycylglycine		132.1	3.15	8.13	—	—
N-2-Hydroxyethylpiperazine-N′-2-ethanesulfonic acid	HEPES	238.3	7.55	—	—	—
N-2-Hydroxyethylpiperazine-N′-3-propanesulfonic acid	HEPPS	252.3	8.0	—	—	—
Histidine		209.7	1.82	6.00	9.17	—
Imidazole		68.1	6.95	—	—	—
2-(N-Morpholino)ethanesulfonic acid	MES	195	6.15	—	—	—
3-(N-Morpholino)propanesulfonic acid	MOPS	209.3	7.20	—	—	—

Compound	Abbreviation	Molecular Weight	pK_1	(20°C) pK_2	pK_3	pK_4
Phosphoric acid		98.0	2.12	7.21	12.32	—
Succinic acid		118.1	4.18	5.60	—	—
3-Tris(hydroxymethyl)aminopropanesulfonic acid	TAPS	243.2	8.40	—	—	—
N-Tris(hydroxymethyl)methyl-2-aminoethanesulfonic acid	TES	229.2	7.50	—	—	—
N-Tris(hydroxymethyl)methylglycine	Tricine	179	8.15	—	—	—
Tris(hydroxymethyl)aminomethane	Tris	121.1	8.30	—	—	—

pK$_a$ Values and pH$_I$ Values of Amino Acids

Name	Abbreviation	pK$_1$ (α-carboxyl)	pK$_2$ (α-amino)	pK$_R$ (side chain)	pH$_I$
Alanine	ala	2.3	9.7	—	6.0
Arginine	arg	2.2	9.0	12.5	10.8
Asparagine	asn	2.0	8.8	—	5.4
Aspartate	asp	2.1	9.8	3.9	3.0
Cysteine	cys	1.7	10.8	8.3	5.0
Glutamate	glu	2.2	9.7	4.3	3.2
Glutamine	gln	2.2	9.1	—	5.7
Glycine	gly	2.3	9.6	—	6.0
Histidine	his	1.8	9.2	6.0	7.6
Isoleucine	ilu	2.4	9.7	—	6.1
Leucine	leu	2.4	9.6	—	6.0
Lysine	lys	2.2	9.0	10.5	9.8
Methionine	met	2.3	9.2	—	5.8
Phenylalanine	phe	1.8	9.1	—	5.5
Proline	pro	2.0	10.6	—	6.3
Serine	ser	2.2	9.2	—	5.7
Threonine	thr	2.6	10.4	—	6.5
Tryptophan	trp	2.4	9.4	—	5.9
Tyrosine	tyr	2.2	9.1	10.1	5.7
Valine	val	2.3	9.6	—	6.0

Molecular Weights of Some Common Proteins

Name (Source)	Molecular Weight
Albumin (bovine serum)	65,400
Albumin (egg white)	45,000
Carboxypeptidase A (bovine pancreas)	35,268
Carboxypeptidase B (porcine)	34,300
Catalase (bovine liver)	250,000
Chymotrypsinogen (bovine pancreas)	23,200
Hemoglobin (bovine)	64,500
Insulin (bovine)	5,700
α-Lactalbumin (bovine milk)	17,400
Lysozyme (egg white)	14,600
Malate dehydrogenase (pig heart mitochondria)	70,000
Myoglobin (horse heart)	16,900
Pepsin (porcine)	35,000
Peroxidase (horseradish)	40,000
Ribonuclease I (bovine pancreas)	12,600
Trypsinogen (bovine pancreas)	24,000
Trypsin (bovine pancreas)	23,800
Tyrosinase (mushroom)	128,000
Uricase (pig liver)	125,000

Appendix *V*

Common Abbreviations used in this Text

AMP, ADP, ATP	Adenosine mono-, di-, or triphosphate
ANS	8-Anilino-1-naphthalenesulfonic acid
BSA	Bovine serum albumin
Ci	Curie
CM cellulose	Carboxymethyl cellulose
DCIP	Dichlorophenolindophenol
DEAE cellulose	Diethylaminoethyl cellulose
DNP	Dinitrophenyl
DOPA	Dihydroxyphenylalanine
DPG	Diphosphoglycerate
EDTA	Ethylenediaminetetraacetic acid
FAD (H_2)	Flavin adenine dinucleotide (reduced form)
FAME	Fatty acid methyl ester
FMN (H_2)	Flavin mononucleotide (reduced form)
GC	Gas chromatography
Hb	Hemoglobin
HDL	High-density lipoprotein
HPLC	High pressure liquid chromatography
IEF	Isoelectric focusing
IHP	Inositol hexaphosphate
IR	Infrared
LDL	Low-density lipoprotein
NAD (H)	Nicotinamide adenine dinucleotide (reduced form)
NADP (H)	Nicotinamide adenine dinucleotide phosphate (reduced form)

NMR	Nuclear magnetic resonance
PAGE	Polyacrylamide gel electrophoresis
PEG	Polyethylene glycol
PMS	Phenazine methosulfate
POPOP	1,4-bis[5-Phenyl-2-oxazolyl]benzene
PPO	2,5-Diphenyloxazole
PTH	Phenylthiohydantoin
RIA	Radioimmunoassay
s	Sedimentation coefficient
SDS	Sodium dodecyl sulfate
TAG	Triacylglycerol
TEMED	N,N,N′,N′-tetramethylene diamine
TLC	Thin-layer chromatography
Tris	Tris(hydroxymethyl)aminomethane
UV	Ultraviolet
VIS	Visible
VLDL	Very low-density lipoprotein

Units of Measurement

Metric Prefixes

Name	Abbreviation	Multiplication Factor
pico	p ($\mu\mu$)	10^{-12}
nano	n (mμ)	10^{-9}
micro	μ	10^{-6}
milli	m	10^{-3}
centi	c	10^{-2}
deca	d	10^{1}
kilo	k	10^{3}

Units of Length

Name	Abbreviation	Multiplication Factor (Relative to Meter)
meter	m	1
centimeter	cm	10^{-2}
millimeter	mm	10^{-3}
micron	μ	10^{-6}
nanometer	nm	10^{-9}
Angstrom	Å	10^{-10}

Units of Weight

Name	Abbreviation	Multiplication Factor (Relative to Gram)
gram	g	1
kilogram	kg	10^3
milligram	mg	10^{-3}
microgram	μg	10^{-6}

Units of Volume

Name	Abbreviation	Multiplication Factor (Relative to Liter)
liter	l	1
milliliter	ml	10^{-3}
microliter	μl	10^{-6}

Table of Atomic Masses*
(Based on Carbon-12)

	Symbol	Atomic No.	Atomic Mass		Symbol	Atomic No.	Atomic Mass
Actinium	Ac	89	227.0278	Mercury	Hg	80	200.59
Aluminum	Al	13	26.98154	Molybdenum	Mo	42	95.94
Americium	Am	95	[243]†	Neodymium	Nd	60	144.24
Antimony	Sb	51	121.75	Neon	Ne	10	20.179
Argon	Ar	18	39.948	Neptunium	Np	93	237.0482
Arsenic	As	33	74.9216	Nickel	Ni	28	58.70
Astatine	At	85	[210]	Niobium	Nb	41	92.9064
Barium	Ba	56	137.33	Nitrogen	N	7	14.0067
Berkelium	Bk	97	[247]	Nobelium	No	102	[259]
Beryllium	Be	4	9.01218	Osmium	Os	76	190.2
Bismuth	Bi	83	208.9804	Oxygen	O	8	15.9994
Boron	B	5	10.81	Palladium	Pd	46	106.4
Bromine	Br	35	79.904	Phosphorus	P	15	30.97376
Cadmium	Cd	48	112.41	Platinum	Pt	78	195.09
Calcium	Ca	20	40.08	Plutonium	Pu	94	[244]
Californium	Cf	98	[251]	Polonium	Po	84	[209]
Carbon	C	6	12.011	Potassium	K	19	39.0983
Cerium	Ce	58	140.12	Praseodymium	Pr	59	140.9077
Cesium	Cs	55	132.9054	Promethium	Pm	61	[145]
Chlorine	Cl	17	35.453	Protactinium	Pa	91	231.0359
Chromium	Cr	24	51.996	Radium	Ra	88	226.0254
Cobalt	Co	27	58.9332	Radon	Rn	86	[222]
Copper	Cu	29	63.546	Rhenium	Re	75	186.207
Curium	Cm	96	[247]	Rhodium	Rh	45	102.9055
Dysprosium	Dy	66	162.50	Rubidium	Rb	37	85.4678
Einsteinium	Es	99	[252]	Ruthenium	Ru	44	101.07

* Atomic masses given here are 1977 IUPAC values

† A value given in brackets denotes the mass number of the longest-lived or best-known isotope.

Erbium	Er	68	167.26	Samarium	Sm	62	150.4
Europium	Eu	63	151.96	Scandium	Sc	21	44.9559
Fermium	Fm	100	[257]	Selenium	Se	34	78.96
Fluorine	F	9	18.998403	Silicon	Si	14	28.0855
Francium	Fr	87	[223]	Silver	Ag	47	107.868
Gadolinium	Gd	64	157.25	Sodium	Na	11	22.98977
Gallium	Ga	31	69.72	Strontium	Sr	38	87.62
Germanium	Ge	32	72.59	Sulfur	S	16	32.06
Gold	Au	79	196.9665	Tantalum	Ta	73	180.9479
Hafnium	Hf	72	178.49	Technetium	Tc	43	[98]
Helium	He	2	4.00260	Tellurium	Te	52	127.60
Holmium	Ho	67	164.9304	Terbium	Tb	65	158.9254
Hydrogen	H	1	1.0079	Thallium	Tl	81	204.37
Indium	In	49	114.82	Thorium	Th	90	232.0381
Iodine	I	53	126.9045	Thulium	Tm	69	168.9342
Iridium	Ir	77	192.22	Tin	Sn	50	118.69
Iron	Fe	26	55.847	Titanium	Ti	22	47.90
Krypton	Kr	36	83.80	Tungsten	W	74	183.85
Lanthanum	La	57	138.9055	Uranium	U	92	238.029
Lawrencium	Lr	103	[260]	Vanadium	V	23	50.9415
Lead	Pb	82	207.2	Xenon	Xe	54	131.30
Lithium	Li	3	6.941	Ytterbium	Yb	70	173.04
Lutetium	Lu	71	174.967	Yttrium	Y	39	88.9059
Magnesium	Mg	12	24.305	Zinc	Zn	30	65.38
Manganese	Mn	25	54.9380	Zirconium	Zr	40	91.22
Mendelevium	Md	101	[258]				

Values of *t* for Analysis of Statistical Confidence Limits

d.f.	Probability of Larger Value of *t*, Sign Ignored			d.f.			
	0.05	0.02	0.01		0.05	0.02	0.01
1	12.706	31.821	63.657	14	2.145	2.624	2.977
2	4.303	6.0965	9.925	15	2.131	2.602	2.947
3	3.182	4.541	5.841	16	2.120	2.583	2.921
4	2.776	3.747	4.604	17	2.110	2.567	2.898
5	2.571	3.365	4.032	18	2.101	2.552	2.878
6	2.447	3.143	3.707	19	2.093	2.539	2.861
7	2.365	2.998	3.499	20	2.086	2.528	2.845
8	2.306	2.896	3.355	21	2.080	2.518	2.831
9	2.262	2.821	3.250	22	2.074	2.508	2.819
10	2.228	2.764	3.169	23	2.069	2.500	2.807
11	2.201	2.718	3.106	24	2.064	2.492	2.797
12	2.179	2.681	3.055	25	2.060	2.485	2.787
13	2.160	2.650	3.012				

From EXPERIMENTAL TECHNIQUES IN BIOCHEMISTRY by J. Brewer, A. Pesce, and R. Ashworth, p. 332. Copyright © 1974. Reprinted by permission of Prentice-Hall, Inc., Englewood Cliffs, N.J.

Appendix *IX*

Solutions to Selected Questions

EXPERIMENT 2

1. *pH 2* · *pH 5* · *pH 11*

	COOH
$H_3\overset{+}{N}$—C—H	
	$(CH_2)_3$
	CH_2
	$^+NH_3$

	COO^-
$H_3\overset{+}{N}$—C—H	
	$(CH_2)_3$
	CH_2
	$^+NH_3$

	COO^-
H_2N—C—H	
	$(CH_2)_3$
	CH_2
	NH_2

2.

	COOH
$H_3\overset{+}{N}$—C—H	
	CH_2
	CH_2
	COOH

	COO^-
$H_3\overset{+}{N}$—C—H	
	CH_2
	CH_2
	COOH

	COO^-
$H_3\overset{+}{N}$—C—H	
	CH_2
	CH_2
	COO^-

$$
\begin{array}{c}
\text{COO}^- \\
| \\
\text{H}_2\text{N}-\text{C}-\text{H} \\
| \\
\text{CH}_2 \\
| \\
\text{CH}_2 \\
| \\
\text{COO}^-
\end{array}
\qquad
\begin{array}{c}
\text{COOH} \\
| \\
\text{H}_2\text{N}-\text{C}-\text{H} \\
| \\
\text{CH}_2 \\
| \\
\text{CH}_2 \\
| \\
\text{COOH}
\end{array}
$$

unionized form

3. The separation of amino acids by paper chromatography is based on polarity. The less polar an amino acid, the greater the R_f. Arrangement in order of decreasing R_f is: leu > val > ala > ser > lysine

4. alanine, pH 3.0: slightly + aspartic acid, pH 4.0: very slightly —

5. *Toward cathode*   *Toward anode*   *Remain stationary*
   lys   glu, phe, ser   val

7. Order of elution: ser > glu > pro > ala > arg

EXPERIMENT 3

1. Edman method vs. dansyl chloride. Because of the cyclization-cleavage reaction that occurs, the Edman method can be used for sequential degradation of a peptide or protein. N-terminal residues can be removed, one at a time, and identified chromatographically. Dansyl chloride can be used only for single N-terminal analysis. Dansyl chloride, however, has some advantages. Dansyl amino acids can be more readily separated by chromatographic techniques, and detection by a UV lamp is more sensitive than for phenylthiohydantoin amino acids (PTH-amino acids). Also, work-up of the dansyl chloride reaction is less cumbersome and, hence, more rapid. Excess phenylisothiocyanate reagent and coupling products must be extracted from the Edman reaction mixture. This requires more work-up steps. In your experiment, this extraction was accomplished with toluene. In the dansyl chloride reaction, excess reagent is hydrolyzed to the corresponding sulfonic acid, which does not interfere with chromatographic analysis.

2. If the solvent level in a chromatography jar is above the application spots, the compounds spotted will be dissolved in the solvent and diffuse throughout the chromatography solvent. The compounds will no longer be highly concentrated at a single area on the plate. Even if there is sufficient compound remaining on the plate for detection, the final developed spot will most likely be too large for an accurate R_f measurement.

3. At least three major factors must be considered in choosing a solvent for the Edman degradation. (1) The solvent must dissolve both the peptide and the phenylisothiocyanate reagent. This problem can best be solved by using an aqueous-organic system. (2) The coupling reaction requires an unprotonated α-amino group of the peptide for nucleophilic attack on the phenylisothiocyanate reagent:

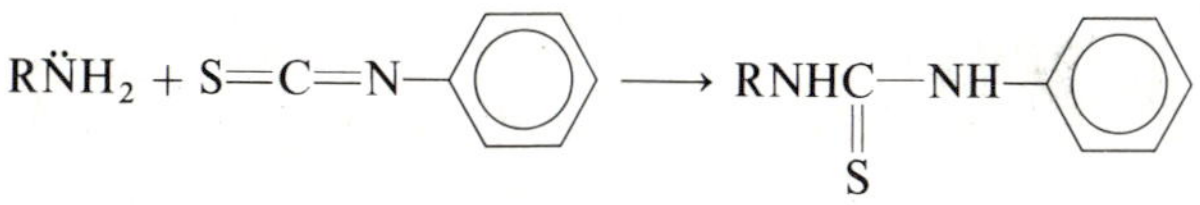

This requires the use of a basic solvent. (3) The solvent must be basic but not highly nucleophilic. A highly nucleophilic solvent, such as alcohol, would react with the thiocyanate reagent, hence reducing the effective labeling of the peptide, and producing side products that must be removed before analysis of the reaction mixture.

Another factor that should also be considered is the volatility of the solvent. Work-up of the reaction mixture calls for evaporation of the solvent; therefore, the lower boiling the solvent the more desirable, as long as the other factors are considered.

4. Analyzing the data given leads to the following conclusions.
    Amino acids present in tripeptide: phe, leu and ala
    N-terminal amino acid: ala
    C-terminal amino acid: phe
The most likely sequence: H_2N-ala-leu-phe-COOH

5. Total acid hydrolysis of a dipeptide would identify the amino acids present. Reaction of a dipeptide with carboxypeptidase would release both amino acids at the same time and it would be impossible to identify the C-terminal amino acid. C-terminal analysis of a dipeptide could be done using a chemical method—hydrazinolysis or reduction by lithium borohydride.

EXPERIMENT 4

2. See Experiment 5.
3. Read Chapter 2, Section E.

EXPERIMENT 5

1. Bromphenol must be the fastest migrating component in the protein solution.

EXPERIMENT 6

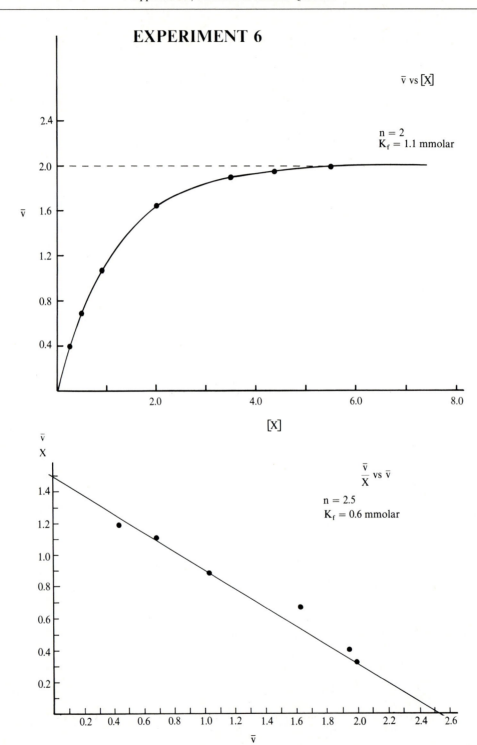

EXPERIMENT 7

1. Derivation of the Eadie-Hofstee equation:

$$\frac{1}{v_0} = \frac{K_{\mathrm{m}}}{V_{\mathrm{max}}} \frac{1}{[S]} + \frac{1}{V_{\mathrm{max}}}$$

Multiply by V_{max}

$$\frac{V_{\mathrm{max}}}{v_0} = \frac{K_{\mathrm{m}}}{[S]} + 1$$

Multiply by v_0

$$V_{\mathrm{max}} = \frac{K_{\mathrm{m}}}{[S]} v_0 + v_0$$

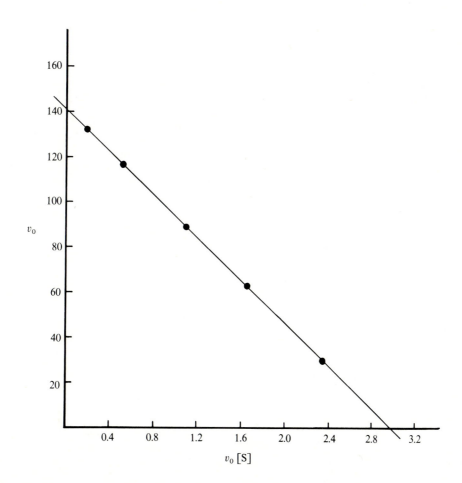

Rearrange

$$v_0 = -\frac{K_m}{[S]} v_0 + V_{max}$$

A plot of v vs. $v/[S]$ yields a straight line as shown in the graph. The intercept on the v axis is V_{max}. The intercept on the $v/[S]$ axis is V_{max}/K_m.

3. The turnover number, k_3, is defined as follows:

$$k_3 = \frac{V_{max}}{[E_T]}$$

From Problem 2, $V_{max} = 140$ μmole/min (see graph in Problem 1).

$$k_3 = \frac{140\ \mu mole/min}{5 \times 10^{-7}\ M} = 280\ min^{-1}$$

6. Type of inhibition: competitive

$$K_I = 5.9 \times 10^{-4}\ molar$$

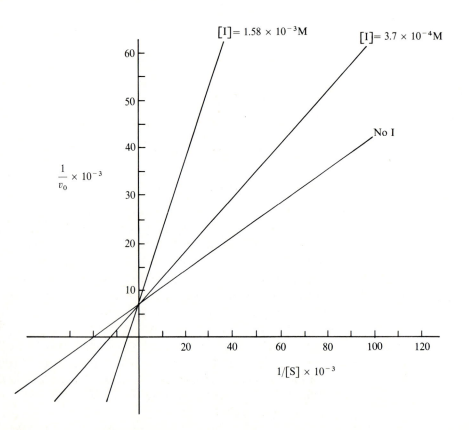

EXPERIMENT 8

1. Fluorescence spectroscopy, ultracentrifugation, or affinity chromatography could be used to detect a complex between α-lactalbumin and galactosyl transferase. See the following references:

   W. Klee and C. Klee, *J. Biol. Chem.*, *247*, 2336–2344 (1972).

   R. Mawal, J. Morrison and K. Ebner, *J. Biol. Chem.*, *246*, 7106–7109 (1971).

   I. Trayer and K. Olsen, *Biochem. J.*, *127*, 9P–10P (1972).

EXPERIMENT 9

1. After completing the experiment you should now recognize that the lipid is a triacylglycerol. The following experimental procedure could be followed to identify the structure.

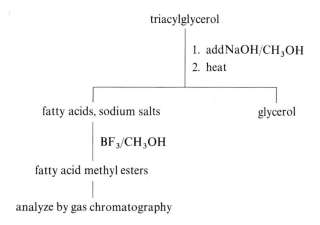

   For a further discussion on this procedure see Experiment 10, "Characterization and Positional Distribution of Fatty Acids in Triacylglycerols."

2. Lipids that have both polar and nonpolar portions (fatty acids, etc.) tend to "streak or tail" during thin layer chromatography development. This leads to long, narrow spots after iodine treatment. Addition of a small amount of a very polar solvent (acetic acid) greatly reduces this "tailing" and results in a more circular, better resolved spot.

3. The order of migration of the standard lipids during chromatography depends primarily on polarity. The less polar the lipid, the greater its affinity for the nonpolar solvent (compared to the polar silica gel) and hence the greater the migration.

4. Iodine interacts with many organic compounds to form pi complexes that are colored. This method of detection is especially useful for lipids containing double bonds. A lipid with several double bonds will give a darker spot with iodine. The darker spots may also be due to a higher concentration of lipid.

5. $R_f = \dfrac{\text{distance of migration of fatty acid}}{\text{distance of migration of solvent}} = \dfrac{10.4 \text{ cm}}{17 \text{ cm}} = 0.61$

EXPERIMENT 10

2. In order for chemical substances to be analyzed by gas chromatography, they must be relatively nonpolar and volatile. Fatty acids are not readily vaporized at temperatures attainable in a gas chromatograph (up to 200°C). FAMEs have much lower boiling points than fatty acids. Polar fatty acids also would have very long retention times unless extremely nonpolar column packings are used.

3. Calcium ion stimulates the action of lipase by forming insoluble salts with liberated fatty acids. The fatty acids are no longer free in solution to inhibit the enzyme. Removal of fatty acid products from solution also retards the reverse reaction (resynthesis of acylglycerols), hence shifting the equilibrium toward hydrolysis.

EXPERIMENT 11

1.

Fischer projection
of D-ribose

Haworth structure
of D-ribose

2. Sucrose contains no anomeric carbons, because the anomeric carbons of the individual monosaccharides (glucose and fructose) are linked together.

3. Invertase catalyzes the hydrolysis of glycoside bonds between fructose and glucose. Note that one of the glycosidic bonds in raffinose is between carbon-2 of fructose and carbon-1 of glucose as in sucrose. The hydrolysis of this bond is facilitated by invertase. The products are fructose plus O-α-D-galactosyl-(1 → 6)-O-α-D-glucose.

4. Specific optical rotation of pure sugar:

$$[\alpha]_D^{20} = \frac{[\alpha]_{obs}}{\ell \times c}$$

$$[\alpha]_D^{20} = \frac{-6.74}{0.05} = -135.2$$

Specific optical rotation of mutarotated sugar:

$$[\alpha]_D^{20} = \frac{-4.55 + 0.01}{0.05} = -91.2$$

The unknown carbohydrate is most likely β-D-fructose.

5. (a) Sample mean = +3.21°
   (b) Standard deviation = ±0.043
   (c) 95% confidence limits = +3.21 ± 0.12, at a probability of 0.05.

EXPERIMENT 12

4. Peripheral proteins usually have some ionic regions and are held in position largely by ionic forces; therefore, they may be extracted with ionic solutions such as salt solutions. The extraction of integral proteins requires the disruption of the hydrophobic interactions between the lipids and proteins. Detergents are convenient reagents to break up these interactions.

5. See Experiments 9 and 10.

EXPERIMENT 13

1. Bio-Rex is a cation exchange resin. (See Chapter 3, Section E.)

3. Glycohemoglobins elute faster than normal hemoglobin from ion-exchange columns.

5. The hemoglobins in this experiment have approximately the same molecular size, so Sephadex G-100 would not lead to effective separation.

8. Heme

EXPERIMENT 14

2. Chlorophyll is a nonpolar molecule, so it is soluble in most organic solvents. A more efficient extraction could be achieved if several extractions were carried out and the extracts pooled.

3. ATP formation may also be measured by monitoring the uptake of inorganic phosphorus. This may be done with radioactive phosphorus (^{32}P) or by colorimetric measurement of P_i.

5. Ion-selective electrodes may be used to measure chloroplast exchange of K^+, Mg^{2+} or Ca^{2+}. For other methods, see Dilley (1972).

EXPERIMENT 15

4. Highly purified DNA may be obtained by repeating the chloroform/isoamyl alcohol extraction several times. The alcohol precipitation step may also be carried out many times. Various chromatographic methods including ion-exchange have been applied to the purification of nucleic acids.

5. 28 $\mu g/ml$

EXPERIMENT 16

1. All of the conditions listed (a–d) will lower the observed T_m as compared with measurement in standard saline citrate, pH 7.0.

3. $\% \, GC = 2.44 \, (T_m - 69.3)$
$60 = 2.44 \, T_m - (2.44)(69.3)$

$$T_m = \frac{60 + (2.44)(69.3)}{2.44}$$

$T_m = 94°C$

6. Dimethylformamide has intense absorption at 260 nm. This would interfere with absorption measurements. The absorption spectrum of DNA shows a broad band between 250 and 280 nm, so a measurement made within this range may be used to evaluate the hyperchromic effect.

EXPERIMENT 17

1. Many plasmids contain genes that carry messages for the synthesis of proteins that protect microorganisms against antibiotics. The presence of certain antibiotics is a signal to the microorganism that more of these proteins are necessary. The microorganism responds by increasing the rate of production of plasmids.

EXPERIMENT 18

2. Preparations of ribonucleases also contain deoxyribonucleases. If not removed or destroyed, DNases would degrade the extracted plasmids. DNases are not stable to high temperatures and are denatured during heating. Ribonucleases remain stable during the heat treatment.

EXPERIMENT 19

1. Restriction endonucleases require the presence of Mg^{2+} for activity. The "quench buffer" contains EDTA, which complexes transition metal ions in the solution. The metal ions are no longer available for binding by the nuclease molecules and enzyme activity is inhibited.

EXPERIMENT 20

1. Properties of a restriction enzyme:
   (a) Ideally, it should cause cleavage at a single or, at most, two well-defined locations on the plasmid DNA.
   (b) The cleavage site should be within a marker gene (insertional marker inactivation).
   (c) The optimal reaction conditions for the enzyme should provide for the stability of the plasmid products.

2. Isopropanol causes the selective precipitation of DNA. Most RNA remains in solution; hence, isopropanol precipitation provides an effective technique for the removal of contaminating RNA from DNA extracts.

3. See Experiment 16.

4. Natural DNA molecules and restriction fragments are too large to penetrate polyacrylamide gels. Even gels with low percentage cross-linking are not useful.

EXPERIMENT 21

1. The following references provide an answer to this question:
   A.R. Morgan, J.S. Lee, D.E. Pulleyblank, N.L. Murray, and D.H. Evans, *Nucleic Acids Research*, 7, 547–569 (1979).
   G.J. Boer, *Anal. Biochem.*, 65, 225–231 (1975).

4. See the answer to question 1 above.

EXPERIMENT 22

1. Mitochondria are most stable in an environment that is relatively polar, but not ionic. Although sucrose is polar, it is not ionic; therefore, it will not disrupt the structure of mitochondria and cause excessive loss of protein material.

4. Electron transport and oxidative phosphorylation may be studied by incubating the mitochondrial fraction with various substrates including malate, succinate, and ascorbate. Oxidative phosphorylation is monitored by measuring the uptake of oxygen, using an oxygen electrode or Warburg manometer. Alternatively, the uptake of inorganic phosphate is measured with a colorimetric assay.

EXPERIMENT 23

1.

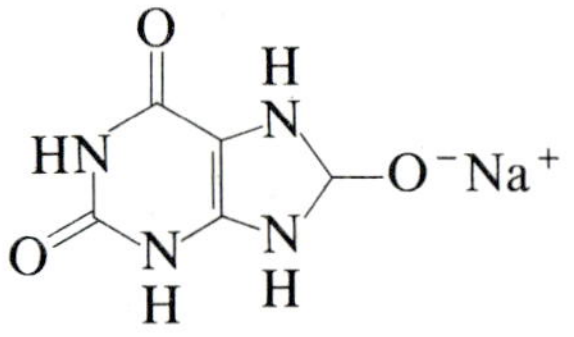

sodium urate

3. 5.4%

4. In order for the single standard method to be valid, there must be a linear relationship between absorbance and cholesterol concentration. The cholesterol assay is linear up to about 500 mg/100 ml.

EXPERIMENT 24

2. 65 mg of vitamin C in 100 ml of orange juice.

3. Compounds containing the sulfhydryl group, shown as R—SH below, react with DCIP:

$$2R—SH + DCIP(oxidized) \longrightarrow R—S—S—R + DCIP(reduced)$$

p-Chloromercuribenzoic acid reacts with sulfhydryl groups, preventing reaction with the dye:

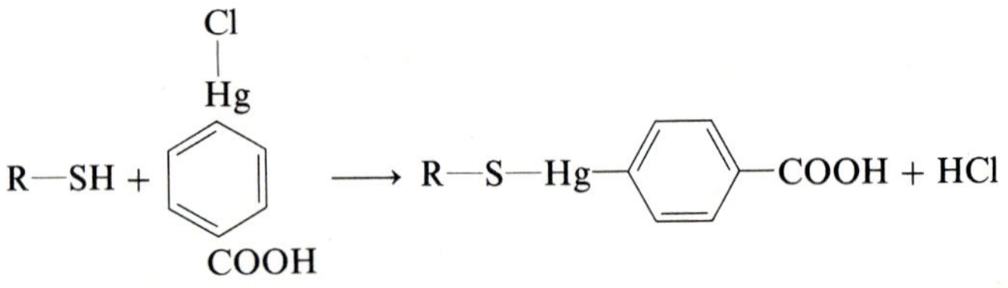

EXPERIMENT 25

2. When the sample is applied to a wet cellulose acetate strip, the spot becomes larger by diffusion and less resolution is attained during the electrophoresis. The dry application is difficult, because the cellulose acetate strip is hard and brittle and the sample does not absorb well.

5. The subunit sizes of the lipoproteins may be determined by gel filtration. First, the lipoproteins must be isolated and partially purified. Second, the subunit structure must be disrupted by treatment with sodium dodecyl sulfate (SDS). Probably a better method for subunit analysis is SDS-gel electrophoresis (Chapter 4, Section B).

EXPERIMENT 26

3. From standard curve of relative % bound vs. log of the amount of estradiol antigen per tube:

      unknown 1   80 pg
      unknown 2   65 pg

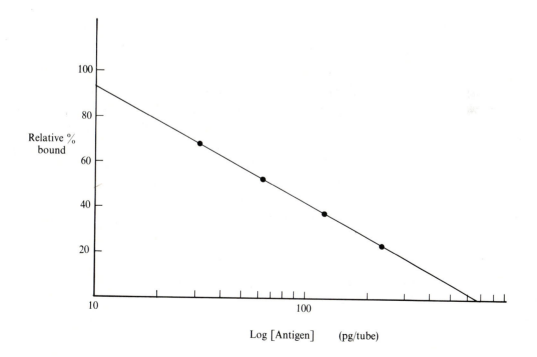

Index